Irago Conference 2012

Journal of Physics: Conference Series Volume 433

Irago, Japan
15-16 November 2012

ISBN: 978-1-62748-310-0
ISSN: 1742-6588

Printed from e-media with permission by:

Curran Associates, Inc.
57 Morehouse Lane
Red Hook, NY 12571

Some format issues inherent in the e-media version may also appear in this print version.

Printed by Curran Associates, Inc. (2013)

For permission requests, please contact the Institute of Physics
at the address below.

Institute of Physics
Dirac House, Temple Back
Bristol BS1 6BE UK

Phone: 44 1 17 929 7481
Fax: 44 1 17 920 0979

techtracking@iop.org

Additional copies of this publication are available from:

Curran Associates, Inc.
57 Morehouse Lane
Red Hook, NY 12571 USA
Phone: 845-758-0400
Fax: 845-758-2634
Email: curran@proceedings.com
Web: www.proceedings.com

TABLE OF CONTENTS

PREFACE

The Irago Conference 2012 – 360 degree outlook on critical scientific and technological challenges for a sustainable society

Organized by the Electronics-Inspired Interdisciplinary Research Institute (EIIRIS) at Toyohashi University of Technology, the Irago Conference, held recently (15–16 November) in Aichi, Japan, aimed to enhance mutual understanding between scientists, engineers and policymakers. Over 180 participants tackled topics ranging from energy and natural resources to public health and disaster prevention.

The 360-degree outlook of the conference impressed speakers and guests. "This conference has been extremely informative," noted Robert Gellar from the University of Tokyo. "A unique conference with experts from a range of backgrounds," agreed Uracha Ruktanonchai from the National Nanotechnology Center (NANOTEC) in Thailand.

Similarly, G P Li, professor of electrical engineering and computer science at the University of California Irvine commented that he had been "able to think the unthinkable" as a range of topics came together. The conference was streamed live on Ustream to ensure that researchers from across the world could benefit from thought-provoking presentations examining global issues such as energy, disaster mitigation and nanotechnology.

"This was wonderful," said Oussama Khatib from Stanford University, "A good recipe of speakers from such a range of backgrounds." Manuscripts submitted to the organizers were peer-reviewed, and the papers in this proceedings were accepted for *Journal of Physics: Conference Series.*

In addition to the formal speaker programme, graduate-student sessions provided a platform for graduate students to describe their latest findings as oral presentations. A series of excursions to relevant locations, such as the Tahara megasolar region under construction and a local car-manufacturing factory, gave participants the opportunity to further consider practical applications of their research in industry.

Irago Conference 2013 is scheduled to be held in October 2013 as a platform for participants from a wide range of backgrounds and specialities to interact and discuss solutions to increasingly important environmental, social, and technological challenges people of the 21st century.

All papers published in this volume of *Journal of Physics: Conference Series* have been peer reviewed through processes administered by the proceedings Editors. Reviews were conducted by expert referees to the professional and scientific standards expected of a proceedings journal published by IOP Publishing.

The Irago Conference 2012 IOP Publishing
Journal of Physics: Conference Series **433** (2013) 012001 doi:10.1088/1742-6596/433/1/012001

Characterization of graphene oxide reduced through chemical and biological processes

M. Boutchich [1*], A. Jaffré [1], D. Alamarguy [1], J. Alvarez [1], A. Barras [2], Y. Tanizawa[3], R. Tero[3], H. Okada[3], T. V. Thu[3], J. P. Kleider [1] and A. Sandhu [3]

[1]LGEP, CNRS UMR8507, SUPELEC, Univ Paris-Sud, UPMC Univ Paris 6, 11 rue Joliot-Curie, Plateau de Moulon, 91192 Gif-sur-Yvette Cedex, France
[2]Institut de Recherche Interdisciplinaire IRI, (USR CNRS 3078), Université Lille 1, Parc de la Haute Borne 50, av de Halley – BP 70478 59658 Villeneuve d'Ascq cedex, France
[3]Toyohashi Tech Graphene Research Group, Toyohashi University of Technology, 1-1 Hibarigaoka, Tempaku-cho, Toyohashi, Aichi 441-8580, Japan.

*corresponding author: mohamed.boutchich@lgep.supelec.fr

Abstract. The study of new materials for transparent electrodes or new heterojunctions made of 2D materials combinations is a very active research topic. Challenges to overcome are the modulation of the optoelectronic properties of such materials to achieve competitive photovoltaic devices. In this work, graphene oxide was reduced into graphene through different chemical (hydrazine, ultraviolet photocatalysis) and biological (microorganisms) processes. We benchmarked the reduction efficiency by probing materials characteristics using various physical characterization techniques. X-ray photoelectron spectroscopy (XPS) analyses were carried out to observe the effectiveness of the reduction processes through the sp^2/sp^3 content. In addition, the homogeneity of the reduction is investigated on micrometer scale sample with micro Raman mapping and extraction of the I_D/I_G ratio. Conductive-probe atomic force microscopy (CP-AFM) was employed to investigate the longitudinal conductivity of the different samples. The results show that hydrazine based reduction remains the most efficient. However, the bacterial procedure demonstrated partial reconstruction of the carbon network and reduced the amount of oxygenated functional groups.

1. Introduction

In order to improve the efficiency and reduce manufacturing costs, corporate and academic research groups are striving to develop new alternative technologies for photovoltaic (PV) devices. The emphasis is on both rigid as well as flexible PV devices opening new opportunities for different class of materials and technologies. 2D materials have attracted tremendous attention worldwide for all kind of nanoelectronic applications including PV. Indeed, graphene [1] is seen as a serious candidate to replace Indium Tin Oxide (ITO) as a transparent electrode [2] but manufacturing cost and optoelectronic properties remain difficult to control. In that respect, reduced graphene oxide (rGO) could possibly be an alternative thanks to its potentially low manufacturing cost [3]. Exfoliated graphene exhibits optoelectronic properties that could potentially outperform any alternative to ITO currently investigated. However, there is no technological pathway to exploit exfoliated graphene. On the other hand, reduced graphene oxide (rGO) is a potential alternative under investigation that could be mass produced at low cost [4]. Nevertheless to envision applications the optoelectronic properties of rGO and CVD-graphene still remain to be tailored to become competitive with current rare earth electrode materials. For that purpose, most of the research work on this topic is devoted to the optimization of the optoelectronic properties through either deposition techniques or reduction for CVD graphene and graphene oxide respectively [5, 6, and 7]. As opposed to pristine exfoliated graphene composed of exclusively sp²-hybridized carbon atoms, a GO sheet is hybrid network of carbon network sp² hybridized with neighboring carbon atoms or with carboxyl and carbonyl groups on the edges as well as sp^3 hybridized carbon atoms that form covalent bonds with epoxyl and hydroxyl groups [8]. The latter renders the GO soluble in various solvents and allows the development of a low cost wafer-scale technology. However, to achieve an electrically active material, GO must be reduced and different reduction methods have emerged [9, 10, 11]. The reduction process partially restores a sp² network by removing a large fraction of these defects. Therefore, the structural disorder decreases and the carrier transport improves. Nevertheless, electrical properties such as conductivity and mobility are orders of magnitude below the ones attainable with exfoliated graphene [12]. Yet, these values allow the fabrication of low cost electronics that usually don't require outmost performance. In this work, we investigate the impact of different reduction methods hydrazine, UV-photocatalysis and microorganisms through physical characterization techniques. Micro-Raman spectroscopy has been utilized to investigate the signatures and the structural disorder with the I_D/I_G ratio before and after reduction. X-ray photoelectron spectroscopy monitors the variation of the sp^2/sp^3 content as well as the oxygen/carbon atomic ratios. At last, conductive CP-AFM set-up was utilized to probe local electrical properties of rGO.

2. Experimental

2.1. General methods

Graphite powder was obtained from SEC Carbon with grade reference SNO-30 [13]. 2D Raman scattering images and Raman spectra were collected at room temperature, using a WiTec alpha 300R confocal Raman microscope with a UHTS300 spectrometer and a DV401A CCD detector. The laser beam spot size was around 900nm. The films were illuminated with 100µW for 2s for each spectrum using a frequency doubled Nd:YAG Laser at 532 nm and Raman signal was collected in a backscattering geometry. A 600 grooves/mm diffraction grating was utilized enabling a spectral resolution of $3cm^{-1}$. The mapping was collected with 250 nm between spectra using a piezoelectric stage.

XPS measurements were performed using a Versaprobe PHI5000 spectrometer (Physical Electronics) operating at a base pressure of $1x10^{-7}$Pa. A monochromatic Al K_α X-ray source (1486.6 eV) was used at 15 kV with take-off angle of 45°. The analysed area of the sample was 50 µm in diameter. High resolution core level spectra were acquired with 23.5 eV pass energy, an energy step of 0.1 eV and a 50 ms dwell time. A dual beam charge neutralisation system consisting of low energy ions and electrons was utilized. The spectrometer was calibrated with Cu $2p_{3/2}$ and Cu LMM of clean copper at 932.7 eV and 918.7 eV respectively. The binding energy scale was calibrated by positioning the reference peak of Si $2p_{3/2}$ corresponding to SiO_2 from the substrate to 103.5 eV. Spectroscopic data were processed by the PHI Multipak software, using a peak fitting routine with symmetrical Gaussian-Lorentzian functions. The Shirley background was subtracted from the spectra.

CP-AFM measurements were performed using a Digital Instruments Nanoscope IIIa Multimode AFM associated with the home-made conducting probe extension called "Resiscope" [14]. This setup allows us to apply a stable DC bias voltage (from -10 to +10 V with 0.01 V resolution) to the device and to measure the resulting current flowing through the tip as the sample surface is scanned in contact mode. Local resistance values can be measured in the range of 10^2-10^{12} Ω, which allows investigations of a variety of materials [15, 16] and devices [17, 18]. Reliable and understandable electrical measurements through CP-AFM setup require a well-characterized conductive tip. Depending on the experimental conditions, the AFM conductive tip should be the most suitable in terms of serial resistance that must be taken into account in the electrical analysis of rGO. PtIr-coated Si cantilever, with a spring constant of about 0.2 N/m, proves to be suitable for our experimental conditions, since measured resistance values are greater than the intrinsic tip resistance that was estimated around 300 Ω.

In this work, we investigated three different reduction methods: hydrazine, UV photocatalysis and bacterial. All samples were extracted from the initial GO source described in the experimental section. GO and rGOs samples were spin coated, on oxidized (300nm) silicon substrate following the procedure described by [19]. This yields to a dispersion of GO flakes ranging from monolayer to multilayer.

2.2. preparation of graphene oxide

GO was synthesized by the modified Hummer's method using natural graphite powder (SEC Carbon SNO-30). The powder was cleaned in a $K_2S_2O_8$ and H_2SO_4 bath and subsequently oxidized in $KMnO_4$ and H_2SO_4. After centrifuging, the resulting graphite oxide was exfoliated into GO by ultra-sonication (100 W, 30 min, 60 °C). At last, a GO aqueous dispersion was produced by centrifugation and dialysis to neutral pH.

2.3. reduction of graphene oxide by hydrazine

$rGO_{hydrazine}$ was obtained by reduction with hydrazine hydrate [9]. A homogeneous GO aqueous suspension (1 $mg.mL^{-1}$) was achieved by dispersing GO in distilled water using ultra sonication and 20 mL were loaded in a 50-mL round bottom flask. Hydrazine monohydrate (200 μL) was then added and the solution heated in an oil bath at 100 °C under a water-cooled condenser for 24 h over which the reduced GO gradually precipitated out as a black solid. This product was isolated by filtration on a 0.45 μm PVDF membrane, washed with water and methanol, and dried on the filter in oven (80 °C) during 24 h. The solid was subsequently dissolved in dimethylformamide (DMF) at 1 $mg.mL^{-1}$.

2.4. reduction of graphene oxide by UV photocatalysis

rGO_{UV} was obtained from a homogeneous GO aqueous suspension (1 $mg.mL^{-1}$) elaborated by dispersing GO in distilled water using ultra sonication. Aqueous graphene oxide solution (20 mL) in a quartz tube was irradiated for 6 h using a home-made photo reactor (λ = 254 nm) under direct irradiation. The dark insoluble graphene was isolated by filtration over Anodisc 0.22 μm, washed copiously with methanol and dried on the filter in oven (80 °C) during 2 minutes. The solid was dissolved in DMF solvent at 1 $mg.mL^{-1}$.

2.5. reduction of graphene oxide by microorganisms

$rGO_{bacterial}$ was reduced using microorganism extracted from a river in Aichi prefecture (Japan) following multiple iterations of transfer and cultivation of river sediments in a medium containing salts and organic materials in anaerobic conditions. Silicon substrates covered with a suspension of GO flakes were immerged into the microbial culture in serum bottles and incubated in the culture at 28°C for 3 days under dark [19].

3. Results and analysis

Figure 1(a) presents the μ-Raman spectra of the different samples acquired at constant power of 100μW and for 2s per spectrum. All samples exhibit a strong D-band (~1350 cm^{-1}) and G-band (~1580 cm^{-1}) usually assigned to structural disorder and the graphitized structure respectively. GO presents a dominant D peak arising from the attachment of functional groups such as hydroxyl, epoxy on the carbon skeleton. The overtone 2D-band does not vary significantly with respect to the reduction processes indicating that graphitization is rather poor. The D+D' peak (~2940 cm^{-1}) represents a defect activated combination of phonons with different momenta [20]. The large D-band and the attenuated 2D-band may be attributed to the small sp^2 domains subsequent to the oxidation process and ultra-sonication [3]. In order to assess the homogeneity of the material, μ-Raman mapping has been carried out on micro scale flakes. The mapping illustrates the variation of the peaks across the sample geometry. The full width half maximum (FWHM) of all samples is large indicating multi-layer material (not shown). Figure 1(b) represents the μ-Raman two dimensional mapping of D, G and 2D peaks for the GO sample.

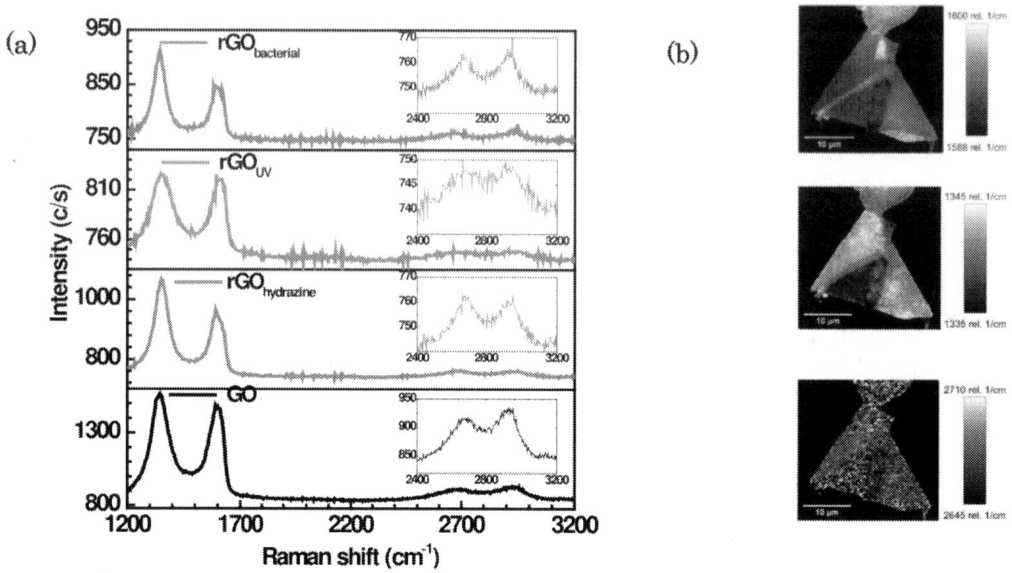

Figure 1: (a) Raman spectra of graphene oxide and reduced graphene oxide samples, (b) μ-Raman mapping of GO sample for D, G and 2D bands. The 2D peak mapping is very noisy due to the poor signal to noise ratio given the low intensities of this peak.

The variation observed on the GO is relatively large as well as the D-band intensity for all samples as displayed in figure 1(a), this indicates the poor crystalline quality of the material as-prepared and reduced. In order to evaluate the disorder in the material, we extracted the I_D/I_G ratio for the various samples, table 1.

	D (cm^{-1})	**G (cm^{-1})**	**2D (cm^{-1})**	**D+D' (cm^{-1})**	**<I_D/I_G> (%)**
GO	1344	1593	2680	2928	1.4
Bacteria	1341	1596	2645	2829	1.55
UV-photocatalysis	1354	1603	2655	2841	1.6
Hydrazine	1351	1595	2683	2932	1.9

Table 1: μ-Raman spectroscopy of GO and reduced GO. <I_D/I_G> represents the average ratio across the mapping.

The I_D/I_G ratios observed present little variation between samples; the slight increase may be related to the removal of the functional groups upon reduction and the subsequent creation of defects [21]. According to [22], this observation implies an inter-defect distance smaller than 3nm and confirms the reduction in size of the sp^2 domains. The graphitic structure could not be completed upon the different reduction processes. Figure 2 illustrates the mapping of the I_D/I_G ratio across the different samples.

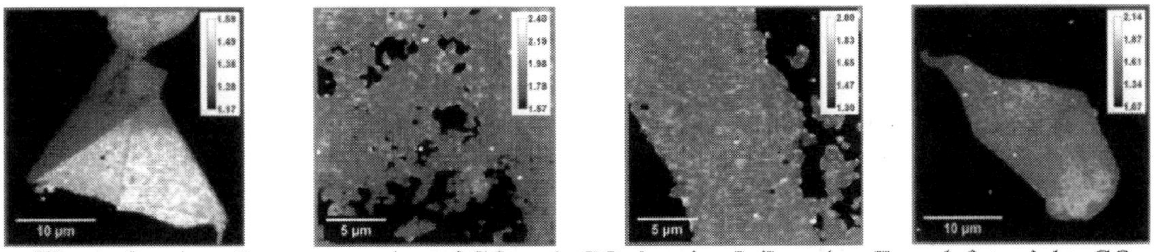

Figure 2: micro-Raman mapping of GO and rGO showing I_D/I_G ratios. From left to right: GO, $rGO_{hydrazyne}$, rGO_{UV}, $rGO_{bacterial}$.

To analyse the impact on the carbon content of the different reduction processes, XPS measurements were carried out for all the samples. Figure 3 shows the deconvoluted C1s and O1s peaks after Gaussian-Lorentzian fitting and Shirley background correction [23]. Figure 3 (a) exhibits C1s spectra acquired from all the samples. The GO sample presents a broad tail towards high energies. This tail is constituted of the various carbon bonding configurations i.e. 2 C components, C-C (for sp^3) and C=C (for sp^2) carbon respectively as well as C-OH, C=O and O=C-OH components, figure 3.

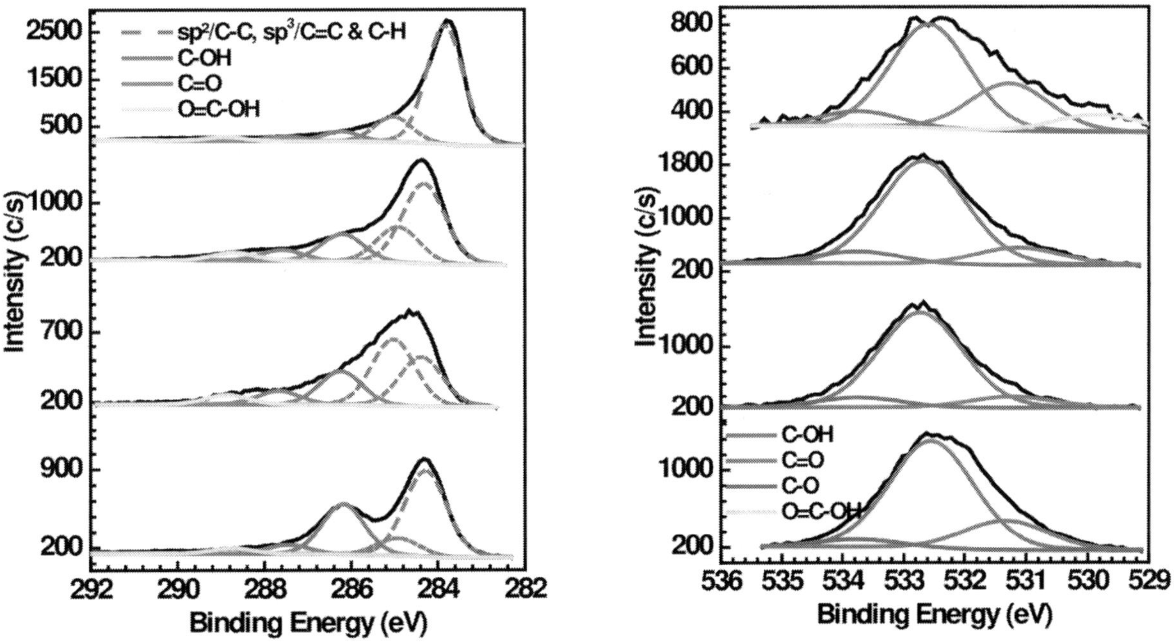

Figure 3: (a) C1s and (b) O1s spectra for GO and reduced GO. From bottom to top : GO, $rGO_{bacterial}$, rGO_{UV}, $rGO_{hydrazine}$.

Upon reduction, the ratios of the different peaks changed, the sp^2 and sp^3 contents increased whereas the contribution of the oxygenated functional groups is diminished as indicated by the atomic ratio C1s/O1s in table 2. The C/C (O) H atomic ratio indicates that $rGO_{hydrazine}$ presents a better reconstruction of the carbon network. The deconvolution of its C1s spectra for each sample reveals the following components: C=C or sp^2 peak ~284.3eV, C-C or sp^3 and C-H peaks ~ 284.9 eV, C-O-H ~ 286.2 eV, C=O ~ 287.5 eV and O=C-OH ~ 288.9 eV. As the reduction improves from GO to $rGO_{hydrazine}$, the sp^2 content increases in intensity and its binding energy shifts to 283.8 eV. The trends are consistent with the decrease in the oxygenated functional groups observed in the O1s spectra for all samples. The bacterial reduction seems to be more efficient in reducing the oxygen content with a

C1s/O1s of 8.1 as opposed to the 3.7 of the UV-photocatalysis but it does not improve the graphitization. However, this interpretation should be confirmed with further experiments as the contribution of the substrates utilized may impact the resulting C1s/O1s ratio. Increasing the contact time with the microorganisms and temperature parameters may improve the process [11]. This latest procedure is of interest as it does not involve any chemical nor radiative process and is fairly simple to implement.

	C1s/O1s (atomic ratios)	**C/C(O)H (atomic ratios)**
GO	3.24	1.5
Bacteria	8.1	1.94
UV-photocatalysis	3.7	2.4
Hydrazine	15.9	6.9

Table 2: XPS spectroscopy of GO and reduced GO. Note that the displayed carbon contents do not discriminate between sp^2 and sp^3 configurations.

CP-AFM measurements were performed on the samples in order to observe the lateral conduction within rGO flakes with respect to the initial GO. 1V was applied between the tip and few μm^2 flakes. The CP-AFM head allows monitoring the change in contact resistance. Note that due to the lack of thickness control during the GO spin coating and subsequent reduction, these measurements only provide qualitative insight on the electrical conductivity of the material. GO sample presents a resistance of 6 GΩ for a monolayer flake estimated from current-voltage measurements exhibiting a linear behaviour in the range -1 to 1 V (not illustrated). This local resistance increases to 170 GΩ for 3 flakes of GO. This is illustrated by the most resistive region on figure 4 (a). Figure 4 (b) represents the mapping for the $rGO_{hydrazine}$. The distribution is rather homogenous and indicates that the reduction is uniform across the flakes. This sample exhibits the lowest resistance in the order of 10-30 kΩ owing to its relatively large carbon content (sp^2 and sp^3 ~6.9compared to GO). The same observation holds for rGO_{UV} but with a higher resistance ~ 1-3 GΩ as expected from the low carbon content (~2.4) compared to $rGO_{hydrazine}$. Interestingly, the $rGO_{bacterial}$ on the other hand presents a lower resistance ~1MΩ on monolayer flake although its carbon content is lower than rGO_{UV} 1.94 versus 2.4%. However, XPS analysis (table 2) showed a more efficient reduction of oxygenated functional groups (8.1 for the $rGO_{bacterial}$ and 3.7 for RGO_{UV}. As a result, the improvement of the electrical conduction implies both the reduction of functional groups i.e. defects as well as the increase in sp^2 carbon domain [24].

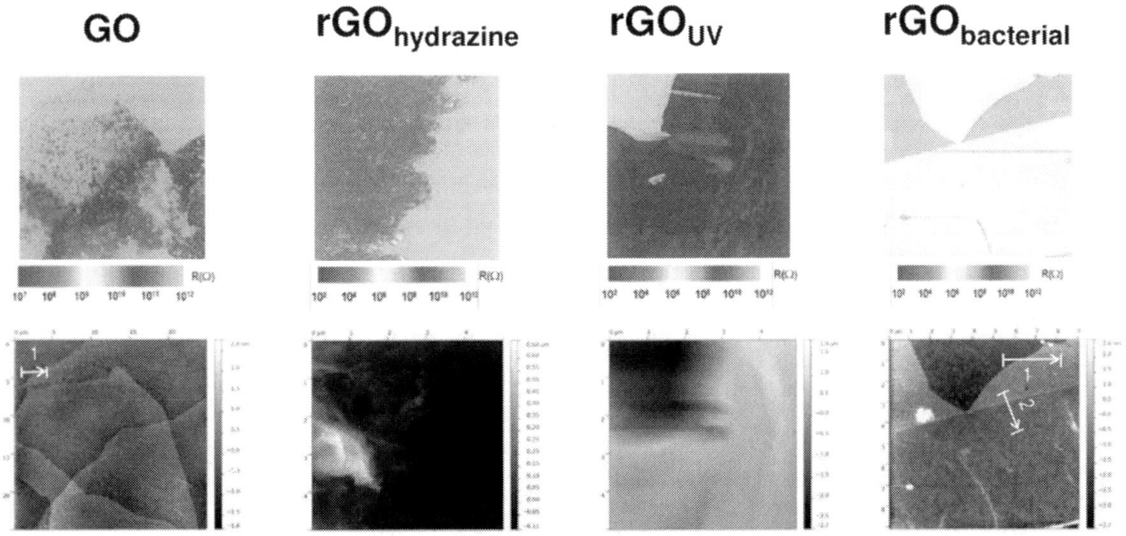

Figure 4: (top) CP-AFM local resistance mapping. **(Bottom)** topography mapping of the samples, the arrows indicate the scanning direction. Arrow 1 for GO represents a scan from substrate to a monolayer flake. Arrows 1 for $rGO_{bacterial}$ indicates a scan from substrate to a 2 layers flake whereas scan 2 is over a one monolayer flake. $RGO_{hydrazine}$ and RGO_{UV} topographies show rather thick deposits.

4. Conclusions

Three different reduction processes of graphene oxide have been characterized using μ-Raman spectroscopy, X-ray photoelectron microscopy as well as conductive probe atomic force microscopy. Among the different processes hydrazine remains the most effective to reduce the content of functional groups as well as to increase the sp^2 domains. On the other hand, the bacterial reduction is rather poor although more efficient than UV-photocatalysis to reduce the amount of oxygenated functional groups. This must be confirmed by further experiments in order to minimize the substrate contribution. Nevertheless, this bacterial procedure may be substantially improved by increasing contact time between GO and microorganisms. This could yield to a chemical free reduction process more amenable than hydrazine and environmentally safe.

Acknowledgements

The research at EIIRIS was partly supported by the Toyohashi Tech Research Program Initiative.

References

[1] Novoselov K S, Geim A K, Morozov S V, Jiang D, Zhang Y, Dubonos S V, Grigorieva I V and Firsov A A 2004 *Science* **306** (5696) 666–9
[2] Liu Z , Li J , Sun Z-H, Tai G , Lau S-P and Yan F 2012 *ACSNanon* **6** 1810–18
[3] Dreyer D R, Park S, Bielawski C W and Ruoff R S 2010 *Chem.Soc.Rev.***39** 228-40
[4] Moon I K, Lee J, Ruoff R S and Lee H 2010 *Nat. Commun.* 1:73 doi:10.1038/ncomms1067
[5] Bae S et al 2010 *Nat Nanotech* **5** 574–8
[6] Salomon A, Labrune M, Damon-Lacoste J, Moreno M and P. Roca i Cabarrocas 2010 *Proc. of the 25th EU-PVSEC* 671
[7] Viel P, Le X T , Huc V, Bar J, Benedetto A, Le Goff A, Filoramo A, Alamarguy A, Noël S , Baraton L and Palacin S 2008 *J. Mater. Chem.***18** 5913-20
[8] Haubner K, Morakowski J, Olk P, Eng L M, Ziegler C, Adolphi B and Jaehne E 2010 *Chem PhysChem* **11** 10 2131–39

[9] Stankovich S, Dikin D A, Piner R D, Kohlhaas K A, Kleinhammes A, Jia Y, Wu Y, Nguyen S T, Ruoff R S 2007 *Carbon* **45** 1558–65
[10] Fellahi O, Das M R, Coffinier Y, Szunerits S, Hadjersi T, Maamache M and Rabah Boukherroub 2011 *Nanoscale* **3** 4662–69
[11] Akhavan O and Ghaderi E 2012 *Carbon* **50** 1853-60
[12] Eda G, Ball J, Mattevi C, Acik M, Artiglia L, Granozzi G et al 2011*Mater Chem* **21** 30 11217-23
[13] Ishikawa R, Bando M, Morimoto Y, Park S Y and Sandhu A 2010 *Jpn. J. Appl. Phys.***49** 06GC02
[14] Houzé F, Meyer R, Schneegans O and Boyer L 1996 *Appl Phys Lett* **69** 1975
[15] Kleider JP, Longeaud C, Brüggemann R and Houzé F 2001 *Thin Solid Films* **57** 383
[16] Planès J, Houzé F, Chrétien P and Schneegans O 2001 *Appl Phys Lett* **79** 2993
[17] Alvarez J, Ngo I, Gueunier-Farret ME, Kleider JP, Yu L, Roca i Cabarrocas P, Perraud S, Rouvière E, Celle C, Mouchet C and Simonato JP 2011 *Nanoscale Research Letters* **6** 110
[18] Alvarez J, Kleider JP, Houze F, Liao MY, Koide Y 2007 *Diamond Relat Mater* **16** 1074
[19] Tanizawa Y, Okamto Y, Tsuzuki K, Nagao Y, Yoshida N, Tero R, Iwasa S, Hiraishi A, Suda A, Takikawa H, Numano R, Okada H, Ishikawa R and Sandhu A 2012 *JPCS* **352** 12011-015
[20] Ferrari A C, Meyer J C, Scardaci V, Casiraghi C, Lazzeri M, Mauri F, Pisanec S, Jiang D, Novoselov K S, Roth S and Geim A K 2006 *PRL* **97** 187401
[21] Eigler S, Dotzer C and Hirsch A 2012 *Carbon* **50** 3666-73
[22] Cançado L G, Jorio A, Martins Ferreira E H, Stavale F, Achete C A, Capaz R B, Moutinho M V O, Lombardo A, Kulmala T S and Ferrari A C 2011 *Nano Lett* **11** 3190-96
[23] S.Kaciullis 2012, *Surf.Interface Anal.* **44** 1155-61
[24] Yang D, Velamakanni A, Bozoklu G L, Park S, Stoller M, Piner R D, Stankovich S, Jung I, Field D A, Ventrice Jr C A, Ruoff R S 2009 *Carbon* **47** 145-152

Journal of Physics: Conference Series **433** (2013) 012002 doi:10.1088/1742-6596/433/1/012002

Simple method to transfer graphene from metallic catalytic substrates to flexible surfaces without chemical etching

P. J. Ko[1], H. Takahashi[1], S. Koide, H. Sakai[1], T. V. Thu[1], H. Okada[1], and A. Sandhu[1,2]

[1]Electronics-Inspired Interdisciplinary Research Institute (EIIRIS), Toyohashi University of Technology, Hibarigaoka, Tempaku-cho, Toyohashi, 441-8580, Japan

[2]Department of Electrical and Electronic Information Engineering, Toyohashi University of Technology, 1-1 Hibarigaoka, Tempaku, Toyohashi, Aichi, 441-8580, Japan

E-mail: ko@eiiris.tut.ac.jp

Abstract. Graphene shows promise for applications in flexible electronics. Here, we describe our procedure to transfer graphene grown on copper substrates by chemical vapor deposition to polydimethylsiloxane (PDMS) and SiO_2/Si surfaces. The transfer of graphene was achieved by a simple, etching-free method onto flexible PDMS substrates.

1. Introduction

Graphene is a single layer of carbon atoms arranged in a honeycomb lattice, and since its first isolation in 2004, the material has attracted tremendous attention due to its novel properties and potential applications [1–3]. In particular, graphene is promising for flexible electronics due to its high optical transparency, physical flexibility, and electrical conductivity [4–6]. Recent developments in the chemical vapor deposition (CVD) method have enabled the successful production of large scale, single layers of graphene (SLG) or few layer graphene (FLG) on metal catalytic surfaces such as Cu, Ni, Pd, Ru or Pt [7–10]. However, for practical flexible-electronic applications, graphene must be transferred to appropriate substrates such as polystyrene (PS), polyvinyl alcohol (PVA), polymethylmetacrylate (PMMA), and polydimethylsiloxane (PDMS) [11–14]. Here, we describe our procedure to transfer graphene deposited on copper (Cu) substrates to PDMS and SiO_2/Si surfaces by chemical etching methods. Furthermore, we also describe the results of experiments on transfer based on an etching-free procedure, where the graphene was transferred by a facile peel-off process for easily exfoliating graphene from Cu surfaces onto flexible surfaces of PDMS substrates. Notably, we demonstrate that the transferred graphene on PDMS by etching-free, not only no significant Raman shift, but also higher 2D/G peak ratio compared to that transferred etching method on the PDMS surface. Optimization for synthesis of the graphene was growth at 1000 °C for 15 min, with growth pressure of 5 Torr, C_2H_2 gas flow rate of 20 sccm and Ar gas flow rate 500 sccm. Furthermore, our approach can be renewable metal substrates, chemical-free, environment-friendly and economy of time. Finally, Transferred graphene on the flexible surface of PDMS can be possible of application in many electronic devices, such as flexible thin-film, solar cells, and touch panels.

2. Experimental

We used chemical vapor deposition (CVD) for the deposition of graphene onto Cu foils using C_2H_2/Ar (20 sccm : 500 sccm), at 5.0 Torr reactor pressure, substrate temperature of 1000 °C followed by rapid cooling to room temperature. The PDMS pre-polymer was prepared by mixing a silicone elastomer base with a curing agent with a ratio of 10:1. The PDMS and curing agent (Silpot 184 and Catalyst Silpot 184) were purchased from Dow Corning, Toray Co., Ltd. A solution of TBAF (tetrabutylammonium fluoride) in NMP (N-Methelpyrrolidone) (5:1; v/v; NMP/75% TBAF in water, TCI, Tokyo, Japan) was used to sufficiently dissolve the PDMS without leaving a residue at room temperature for 24 hours. All the Raman spectra were measured by a Nihon Bunko NRS-7100 system with excitation light of a 532 nm laser, with an estimated laser spot of less 1 μm.

3. Results and Discussion

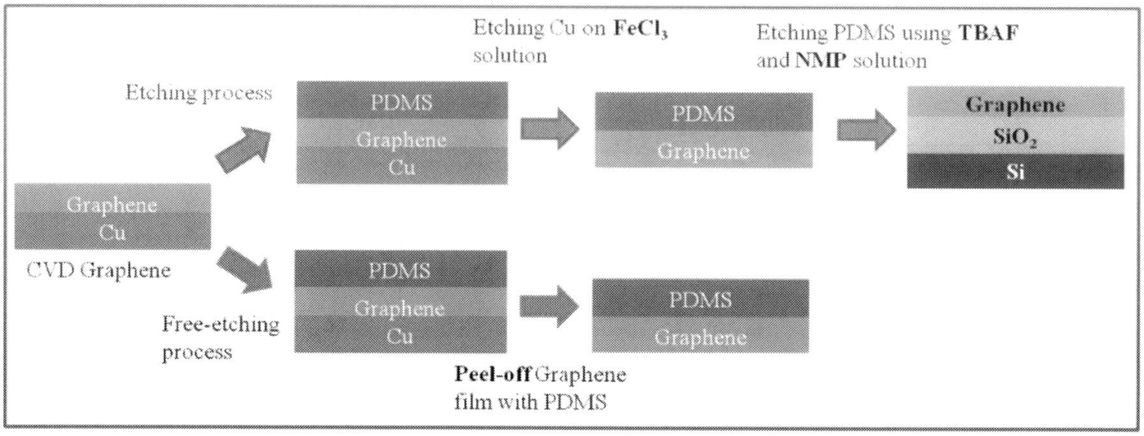

Figure 1. Schematic diagram of the processes used for transferring graphene from copper foils to target substrates using etching and etching-free methods.

Figure 1 is a schematic of the steps used to transfer graphene from copper foil substrates to flexible PDMS substrates using $FeCl_3$ etching and the etching-free, direct peel-off method. We deposited graphene on copper foils by CVD at 1000 °C in a gas mixture of C_2H_2 : Ar = 20 : 500 sccm. The liquid mixture of PDMS and curing agent (PDMS : curing agent = 10 : 1) was poured on the surface of the graphene-Cu foils. Next, the PDMS mixture was placed into a vacuum chamber for 30 min to remove bubbles. We prepared same substrate of graphene-Cu on solid PDMS for using chemical-etching and etching-free experiments.

In the chemical etching process the Cu foil was etched using $FeCl_3$ for 24 h and the graphene transferred using PDMS. The PDMS-graphene was attached onto a SiO_2/Si substrate and the PDMS dissolved in a solution of TBAF : NMP (mixing ratio of 10 : 1) at room temperature for 24 hours.

In the etching-free process, a solid layer of PDMS was slowly removed from the Cu substrate. This resulted in the graphene being transferred to the PDMS surface because the adhesion force of PDMS-graphene is higher than that between the graphene-graphene. We noticed that the ratio of the PDMS and curing agent determined the adhesion force between PDMS and target substrate.

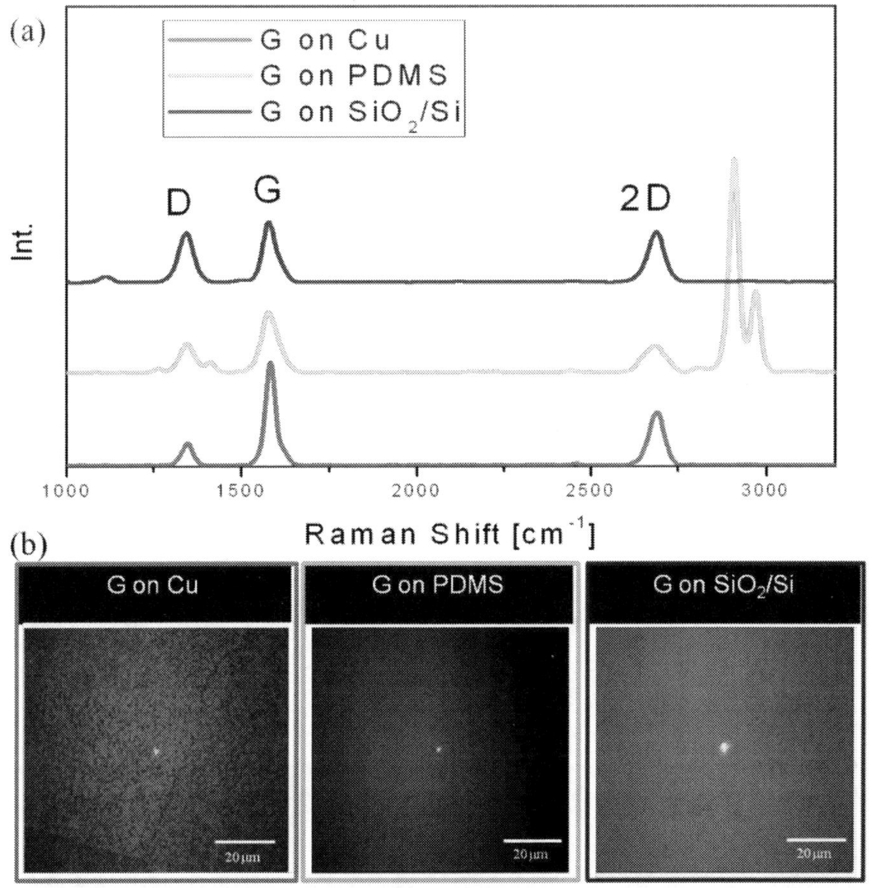

Figure 2. (a) Raman spectra and (b) optical microscope images show graphene on Cu, PDMS, and SiO_2/Si substrates.

Figure 2 (a) shows Raman spectra of graphene on Cu, PDMS, and SiO_2/Si substrates. Few-layered graphene resulted by CVD growth at 1000 °C (C_2H_2 : Ar = 20 : 500 sccm). The sharp peaks of graphene-Cu spectrum are due to graphene with the three signature peaks at 1345 cm^{-1} (D peak), 1576 cm^{-1} (G peak), and 2681 cm^{-1} (2D peak). The PDMS graphene was transferred from the CVD graphene-Cu by etching Cu with $FeCl_3$ solution. The graphene transferred to PDMS showed D, G, and 2D peaks and no significant changes. However, the Raman spectra showed an increase of the intensity of G to D peak (I_G/I_D) after chemical etching with $FeCl_3$, indicating the possibility of damage of the graphene surface layer [15]. Furthermore, the intensity of (I_G/I_D) of the graphene transferred to SiO_2/Si by dissolving in a solution of TBAF/NMP increased. Figure 2(b) shows optical microscopic images of graphene on Cu, PDMS, and SiO_2/Si substrates.

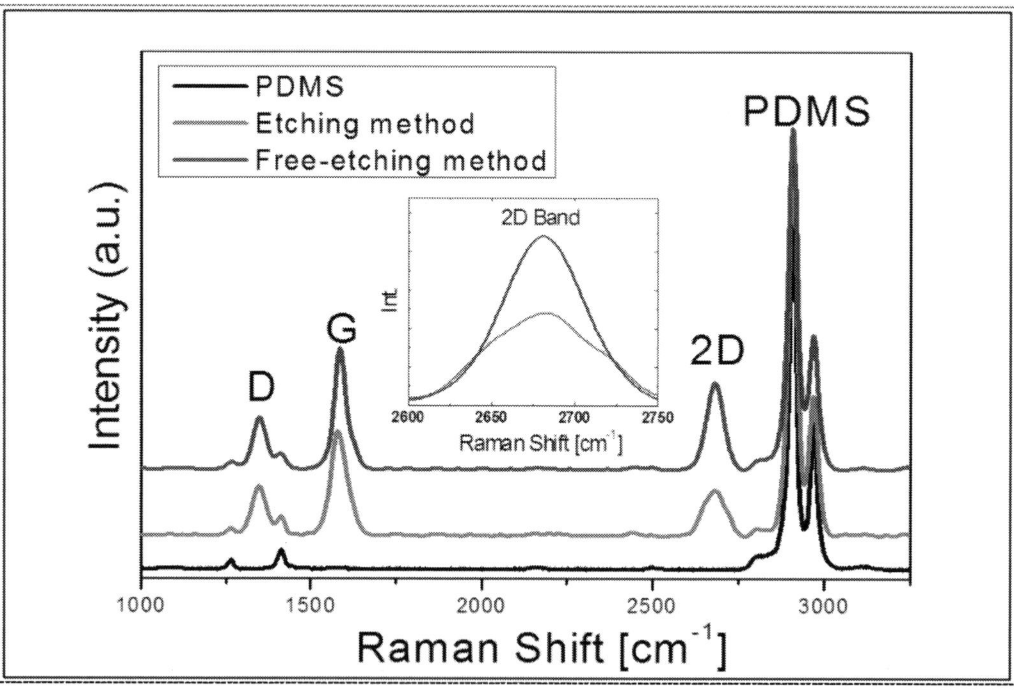

Figure 3. Comparison of Raman spectra of the graphene transferred on PDMS by different methods.

Figure 3 is a comparison of Raman spectra of graphene transferred onto PDMS by etching copper, etching-free, and PDMS based methods. The PDMS (black line) peaks were observed for all transferred samples by etching PDMS (red line) and free-etching PDMS (blue line). The D 1345 $cm^{-1}$and G 1576 cm^{-1} band positions and values of the G-to-D peak ratio (I_G/I_D)

were similar for etching (~ 2.06) and etching-free (~2.15) samples. However, the full width at half maximum (FWHM) of the 2D Raman band of the etching-free sample was narrower than the etching sample, with the etching sample being 80.4 cm^{-1} and etching-free 55.9 cm^{-1}, respectively. The shape, 2D to G peak ratio (I_{2D}/I_G), and FWHM of the 2D band was used to determine the number of graphene layers because the 2D band originates from the two phonon double resonance process and is closely related to the band structure of graphene layers [16]. Therefore, the narrow 2D peaks imply that the etching-free method can transfer few-layered graphene.

4. Conclusion

We described two simple methods etching and etching-free for transferring graphene grown on Cu substrates onto flexible surfaces of PDMS and silicon. We successfully transferred graphene onto flexible surfaces from Cu substrates without chemical etching. Chemical etching with $FeCl_3$ led to an increase in the (I_G/I_D) ratio indicating the formation of chemical induced damage in the graphene layer. The observation of a narrow 2D peak from graphene transferred by a etching-free implied the possibility of transferring few-layered graphene onto flexible surfaces. The graphene on flexible PDMS opens up the possibility of applications for electronic devices such as solar cells and touch panels.

Acknowledgements: This work was supported in part by a grant from the Ministry of Education, Culture, Sports, Science, and Technology, Japan in the form of the Special Coordination Funds for Promoting Science and Technology for the Integrated Research Institute. Pil Ju Ko acknowledges the support of the Toyohashi Tech Venture Business Laboratory (VBL) for financial support from April 2012.

References

[1] A. K. Geim. & K. S. Novoselov: *Nat. Mater.* **6** (2007) 183.

[2] C. Lee, X. D. Wei, J. W. Kysar, and J. Hone: *Science* **321** (2008) 385.

[3] A. A. Balandin, S. Ghosh, W. Z. Bao, I. Calizo, D. Teweldebrhan, F. Miao, and C. N. Lau: *Nano Lett.* **8**, (2008) 902.

[4] X. Wang, L. J. Zhi, and K. Mullen: *Nano Lett.* **8** (2008) 323.

[5] Y. M. Lin, K. A. Jenkins, A. Valdes-Garcia, J. P. Small, D. B. Farmer, and P. Avouris: *Nano Lett.* **9** (2009) 422.

[6] K. S. Kim, Y. Zhao, H. Jang, S. Y. Lee, J. M. Kim, K. S. Kim, J. H. Ahn, P. Kim, J. Y. Choi, and B. H. Hong: *Nature* **457** (2009) 706.

[7] X. Li, W. Cai, J. An, S. Kim, J. Nah, D. Yang, R. Piner, A. Velamakanni, I. Jung, E. Tutuc, S. K. Banerjee, L. Colombo and R. S. Ruoff: *Science* **324** (2009) 1312.

[8] P. W. Sutter, J.-I. Flege and E. A. Sutter: *Nat. Mater.* **7** (2008) 406.

[9] C. Mattevi, H. Kim, and M. Chhowalla: *J. Mater. Chem.* **21** (2011) 3324.

[10] T. Gao, S. Xie, Y. Gao, M. Liu, Y. Chen, Y. Zhang, and Z. Liu: *ACS Nano* **5** (2011) 9194.

[11] H. Hu, X. Wang, J. Wang, L. Wan, F. Liu, H. Zheng, R. Chen , C. Xu: *Chem. Phys. Lett.* **484** (2010) 247.

[12] J. H. Wu, Q. W. Tang, H. Sun, J. M. Lin, H. Y. Ao, M. L. Huang, Y. F. Huang: *Langmuir* **24** (2008) 4800.

[13] J. Wu, M. Agrawal, H. A. Becerril, Z. Bao, Z. Liu, Y. Chen, P. Peumans: *ACS Nano* 4 (2009) 43.

[14] W. Cai, Y. Zhu, X. Li, R. D. Piner, and D. S. Ruoff: *Appl. Phys. Lett.* **95** (2009) 123115.

[15] W. Regan, N. Alem, B. Aleman, B. Geng, C. Girit, L. Maserati, F. Wang, M. Crommie, A. Zettl: *Appl. Phys. Lett.* **96** (2010) 113102.

[16] C. Thomsen, S. Reich: *Phys. Rev. Lett.* **85** (2000) 5214.

Synthesis and characterization of graphite nanoplatelets

T V Thu[1], **Y Tanizawa**[1], **N H H Phuc**[2], **P J Ko**[1] and **A Sandhu**[1,2]
[1] Electronics-Inspired Interdisciplinary Research Institute, Toyohashi University of Technology, 1-1 Hibarigaoka, Tempaku, Toyohashi, Aichi 441-8580, Japan
[2] Department of Electrical and Electronic Information Engineering, Toyohashi University of Technology, 1-1 Hibarigaoka, Tempaku, Toyohashi, Aichi, 441-8580, Japan

E-mail: thu@eiiris.tut.ac.jp, and tranvietthu@gmail.com

Abstract. We report a facile method for the synthesis of graphite nanoplatelets (GNPs) from a carbon rod. GNPs were produced by vacuum filtration of a suspension obtained from electrochemically exfoliated carbon rods. The resulting GNPs were characterized using scanning electron microscopy, atomic force microscopy, ultraviolet-visible spectroscopy, and Raman spectroscopy. GNPs had sizes of 40–170 nm, thickness of 2.8–5.8 nm and were well dispersed in dimethyl sulfoxide. We discuss a possible mechanism for the formation of GNPs in our experiments.

1. Introduction

Graphite nanoplatelets (GNPs) have recently attracted considerable attention as a cost effective substitute for carbon nanotubes in many applications [1]. GNPs combine beneficial properties of graphite such as thermal and electrical conductivity with those derived from their unique size and shape [1]. However, in spite of the intriguing properties of these materials there are few reports about the synthesis of such materials. Therefore, it is important to explore ways of synthesing them and understanding their properties. Electrochemical exfoliation is a promising strategy for the preparation of graphene nanosheets [2,3,4] yet its potential for producing GNPs has not been explored. In addition, electrochemical exfoliation of graphite-based materials has heavily relied on ionic liquids, which are very expensive and thus not attractive for the large scale production of GNPs.

Here, we produced GNPs by electrochemical exfoliation of carbon rods in an electrolyte solution. We utilized an aqueous solution of sulphuric acid (H_2SO_4) as an electrolyte, which is low-cost but highly efficient in exfoliating GNPs from carbon rods. GNPs were successfully separated from the reaction solution by vacuum filtration. GNPs exhibited uniform shapes, a narrow size distribution, and unique optical properties. Our method enables the low cost, large scale production of GNPs.

2. Experimental

2.1. Synthesis of GNPs by electrochemical exfoliation of a carbon rod

Figure 1 is a schematic illustration of the electrochemical exfoliation of a carbon rod by our method. The carbon rod was taken from an exhausted battery and employed as the source of GNPs. A Pt wire was placed in parallel to the carbon rod at a distance of 3 cm. The two electrodes were immersed in an aqueous solution of H_2SO_4 (concentration 0.722 M). To initiate the electrochemical exfoliation, a DC bias of +15 V was applied to the carbon rod with the Pt wire grounded. The resulting suspension was filtered and repeatedly washed in distilled water. The solid material that remained on the filter membrane was collected and ultrasonically dispersed in dimethyl sulfoxide (DMSO) for 5 min. Finally, the DMSO dispersion was vacuum-filtered through a 0.2 μm polytetrafluoroethylene (PTFE) membrane, and the filtration was collected for characterization.

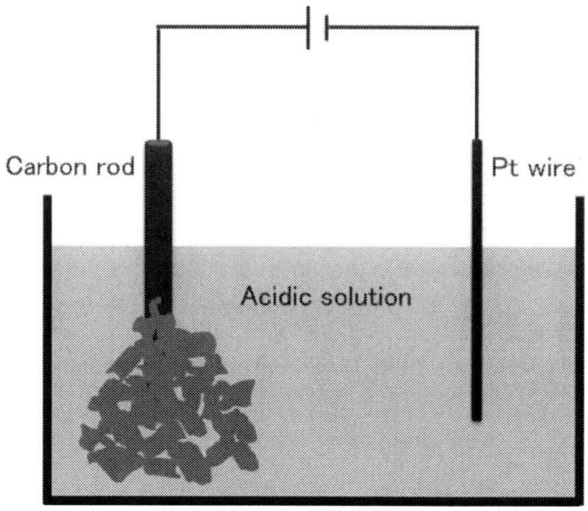

Figure 1. Schematic illustration of the electrochemical exfoliation of a carbon rod for synthesis of graphite nanoplatelets (GNPs).

2.2. Characterization

Scanning electron microscopy (SEM) images of GNPs were observed in an Hitachi S-4800 field-emission scanning electron microscope. Atomic force microscopy (AFM) images were performed in tapping mode using a PicoScan2500 atomic force microscope (Agilent). AFM data were visualized and analysed using Gwyddion software package (http://gwyddion.net/). UV-VIS spectroscopy was performed on a V-670 spectrophotometer (JASCO). Raman spectra of GNPs were collected with a NRS-7100 Raman system (JASCO) with the 532 nm green laser line. In UV-VIS measurements, filtration was directly used with a DMSO reference solution. For other characterization, the samples were prepared by dropping the filtration solution onto a silicon substrate and drying in air.

3. Results and discussion

As observed by naked eyes, the carbon rod corroded continuously during electrochemical exfoliation. As a result, the electrolyte solution was increasingly filled with black particles and sheets, which might contain graphite-based particles of varying sizes. In addition, we expected some undesired components such as binders (from the carbon rod) to be incorporated in the solution. Therefore, appropriate processing of the resulting solution was required to produce a suspension of high purity GNPs . Thus, we have devised an efficient protocol consisting of three steps: washing, ultrasonication, and filtering. The solution was first filtered and repeatedly washed with distilled water and ethanol to eliminate impurities, and then the solid material on filter paper was ultrasonicated in DMSO to dissolve soluble particles. Finally, the DMSO suspension was filtered through a 0.2μm PTFE membrane in order to remove particles larger than the membrane pore size. Only the filtrated solution was collected for subsequent characterization. It should be noted that in most previous studies on electrochemical exfoliation of graphite-based materials, the filtration is normally treated as waste, and only the solid material on the filter membrane is collected for characterization. However, we found that the filtrated solution obtained using the 0.2μm PTFE membrane contained large quantities of GNPs, which are the focus in this study.

Figure 2(a) shows a typical SEM image of GNPs obtained in our experiments, and Fig. 2(b) shows our statistical analysis of this image, which contains about 720 GNPs. The statistical analysis shows that the size of GNPs varied from 50 nm to 170 nm, with the majority is found in the range 70–140 nm.

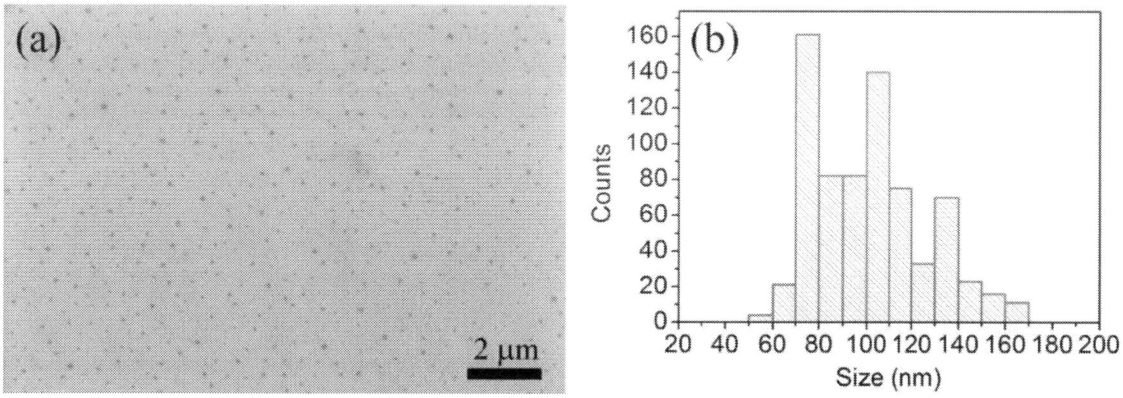

Figure 2. FE-SEM image (a) and size distribution (b) of GNPs

The morphology of GNPs was further investigated by using AFM. Figure 3(a) shows the AFM topography of GNPs, with data levelled by mean plane subtraction for easy visualization. Figure 3(b) shows the two corresponding cross section profiles. From these profiles, the thicknesses of GNPs were estimated to be from 2.8 nm (few-layer graphene) to 5.8 nm (more than 10 layers of graphene), assuming the graphite interlayer spacing to be about 0.35 nm. We did not observe any GNPs thicker than 10 nm. Furthermore, as shown in Fig. 3(b), the sizes of GNPs was within 40–90 nm, which is in approximate agreement with SEM observations.

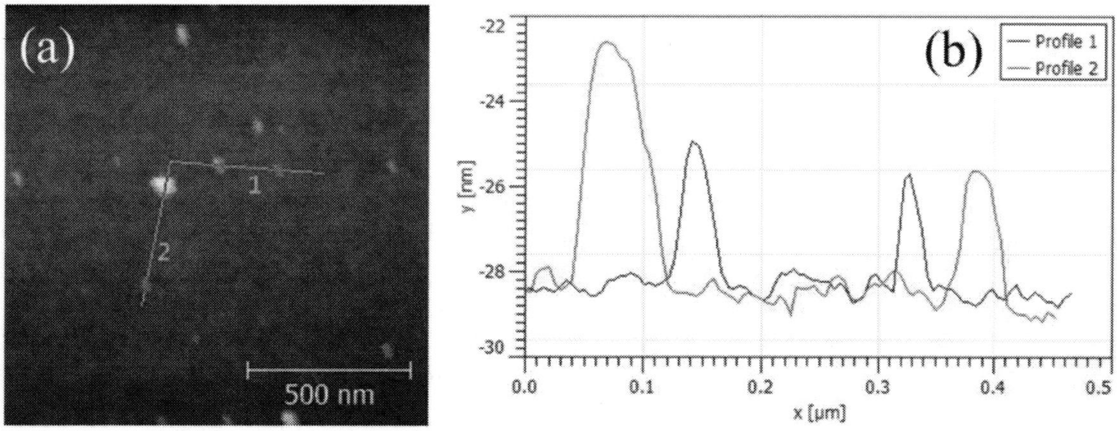

Figure 3. AFM tomography (a) and cross section profiles (b) of GNPs

The absorbance spectrum of GNPs is shown in Fig. 4. A single strong and sharp adsorption peak appeared at 259 nm was observed. We assign this feature to the $\pi \rightarrow \pi^*$ transition of C = C double bonds in GNPs [5]. The C = C double bonds are formed because graphite is partly oxidized in acidic solution due to the applied voltage. The formation of oxidized carbon in GNPs results in electrostatic repulsion among GNPs when they are dispersed in a polar solvent such as DMSO. In agreement with this, our GNPs are stable in DMSO for months after preparation.

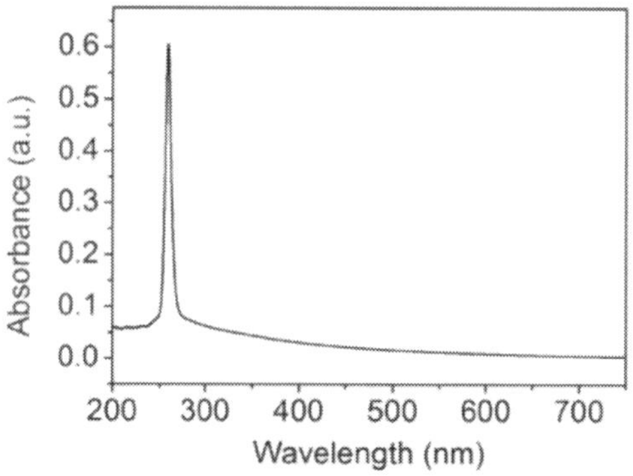

Figure 4. Absorption spectrum of GNPs

If the oxidation takes place at the basal plane, then C – O – C bondings will be formed. As a result, a broad shoulder peak is expected to appear at around 300 nm due to n $\rightarrow \pi^*$ transition [6]. However, such a peak was not observed. This suggests that GNPs are only partly oxidized at the edge of GNPs.

Figure 5 shows the Raman spectrum of GNPs. Two strongest peaks centered at 1351 cm^{-1} and 1602 cm^{-1} correspond to D band and G band vibrations of graphite, with a I_D/I_G ratio of 0.95. The D band is caused by structural imperfections such as oxidized carbon at the edge [7]. The G band corresponds to the in-phase vibration of the graphite lattice [8].

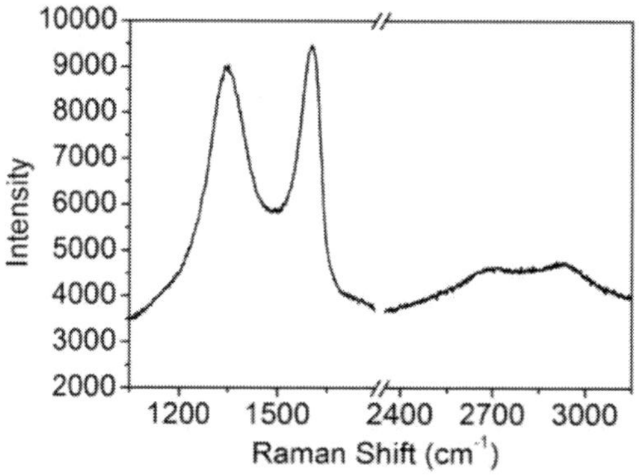

Figure 5. Raman spectrum of GNPs

We suggest that there are two main factors involved in the formation of GNPs in our experiment. The first one is the role of water. Water molecules intercalate into interlayer space of graphite domains in the carbon rod and donate their electrons when a positive voltage bias is applied to the rod. This process leads to the formation of oxygen with increasing density, and finally gaseous oxygen is liberated, resulting in the exfoliation of graphite domains. This factor is, however, responsible for the formation of graphene sheets rather than GNPs. Therefore the second factor must be associated with the composition and structure of the carbon rod employed in this work. In high purity graphite, the size of graphene layers is on the micrometer scale. Our carbon rod was produced by graphitizing a mixture of graphite powder and binders at high temperature, therefore, graphite domains at nanoscale do exist. After size screening by vacuum filtration, GNPs could be obtained.

4. Conclusions

We demonstrated that electrochemical exfoliation is a promising route for preparing GNPs. The successful preparation of GO from waste carbon rods may offer an alternative and inexpensive way of producing high-quality GO for both fundamental research and practical applications.

Acknowledgments

This work was supported by a start-up grant from the Electronics-Inspired Interdisciplinary Research Institute (EIIRIS) at Toyohashi University of Technology, Japan. The authors would like to thank Prof. A. Matsuda and Dr. R. Tero of Toyohashi University of Technology for their support in the characterization of the materials described in this report.

References

[1] Kalaitzidou K, Fukushima H and Drzal L T 2007 *Carbon* **45** 1446
[2] Liu N, Luo F, Wu H, Liu Y, Zhang C and Chen J 2008 *Adv. Funct. Mater.* **18** 1518
[3] Su C Y, Lu A Y, Xu Y, Chen F R, Khlobystov A N and Li L J 2011 *ACS Nano* **5** 2332
[4] Lu J, Yang J X, Wang J, Lim A, Wang S and Loh K P 2009 *Nano Lett.* **3** 2367
[5] Li D, Müller M B, Gilje S, Kaner R B and Wallace G G 2008 *Nat. Nanotech.* **3** 101
[6] Saxena S, Tyson T A, Shukla S, Negusse E, Chen H and Bai J 2011 *J. Appl. Phys.* **99** 0131041
[7] Yang D, Velamakanni A, Bozoklu G, Park S, Stoller M, Piner R, Stankovich S, Jung I, Field D and Ventricejr C 2009 *Carbon* **47** 145–52
[8] Kudin K N, Ozbas B, Schniepp H C, Prud'homme R K, Aksay I A and Car R 2008 *Nano Lett.* **8** 36

Electronic structure and magnetic properties of endohedral metallofullerenes based on mixed metal cluster within fullerene cage with trifluoromethyl groups

A Suzuki and T Oku

Department of Materials Science, The University of Shiga Prefecture
2500 Hassaka, Hikone, Shiga 522-8533, Japan

E-mail: suzuki@mat.usp.ac.jp

Abstract. Endohedral metallofullerenes based on encapsulated nitride scandium cluster within C_{80} fullerene cage with trifluoromethyl group of CF_3 were characterized as a candidate model for controlling nuclear spin gate in NMR quantum computers. Effects of yttrium in nitride scandium cluster of Sc_3N within the fullerene cage with trifluoromethyl group on electronic structures and magnetic properties were investigated. Electronic structure and chemical shifts in ^{45}Sc, ^{89}Y, ^{19}F, ^{13}C and ^{14}N atoms were calculated by quantum calculation using density functional theory. Addition of yttrium in nitride scandium cluster of Sc_3N within the fullerene cage provided a different position of chemical shift in ^{45}Sc, ^{89}Y, ^{19}F, ^{13}C and ^{14}N atoms. This result was originated in magnetic interaction with spin-local interaction, dipole-dipole interaction and nuclear quadrupole interaction based on nuclear quadrupole moment and asymmetry of electron field gradient with a bias value of charge and spin density distribution of spd-hybrid orbital by inductive effect.

1. Introduction

Electronic structure and magnetic parameters of principle g-tensor, A-tensor for hyperfine coupling constant (hfc) of nitrogen, phosphorus, scandium atoms, and isotropic chemical shift of ^{13}C in $^{14}N@C_{60}$, $^{31}P@C_{60}$ and $^{45}Sc@C_{82}$ within SWCNT as peapods for chiral index had been studied by electron spin resonance (ESR), electron nuclear double resonance (ENDOR), NMR and density functional theory (DFT) [1-4]. Recently, electronic structure and magnetic properties of endohedral fullerene based on encapsulating mixed metal cluster with several varieties of the modified fullerene cage and the mixed metal cluster have considerably been studied [5-8]. Trimetallic nitride endohedral fullerenes of $Sc_3N@C_{80}$ and $Sc_{3-x}Y_xN@C_{80}$ based on central atom of nitrogen atom combined with alterable encaged metal atoms have a great potential to be a candidate model for spin electronic devices with controlling nuclear spin gate as NMR quantum computer. Characterization and electronic structure of endohedral metallofullerene derivatives based on asymmetric nitride scandium clusters within modified fullerene cage has been investigated by NMR, ESR and ultra-visible-near-infrared (UV-vis-NIR) absorption spectrometry using DFT [9-11]. For instant, synthesis and spectroscopic studies of a family of $Sc_3N@C_{80}(CF_3)_n$ (n=2-16) and their radical anion have been reported [9]. Spectroscopic (Mass spectrometry, ^{19}F-NMR, UV-vis-NIR, ESR), structural, electrochemical studies revealed that addition of substitution CF_3 groups in $Sc_3N@C_{80}(CF_3)_n$ leaded to be easier to reduce than $Sc_3N@C_{80}$ [10]. The experimental and computational studies of the CF_3-addition motifs to $Sc_3N@C_{80}$

Journal of Physics: Conference Series **433** (2013) 012004 doi:10.1088/1742-6596/433/1/012004

reported a special role played by the endohedral metal cluster in determining electronic structure of $Sc_3N@C_{80}$ [11]. In addition, effect of encapsulating mixed metal cluster in a series of endohedral fullerenes, $ScYErN@C_{80}$, $ScLuN@C_{80}$ and $Sc_{3-x}Y_xN@C_{80}$ (x=0-3) on their electronic structure and reactivity have been investigated [12, 13]. The size of the endohedral moiety of mixed metal cluster in $Sc_2YN@C_{80}$ and $ScY_2N@C_{80}$ as similar geometric structure played an important role in affecting the electronic properties and chemical properties of endohedral fullerenes. In this work, effect of yttrium in nitride scandium cluster within modified fullerene cage of trifluoromethylated metallofullerene, $Sc_3N@C_{80}(CF_3)_2$ on electron structure and magnetic properties will be investigated by quantum calculation using DFT. The electronic structure and magnetic interaction will be discussed on the basis of quantum consideration using spin density distribution. Theoretical consideration based on Ramsey theory will be performed to reveal the factors determining the magnitude of chemical shifts in nitride mixed metal cluster fullerenes.

2. Quantum calculation

Isolated molecular structures were optimized by *ab-initio* quantum calculation using restricted Hartree–Fock (RHF) and DFT using B3LYP with hybrid function LANL2DZ and STO-3G*, 3-31G* and 6-31G* as basis set because of computational limitations (Gaussian 03 Inc.). The highest occupied molecular orbital (HOMO), HOMO-1, the lowest unoccupied molecular orbital (LUMO), LUMO+1 and the HOMO-LUMO band gap (E_g) were calculated. Mulliken atomic charges and spin density distribution were estimated by Mulliken population analysis. Continuously, isotropic chemical shift (δ) of ^{13}C, ^{14}N and ^{45}Sc, ^{89}Y and ^{19}F atoms were calculated by DFT using NMR/GIAO with hybrid function B3LYP and 3-31G* as basis set. Nuclear quadruple interactions were estimated on basis of multiplication of nuclear quadruple moment and the electro field gradient (EFG) tensor generated by charge distribution around nucleus. Excited energy level at state and wavelength in UV-vis-NIR spectra of $Sc_3N@C_{80}(CF_3)_2$, $Sc_2YN@C_{80}(CF_3)_2$ and $ScY_2N@C_{80}(CF_3)_2$ as referred to $Sc_3N@C_{80}$, $Sc_2YN@C_{80}$, $ScY_2N@C_{80}$, Sc_3N, Sc_2YN and ScY_2N were calculated by time-dependent DFT (TD-DFT) with B3LYP and RHF using 3-21G* as basis set.

3. Results and Discussion

Figure 1 show DFT-computed spin densities at HOMO and LUMO of trifluoromethylated metallofullerenes: (a) $Sc_3N@(C_{80}\text{-}I_h)(CF_3)_2$, (b) $Sc_2YN@C_{80}(CF_3)_2$ and (c) $ScY_2N@C_{80}(CF_3)_2$. Electron spin density distribution of s, p and 3d-orbital in ^{14}N, ^{45}Sc and ^{13}C atoms at HOMO was delocalized on the carbon cage around trifluoromethyl group. The electron spin density distribution at LUMO was delocalized on surface of carbon cage near trifluoromethyl group. Addition of yttrium in scandium cluster of Sc_3N as mixed metal cluster within fullerenes cage influenced orbital-phase with electron spin density distribution. Calculated energy levels and band gap of metallofullerenes and metal clusters are listed in Table 1. In the case of $Sc_3N@(C_{80}\text{-}I_h)(CF_3)_2$, energy levels at LUMO and HOMO and the energy gaps (E_g) between HOMO and LUMO were estimated to be -3.7 eV, -5.9 eV and 2.2 eV, which was corresponded to $\pi \rightarrow \pi^*$ excited transition. Addition of only two CF_3 groups as electron-withdrawing group caused inductive effect to result a considerable bias of hybrid spin density distribution on the fullerene surface cage encapsulated with homo-metallic cluster. Addition of trifluoromethyl group on the fullerene cage decreased the energy levels on LUMO, providing a narrow band gap. Trifluoromethylation stabilized electronic structure with a narrow band gap, yielding expectation with a long wavelength. Similarly, energy gaps of $Sc_2YN@C_{80}(CF_3)_2$ and $ScY_2N@C_{80}(CF_3)_2$ were slightly decreased to be 2.1 eV and 2.2 eV, as compared with those of $Sc_2YN@C_{80}$ and $ScY_2N@C_{80}$ to be 2.5 eV and 2.6 eV, respectively. As reference, energy gaps of metal cluster using Sc_3N, Sc_2YN and ScY_3N were strongly increased to be 6.0 eV, 5.5 eV, and 5.3 eV, which were corresponded to $d \rightarrow d^*$ excited transition.

Addition of yttrium in nitride scandium cluster of Sc_3N within the fullerene cage will provide expect of a slight different position of chemical shift, which based on nuclear quadrupole interaction with nuclear quadrupole moment and asymmetry of electronic field graduate (EFG) with a bias value of charge and spin density distribution of spd-hybrid orbital. For instant, the nuclear quadrupole interaction in $ScY_2N@C_{80}(CF_3)_2$ is depending on nuclear quadrupole moment to be -588~-593 Debye-Ang. with a bias value of EFG of ^{89}Y, ^{45}Sc and ^{14}N atoms, asymmetry parameter (η) to be 0.36~0.38, 0.58 and 0.20, Mulliken atomic charge of ^{89}Y, ^{45}Sc and ^{14}N atoms to be 2.03~2.04 e, 2.02 e, -0.92 e. The nuclear quadrupole interaction in $Sc_3N@C_{80}(CF_3)_2$ is caused by nuclear quadrupole moment to be -573~-577 Debye-Ang. with a bias value of EFG of ^{45}Sc and ^{14}N atoms with η to be 0.59~0.64 and 0.11, Mulliken atomic charge of ^{45}Sc and ^{14}N atoms to be 1.85~1.92 e and -0.87 e. The nuclear quadrupole interaction of ^{45}Sc and ^{14}N based on nuclear quadrupole moment with a bias of EFG will provide a different position of the chemical shift in ^{45}Sc and ^{14}N atom with multicity at I=7/2 and 1.

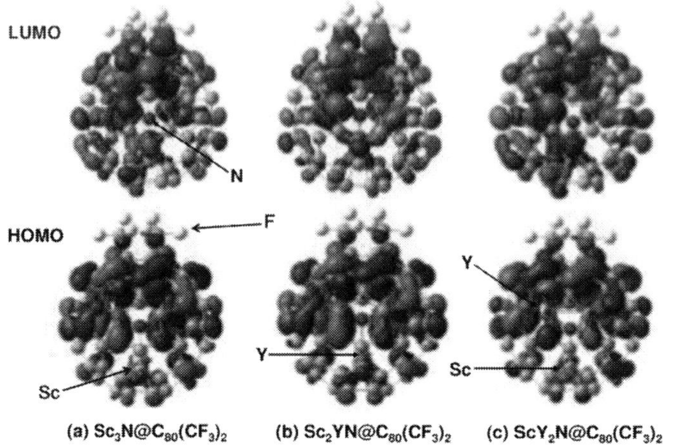

Figure 1 DFT-computed spin densities at HOMO and LUMO of trifluoromethyl metallofullerenes: (a) $Sc_3N@(C_{80}\text{-}I_h)(CF_3)_2$, (b) $Sc_2YN@C_{80}(CF_3)_2$ and (c) $ScY_2N@C_{80}(CF_3)_2$.

Calculated excited states and wavelength of mixed metallofullerenes of $Sc_3N@(C_{80}\text{-}I_h)(CF_3)_2$, $Sc_2YN@C_{80}(CF_3)_2$ and $ScY_2N@C_{80}(CF_3)_2$ as calculated by TD-DFT with B3LYP using 3-21G* as basis set are shown in Fig. 2 and Table 2. Energy levels and wavelengths of mixed metallofullerenes were calculated to 2.28 eV, 2.43 eV and 2.36 eV, which converted to be 544 nm, 511 nm and 526 nm. The calculated excited states corresponded to HOMO to LUMO+3 excited transition. As shown in Fig. 2, there were multi levels of energy between HOMO, LUMO, LUMO+1, LUMO+3 and band gaps, which would be assigned as multi states as exited transition, suggesting a wide range of wavelength. This result would be originated in s, p, and d-hybridization at d–d* and π-π* exited-transitions near HOMO and LUMO+3. Addition of yttrium in nitride scandium cluster within trifluoromethylated fullerene cage of $C_{80}(CF_3)_2$ provided a slight change of band gap with a slight blue shift of wavelength, as compared with those of $Sc_2Y@C_{80}$, $ScY_2@C_{80}$, Sc_2Y and ScY_2 as listed in Table 1 [13].

Table 1 Calculated energy levels and band gaps of metallofullerenes and nitride metal clusters.

	LUMO+1 (eV)	LUMO (eV)	HOMO (eV)	HOMO-1 (eV)	E_g (eV)
$Sc_3N@C_{80}(CF_3)_2$	-3.4	-3.7	-5.9	-6.2	2.2
$Sc_2YN@C_{80}(CF_3)_2$	-3.5	-3.9	-6.0	-6.2	2.1
$ScY_2N@C_{80}(CF_3)_2$	-3.5	-3.7	-5.9	-6.2	2.2
$Sc_3N@C_{80}$	-3.5	-3.6	-6.1	-6.1	2.5
$Sc_2YN@C_{80}$	-3.3	-3.5	-6.1	-6.2	2.6
$ScY_2N@C_{80}$	-3.2	-3.5	-6.1	-6.2	2.7
Sc_3N	0.4	0.4	-5.6	-7.1	6.0
Sc_2YN	0.4	0.3	-5.2	-5.5	5.5
ScY_2N	0.4	0.3	-5.0	-5.3	5.3

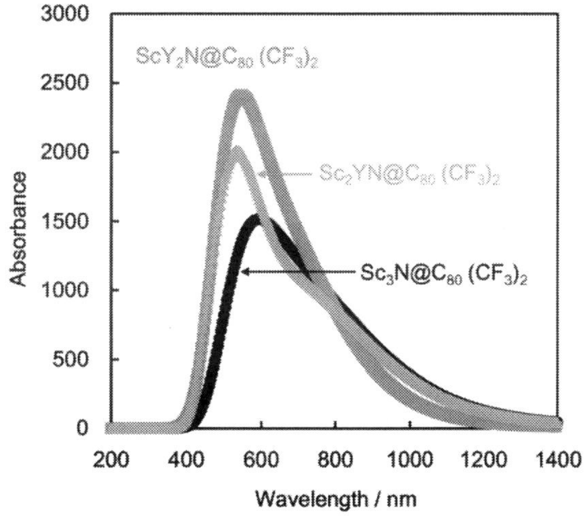

Figure 2 UV-vis-NIR spectra of metallofullerenes.

Table 2 Calculated excited states and wavelength of metallofullerenes.

Sample	Excited states	Energy (eV)	Wavelength (nm)
$Sc_3N@C_{80}(CF_3)_2$	HOMO → LUMO+3	2.28	544
$Sc_2YN@C_{80}(CF_3)_2$	HOMO → LUMO+3	2.43	511
$ScY_2N@C_{80}(CF_3)_2$	HOMO → LUMO+3	2.36	526

Comparison between chemical shifts of ^{45}Sc, ^{89}Y, ^{14}N, ^{13}C and ^{19}F in $Sc_3N@C_{80}(CF_3)_2$, $Sc_2YN@C_{80}$ $(CF_3)_2$ and $ScY_2N@C_{80}(CF_3)_2$ were investigated for accumulating on the principle of quantum superposition of nuclear spin, ^{45}Sc, ^{89}Y and ^{14}N with multiplicity at I=7/2, 1/2 and 1. As shown in Fig. 3 (a), chemical shift of ^{45}Sc, ^{14}N, ^{13}C and ^{19}F atoms inside $Sc_3N@C_{80}(CF_3)_2$ was individually separated and identified on a high magnetic field in a wide range from -300 ppm to 1000 ppm. As shown in Fig. 3 (a)-(c), chemical shifts of ^{45}Sc in nitride scandium cluster within the fullerene cage were placed near 950 ppm. A detail of chemical shift of ^{45}Sc in $Sc_3N@C_{80}(CF_3)_2$ is listed in Table 3. This result indicates anisotropic behavior as similar to those of ^{45}Sc in $S_2YN@C_{80}(CF_3)_2$ and $ScY_2N@C_{80}(CF_3)_2$. The chemical shifts of ^{45}Sc in $Sc_3N@C_{80}(CF_3)_2$, $S_2YN@C_{80}(CF_3)_2$ and $ScY_2N@C_{80}(CF_3)_2$ compared with nitride scandium cluster of Sc_3N to be -1831 ppm was considerably shifted by cage effect on magnetic interaction in ^{45}Sc with multicity at I=7/2. A slight change of chemical shifts of ^{45}Sc in the nitride scandium cluster with addition of ^{89}Y would be explained by asymmetrical spin density distribution with synergistic effect on nuclear quadrupole interaction with nuclear quadrupole moment and EFG.

The chemical shift of ^{89}Y atom in mixed metal cluster within the trifluoromethylated fullerene cage of $Sc_2YN@C_{80}N(CF_3)_2$ and $ScY_2N@C_{80}N(CF_3)_2$ was considerably shifted in a low magnetic field near 3000 ppm as shown in Fig. 3 (b) and (c). Influence of ^{89}Y atom in mixed metal cluster of Sc_3N on chemical shift of ^{14}N is listed in Table 3. A slight change of the ^{14}N chemical shifts in the series of nitride cluster would be explained by magnetic interaction in the nitrogen nuclei with metal nuclei of ^{45}Sc and ^{89}Y atoms in Sc_3N^{6+} clusters within the fullerene cage. Encapsulation of the nitride metal cluster into the carbon cage dramatically influenced nitrogen shielding constants. Similarly, chemical shift in ^{89}Y atoms was based on magnetic interaction between ^{45}Sc, ^{89}Y and ^{14}N atoms with nuclear quadrupole moment of ^{45}Sc and ^{14}N atoms with asymmetry of EFG based on a bias value of charge and spin density distribution of ^{45}Sc, ^{89}Y and ^{14}N atoms by inductive effect.

The chemical shifts of ^{13}C atom with splitting in ^{13}C-NMR spectra are shown in Fig. 3 (d)-(f). The chemical shift of ^{13}C atom in $Sc_2YN@C_{80}(CF_3)_2$ were similarly identified to those of

$ScY_2N@C_{80}(CF_3)_2$ near 100 ppm. A detail of chemical shifts of ^{13}C on the modified fullerene cages of (a) $Sc_3N@C_{80}(CF_3)_2$, (b) $Sc_2YN@C_{80}(CF_3)_2$ and (c) $ScY_2N@C_{80}(CF_3)_2$ are shown as remarked red in Fig. 4. The chemical shifts of ^{13}C in pentagon and hexagon aromatic ring near nitrogen atom on the modified fullerene cages in $Sc_3N@C_{80}(CF_3)_2$, $Sc_2YN@C_{80}(CF_3)_2$ and $ScY_2@C_{80}(CF_3)_2$ were differed slightly. Addition of yttrium in scandium cluster within the fullerene cage influenced the chemical shift of ^{13}C around ^{89}Y atom. This result would be explained by a magnitude of magnetic interaction based on s, p, and 3d hybrid orbital with a bias value of charge and spin density delocalization on hexagon aromatic ring on the fullerene cages by induction effect. Addition of yttrium in nitride mixed metal cluster within the modified fullerene cage with trifluoromethyl groups played an important role in affecting the electron structure and magnetic interaction.

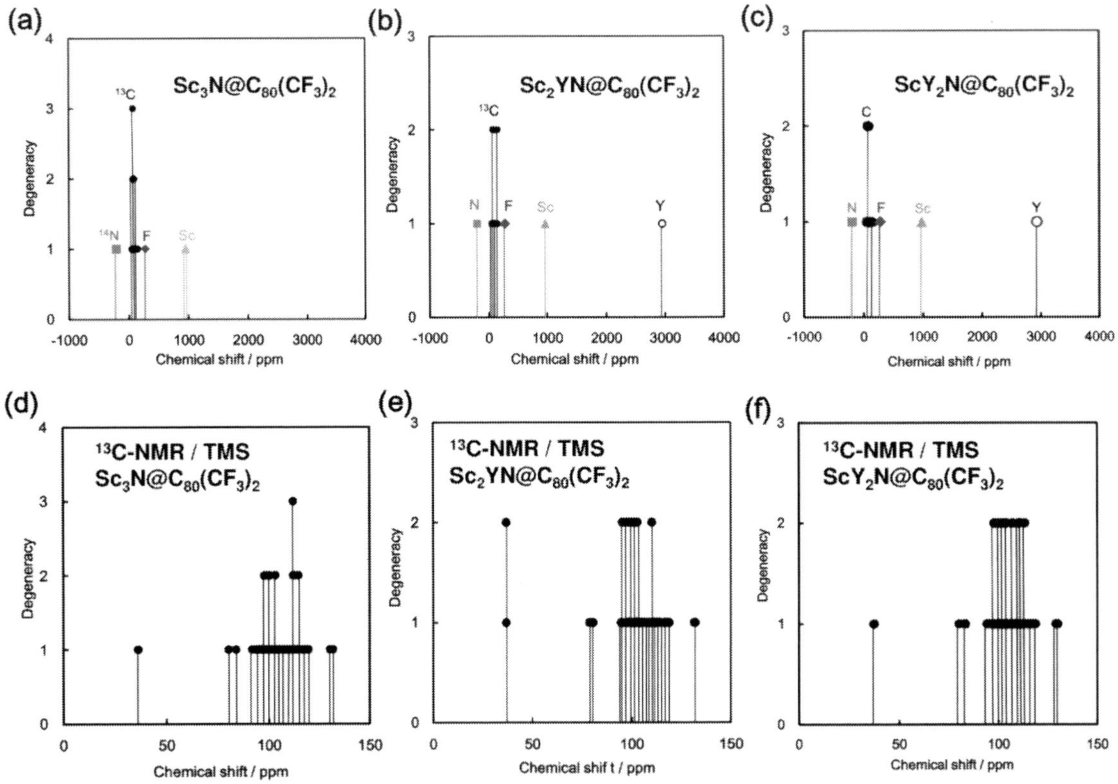

Figure 3 NMR spectra of chemical shifts in (a)-(c) ^{14}N, ^{19}F, ^{45}Sc and ^{89}Y atoms in nitride scandium cluster within (d)-(f) ^{13}C fullerene cage: $Sc_3N@C_{80}(CF_3)_2$, $Sc_2YN@C_{80}(CF_3)_2$ and $ScY_2N@C_{80}(CF_3)_2$.

Table 3 Chemical shifts of ^{14}N, ^{45}Sc and ^{89}Y atoms in mixed metal cluster within the fullerene cage.

Chemical shift (ppm)	$Sc_3N@C_{80}(CF_3)_2$	$Sc_2YN@C_{80}(CF_3)_2$	$ScY_2N@C_{80}(CF_3)_2$
^{14}N	-209	-198	-191
^{45}Sc	944	966	979
	953	971	-
	960	-	-
^{89}Y	-	2952	2921
	-	-	2944

In Ramsey's theory for nuclear magnetic shielding, the components of the shielding tensor are evolved as a sum of two terms: diamagnetic and paramagnetic terms [14]. The chemical shift will be mainly in dependence on paramagnetic effect based on spin-local interaction, the spin-dipole interaction and hybrization of molecular orbital between excited state and ground states. The chemical shift and multicity of ^{13}C for optimized structure with paramagnetic moment could be explained by Karplus-Pople equation [15]. In this case, the chemical shift mainly depends on spin-local interaction, dipole-dipole interaction with nuclear quadrupole interaction in ^{45}Sc and ^{14}N atoms at nuclear multicity state, I=7/2 and I=1 on the basis of quantum consideration. The magnetic interactions determining the magnitude of chemical shifts in nitride metallofullerenes were related on hybrization of molecular orbital of s, p and 3d spin orbital with a magnitude of spin density distribution at multi-energy state. Addition of yttrium in nitride scandium cluster of Sc_3N within the modified fullerene cage with trifluoromethyl groups plays an important role in affecting the electronic structure and magnetic parameters in NMR, ESR and ENDOR spectra. The quantum consideration expect that the chemical shift of mixed metal cluster within the modified fullerene cage can be manipulated to control quantum spin gate with decoupling pulses in the NMR quantum computer. Molecular design of endohedral metallofullerene based on mixed metal cluster within fullerene cage has a great advantage to quantum-calculate in the NMR quantum computer.

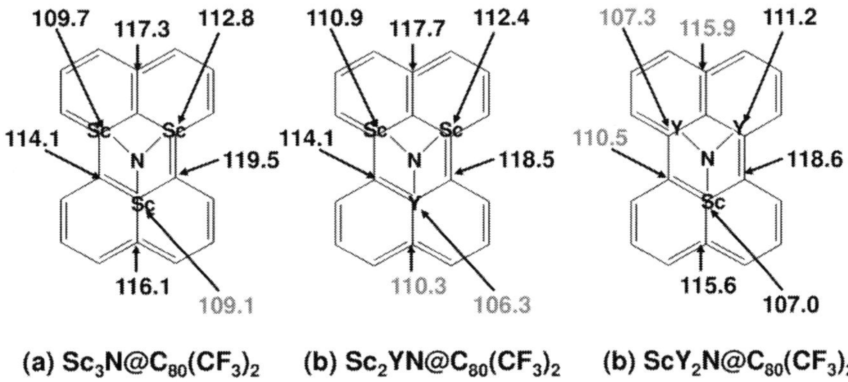

Figure 4 Chemical shift of ^{13}C on fullerene cages of (a) $Sc_3N@C_{80}(CF_3)_2$, (b) $Sc_2YN@C_{80}(CF_3)_2$ and (c) $ScY_2N@C_{80}(CF_3)_2$, referred by TMS.

4. Conclusion

Electronic structure and magnetic properties of endohedral metallofullerenes based on nitride mixed metal cluster within the fullerene cage were studied for design of spin gate in the NMR quantum computer. Endohedral metallofullerenes based on encapsulated nitride mixed metal clusters within the fullerene cage with trifluoromethyl groups were characterized. Effects of modified fullerene cage and yttrium in scandium cluster on electronic structures and magnetic properties were investigated by quantum calculation using DFT. Addition of yttrium in scandium cluster within the fullerene cage resulted a slight different position of chemical shift in ^{45}Sc, ^{89}Y, ^{14}N and ^{13}C atoms. This result was originated in magnetic interaction with spin-local interaction, dipole-dipole interaction and nuclear quadrupole interaction based on nuclear quadrupole moment and asymmetry of EFG with a bias value of charge and spin densities distribution of spd-hybrid orbital by inductive effect. Addition of yttrium in nitride scandium cluster within the modified fullerene cage with trifluoromethyl groups as electron-withdrawn groups plays an important role in affecting chemical shift. Molecular design of mixed metallic nitride endohedral fullerenes has a great advantage to be a candidate model for controlling nuclear spin gate in the NMR quantum computers.

References

[1] Ge L, Montanari B, Jefferson J H, Pettifor D G, Harrison N M and Briggs G A D, 2008 *Phys. Rev. B* **77** 235416
[2] Morton J J L, Tyryshkin A M, Ardavan A, Benjamin S C, Porfyrakis K, Lyon S A and Briggs G A D, 2006 *Phys. Stat. Solid. B* **243** 3028
[3] Suzuki A, Oku T, 2011 *Physica B* **406** 3274
[4] Suzuki A, Oku T, 2012 *J. Phys. Conf. Series* **352** 012012
[5] Wu J, Hagelberg F, 2008 *J. Phys. Chem. C* **112** 5770
[6] Popov A A, Zhang L, Dunsch L, 2010 *ACS Nano* **4** 795
[7] Ling Y, Stevenson S, Zhang Y, 2011 *Chem. Phys. Lett.* **508** 121
[8] Chen N, Zhang E Y, Wang C R, 2006 *J. Phys. Chem. B* **110** 13322
[9] Shustova N B, Popov A A, Mackey M A, Coumbe C E, Philips J P, Stevenson S, Strauss S. H, Boltalina O V, 2007 *J. Am. Chem. Soc.* **129** 11676
[10] Popov A A, Dunsch L, 2011 *J. Phys. Chem. Lett.* **2** 786
[11] Shustova N B, Peryshkov D V, Kuvychko I V, Chen Y S, Mackey M A, Coumbe C E, Heaps D T, Confait B S, Heine T, Phillips J P, Stevenson S, Dunsch L, Popov A A, Strauss S H, Boltalina O V, 2011 *J. Am. Chem. Soc.* **133** 2672
[12] Popov A A, Schiemenz S, Avdoshenko S M, Yang S, Cuniberti G, Dunsch L, 2011 *J. Phys. Chem. C* **115** 15257
[13] Chen N, Fan L Z, Tan K, Wu Y Q, Shu C Y, Lu X, Wang C R, 2007 *J. Phys. Chem. C* **111** 11823
[14] Ramsey N F, 1950 *Phys. Rev.* **78** 699
[15] Pople J A, 1957 *Proc. Roy. Soc. A* **239** 541, 550

Magnetic properties of endohedral metallofullerenes based on asymmetric nitride yttrium cluster within azafullerenes

Hitoshi Nakagawa, Atsushi Suzuki and Takeo Oku

Department of Material Science, The University of Shiga Prefecture
2500 Hassaka, Hikone, Shiga 522-8533 Japan.

E-mail: suzuki@mat.usp.ac.jp

Abstract. Magnetic properties of endohedral metallofullerene based on asymmetric triyttrium cluster of Y_3N within azafullerene cage of $C_{79}N$ were studied to control spin-flop of nuclear magnetic resonance quantum computer. Difference of nuclear magnetic resonance peaks were exerted by symmetry of structure, electron density distribution and magnetic interaction. In the present calculation, magnetic properties of $Y_3N@C_{80}$, $(Y_3N@C_{80})^{-1}$, $Y_3N@C_{80}$ $(CF_3)_2$, $Y_3N@C_{79}N$ and $(C_{79}N)^{+1}$ were compared for spin controlling in nuclear magnetic resonance quantum computers. The chemical shift of $Y_3N@C_{79}N$ appeared the highest magnetic field side. Difference of electron density and magnetic parameters are cause of this result. The chemical shift of $Y_3N@C_{80}$ $(CF_3)_2$ had the widest peak in all of samples. Substituent groups of $(CF_3)_2$ caused the wide peaks by breaking symmetry of the an fullerene cages. The effect of difference of fullerene cage was shown in NMR chemical shift.

1. Introduction

Endohedral metalofullerenes have various magnetic properties [1-3]. Those are used for monomolecular transistor, spin electronics, nuclear magnetic resonance (NMR) quantum computer [4-6], molecular spintronics, organic field-effect transistor (FET) and organic electroluminescence (EL) displays. Recently, the endohedral magnetic fullerene within single walled carbon nanotube (SWCNT) called carbonpeapods has been studied by spectroscopic analysis measures like electron spin resonance (ESR), NMR, Raman and ultraviolet (UV)-visible-near-infrared (NIR) to make a valuation of those electron structure and magnetic properties. Moreover, experimental results of magnetic properties and theoretical analysis by density functional theory (DFT) calculation evinced that endohedral metallofullerene and carbon peapods has a great potential to do spin controlling of NMR quantum computer. The magnetic interaction based on spin-orbit interaction and dipole-dipole interaction and nuclear quadroouple interaction of $^{14}N@C_{60}$-SWCNT has been reported by experimental result in late years [7, 8].

In the present work, electronic structures and magnetic properties of encapsulated nitride yttrium cluster of $Y_{.3}N$ within azafullerene [9] cage were studied for spin controlling of NMR quantum computer. Magnetic interactions of spin-orbit interaction, spin-spin interaction and dipole-dipole interaction are discussed on the basis of magnetic parameters and excitation energies. A yttrium belongs to the rare earth. Both electron spin and nuclear spin is 1/2. Electronic configuration is $4d^15s^2$. The yttrium has strange behavior in their orbital [10].

2. Quantum chemical calculation

Isolated molecular structures were optimized by calculation using Hantree-Fock and DFT of UB3LPY with STO, 3-21G* and 6-31G* as basic set because of computational limitations (Gaussian 03 and 09). Highest occupied molecular orbital (HOMO), lowest unoccupied molecular orbital (LUMO) and the HOMO-LUMO energy gap (E_g) were obtained by calculation. Mulliken charge and spin density distribution were estimated by Mulliken population analysis. DFT calculation of NMR/GIAO with hybrid function UB3LYP and 3-21G* was performed to calculate isotropic chemical shift (δ), principle g-tensor (g_{xx}, g_{yy}, g_{zz}), A-tensor (A_{xx}, A_{yy}, A_{zz}) in hyperfine coupling constant (hfc) and Q-tensor (Q_{xx}, Q_{yy}, Q_{zz}) of ^{13}C, ^{14}N and ^{38}Y atoms. Nuclear quadruple interactions were estimated on basis of multiplication of nuclear quadruple moment Q and the electro field gradient (EFG) tensor that is generated by charge distribution around nucleus. Excited energy and wavelength in UV-vis-NIR spectra of $Y_3N@C_{79}N$, $(Y_3N@C_{80})^{-1}$ and Y_3N were calculated by TD-DFT with B3LYP using 3-21G* as basis set.

3. Results and Discussion

Electronic structures of molecular orbital at HOMO, LUMO and energy levels of $Y_3N@C_{79}N$ were calculated by DFT with UB3LYP using 3-21G*. Figure 1 shows (a) molecular structure, (b) HOMO and LUMO, (c) energy levels of $Y_3N@C_{79}N$ and (d) HOMO of $Y_3N@C_{80}$, $(Y_3N@C_{80})^{-1}$, $Y_3N@C_{80}(CF_3)_2$, $Y_3N@C_{79}N$ and $(C_{79}N)^{+1}$. Electron spin density distribution of $Y_3N@C_{79}N$ was localized on surface of azafullerene cage. The spd-hybrid obital was identified around ^{89}Y in nitride yttrium cluster as shown in Fig 1 (b). Energy gap of $Y_3N@C_{79}N$ is 1.0 eV as shown in Fig 1 (c). As shown in Fig 1 (d), electron spin density distribution of $(Y_3N@C_{80})^{-1}$, $Y_3N@C_{79}N$ and $Y_3N@C_{80}(CF_3)_2$ were localized on carbon surface of azafullerene cage. Existence of radical in substance and substituent of trifluoromethyl group of CF_3 as drawn in $Y_3N@C_{80}(CF_3)_2$ caused a bias of spin density distribution. In the cage of $Y_3N@C_{80}$ and $(C_{79}N)^{+1}$, electron spin density distribution were delocalized on surface of fullerene cage. The effects of fullerene cage and nitride yttrium cluster within fullerene cage were suggested.

The calculated energy levels and band gaps of endohedralmetallofullerenes were listed in Table 1. The energy gaps of $Y_3N@C_{79}N$ and $(Y_3N@C_{80})^{-1}$ were both 1.00 eV. Those E_g were smaller than $Y_3N@C_{80}$ (2.61 eV). The asymmetrical structure of modified metallofullerene with trifluoromethyl group and azafullerene has a tendency to arise energy level at HOMO, with a narrow energy gap.

Table 1. Calculated energy levels and band gap of endohedral metallofullerenes.

Sample	HOMO	LUMO	Eg(eV)
$Y_3N@C_{79}N$	-4.23	-3.23	1.00
$Y_3N@C_{80}(CF_3)_2$	-5.76	-3.51	2.25
$(Y_3N@C_{80})^{-1}$	-1.33	-0.33	1.00
$Y_3N@C_{80}$	-6.13	-3.5	2.61
C_{80}	-5.76	-4.98	0.78
Y_3N	-3.73	-1.82	1.91

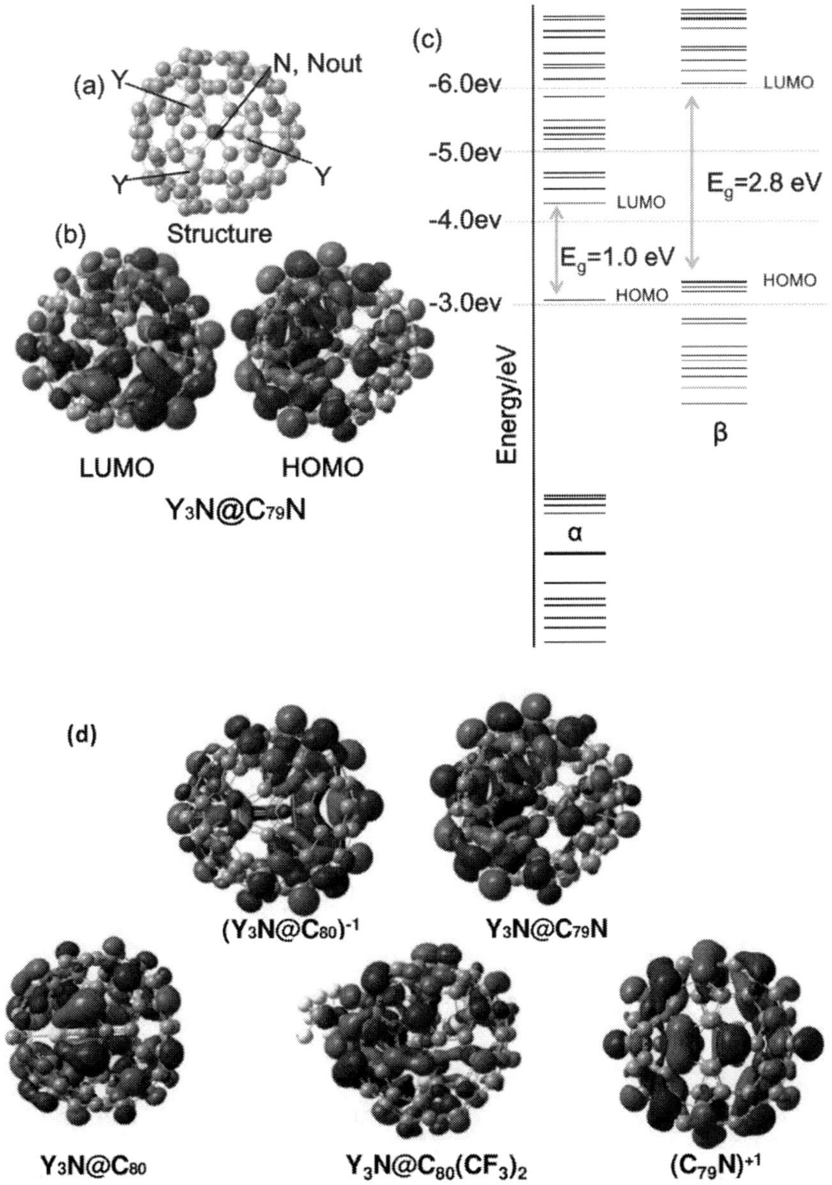

Figure 1 (a) Molecular structure, (b) HOMO, LUMO, (c) Energy levels of $Y_3N@C_{79}N$, (d) HOMO of $Y_3N@C_{80}$, $(Y_3N@C_{80})^{-1}$, $Y_3N@C_{80}(CF_3)_2$, $Y_3N@C_{79}N$ and $(C_{79}N)^{+1}$.

Table 2 shows magnetic parameters of principle g-tensor, A-tensor in hfs, Q-tensor in $Y_3N@C_{79}N$. Atomic charge of Y and N atom, which were 2.1 e and -0.9 e, confirmed charge transfer Approximately 6 electrons moved from yttrium atom to nitrogen atom on azafullerene cage. A slight charge of Milliken atomic spin density of Y and N was caused by inductive effect. The chemical shift of NMR was represented by Karplus-Pople equation as written by following equation (1).

Journal of Physics: Conference Series **433** (2013) 012005 doi:10.1088/1742-6596/433/1/012005

$$\sigma = \sigma_d + \sigma_p$$

$$\sigma = \frac{e^2}{3mc^2}\left\langle 0\left|\sum_k \frac{1}{r_k}\right|0\right\rangle$$

$$-\frac{2}{3}\left(\frac{e\hbar}{2mc}\right)^2 \sum_n \frac{1}{E_n - E_0}\left\{\left\langle 0\left|\sum_k L_k\right|n\right\rangle\left\langle n\left|\sum_k \frac{L_k}{r_k^3}\right|0\right\rangle + \left\langle 0\left|\sum_k \frac{L_k}{r_k^3}\right|n\right\rangle\left\langle n\left|\sum_k L_k\right|0\right\rangle\right\} \quad (1)$$

Karpulus-Pople equation was divided diamagnetic term (σ_d) and paramagnetic term (σ_p). The paramagnetic term was under the control of spin-local interaction and mixed hybrid orbital between excited and grand states. In the Karpulus-Pople equation, the e, m, c and L represent charge of electoron, mass of elrctron, the speed of light and the operator of orbital angular momentum, respectively.

Spin Hamiltonian was written by following equations from (2) to (5).

$$H = g_s\beta SH - g_N\beta IH + SAI + IQI \quad (2)$$

$$g_{ii} = g_e\left\{1 + \sum_m \frac{|\varphi_m|Li|\varphi_p|}{\varepsilon_p - \varepsilon_m}\right\} \quad (3)$$

$$A_{iso} = \frac{8\prod}{3g_n\mu_N\rho(0)} \quad (4)$$

$$A'_{ii} = g_N\mu_N\int\frac{\cos^2\theta - 1}{e^3}\rho(r)dv \quad (5)$$

Spin Hamiltonian is shown in eq. (2). In the spin Hamiltonian, the g_e, ψ, L, ε_m, ε_p, g_n, μ and ρ represent g factor of free electron spin, the wave function, the operator of orbital angular momentum, energy of each orbital, energy of ground state, g factor of nuclear, magnetism moment of nuclear and density of electrons, respectively. Formulas are configured by a sum of free radical Zieman interaction, dipole-dipole interaction and nuclear quadrupole interaction. Norm of g-tensor is approximately 2.0023. A_{iso} means Isotropic Fermi constant. A'_{ii} is same with A_{aniso} and it means anisotropic spin coupling constant in hyperfine structure (hfs). Both A_{iso} and A_{aniso} contain electron density (ρ). A_{iso} pertains to s orbital. A_{aniso} has angle dependence of p and d orbital. Magnetic parametyers of g-tensor and A-tensor (both A_{iso} and A_{aniso}) in hfs and nuclear quadropole constant (Q_{XX}, Q_{YY}, Q_{ZZ}) were listed in Table 2. The g-tensor, (g_{xx}, g_{yy}, g_{zz}) were g_{xx} = 1.859, g_{yy} = 1,959 and g_{zz} = 1.971, respectively. Anisotoropic A-tensors of yttrium have prominent value around 10 MHz. In the case of central nitrogen atom, anisotropic A-tensor in hfs had angle dependence of p spin distribution. The nuclear guadrapole interaction depends on the nuclear guadraple moment and electron field graduate (EFG) with a bias of charge around nuclei. The anisotropic A-tensor (A'_{XX}, A'_{YY}, A'_{ZZ}) of ^{14}N atom in the fullerene cage were A'_{XX} = -0.514, A'_{YY} = -0.479 and A'_{ZZ} = 0.993, respectively. This result indicates a large extent of density distribution of p-orbital in N atom in fullerene cage.

Table 2 Magnetic parameters of principle *g*-tensor, *A*-tensor in *hfc*, *Q*-tensor in $Y_3N@C_{79}N$.

ESR parameters: g_{XX}=1.859, g_{YY}=1.959, g_{ZZ}=1.971

	Atomic charge (e)	Mulliken atomic spin density	A_{iso}	A'_{xx}	A'_{yy}	A'_{zz}	Q_{xx}	Q_{yy}	Q_{zz}
				(MHz)				(MHz)	
N	-0.937	-0.011	-0.919	-0.273	0.016	0.257	-0.719	-0.216	0.935
Y_1	2.119	0.011	1.027	0.498	-0.062	0.257	0	0	0
Y_2	2.092	0.132	-11.857	1.452	1.098	-0.436	0	0	0
Y_3	2.055	0.128	-13.771	1.703	0.935	-2.55	0	0	0
$N_{(out)}$	-0.851	0.007	0.862	-0.514	-0.479	0.993	1.518	1.206	-2.724

A_{iso} : Isotropic Fermi constant, A_{aniso}: Anisotropic spin coupling constant in hfc,
Q: Nuclear quadrupole constant.

The calculated chemical shifts of (a) $Y_3N@C_{79}N$, (b) ^{13}C-NMR in $Y_3N@C_{79}N$ and (c) $Y_3N@C_{80}(CF_3)_2$ are shown in Fig 2. The chemical shift of ^{14}N, ^{89}Y and ^{13}C in $Y_3N@C_{79}N$ were appeared separately in Fig 2 (a). Especially, yttrium could be around 2500 pmm. And chemical shifts of ^{13}C atom in ^{13}C-NMR had wide peak between around 70 ppm and 95 ppm in Fig 2 (b). As shown in Fig 2 (c), the chemical shift of ^{14}N, ^{89}Y, ^{19}F and ^{13}C in $Y_3N@C_{80}(CF_3)_2$ also appeared separately. But the chemical shifts of ^{89}Y in $Y_3N@C_{80}(CF_3)_2$ appeared around 30 ppm, which was completely different position as compared to those of ^{89}Y in $Y_3N@C_{79}N$. The reason of this difference was thought that a different extent of electron density of spd-hybrid orbital. Substituent group of $(CF_3)_2$ played a role of falling spin density distribution on the fullerune cage.

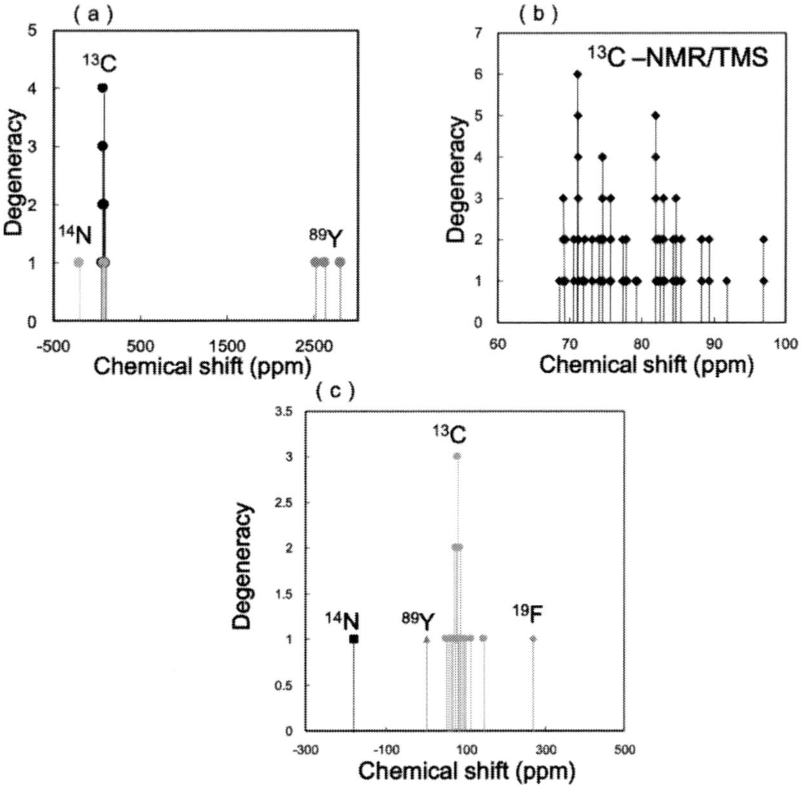

Figure 2 Calculated chemical shifts of (a)$Y_3N@C_{79}N$, (b) ^{13}C-NMR in $Y_3N@C_{79}N$, (c) $Y_3N@C_{80}(CF_3)_2$.

Figure 3 shows Calculated chemical shifts of ^{13}C-NMR in $Y_3N@C_{80}$, $(Y_3N@C_{80})^{-1}$, $Y_3N@C_{80}(CF_3)_2$ and $Y_3N@C_{79}N$. Those samples had different chemical shifts. The chemical shift of $Y_3N@C_{79}N$ appeared the highest magnetic field side. Difference of electron density as shown in Fig 1 and Magnetic parameters in table 2 are cause of this results. The chemical shift of $Y_3N@C_{80}(CF_3)_2$ had the widest peak in all of samples. The range of chemical shift of $Y_3N@C_{80}(CF_3)_2$ was between 90 ppm and 150 ppm. The substituent group of $(CF_3)_2$ cause this wide peak by breaking symmetry of fullerene cage. The effect of difference of fullerene cage was shown obviously in NMR chemical shift.

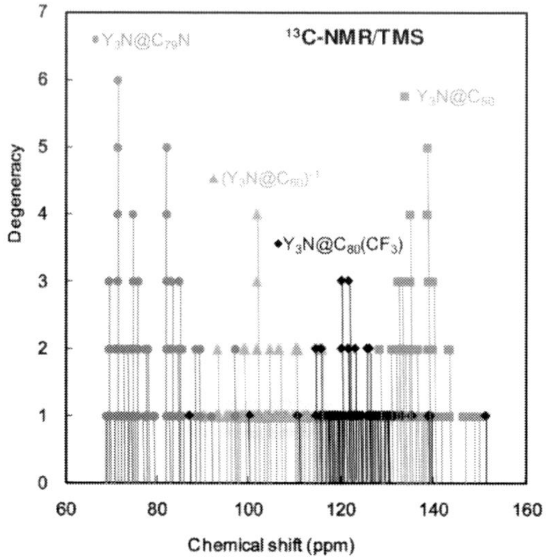

Figure 3 Calculated chemical shifts of ^{13}C-NMR in $Y_3N@C_{80}$, $(Y_3N@C_{80})^{-1}$, $Y_3N@C_{80}(CF_3)_2$ and $Y_3N@C_{79}N$.

Details of chemical shifts of ^{13}C on fullerene cage in $Y_3N@C_{79}N$, $Y_3N@C_{80}$ and $(Y_3N@C_{80})^{-1}$ are shown in Fig 4. In the ^{13}C-NMR chemical shift of $Y_3N@C_{79}N$, ^{14}N on fullerene cage was pointed. The numerical vales of ^{13}C-NMR around ^{14}N on the fullerene cage were displayed.The ^{13}C-NMR chemical shift of $Y_3N@C_{80}$ and $(Y_3N@C_{80})^{-1}$ were displayed exactly same point on the fullerene cage compared with the ^{13}C-NMR chemical shift of $Y_3N@C_{79}N$. The numerical value of ^{13}C-NMR of $Y_3N@C_{79}N$ was marked low values compared with $Y_3N@C_{80}$ and $(Y_3N@C_{80})^{-1}$. The ^{14}N on fullerene cage could be localized factor in $Y_3N@C_{79}N$. Therefore the ^{13}C-NMR chemical shift of $Y_3N@C_{79}N$ appeared higher side of magnetic field than others.

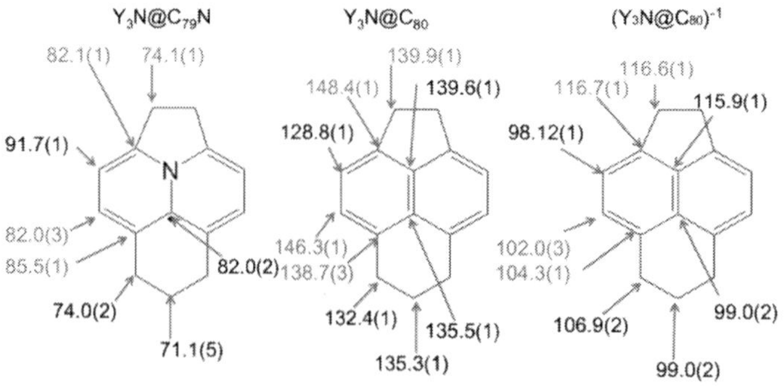

Figure 4 Calculated chemical shifts of ^{13}C on fullerene cage in $Y_3N@C_{79}N$, $Y_3N@C_{80}$ and $(Y_3N@C_{80})^{-1}$.

4. Conclusion

Electronic structures and magnetic properties of endohedral metallofullerenes based on asymmetric nitride yttrium cluster within fullerene cage are studied to control the spin flopping of NMR quantum computer. The effects of fullerene cage and inside nitride yttrium cluster were investigated by quantum chemical calculation. Characterization of NMR chemical shifts, g-tensor, A-tensor, Q-tensor of ^{89}Y and ^{14}N atoms with magnetic interaction between electronic spin and nuclear spin were calculated. The significance of sorts of fullerene cages to control the NMR chemical shift were explained by Spin Hamiltonian. The Karplus-Pople equation exhibited the effect of radical in metallofullerenes. Substituent group of $(CF_3)_2$ and inside metal cluster within fullerene cage could control a variable position of chemical shift. The endohedral metallofullerenes based on asymmetric nitride yttrium cluster within azafullerenes should have a great potential for NMR quantum computer to quantum-calculate at multiple-state.

References

[1] Jing L, Xinwei Z, Xiangeng Z, 2000 *J. Phys. Chem.*.**332** 51
[2] Shustova N B, Peryshkov D V, 2011 *J. Am. Chem. Soc.* **133** 2672
[3] Schiemenz S, Avdoshenko S M, Yangv S, Cuniberti G, Dunsch L, 2011 *J. Phys. Chem. C* **115** 15257.
[4] Chen N, Fan L Z, Tan K, Wu Y Q, Shu C H, Lu X, and Wang C R, 2007 *J. Phys. Chem.* C **111** 11823
[5] Tiaming Z, Liasa X, Christine M B, Marilyn M O, 2008 *J. Am. Chem. Soc* **130** 12992
[6] Lieven M K, Matthias S, Gregory B, Costantino S, 2001 *Nature* **414** 883
[7] Suzuki A, Oku T, 2011 *Physica B* **406** 3274
[8] Suzuki A, Oku T, 2012 *J. Phys. Conf. Series* **352** 012012
[9] Van D S T, Wang Z H, Blok M S, Bernien H, Taminial T H, 2012 *Nature* **485** 82
[10] Popov A A, Zhang L, Dunsch L, 2010 *ASC Nano* **4** 795

Optical waveguiding in an electrospun polymer nanofiber

Yuya Ishii[1]*, Ryohei Kaminose[1], and Mitsuo Fukuda[1]
[1]Department of Electrical and Electronic Information Engineering, Toyohashi University of Technology, 1-1 Hibarigaoka, Tempaku-cho, Toyohashi, Aichi 441-8580, Japan

E-mail: yishii@ee.tut.ac.jp

Abstract. Optical waveguides with nanometer diameters are needed to realize nanometer-scaled optical circuits, which promise high-speed and high-capacity information processing. In this work, we fabricate polymer nanofibers with uniform diameters, containing an organic dye. Fluorescence microscopy observations and photoluminescence measurements demonstrate optical waveguiding in the nanofiber.

1. Introduction

To achieve high-speed and high-capacity information processing, optical circuits are attracting much attention as a substitute for electronic circuits [1, 2]. An optical waveguide is an essential element in optical circuits, and several nanometer ranged optical waveguides have been developed, such as silicon waveguides [1, 3], highly-drawn silica wires [4], and lithographic patterned polymer waveguides [2]. Electrospun polymer fibers have diameters in the nanometer range, high aspect ratios [5, 6], and show mechanical flexibility. Thus, the fibers are prime candidates for small waveguides. To date, only several groups have reported waveguiding properties inside electrospun nanofibers [7, 8].

In this work, we fabricate electrospun polymer nanofibers containing an organic dye, and demonstrate optical waveguiding in the fiber.

2. Experimental details

2.1. Materials

Nile Blue A perchlorate (NBA, Fig. 1) and poly(methyl methacrylate) (PMMA, average *M*w ~ 350,000) was purchased from Sigma-Aldrich and used as received. *N,N*-dimethylformamide (DMF) and chloroform were purchased from Wako Pure Chemical Industries, Ltd. and used as received.

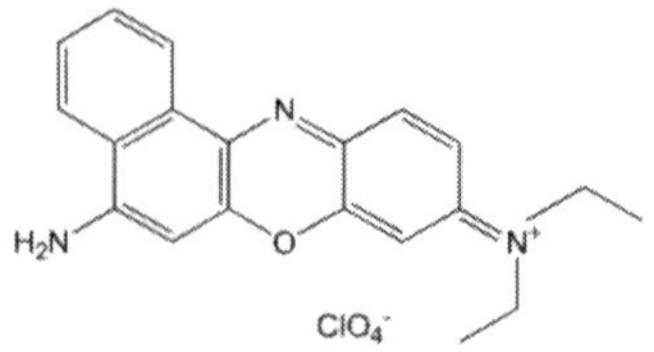

Figure 1. Molecular structure of NBA.

* To whom any correspondence should be addressed.

2.2. Electrospun fibers

To produce composite nanofibers by electrospinning, the different materials first need to be dissolved [9]. PMMA and NBA were dissolved in a mixture of DMF and chloroform (DMF:chloroform = 7:3 volume ratio). The PMMA concentration is varied to give 6, 8, and 10 wt% solutions and the NBA concentration is 1×10^{-3} mol·l^{-1} for the PMMA volume. The solution of PMMA/NBA was loaded into a glass syringe equipped with a stainless steel needle (0.3 mm in diameter) and syringe pump (KDS-100, KD Scientific Inc.). The PMMA/NBA solutions are ejected from the needle tip at constant rates (Table 1). A high-voltage power supply (HVU-30P100, Mecc Co., Ltd.) applied a high voltage to the needle, and a grounded silicon plate collector was placed 15 cm below the tip of the needle to collect the fibers. All electrospinning experiments were performed in air at room temperature with a humidity of 35–48%.

Table 1. Electrospinning conditions for each PMMA/NBA solution.

PMMA concentration (wt%)	Feed rate (ml·h^{-1})	Applied voltage (kV)
6	0.30	4.5
8	0.32	3.8
10	0.35	3.5

2.3. Characterization method

Shapes and diameters of the fibers were characterized using field-emission scanning electron microscopy (FESEM, SU8000, Hitachi) and fluorescence microscopy (BX-51, Olympus). The average fiber diameter was calculated from 100 SEM images of individual fibers (2 images per 50 individual fibers). Note that the standard deviation in fiber diameters among electrospun fibres is about two times larger than that within an individual fiber [10].

Photoluminescence (PL) spectra were measured with a spectral analyzer (USB4000, Ocean Optics), which was connected to the fluorescence microscope via an optical fiber. Excitation light was supplied by a halogen light source (U-LH100-3, Olympus) through a 510–550 nm bandpass filter. The PL spectra from the samples were collected through a sharp-cut filter ($\lambda > 590$ nm). The circular detection area had a diameter of 20 μm.

The absorption spectra of dilute solutions were measured with a UV-Vis spectrophotometer (U-4100, Hitachi). To prepare a dilute solution, NBA was dissolved in ethanol at a concentration of 4.26 mg·l^{-1}.

3. Results and discussion

The optical microscopy images of the electrospun fibers deposited on silicon plate collectors are shown in Fig. 1. The PMMA/NBA fibers electrospun from the 6 wt% solutions have a beaded-fiber morphology. With increasing concentration, the number and size of the beads on the nanofibers is reduced, and uniform fibers are obtained from the 10 wt% solution. This homogenization of fiber

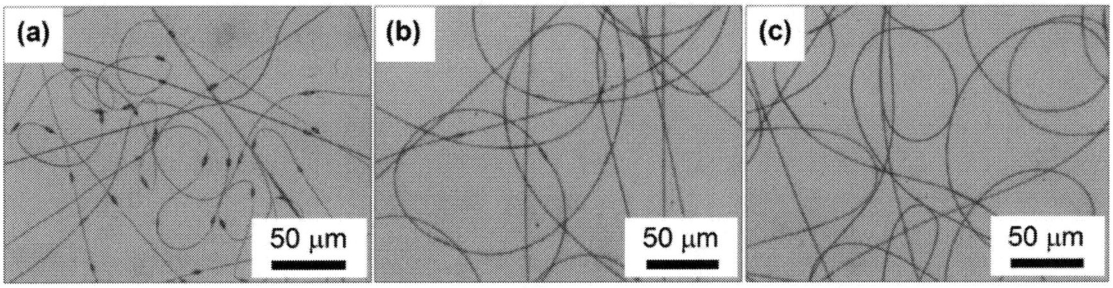

Figure 2. Optical microscopy images of NBA/PMMA fibers electrospun from NBA/PMMA solutions with different concentrations: (a) 6 wt%, (b) 8 wt%, and (c) 10 wt%.

diameters results from increased solvent distribution over the polymer molecules [11]. To suppress the loss of light from inside the fiber, uniform fibers are required. Thus, we use the fibers spun from the 10 wt% solutions to probe the waveguiding properties inside these fibers.

The inset in Fig. 3 shows an FESEM image of PMMA/NBA electrospun fibers. Uniform fibers with nanometer diameters are obtained. We measured the fiber diameters from 100 SEM images (Fig. 3). The average diameter of the fibers was 600 nm with a standard deviation of 70 nm.

We cut some nanofibers with a knife, and obtained the optical and fluorescence microscopy images (Fig. 4). The fibers exhibit bright and uniform PL [Fig. 4(b)], which demonstrates the homogeneous incorporation of the NBA molecules. The isolated fiber shown in Fig. 4(c) bends upwards and the end face of the fiber is clearly observed. PL from the end face of the fiber [point B in Fig. 4(c) and (d)] is more intense than that from the middle of the fiber [point A in Fig. 4(a) and (b)]. We measured the PL of the fiber at points A and B (Fig. 5). The PL intensity at point B is approximately 9.5-times higher than that at point A. This high PL intensity at the end face of the fiber is due to the integrated NBA PL and clearly demonstrates that waveguiding occurs in the nanofiber. Figure 6 shows the normalized PL spectra of the fiber at points A and B. The PL at point B is red-shifted compared to that at point A. This is due to re-absorption effects. The NBA absorption spectrum overlaps with the PL spectrum in the range 590–690 nm (Fig. 6). Although the average diameter of the fibers is 600 nm, the NBA PL occurs at wavelengths >600 nm, and waveguiding occurs inside the fiber because of the higher refractive index of PMMA (n=1.49 at 633 nm) [12] compared to air.

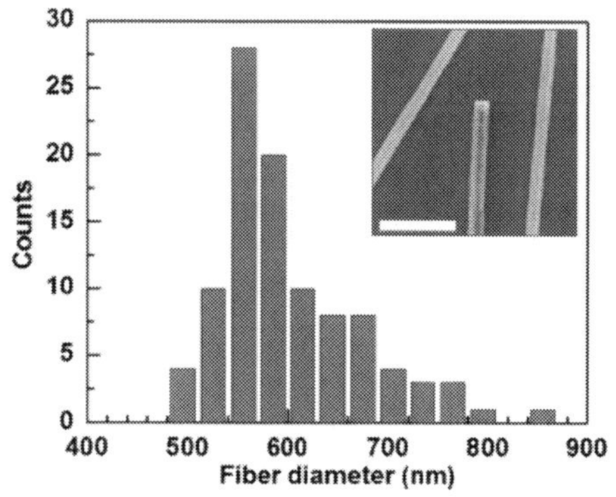

Figure 3. Distribution of diameters in NBA/PMMA fibers electrospun from the 10 wt% NBA/PMMA solution. We measured the diameters from 100 different fibers. The diameter interval for fibers is 5% of the average fiber diameter. Inset: FESEM image of the fibers. The scale bar is 3μm.

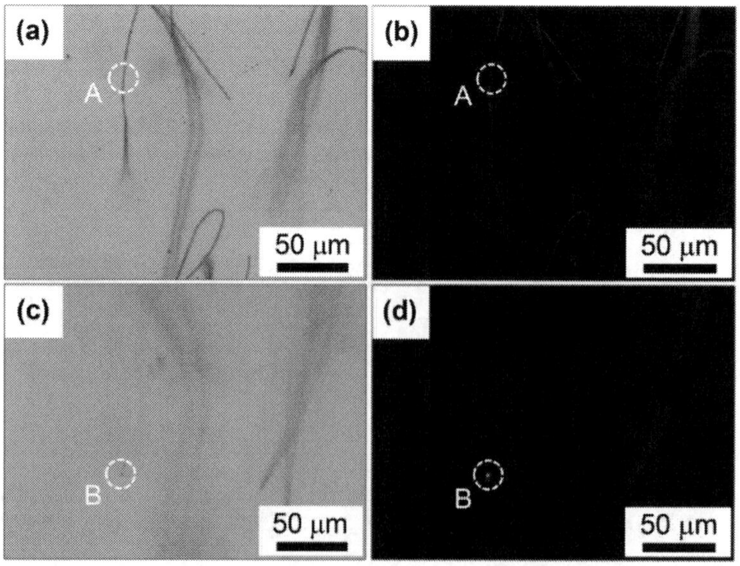

Figure 4. Optical [(a) and (c)] and fluorescence [(b) and (d)] microscopy images of PMMA/NBA electrospun nanofibers. Middle [(a) and (b)] and end face [(c) and (d)] fibers are shown. At points A and B, PL spectra are measured.

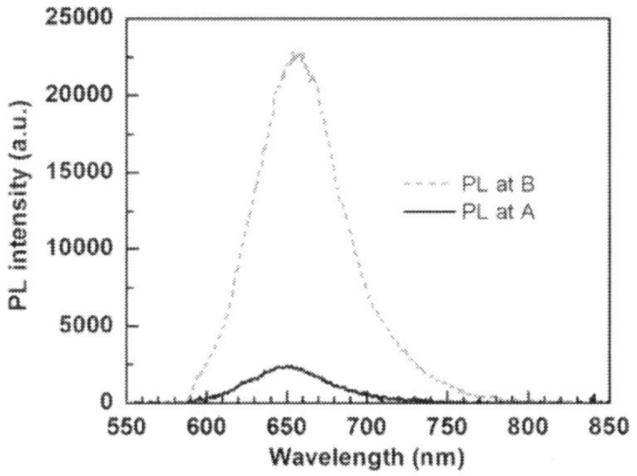

Figure 5. PL spectra of the nanofiber at points A and B marked in Fig. 4.

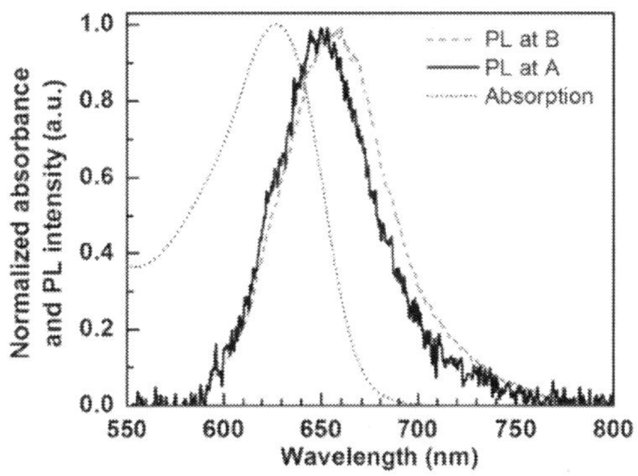

Figure 6. Normalized PL spectra of the nanofiber at points A and B marked in Fig. 4 and the normalized absorption spectrum of the NBA dilute solution.

4. Summary

We have fabricated uniform PMMA/NBA electrospun nanofibers by adjusting the concentration of PMMA/NBA solutions. Fluorescence microscopy images and PL spectra of the fiber demonstrate optical waveguiding in the nanofibers.

Acknowledgments

This work was partially supported by JSPS KAKENHI Grant Number 24810013.

References

[1] Jalali B and Fathpour S 2006 *Journal of Lightwave Technology* **24** 4600
[2] Ma H, Jen A K Y, and Dalton L R 2002 *Adv. Mater.* **14** 1339
[3] Liang D and Bowers J E. 2010 *Nat. Photon.* **4** 511
[4] Tong L, Gattass R R, Ashcom J B, He S, Lou J, Shen M, Maxwell I, and Mazur E 2003 *Nature* **426** 816
[5] Li D and Xia Y 2004 *Adv. Mater.* **16** 1151
[6] Ishii Y, Sakai H, and Murata H 2011 *Nanotechnology* **22** 205202
[7] Liu H, Edel J B, Bellan L M, and Craighead H G 2006 *Small* **2** 495
[8] Camposeo A, Di Benedetto F, Stabile R, Neves A A R, Cingolani R, and Pisignano D 2009 *Small* **5** 562
[9] Ishii Y and Murata H, 2012 *J. Mater. Chem.* **22** 4695
[10] Ishii Y, Sakai H, and Murata H 2009 *Thin Solid Films* **518** 647
[11] Ramakrishna S, Fujihara K, Teo W-E, Lim T-C, and Ma Z 2005 *An Introduction to Electrospinning and Nanofibers* (Word Scientific)
[12] Tanio N and Irie M 1994 *Jpn. J. Appl. Phys.* **33** 3942

Fabrication of C_{60} assembly films via an fullerene-amine addition reaction by using stepwise immersion

Shoto Banya, Taisuke Matsumoto, Takeo Oku and Tsuyoshi Akiyama*

Department of Materials Science, The University of Shiga Prefecture, 2500 Hassaka, Hikone, Shiga 522-8533, Japan

E-mail: akiyama.t@mat.usp.ac.jp

Abstract. Fullerene C_{60} thin films were successfully fabricated on the surface of amino-substituted substrates via a reaction between C_{60} and an NH_2 group of 1,2-diaminoethane (DAE) using simple immersion processes. The maximum thicknesses of the fullerene assembled films increased with the number of immersions into the fullerene solution. Layered fullerenes were obtained by a series of repeated processes. The aggregated C_{60} thin films were evaluated using transmission absorption measurements, transmission electron microscopy, and atomic force microscopy. These analyses suggest the presence of C_{60}-DAE aggregates on the substrates.

1. Introduction

Fullerenes are highly attractive materials as both electron acceptors and organic n-type semiconductors. Indeed, fullerenes are widely used in artificial photosynthesis models [1, 2], organic solar cells [3], and organic transistors [4]. In organic solar cells (OSCs), the carrier mobility of the organic films is an important factor in the photoelectric conversion performance. Adjacent or aggregated fullerene structures feature efficient electron transport [5]; therefore, the development of a facile method for the preparation of fullerene aggregates or wire structures is important.

Chemical modification and assembling of fullerenes are fundamentally important for the application of fullerenes as novel functional molecules and in molecular devices and organic electronic devices. One of the simplest chemical modification methods is an addition reaction between amines and fullerenes [6-9]. Recently, we discovered that the reaction of fullerenes (C_{60} and C_{70}) with a large excess of 1,2-diaminoethane (DAE) affords uniform sub-micrometer particles consisting of fullerenes and DAE [10-12]. Using these molecules, it was determined that the incorporation of the microparticles substantially enhanced the photocurrent generation of a polythiophene film [10].

The addition reaction between fullerenes and amines also occurs at the liquid/solid interface; this reaction can therefore be used to prepare ultra-thin fullerene films [13-15]. Furthermore, we found that fullerenes contribute to photocurrent generation as electron acceptors via photo-induced electron transfer even in ultra-thin films [16, 17].

The fabrication of assembled fullerene films with various morphologies should also be useful for the development of organic electronics materials. Accordingly, we fabricated C_{60} assembly films by repeated fullerene-amine addition reactions. The structures of the resultant C_{60} assembly films were investigated.

2. Experimental procedure

All chemicals and materials were used as received. Milli-Q water (18.2 MΩ cm) was used. All immersions of the substrates into the fullerene and amine solution were performed using hand-made alternating dip-coating equipment.

The procedure for the preparation of C_{60} assembly films is as follows (Figure 1): The pre-cleaned quartz glass substrate was immersed into a toluene solution of 3-aminotrimethoxysilane. After the solution was refluxed for 12 h, the substrate was removed and washed with sufficient amounts of toluene, ethanol, aqueous $NaHCO_3$, and water to obtain a NH_2-modified substrate, which is denoted as Quartz/NH_2, where / represents the interface. The presence of amino groups on the surface of Quartz/NH_2 was confirmed via the formation of imine bonds between amino groups on the surface and 4-nitrobenzaldehyde [18].

After drying, Quartz/NH_2 was immersed in a toluene solution of C_{60} (2.0×10^{-3} mol dm^{-3}) for 5 min at room temperature and washed with enough amounts of toluene to obtain a C_{60}-modified quartz substrate, which is denoted as Quartz/$[C_{60}]_1$ (Step 1).

Then, Quartz/$[C_{60}]_1$ was immersed in a toluene solution of 1,2-diaminoethane (2.0×10^{-2} mol dm^{-3}) for 5 min to substitute the amino groups on the surface of the C_{60} layer of Quartz/$[C_{60}]_1$ (Step 2). After washing this modified quartz substrate, it was immersed in the same C_{60} toluene solution used in Step 1 to obtain a double layer of C_{60}; the resultant substrate is denoted as Quartz/$[C_{60}]_2$ (Step 3). Thus, through a similar process of alternating immersion in the toluene solutions of fullerene and amine (Steps 4 and 5), we obtained aggregated C_{60} assembly films, which are denoted as Quartz/$[C_{60}]_n$, where n is the number of immersions into the C_{60} toluene solution.

For the transmission electron microscopy measurements, a gold microgrid (400 mesh, 3.0 mm) was immersed in an ethanol solution of 2-aminoethanethiol (1.0×10^{-3} mol dm^{-3}) for 1 day at room temperature. After withdrawal, the grid was washed with sufficient amounts of ethanol and then dried in order to obtain an amino-modified microgrid, which is denoted as Au/NH_2. C_{60} and diamine modifications were also performed similarly to the preparation of Quartz/$[C_{60}]_n$, to obtain C_{60}-assembly-film-modified gold grids, which are denoted as Au/$[C_{60}]_n$.

The UV-Vis spectra of the C_{60} assembly films were measured using a UV-Vis spectrophotometer (V-670, JASCO). The surface morphologies of the modified thin films were determined using atomic force microscopy (AFM; SPA400, SII). The thicknesses and cross-sectional images of C_{60} assembly films were measured using transmission electron microscopy (TEM; H-8100, HITACHI).

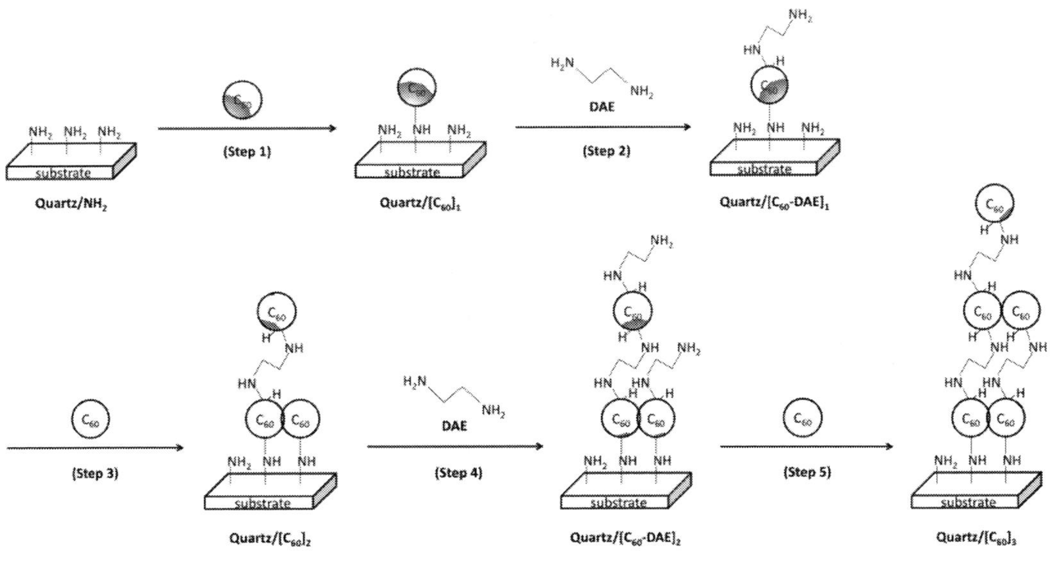

Figure 1. Schematic illustration of the preparation of the C_{60} assembly films.

3. Results and discussion

The surface coverage of Quartz/NH_2 with amino groups was ~1.5 - 2.0 × 10^{-10} mol cm^{-2}, as determined via comparison of the transmission absorption spectra before and after modification of Quartz/NH_2 with 4-nitrobenzaldehyde. The surface coverage was similar to that predicted for a monolayer of C_{60} using a model based on crystallographic data (fcc; 1.9 × 10^{-10} mol cm^{-2}) [13].

Figure 2(a) shows transmission absorption spectra of Quartz/$[C_{60}]_n$ (n = 1, 2, and 3). In each assembly film, the absorption profile decayed from shorter to longer wavelengths and the baseline increased from n = 1 to 3. These absorption characteristics are attributable to the sum of the π-π* absorption band of the C_{60} structure and light scattering of the C_{60} assemblies.

In Figure 2(b), the intensity of the absorbance at 350 nm, which is mainly attributed to the π-π* absorption, is plotted as a function of the number of immersions into the C_{60} toluene solution: An approximately straight line results with an increasing slope from n = 1 to 3. This suggests that the amount of C_{60} absorbed increases with increasing number of C_{60} modification processes. The estimated absorbed amount of C_{60} after one immersion is on the sub-monolayer level, as determined from the absorbance of Quartz/$[C_{60}]_n$ (n = 1, 2, and 3) at 350 nm.

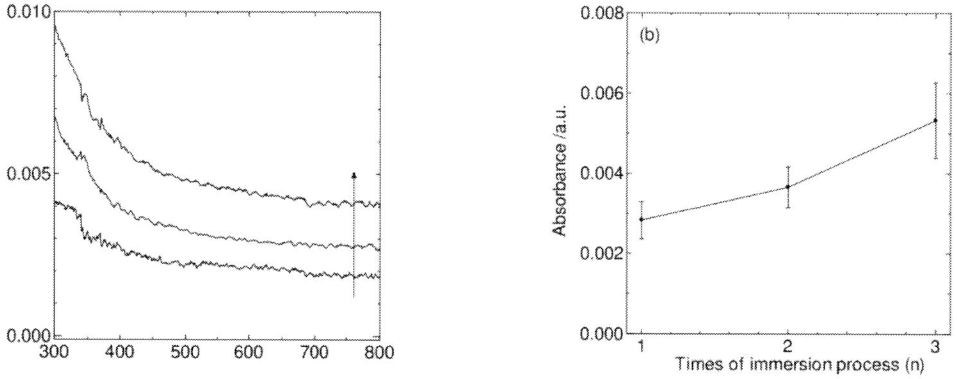

Figure 2. UV-Vis absorption spectra of (a) Quartz/$[C_{60}]_n$ (n = 1, 2, and 3) on both sides, (b) relationship between the absorbance of Quartz/$[C_{60}]_n$ (n = 1, 2, and 3) at 350 nm and the number of immersions (n) into the C_{60} toluene solution.

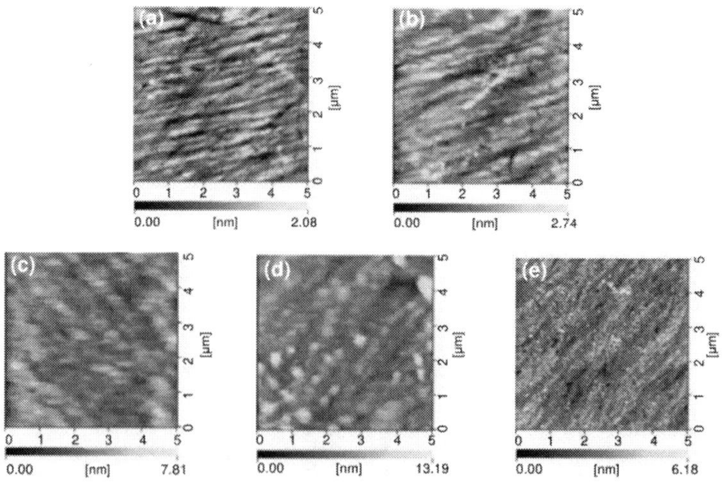

Figure 3. Contact-mode AFM images of (a) bare quartz, (b) Quartz/NH_2, (c) Quartz/$[C_{60}]_1$, (d) Quartz/$[C_{60}]_2$, (e) Quartz/$[C_{60}]_3$. All images display the top view.

AFM images of bare quartz, Quartz/NH_2, and Quartz/$[C_{60}]_n$ (n = 1, 2 and 3) are shown in Figure 3. Very similar surface structures are observed for quartz and Quartz/NH_2, and the differences between the maximum and minimum heights in both samples are quite comparable. This is reasonable because the molecular length of 3-aminotrimethoxysilane is relatively short with respect to the surface roughness of quartz and Quartz/NH_2. In contrast, the surface morphology of Quartz/$[C_{60}]_1$ is dramatically different. Modification of C_{60} increased the surface roughness from ~3 to ~8 nm. In addition, sub-micrometer aggregated structures were evident on the surface of Quartz/$[C_{60}]_1$. The 5 nm increase in the surface roughness is considerably larger than the diameter of C_{60}, which suggests that C_{60} adsorbs onto the NH_2 surface of Quartz/NH_2 as aggregates or clusters [9]. The surface roughness increased to ~13 nm for Quartz/$[C_{60}]_2$, but was only ~6 nm for Quartz/$[C_{60}]_3$. The detailed mechanism for these changes in surface roughness has not yet been elucidated; one possibility is the formation of an amino-group domain in Quartz/NH_2, which may cause morphemic changes, as follows (Figure 4): 1) Oligomers of 3-aminotrimethoxysilane are formed on the surface of the quartz, 2) an uncoated quartz surface remains, 3) C_{60} molecules attach to the amino groups of Quartz/NH_2 to give Quartz/$[C_{60}]_1$, and 4) C_{60} assemblies grow in all directions.

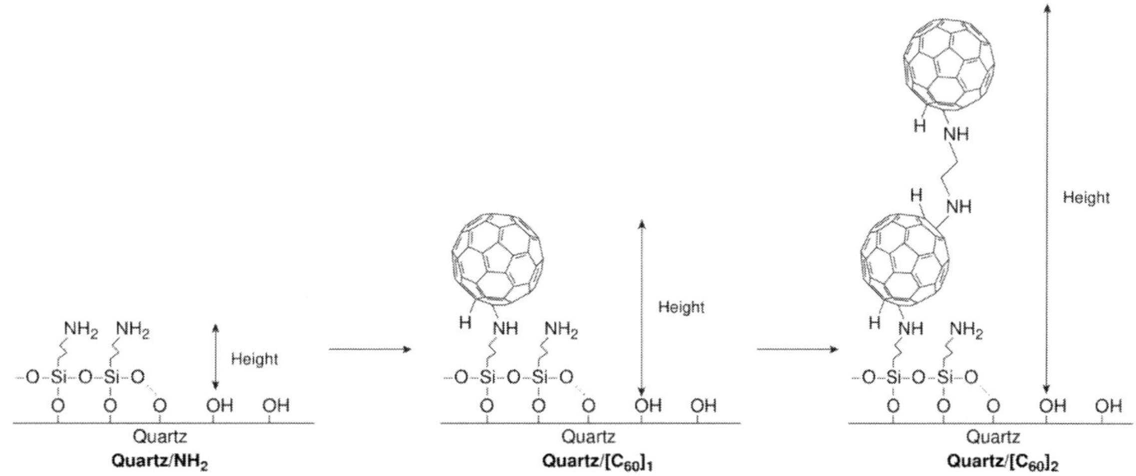

Figure 4. Proposed reaction scheme and structural changes on the quartz surface following silanization, and the attachment of C_{60} and DAE.

Typical cross-sectional TEM images of Quartz/$[C_{60}]_1$ and Quartz/$[C_{60}]_3$ are shown in Figure 5. The TEM images indicate that the densities of the organic layers are not equal, which is reasonable because C_{60} assembly films have considerably larger surface roughness. The results of statistical analyses, including the average C_{60} film thicknesses and standard deviations of the maximum thicknesses of the films, are summarized in Table 1.

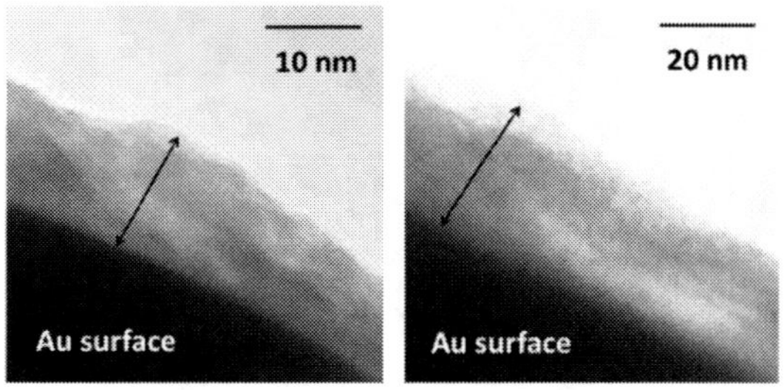

Figure 5. TEM images of (a) Au/$[C_{60}]_1$ and (b) Au/$[C_{60}]_3$.

Table 1. Thicknesses of the $[C_{60}]_n$ layers.

Sample	Average film thickness / nm	Standard deviation / nm
$Au/[C_{60}]_1$	14	5
$Au/[C_{60}]_3$	32	11

These results indicate that the facile stepwise chemical modification process yielded 10–30 nm thick C_{60} assembly films. In addition, the obtained C_{60} assembly films have rough surfaces. C_{60} assembly films with these morphological characteristics are likely particularly suitable for electronic applications. Further investigations into controlling and expanding the C_{60} assembly films are currently underway. In the near future, we will apply these C_{60} assembly films as n-type semiconductors for organic electronics such as organic field effect transistors and organic solar cells.

4. Conclusion

C_{60} assembly films were attached onto a Quartz/NH_2 surface via addition reactions between fullerenes and amines across double bonds of the C_{60} molecules. The absorption spectra of the C_{60} assembly films showed the sum of the π-π* absorption band from the C_{60} structure and light scattering of the C_{60} assemblies. Via AFM, it was determined that a homogeneous thin C_{60} assembly film was obtained. The TEM images indicated that the film thickness increased with repeated immersions into the C_{60} toluene solution.

Acknowledgement

This research was partially supported by a Grant-in-Aid for Scientific Research (C) (No. 22550123) from the Japan Society for the Promotion of Science (JSPS).

References

[1] Imahori H, Sakata Y, 1999 *Eur. J. Org. Chem.* **10** 2445
[2] Imahori H, 2007 *Bull. Chem. Soc. Jpn.* **80** 621
[3] Hoppe H, Sariciftci N S, 2004 *J. Mater. Res.* **19** 1924
[4] Kobayashi S, Takenobu T, S. M, Fujiwara A, Iwasa T, 2003 *Appl. Phys. Lett.* **82** 4581
[5] Tezuka N, Umeyama T, Seki S, Matano Y, Nishi M, Hirao K, Imahori H, 2010 *J. Phys. Chem. C* **114** 3235
[6] Hirsch A, Qiaoying L, Wudl F, 1991 *Angew. Chem. Int. Ed.* **30** 1309
[7] Wudl F, Hirsch A, Khemani K C, Suzuki T, Allemand P-M, Koch A, Eckert H, Srdanov G, Webb H M, 1992 *ACS Symposium Series* **481** 161
[8] Isobe H, Tomita N, Nakamura E, 2000 *Org. Lett.* **2** 3663
[9] Wang Q and Moriyama H, 2009 *Langmuir* **25** 10834
[10] Matsuoka K-i, Seo H, Akiyama T, Yamada S, 2007 *Chem. Lett.* **36** 934
[11] Matsuoka K-i, Matsumura S, Akiyama T, Yamada S, 2008 *Chem. Lett.* **37** 932
[12] Matsuoka K-i, Akiyama T, Yamada S, 2009 *Langmuir* **26** 4274
[13] Caldwell W B, Chen K, Mirkin C A, Babinec S J, 1993 *Langmuir* **9** 1945
[14] Chen K, Caldwell W B, Mirkin C A, 1993 *J. Am. Chem. Soc.* **115** 1193
[15] Ozawa S, 2004 *Hyomenkagaku* **25** 428
[16] Akiyama T, Matsuoka K-i, Arakawa T, Kakutani K, Miyazaki A, Yamada S, 2006 *Jpn. J. Appl. Phys.* **45** 3758
[17] Matsuoka K-i, Akiyama T, Yamada S, 2008 *J. Phys. Chem. C* **112** 7015
[18] Moon J H, Shin J W, Youl Kim S Y, Park J W, 1996 *Langmuir* **12** 4621

Supporting PtRu catalysts on various types of carbon nanomaterials for fuel cell applications

Yoshiyuki Suda[1, *], Masahiro Ozaki[1], Hideto Tanoue[1], Hirofumi Takikawa[1], Hitoshi Ue[2], Kazuki Shimizu[3], Hirokazu Muramoto[4]

[1] Department of Electrical and Electronic Information Engineering, Toyohashi University of Technology (1-1 Hibarigaoka, Tempaku, Toyohashi, Aichi 441-8580, Japan)

[2] Tokai Carbon Co., Ltd. (394-1 Subashiri, Oyama, Sunto, Shizuoka 410-1431, Japan)

[3] Shonan Plastic Mfg. Co., Ltd. (31-27 Daikan, Hiratsuka, Kanagawa 254-0807, Japan)

[4] Cooperative Research Facility Center, Toyohashi University of Technology (1-1 Hibarigaoka, Tempaku, Toyohashi, Aichi 441-8580, Japan)

E-mail: suda@ee.tut.ac.jp

Abstract: PtRu catalysts were supported on five types of carbon nanomaterials of various shapes, sizes, and graphitic properties and the catalyst supports evaluated. The carbon nanomaterial used included three types of nanoparticles: Arc Black (AcB), Vulcan XC-72 (Vulcan) and graphene oxide (GO), and two types of nanofibers: carbon nanocoil (CNC) and carbon nanotube (CNT). Pt and Ru were supported by the reduction method using sodium borohydride. The metal catalyst loading was confirmed by thermo-gravimetric analysis (TGA), electron microscopy, and X-ray diffraction (XRD). Transmission electron microscopy (TEM) and XRD revealed that the diameter of PtRu catalyst nanoparticles loaded on reduced GO (rGO) and AcB were ~2 nm and was the smallest among all the samples. Shifts in Pt (111) XRD peaks of CNC and CNT were larger than those of AcB, Vulcan, and rGO. These results suggest that the diameters of catalyst nanoparticles became smaller by loading on the carbon nanoparticles with a large surface area including rGO, AcB, and Vulcan. Loading onto the carbon nanofibers enhanced the degree of PtRu alloying.

1. Introduction

Direct methanol fuel cells (DMFC) are applicable to portable electric power supply because the liquid fuel has a high energy density and DMFC is operated in a low temperature range (room temperature–100°C) [1]. In order to apply DMFCs to commercial electronic devices, low cost, miniaturization, and high efficiency issues needs to be solved [2-7]. As fuel cells employ precious metals including platinum (Pt) and ruthenium (Ru) for the oxidation of fuel and reduction of oxygen, these catalyst metals are required to be used effectively. This can be achieved by using their nanoparticulate form. Carbon nanomaterial is efficient for supporting nanometer-sized metal catalysts and has been used as a catalyst support in energy devices. There are various types of carbon nanomaterials including carbon nano-onions, carbon nanocapsules, carbon nanohorns, and carbon nanotubes [8-10]. Recently, fibrous carbon nanomaterials including carbon nanofibers (CNFs) [11, 12] and CNTs [13, 14] have been studied for applications as catalyst supports in fuel cells. These fibrous carbon nanomaterials are capable of having minute structures as well as high electric conductivity in the fuel cell electrode. In contrast, other type fibrous CNFs have a helical shape. These are called carbon nanocoils (CNCs) [15-18]. It was shown that CNCs could also be used as a catalyst support for fuel cells [19-21].

The properties of carbon nanomaterials as a catalyst support have been studied for the following parameters: (1) electric conductivity and graphitic property of carbon nanomaterials, (2) dispersion of the catalyst nanoparticles on carbon nanomaterials with the aim of having a narrow size distribution [22, 23], (3) catalytic activity by electrochemical characterization, and (4) fuel cell performance of the anode/cathode electrode of the catalyst-supported carbon nanomaterials [24]. In this study, we supported PtRu catalysts on five types of carbon nanomaterials with different shapes, sizes, and graphitic properties and evaluated the state of the catalyst loading. The carbon nanomaterials used included the following nanoparticles: Arc Black (AcB), Vulcan XC-72 (Vulcan) and graphene oxide (GO), and two types of nanofibers: carbon nanocoil (CNC) and carbon nanotube (CNT). A quantity synthesis apparatus of AcB [23, 24] and CNC was developed [15]. The properties of Vulcan, CNT, and GO were compared with those of AcB and CNC.

2. Experimental

2-1. Preparation of carbon nanomaterials

2-1-1. Carbon nanocoil (CNC)

Carbon nanofiber with a helical shape is called helical carbon nanofiber (HCNF). HCNFs are classified into four types: angular-type CNC and carbon nanotwist (CNTw), and round-type CNC and CNTw by their coil diameter and surface morphology of the fiber. These classifications are determined by their coil diameter and by the surface morphology of the fiber [15]. Angular-type CNC with fiber diameters of 120–400 nm and coil diameters of 400–1000 nm were used in this study. This angular-type CNC is composed of two, or more, thin helical carbon nanofibers with an edged surface [15]. CNC was synthesized on a Fe/Sn catalyst which was deposited on a graphite substrate using an automatic chemical vapor deposition (CVD) system with consecutive substrate transfer mechanism [15] that was developed in our laboratory. Conventional CVD apparatus requires cooling down after synthesis in every operation, resulting in a longer process time. This apparatus employs two gate valves to separate the load lock, synthesis, and cooling chambers. The sample synthesized can be replaced by a new catalyst while keeping the synthesis chamber heated. By using this apparatus, CNC was obtained at a production rate of 2 g/h. The CVD conditions to produce CNC are listed in Table 1.

Table 1 CNC growth conditions

Substrate	Graphite
Catalyst	Fe/Sn composite catalyst
Dilution gas	N_2 (1400 ml/min)
Feedstock	C_2H_2 (350 ml/min)
Reaction temperature	780°C
Catalyst annealing time	10 min
Reaction time	10 min

2-1-2. Graphene oxide (GO)

Graphene oxide (GO) is a sheet-like graphite and is obtained by treating with strong oxidizers. GO was synthesized from graphite powder (JEOF carbon SNO-3) by a modification of the Hummers method. 2.5 g of potassium peroxodisulfate ($K_2S_2O_8$) and 2.5 g of phosphorus pentoxide (P_2O_5) were added to 12 mL of sulfuric acid (H_2SO_4) aqueous solution, which was heated at 90°C. This mixture was then cooled to 80°C. Then 3 g of graphite powder was added to the mixture and stirred at 80°C for 4.5 h. After cooling to room temperature, 500 mL of ultrapure water was added for dilution. The solution was filtered and dried in clock glass overnight. 15 g of potassium permanganate ($KMnO_4$) was slowly added while stirring. The rate of addition was carefully controlled to avoid sudden increases in temperature. The mixture was then maintained at 35°C for 2 hours. 250 mL of deionized water was gradually added; causing an increase in temperature that was kept below 50°C. After 2 hours, the mixture was further treated with 700 mL of deionized water and 20 mL of 30% hydrogen peroxide (H_2O_2) solution. The mixture was washed with a 1:10 hydrochloric (HCl) and water solution by repeated centrifugation. The solution was filtered and further washed with deionized water.

2-1-3. Arc-black

Arc-black (AcB) is a spherical particulate of carbon nanomaterial with an amorphous structure. AcB was synthesized by the twin-torch arc discharge apparatus developed in our laboratory [23, 24]. In the discharge chamber, two graphite rod electrodes are mounted at sharp angles, and an arc discharge with a AC current of 200 A occurred between the electrodes in an 80-kPa N_2 gas environment. During the discharge, the anode electrode was vaporized and the generated carbon particles of atoms, molecules, and clusters were cooled down by inelastic collisions with ambient N_2 gas molecules, forming AcB.

2-1-4. Other Materials

Commercial Vulcan XC-72R (Cabot Corp) and vapor grown carbon fiber (VGCF; SHOWA DENKO K.K.) were used as a Vulcan and carbon nanotubes (CNTs), respectively.

2-2. Catalyst preparation

PtRu catalysts were loaded by the reduction method using sodium boron hydrate ($NaBH_4$). Hydrogen hexachloroplatinate (IV) hexahydrate ($H_2PtCl_6{\bullet}6H_2O$) and Ruthenium trichloride ($RuCl_3$) were respectively used as precursors of Pt and Ru. The molar ratio of Pt and Ru was set at 1:1 and the precursors were balanced between 20wt.% and 30wt.% with respect to the weight of the carbon nanomaterials. The carbon nanomaterial was dispersed in 500 mL of deionized water by sonication for 20 min. $H_2PtCl_6{\bullet}6H_2O$ and $RuCl_3$ were stirred in 50 mL of deionized water at 60 rpm for 10 min. This metal precursor solution was added to the carbon nanomaterial solution and the mixture stirred at 600 rpm for 10 min. $NaBH_4$, with 30-fold mole of the metal precursors, was added to 400 mL deionized water. This $NaBH_4$ solution was added to the metal precursor solution and stirred at 600 rpm for 2 hours. The solution was filtered by a membrane and washed repeatedly with deionized water, and the catalyst was obtained after drying.

It is reported that GO was reduced during the catalyst preparation [25, 26]. The reduced GO shall be referred to as rGO.

2-3. Characterization

The carbon nanomaterials were characterized using a scanning electron microscope (SEM; Hitachi, S-4500II), a compressive resistivity meter, and a laser Raman spectroscope (JASCO NRS-1000, excitation wavelength = 532 nm). The crystalline structure of the carbon nanomaterials was analyzed by a Raman spectrometer and the intensity ratio of the G-peak (1581 cm^{-1}) to the D-peak (1360 cm^{-1}) (I_G/I_D) in the Raman spectra was noted. The morphological characteristics of the prepared catalysts were analyzed using a transmission electron microscope (TEM; JEOL, JEM-2100F). Before and after catalyst loading, the carbon nanomaterials were analyzed by a thermo-gravimetric analyzer (TGA; Shimadzu, DTG-60) in a temperature range of 100–1000°C. The rate of temperature increase was 10°C/min under a flow rate of 50 ml/min of ambient air. By determining the residual weight after heating, the catalyst load was measured. The catalytic activities of the catalysts loaded on carbon nanomaterials were measured by a three-electrode cell with an electrolyte of 0.5 M H_2SO_4 and a voltage scan rate of 50 mV/s.

3. Results and discussion

3-1. Configuration and graphitic structure of carbon nanomaterials

Scanning electron microscopy (SEM) micrographs of the carbon nanomaterials are shown in Fig. 1. AcB and Vulcan are spherical particles with diameters of 50 nm and 30 nm, respectively. CNC and CNT are confirmed to be of a fibrous configuration. The fiber diameters of CNC and CNT are similar at 150 nm, and the coil diameter of CNC ranges between 400 and 1000 nm. GO is applied in the form of thin sheets with a thickness of 3 μm.

Journal of Physics: Conference Series **433** (2013) 012008 doi:10.1088/1742-6596/433/1/012008

Fig. 1. SEM micrographs of carbon nanomaterials: (a) AcB, (b) Vulcan, (c) CNC, (d) CNT (VGCF), and (e) GO.

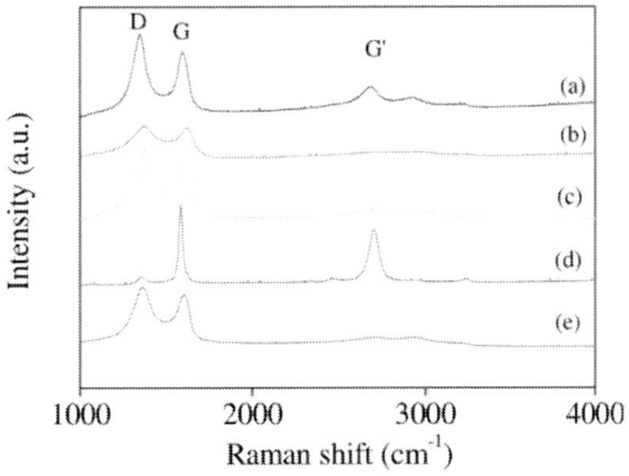

Fig. 2. Raman spectra of (a) AcB, (b) Vulcan, (c) CNC, (d) CNT, and (e) GO.

Table 2 I_G/I_D ratios in the Raman spectra of the carbon nanomaterials

	AcB	Vulcan	CNC	CNT	GO
I_G/I_D	0.86	0.90	0.90	2.61	0.91

Raman spectra of the carbon nanomaterials are shown in Fig. 2. The G-peak (1581 cm^{-1}), which represents the graphite crystalline structure, and the D-peak (1360 cm^{-1}), which indicates the defect in crystalline graphite, can be seen in all the spectra. G'-peak (2700 cm^{-1}) which appears in crystalline graphite is also seen in the spectra of AcB, CNC, and CNT. The intensity ratios of the G-peak to the D-peak (I_G/I_D) in the Raman spectra was evaluated and shown in Table 2. CNT has the highest I_G/I_D ratio among the carbon nanomaterials, while those of the other four are more similar.

3-2. TEM observation of catalyst loaded on carbon nanomaterials

Fig. 3 shows the TEM micrographs of catalyst nanoparticles loaded on the carbon nanomaterials. Nanometer scale particles are well dispersed on all the carbon nanomaterials. Energy-dispersive X-ray analyzer equipped with TEM revealed that the nanoparticles loaded are of Pt and Ru.

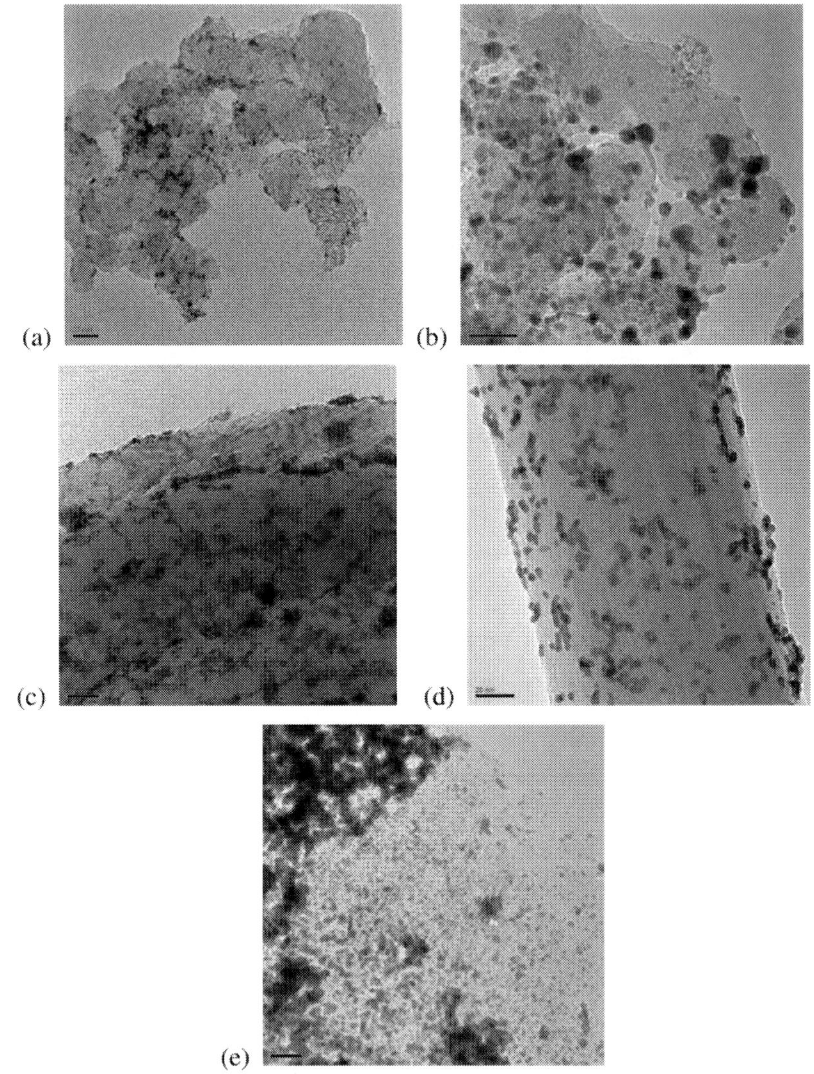

Fig. 3. TEM micrographs of catalyst nanoparticles loaded on carbon nanomaterials: (a) AcB, (b) Vulcan, (c) CNC, (d) CNT, and (e) rGO. The weight ratio of Pt and Ru to carbon nanomaterial was 30wt.%. Scale bar = 20 nm.

For each sample, 70 nanoparticles in the TEM micrographs were counted and their size distributions shown in Fig. 4. The catalyst nanoparticles loaded were in the range of 1 and 6 nm. The average diameters of catalyst nanoparticles loaded on rGO and AcB were ~2 nm and were the smallest among those on all the carbon nanomaterials. These results show that the form of nanofibers including CNC and CNT tended to load larger catalyst nanoparticles than the form of particulates including AcB, Vulcan, and rGO. The surface areas of the carbon nanomaterials without catalysts are listed in Table 3.

3-3. Thermo-gravimetric analysis (TGA)

Fig. 5 shows the TGA results of the carbon nanomaterials without PtRu catalysts. CNT has the highest combustion temperature among all the carbon nanomaterials because of the highest crystalline order of graphite. However, the combustion temperature of CNC was very low. The combustion temperature of GO was the lowest and this is probably because of the high surface area.

The TGA results of the PtRu-catalyst-loaded carbon nanomaterials with two different catalyst weight ratios are shown in Fig. 6. After heating these catalyst-loaded carbon nanomaterials to 1000°C, the error of the residue weights were within 5% with respect to the weight ratio of the PtRu precursor to carbon nanomaterial. When PtRu catalysts were supported, the combustion temperatures of AcB and Vulcan were lower than those of CNT and CNC.

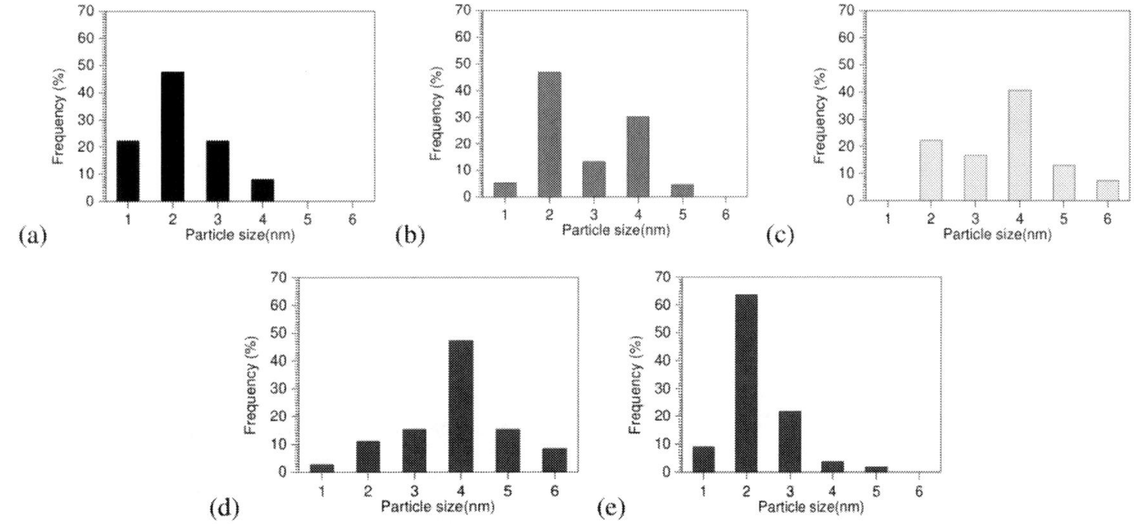

Fig. 4. Size distributions of PtRu catalyst nanoparticles observed in Fig. 3. Carbon nanomaterial and the average diameter are as follows: (a) PtRu/AcB (2.16 nm), (b) PtRu/Vulcan (2.82 nm), (c) PtRu/CNC (3.67 nm), (d) PtRu/CNT (3.86 nm), (e) PtRu/rGO (2.25 nm).

Table 3 Surface area of carbon nanomaterials measured by Brunauer, Emmett and Teller (BET) method

	AcB	Vulcan	CNC	CNT	GO
Surface area (m^2 / g)	153	237	115	13	347

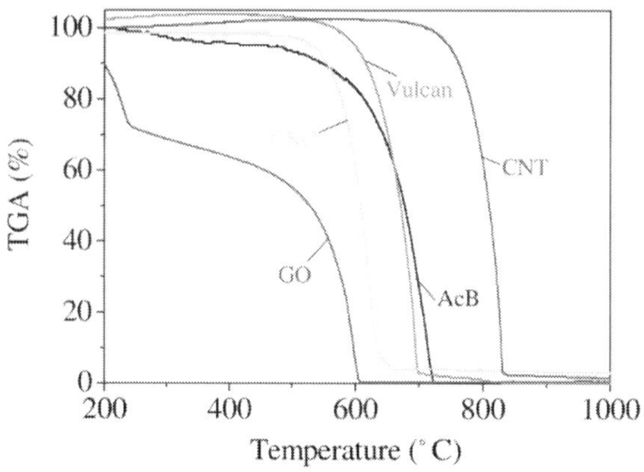

Fig. 5. TGA results of carbon nanomaterials. (black) AcB, (red) Vulcan, (green), CNC, (blue) CNT, and (purple) GO. Temperature range: 100–1000°C; rate of temperature increase: 10°C/min.

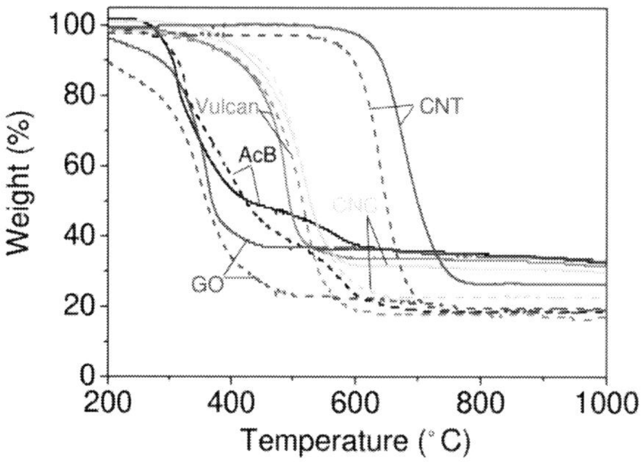

Fig. 6. TGA results of PtRu-loaded carbon nanomaterials (solid line, 30wt.%; dashed line, 20wt.%). (black) AcB, (red) Vulcan, (green) CNC, (blue) CNT, and (purple) GO.

3-4. X-ray diffraction (XRD)

XRD patterns of PtRu-catalyst-loaded carbon nanomaterials are shown in Fig. 7. All the catalysts show typical diffraction peaks of Pt (111), Pt (200), Pt (220), and Pt (311). C (200) peak is also seen in the carbon nanomaterials of AcB, CNC and CNT, which have a high G' intensity in the Raman spectra. The average diameter, lattice constant, and surface area of the Pt nanoparticles are evaluated by the following formula [27, 28]:

$$d = \frac{0.9\lambda_{k\alpha1}}{\beta \cos\theta_{max}} \quad (1)$$

$$\alpha_{fcc} = \frac{\sqrt{2}\lambda_{k\alpha1}}{\sin\theta_{max}} \quad (2)$$

$$S = \frac{6000}{\rho d} \tag{3}$$

Where d is the average diameter (nm), $\lambda_{K\alpha 1}$ is the wavelength of the X-ray (0.154 nm), β is the full width of the diffraction peak at half maximum (rad), θ_{max} is the angle at the position of the peak maximum (deg), a_{fcc} is the lattice constant (nm), S is the metal surface area (m^2/g), and ρ is the density of Pt (21.4 g/cm^3). The calculated values are presented in Table 4. The average diameter of catalyst nanoparticles was in the order of: rGO, Vulcan, AcB, CNC, and CNT. This tendency corresponds to the results from TEM observation. The 2θ value of Pt (111) in single crystalline Pt is 39.8° [27]. Shifts in the Pt (111) peak in XRD patterns, which indicate the alloying of Pt with Ru, were seen in all the samples. The shifts in the samples of CNC and CNT were larger than those of AcB, Vulcan, and rGO.

From the analysis above, it suggests that the diameter of catalyst nanoparticles became smaller by loading on the carbon nanoparticles with a large surface area including rGO, AcB, and Vulcan, and that the degree of PtRu alloying was enhanced by loading on the carbon nanofibers including CNC and CNT.

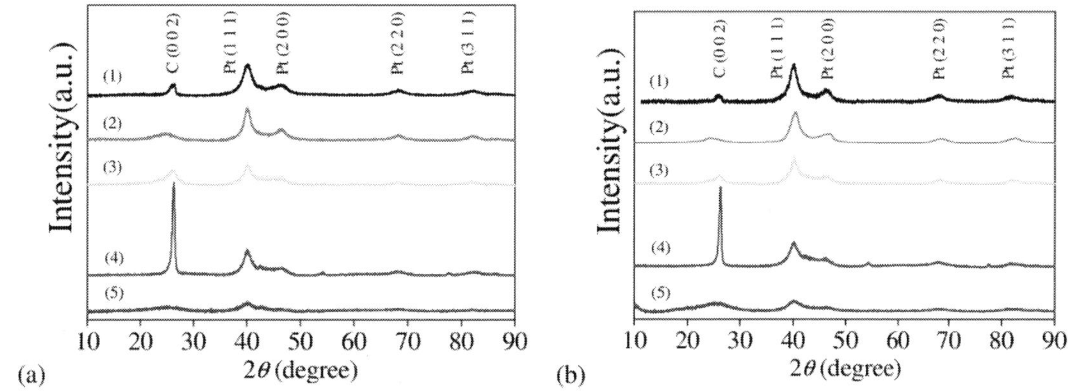

Fig. 7. XRD patterns of (a) 20wt.% PtRu- and (b) 30wt.% PtRu-loaded carbon nanomaterials. (1) AcB, (2) Vulcan, (3) CNC, (4) CNT, and (5) GO.

Table 4 Results of XRD analysis of PtRu catalysts loaded on carbon nanomaterials (20wt.%, 30wt.%)

Catalyst	Peak of Pt (111) (degree)	Lattice constant (Å)	Particle size (nm)	Pt surface area (m^2/g)
PtRu/AcB	40.1, 40,1	3.90, 3.90	2.98, 3.34	94.2, 84.0
PtRu/Vulcan	40.1, 40.2	3.89, 3.89	3.07, 3.09	91.2, 90.6
PtRu/CNC	40.2, 40.3	3.89, 3.88	3.20, 3.69	87.7, 75.9
PtRu/CNT	40.2, 40.3	3.89, 3.89	3.69, 3.84	76.0, 73.0
PtRu/GO	40.1, 40.1	3.90, 3.89	2.59, 2.90	108.2, 96.4

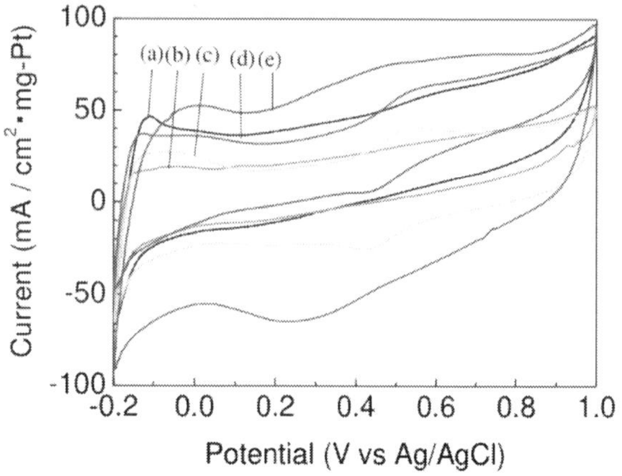

Fig. 8. CV curves of (a) PtRu/AcB, (b) PtRu/Vulcan, (c) PtRu/CNC, (d) PtRu/CNT, and (e) PtRu/rGO electrodes in Ar saturated 0.5 mol L^{-1} H_2SO_4 solution at a potential scan rate of 50 mV s^{-1}.

Table 5 Electrochemically active surface areas of PtRu-loaded carbon nanomaterials.

	PtRu/AcB	PtRu/Vulcan	PtRu/CNC	PtRu/CNT	PtRu/rGO
Electrochemically active Surface area (m^2 g^{-1})	8.14	6.46	16.45	9.20	3.52

3-5.Electrochemical measurement

The cyclic voltammograms (CVs) of PtRu-loaded carbon nanomaterials (PtRu/C) were obtained in Ar-purged 0.5M H_2SO_4 solution (Fig. 8). The electrochemically active surface area of PtRu/C was obtained by integrating the total charge corresponding to the desorption peak of hydrogen, and normalizing with scan rate, Pt loading, and the charge value of 210 μC cm^{-2} for Pt surface [28,29]. Thus , the obtained active surface areas of PtRu/C catalysts were 8.14 m^2 g^{-1} for PtRu/AcB , 6.46 m^2 g^{-1} for PtRu/Vulcan, 16.45 m^2 g^{-1} for PtRu/CNC, 9.20 m^2 g^{-1} for PtRu/CNT, and 3.52 m^2 g^{-1} for PtRu/rGO, as listed in Table 5.

4. Conclusion

Various carbon nanomaterials with different configurations and degree of crystalline graphite were used for PtRu catalyst nanoparticle support, which can be applied in fuel cells. When the weight ratio of the catalyst precursor to carbon nanomaterial is constant, the carbon nanomaterials with higher surface area can be loaded to form smaller catalyst nanoparticles. The degree of alloying of Pt with Ru was evaluated by X-ray diffraction pattern, and the form of nanofibers influenced a higher degree of alloying. We initially evaluated the catalytic activity of these catalysts by using a three-electrode cell. The direct methanol fuel cell performance of the anode and cathode electrodes using these catalysts will be presented in future work.

Acknowledgements

We would like to thank Prof. A. Sandhu, Dr. R. Tero, and Mr. Y. Okamoto at EIIRIS, Toyohashi University of Technology for GO preparation. This work has been partly supported by the Research Project of the Venture Business Laboratory from Toyohashi University of Technology (TUT); the Core University Programs (JSPS-CAS program in the field of "Plasma and Nuclear Fusion") from the Japan Society for the Promotion of Science (JSPS); and a Grand-in-Aid for Scientific Research from the Ministry of Education, Culture, Sport, Science and Technology (MEXT).

References

[1] Mao Q, Sun G Q, Wang S L, Sun H, Wang G X and Gao Y 2007 *Electrochimica Acta* **52** (24) p6763–6770.

[2] Gallo J M R, Andrea Icardi U, Baglio V, Coralli A and Graizzaro A 2011 *Int. J. Hydrogen Energy* **36** (13) p8082–8087.

[3] Yao K, Chen Y, Chao C, Wang W, Lien S, Shih H, Chen T and Weng K 2010 *Thin Solid Films* **518** (24) p7225–7228

[4] Kamarudin S K, Daud W R W, Ho S L and Hasran U A 2007 *J. Power Sources* **163** (2) p743–754

[5] Park Y, Peck D, Kim S, Lim S, Jung D, Jang J and Lee D 2010 *J. Power Sources* **195** (13) p4080–4089

[6] Chu D, Jiang R, *Electrochimica Acta* 2006 **51** (26) p5829–5835

[7] Li X, Faghri A, Xu C 2010 *Int. J. Hydrogen Energy* **35** (16) p8690–8698

[8] Ando Y, Zhao X, Hirahara K, Suenaga K, Bandow S and Iijima S 2001 *Diamond Relat. Mater.* **10** p1185

[9] Thess A, Lee R, Nikoleav P, Dai H, Petit P, Rober J T, Xu C, Lee Y H, Kim S G, Rinzler A G, Colbert D T, Scuseria G E, Tomanek D, Fischer J E and Smalley R E 1996 *Science* **273** p483

[10] Yacaman J M, Yoshida M M, Rendon L and Santiesteban J G 1993 *Appl. Phys. Lett.* **62** p202

[11] Hamoudi Z, Khakani M A E and Mohamedi M 2011 *Int. J. Hydrogen Energy* **36** (8) p4682–4688

[12] Maiyalagan T 2009 *Int. J. Hydrogen Energy* **34** (7) p2874–2879

[13] Maiyalagan T 2008 *Appl. Catal. B: Environ.* **80** (3-4) p286–295

[14] Drillet J F, Bueb H, Dettlaff-Weglikowska U, Dittmeyer R and Roth S 2010 *J. Power Sources* 195 (24) p8084–8088

[15] Yokota M, Shinohara Y, Kawabata T, Takimoto K, Suda Y, Oke S, Takikawa H, Fujimura Y, Yamaura T, Itoh S, Miura K and Morioki M 2010 *J. Nanosci. Nanotechnol.* **10** (6) p3910–3914

[16] Yokota M, Suda Y, Takikawa H, Ue H, Shimizu K and Umeda Y 2011 *J. Nanosci. Nanotechnol.* **11** p2344–2348

[17] Takikawa H, Yatsuki M, Miyano R, Nagayama M, Sakakibara T, Itoh S and Ando Y 2000 *Jpn. J.*

Appl. Phys. **39** p5177–5179

[18] Xu G, Chen B, Shiki H, Katsumata T, Takikawa H, Sakakibara T, Itoh S and Ina T 2005 *Jpn. J. Appl. Phys.* **44** p1569–1576

[19] Hyeon T, Han S, Sung Y E, Park K W and Kim Y W 2003 *Angewandte Chemie* **115** (36) p4488–4492

[20] Park K W, Sung Y E, Han S, Yun Y K, Hyeon T 2004 *J. Phys. Chem. B* **108** (3) p939–944

[21] Sevilla M, Sanchís C, Valdés-Solís T, Morallón E and Fuertes A B 2009 *Electrochimica Acta* **54** (8) p2234–2238

[22] Kang S, Lim S, Peck D H, Kim S K, Jung D H, Hong S H, Jung H G and Shul Y 2012 *Int. J. Hydrogen Energy* **37** p4685–4693

[23] Oke S, Higashi K, Shinohara K, Izumi Y, Takikawa H, Sakakibara T, Itoh S, Yamaura T, Xu G, Miura K, Yoshikawa K, Sakakibara T, Sugawara S, Okawa T and Aoyagi N 2008 *Chem. Eng. J.* **143** p225–229

[24] Ikeda T, Kaida S, Satou T, Suda Y, Takikawa H, Tanoue H, Oke S, Ue H, Okawa T, Aoyagi N and Shimizu K 2011 *Jpn. J. Appl. Phys.* **50** 01AF13

[25] Shen J, Yan B, Shi M, Ma H, Li N and Ye M 2012 *Mater. Res. Bull.* **47** p1486–1493

[26] Xin Y, Liu J G, Zhou Y, Liu W M, Gao J, Xie Y, Yin Y and Zou Z 2011 *J. Power Sources* **196** p1012–1018

[27] Antonucci P L, Arico A S, Cret`ı P, Ramunni E and Antonucci V 1999 *Solid State Ionics* **125** p431–437

[28] Lu Y, Reddy R G 2007 *Electrochimica Acta* **52** p2562–2569

[29] Pozio A, De Francesco M, Cemmi A, Cardellini F and Giorgi L 2002 *J. Power Sources* **105** p13–19

Study on Evaluation Methods for Mechanical Properties of Organic Semiconductor Materials

T. Kobayashi*[1], T. Yokoyama[1], Y. Utsumi[2], H. Kanematsu[3] and T. Masuda[4]

[1]Department of Electronics & Control Engineering, Tsuyama National College of Technology, Okayama, Japan
[2] Laboratory of Advanced Science and Technology for Industry, University of Hyogo, Hyogo, Japan
[3]Department of Materials Science & Engineering, Suzuka National College of Technology, Mie, Japan
[4]Q-Light Co., Ltd., Iwate, Japan

E-mail: t-koba@tsuyama-ct.ac.jp

Abstract. This paper describes the evaluation method of the mechanical properties of the materials constituting organic semiconductor, and the test result of the relation between applied strain and the fracture of thin films. The final target of this work is the improvement of flexibility of organic light emitting diode(OLED), the tensile test of the thin films coated on flexible substrate is conducted, and the vulnerable parts of the constituent material of the OLED is quantitatively understood, further the guideline for designing OLED structure will be obtained. In the present paper, tensile test of an aluminium oxide thin films deposited on a poly-ethylene-tere-phtalate (PET) substrate was carried out under constant conditions, the following results were obtained:(1)Cracking of the aluminium oxide thin films was observed using an optical transparent formula microscope at more than 40 times magnification; (2)Cracking was initiated at a strain of about 3%; (3)the number of cracks increased proportional to the strain, and saturated at about 9% strain; (4)Organic thin films α-NPD caused the same cracking as oxide thin films.

1. Introduction

Since organic semiconductors can be manufactured by printing, and have higher flexibility compared to inorganic semiconductors, such as type silicon, their future is expected and the research and development of organic semiconductor including solar cells has been extensively conducted regardless of inferior electrical characteristics. In addition, organic light emitting diode (OLED) which is one of the organic semiconductors, has been practically applied as displays and lighting due to the features thin, self luminous and fast response.

It is possible to obtain flexibility by applying a polymer sheet or films, such as PEM or PET, for the substrate of organic semiconductor in stead of a conventional glass substrate. However, transparent conductive oxide (TCO) such as Indium Tin oxide (ITO) or Indium Zinc oxide (IZO) films such as SiNx should be coated on the flexible polymer sheet substrate which easily transmitted through the vapour, in order to prevent deterioration of the organic semiconductor element due to moisture. These transparent electrodes and SiNx may impair the flexibility because they are brittle inorganic materials therefore the tensile and bending tests of ITO and SiNx have been carried out to evaluate the cracking due to distortion. (1) Recently, it has been reported that transparent conductor organic material PEDUT:PSS (polyethylene dioxythiophene doped with polystyrene sulfonate) does not show cracking even in a tensile strain of 60% although its electric conductivity is not so low. (2) More recently, it has been also reported that the multilayer transparent electrodes having a structure of ZnS/Ag/Wo_3 shows much higher flexibility than ITO and comparable electric properties. (3) Therefore, the improvement of flexibility of OLED can be expected.

As described above, the establishment of a method of assessing the flexibility of the thin film constituting organic semiconductor has become an important issue, however systematic consideration has not been found. Therefore, in the present work, the observation methods for the cracking of TCO and transparent organic light emitting materials were comparatively studied, and the observation results for the cracking of inorganic thin films, such as aluminium oxide and IZO, and typical organic thin film used for OLEDs were reported.

2. Experimental

2.1 Test specimens

The test specimens of inorganic thin films were the aluminium oxide with an 80 nm thickness coated on a PET substrate having a thickness of 25 μ m and IZO with a 200 nm thickness coated on a PEN substrate with a 250 μ m thickness. The test specimen of organic thin film was the NPD with a 200nm thickness deposited on the PEN substrate with a thickness of 250nm. The aluminium oxide thin film was vacuum deposited by a roll to roll web coater. The IZO thin film was deposited by a batch type magnetron sputtering apparatus. The deposition of the NPD thin film was carried out by a multi chamber type vacuum deposition system for organic thin films.

2.2 Tensile test

The above test pieces were cut to 4 mm in width, and the occurrence of film cracking while providing tensile strain was observed under an optical microscope. In a manner to clamp the specimen to impart tensile strain with the micromere, the gauge length was 20 mm, Therefore, assuming an error of $\pm 5\mu$ m in micrometer and mechanical system, and error of 0.025% was estimated at a strain of 1%, namely the relative error results in 2.5%. Moreover, in order to correspond to the transmission type optical microscope, the parts under test piece were composed of transparent acryl and opening.

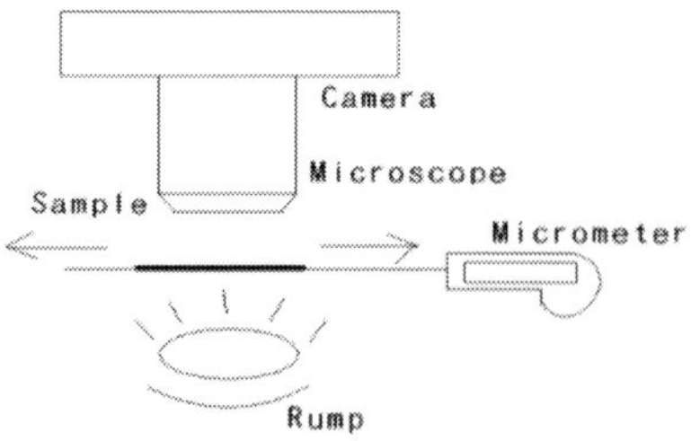

Fig.1Schematic diagram of the experimental method

Fig.2 Tensile test apparatus

2.3 Observation of crack

Digital photos were taken while specimens were observed by the optical microscope, to understand the relationship between cracking and applied strain. Thin films observed in the present study had a high light transmittance, and the substrate was transparent plastic sheet, thus it was not clearly visible by a reflection type optical microscope in which light was irradiated from the objective lens.

Therefore, transmission type optical microscope where light was introduced from blow the transparent specimen to the objective lens was employed. In addition, since the cracking of the organic thin films was too fine to observe clearly even with the transmission type optical microscope, a laser microscope (manufactured by JOEL, JSM 6700F) were used in combination for observing the occurrence and the configuration of the cracking. In the observation of organic film, gold was deposited in a thickness of 3nm on the specimen for the prevention of charge up because of the high electrical resistance materials.

3. Result and Discussion

3.1. Inorganic thin films

Firstly, Fig3 shows an example of the cracking occurred in the aluminium oxide film. Cracking was easily observed by the transmission type optical microscope. The relation between the tensile strain and the number of cracks per unit length in the direction of the tensile is show in Fig.4. The number of cracks increased with the increase of strain, then strain of about 8% or more, number of cracks was saturated to a constant value. Next, Fig.5 shows an example of the cracking in the IZO film. As with the aluminium oxide film, cracking occurred in approximately 3% applied strain, the number of cracks increased with the increasing strain, and it showed a constant value.

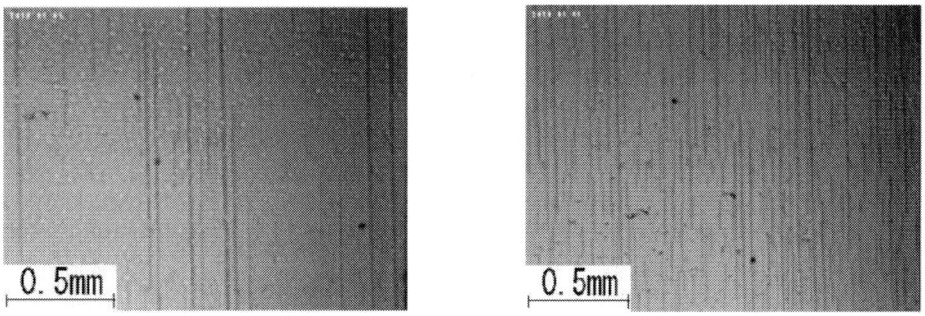

(a) Strain 4% (b) Strain 8%

Fig.3 An example of the cracking in aluminium oxide film

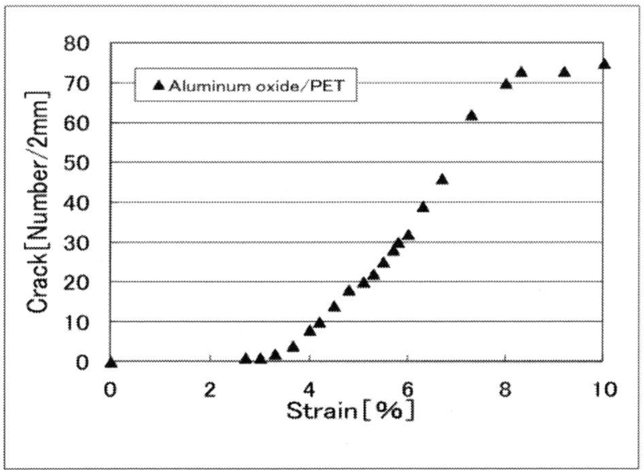

Fig.4 Relation between strain the number of cracks per unit area in aluminium oxide film on PET substrate

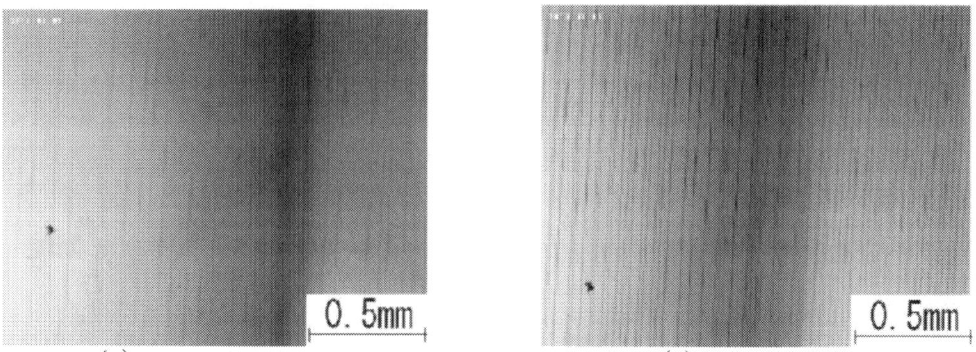

(a)Strain 3% (4) Strain 6%

Fig.5 An example of cracking of IZO

3.2. Organic film

3.2.1 Observation result using optical microscope

Figure 6(a) shows a transmission type optical micrograph of α-NPD film. Cracking could not be recognized in 40 times magnification which was different from the cracking of aluminum oxide and IZO films, and fine cracking in α-NPD film was confirmed in 400 times magnification. The cracks narrow spacing and unclear in the magnification.

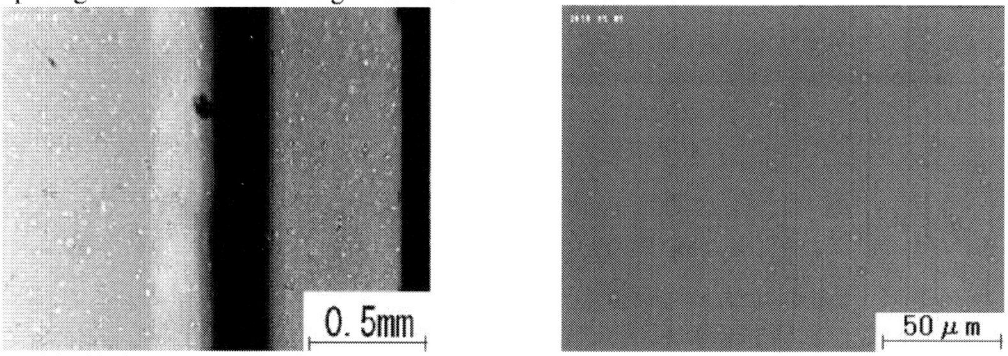

(a) 40 times magnification (b) 400 times magnification

Fig.6 A transmission type optical micrograph of α-NPD film

As shown in Fig.7, the cracking of the test piece applied around 8% strain was clearly observed at 400 times magnification with the optical microscope.

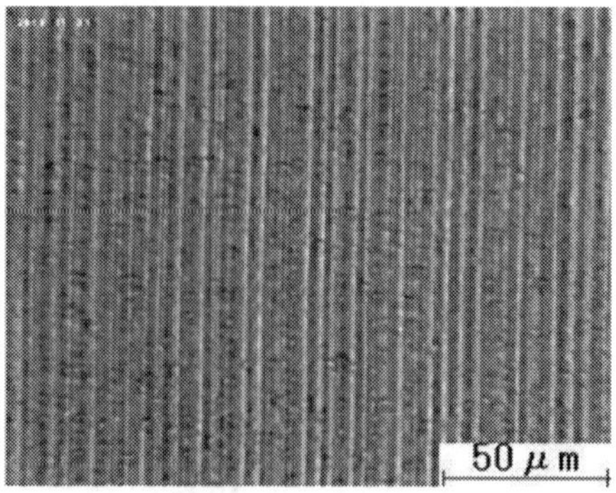

Fig.7 A transmission type optical micrograph of α-NPD film applied around 8% strain

3.2.2 Observation result using laser microscope

In order to confirm the presence or absence of cracking in the specimens applied low strain, specimens were observed after tensile test using laser microscope and a field emission type scanning electron microscope (FE-SEM). First, Fig.8 shows the result observed specimen shown in Fig.6 with the laser microscope. The optical microscope images were taken with a reflection type optical microscope. It was found that the recognition of cracks, which was not possible at 400 magnifications with transmission type microscope, was possible at 4500 times magnification even with reflection type optical microscope, although the image was not so clear. On the other hand, crack was clearly recognized by the reflection of the laser light image. In addition, it was confirmed by the height measurement that something that looks like points were the projections of about 2 μ m in height, thus it has been suggested that the projections are the adhesion of splash or foreign material on the substrate. Since cracks did not pass through the toe of the projection, it has been suggested that the projections are not the origin or the passage.

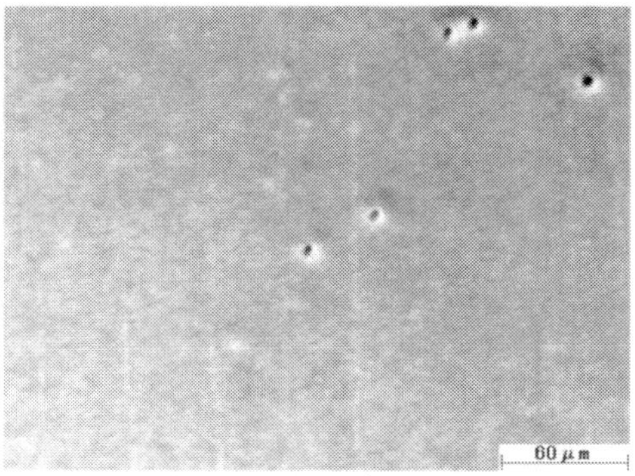

(a) Laser + Colour

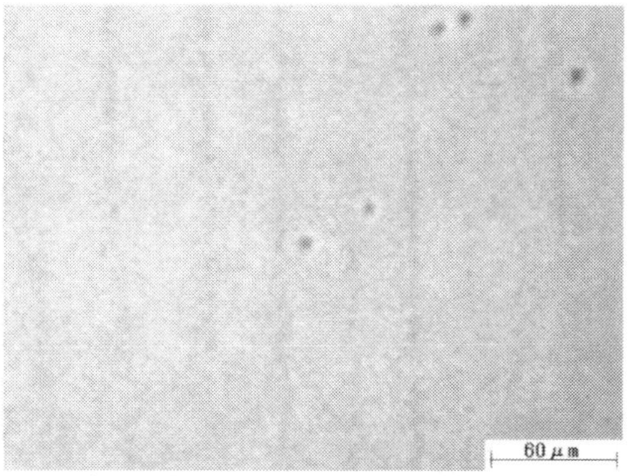

(b) Colour

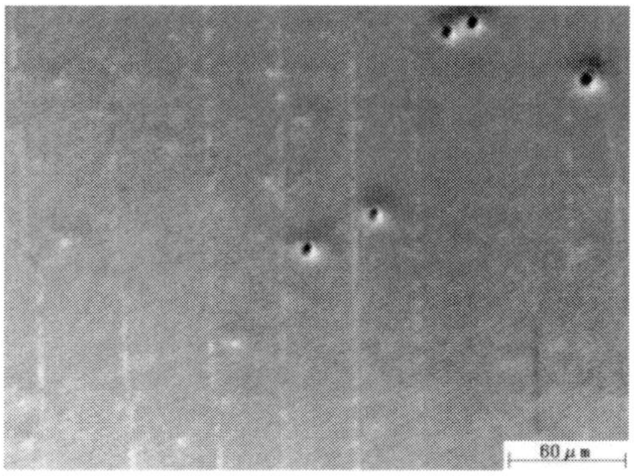

(c)Laser

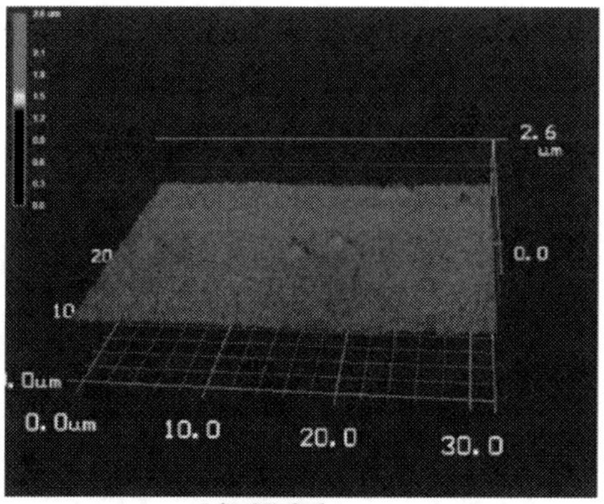

(d) Height

Fig.8 Images of the specimen shown in Fig.6 observed by a laser microscope at 4500 times magnification

Secondly, Fig.9 shows the result observed the specimen shown in Fig7. In addition to the vertical line-shaped cracks, granular or short horizontal stripes were observed. From the result of height measurement by laser, the longitudinal line and the granular or lateral stripes had a convex height of around more than 9 μ m, which was different from the concave line indicating crack that was expected. Therefore, further observation was conducted using the FE-SEM at higher magnification to investigate the configuration, as the following section.

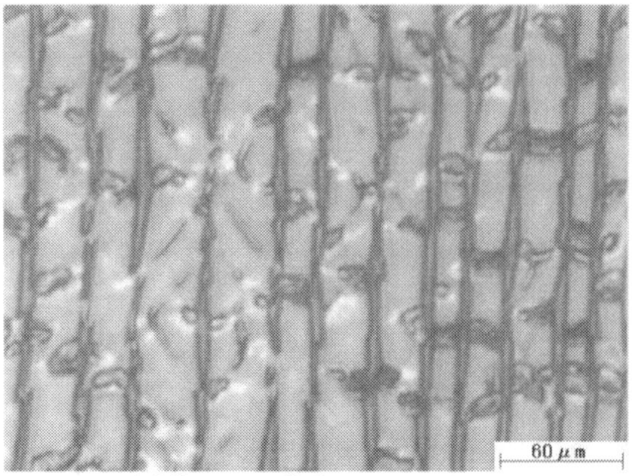

(a) Laser + Colour

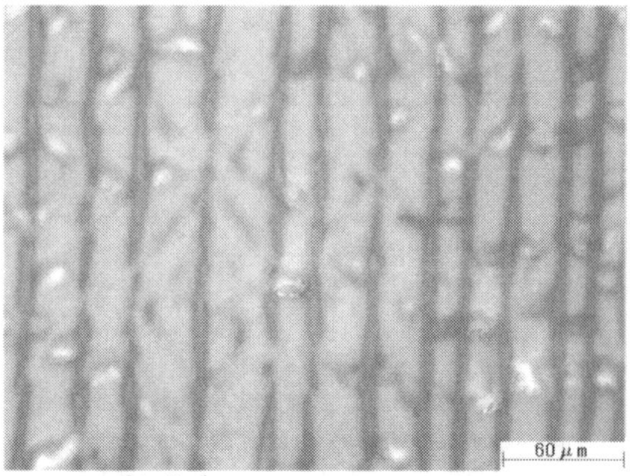

(b) Colour

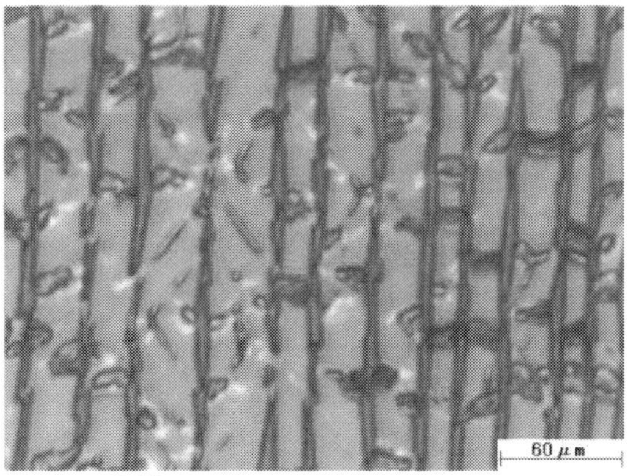

(c)Laser

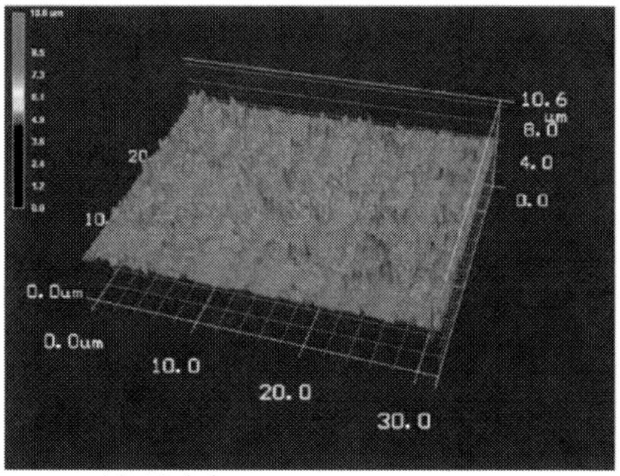

(d) Height

Fig.9 Images of the specimen shown in Fig.7 observed by a laser microscope at 4500 times magnification

3.2.3 Observation result using laser microscope

Figure10 shows the image of the specimen, shown in Fig.8 (b), by FESEM. From observation at a magnification of 10000, it was estimated that the vertical stripes were cracks and they were concave. The protrusion having a height of 6 μ m in Fig.8 (b) was assumed that the end of the crack in the film is peeled from the substrate and curled. It was also estimated that the convex shape, which was recognized as a short horizontal lines or granular, unlike the vertical protrusion, was an overlapping or bridged shape which was created by buckling lifted the coating film to peel resulted from the compressive stress. The compressive stress was produced by the shrinkage in the longitudinal direction due to the tensile strain in the horizontal direction. For reference, the specimen in Fig.6 (a) was observed by FESEM, the crack was not recognized. It was considered that this was because the crack had not opened almost no irregularities were formed.

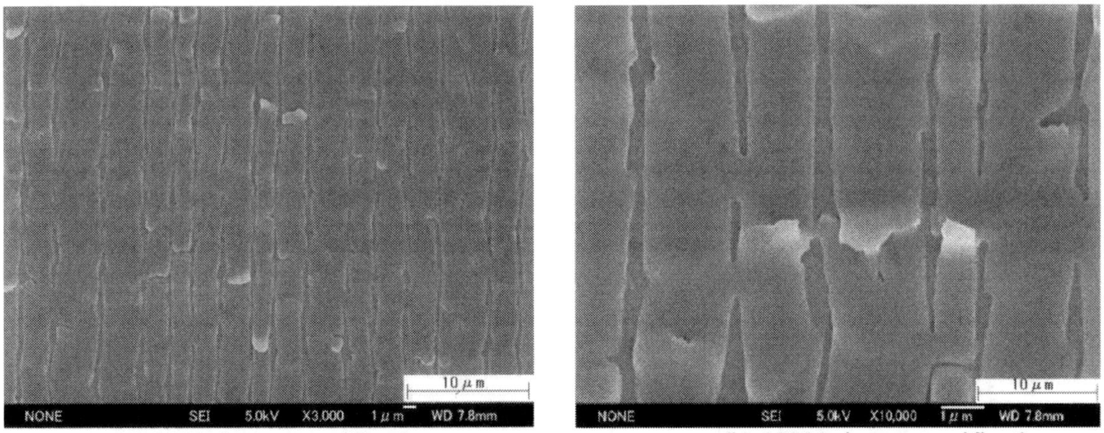

(a) 3000 times magnification (b) 10000 times magnification

Fig10 Images of specimen shown in Fig.9 by FE-SEM

3.2.4 Summary of observation method

・Regarding transparent films and substrate, transmission type optical microscope was suitable for observing crack in coated film, and crack was observed at a magnification of 40 times or more.

・When the interval of crack was narrow and crack was unclear, the magnification of the optical microscope needed to be changed to 400 times, or laser reflection microscope was recommended.

・Less information was obtained by using SEM for small cracks in the low strain condition where convex-concave was not created.

4. Conclusion

(1)Cracking of the aluminium oxide thin films was observed using an optical transparent formula microscope at more than 40 times magnification.
(2)Cracking was initiated at a strain of about 3%.
(3)The number of cracks increased proportional to the strain and saturated at about 9%.
(4)Organic thin film α -NPD caused same cracking as oxide thin films.

References

[1] Leteir Y, Medico L, Demarco F, Manson J A E, Betz U, Escola M E, Olsson M K and Atamny F 2004 Mechanical integrity of transparent conductive oxide films for flexible polymer-based display *Thin Solid Films* **460** pp 156-166

[2] Kim W H and Kafafi Z H 2002 Flexible organic light emitting devices using conductive polymeric anodes *Soc. Inf. Display. Dig. Tech. Papers* **XXXIII** pp 1090-1091

[3] Hyunsu C, Changhun Y, Jae-Woo P and Seunghyup Y 2009 Highly flexible organic light emitting diodes base on ZnS/Ag/WO_3 multilayer transparent electrodes *Organic Electronics* **10** pp 1163-1169

Photochemical half-cells using mixture films of fullerene-ethylenediamine adduct microparticles and polythiophene

Tsuyoshi Akiyama[1*], Satoshi Matsumura[2], Ken-ichi Matsuoka[2], Takeo Oku[1], Sunao Yamada[3]

[1]Department of Materials Science, School of Engineering, The University of Shiga Prefecture, 2500 Hikone, Hikone, Shiga, 522-8533, Japan

[2]Department of Materials Physics and Chemistry, Graduate School of Engineering, Kyushu University, 744, Moto-oka, Nishi-ku, Fukuoka 819-0395, Japan

[3]Department of Applied Chemistry, Faculty of Engineering, Kyushu University, 744, Moto-oka, Nishi-ku, Fukuoka 819-0395, Japan

E-mail: akiyama.t@mat.usp.ac.jp

Abstract. In this study, C_{60} fullerene–ethylenediamine adduct microparticles were prepared. Mixture films of these microparticles and polythiophene were fabricated on indium–tin-oxide transparent electrodes by spin-coating. Incorporation of C_{60}–ethylenediamine microparticles was verified by scanning electron microscopy (SEM) measurements. The coverage values of these microparticles were approximately 3–17%, which were calculated from SEM images of modified electrodes. Fluorescence spectra of modified electrodes indicated that the emission intensity of polythiophene in these mixture films was apparently quenched by these C_{60}–ethylenediamine microparticles as compared with a polythiophene film without these microparticles. In the presence of methylviologen, these modified electrodes generated stable photocurrent. The photoexciting species was polythiophene, which was verified by profiles of photocurrent action spectra. The C_{60}–ethylenediamine microparticles substantially enhanced the photocurrent signals generated by the polythiophene-modified electrode.

1. Introduction

Fullerenes are highly symmetrical aromatic molecules; they are known as good electron acceptors and n-type organic semiconductors because of the unique structure of their electron orbital. These unique electronic characteristics of fullerenes have attracted great attention. Recently, applications of fullerenes to a photoelectric conversion device [1-3], an organic solar cell [4-6], and an organic transistor [7] have been investigated intensively.

The chemical modification and clustering of fullerenes are essentially important for a wide range of applications in fullerene-based science and technology. Many organic chemists have reported the chemical modification of fullerenes. One of the simplest examples is the addition reaction between fullerene and aliphatic amine [8]. This addition reaction can occur even on an amino-substituted solid surface, which is useful in preparing fullerene-modified substrates [9-11].

Further, Kamat et al. reported the clustering of fullerenes and their applications to photocurrent generation [12-15]. The fullerene cluster is generated by sedimentation, generally by the addition of a poor solvent to a solution of fullerene. However, controlling the sizes and shapes of the clusters in the sedimentation method is not very easy.

With this background, fullerene clusters consisting of fullerenes and aliphatic diamines are obtained by the addition reaction between fullerenes and aliphatic amines. Formation of a micropowder by mixing of C_{60} fullerene and diamines was already reported [16, 17]. We discovered a sub-micrometer-sized C_{60} fullerene–diamine microparticle ($C_{60}P$) with an almost sphere shape under optimized conditions of the addition reaction [18]. We also expanded the addition reaction to C_{70} fullerene and diamines [19].

In our previous research, we succeeded in depositing $C_{60}P$ on the surface of a poly-3-dodecylthiophene(P3DT)-modified indium–tin-oxide (ITO) electrode using capillary action. We also determined the $C_{60}P$-enhanced photocurrent efficiency from the P3DT film in the presence of a sacrificial reagent [18]. However, it is difficult to control the surface coverage of $C_{60}P$ on P3DT for a $C_{60}P$-deposited P3DT film using capillary action.

In this study, a mixed solution of P3DT and dispersed $C_{60}P$ was modified on the ITO electrode by spin-coating. The photoexcited state of P3DT was quenched by the existence of $C_{60}P$ in the mixed film. The photocurrent generation properties of the mixture film and a polythiophene film as a corresponding reference were investigated.

2. Experimental Procedures

C_{60} fullerenes, ethylenediamine, P3DT, and other chemicals were used as received. The ITO electrode was obtained from Geomatec. The ITO electrode was cleaned by ozone atmosphere before being used. Preparation of $C_{60}P$ has been reported elsewhere [18]. A toluene solution of C_{60} fullerene (2.0×10^{-3} mol dm^{-3}) and a toluene solution of ethylenediamine (2 mol dm^{-3}) were mixed under sonication. The reaction mixture was left for 30 min; then, $C_{60}P$ was obtained as a powder. The $C_{60}P$ powder was collected by vacuum filtration and dried in vacuo.

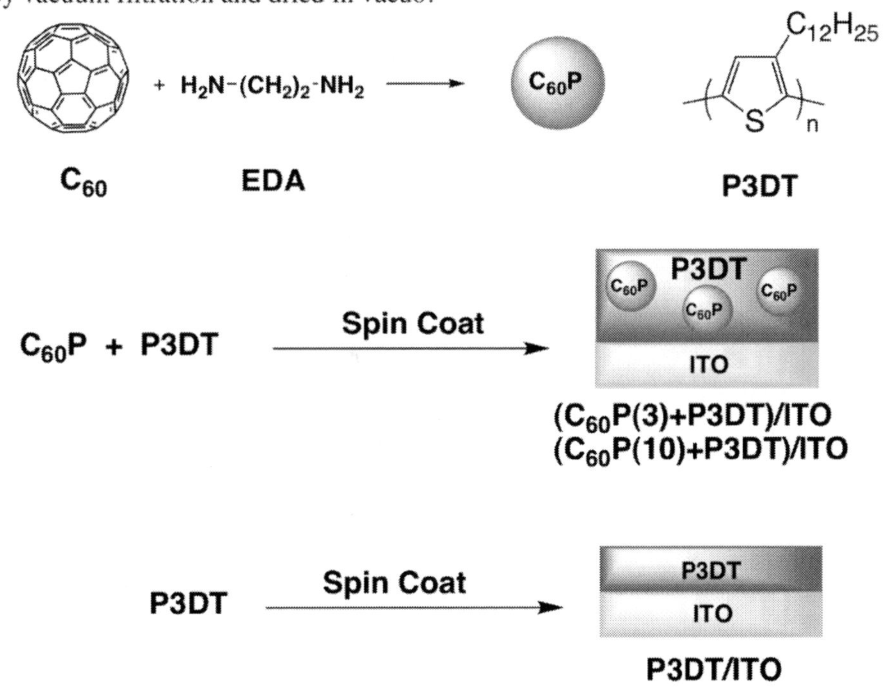

Figure 1. Schematic illustration of fabrication procedure of modified electrodes.

Figure 1 shows the preparation procedure of mixture films consisting of $C_{60}P$ and P3DT. First, $C_{60}P$ (3 or 10 mg mL^{-1}) and P3DT (2.3×10^{-2} mol dm^{-3}) were mixed in toluene. Next, the mixture films were prepared by spin-coating of the toluene solutions consisting of P3DT and dispersed $C_{60}P$ on the ITO electrode, denoted as (C_{60}P(3)+P3DT)/ITO and (C_{60}P(10)+P3DT)/ITO, respectively. A P3DT film without C_{60}P was also prepared by spin-coating of a toluene solution of P3DT (2.3×10^{-2} mol dm^{-3}) on the ITO electrode, as a corresponding reference film, denoted as P3DT/ITO.

Scanning electron microscopy (SEM) measurements were performed using a Hitachi S-5000 microscope. The absorption spectra and fluorescence spectra of the mixture films and the P3DT film were measured using a Shimadzu UV-2500PC spectrophotometer and a JASCO FP-6500 spectrofluorometer, respectively. Photocurrent measurements were carried out using a three-electrode electrochemical cell containing 0.1 mol dm^{-3} $NaClO_4$ in aqueous solution at room temperature. The sacrificial reagent was methylviologen (5×10^{-3} mol dm^{-3}). Light from a xenon lamp was passed through a monochromator that irradiated the modified electrode. Photocurrent action spectra were measured by changing the excitation wavelength at an applied potential (*E*) of 0 V vs. Ag/AgCl (sat. KCl). Dependence of the photocurrent on the applied potential was measured under irradiation of 520 nm light. All photocurrents were recorded using a Huso HECS-318C potentiostat. The electrolyte solution was degassed by N_2 bubbling (30 min) before photocurrent measurement.

3. Results and discussion

SEM images of the modified electrodes ((C_{60}P(3 or 10)+P3DT)/ITO and P3DT/ITO) are shown in Figure 2. The spherical particles in the SEM images are C_{60}Ps. The SEM images of C_{60}P(3 or 10)+P3DT)/ITO show that the amount of C_{60}P on the surface of P3DT changes by the ratio of C_{60}P to P3DT in the mixed solution. The mean fractional coverage of C_{60}Ps in the mixture films is ~3% for (C_{60}P(3)+P3DT)/ITO and ~17% for (C_{60}P(10)+P3DT)/ITO. These coverage values of C_{60}Ps are considerably smaller than that in the C_{60}P-deposited P3DT film prepared by capillary action (~30%) [18]. As observed from the cross-sectional SEM images of C_{60}P(3 or 10)+P3DT)/ITO, C_{60}P is partly buried into the P3DT film. In this case, C_{60}P and P3DT appear to be in considerably greater contact with each other as compared to the case of the C_{60}P-deposited P3DT film [18].

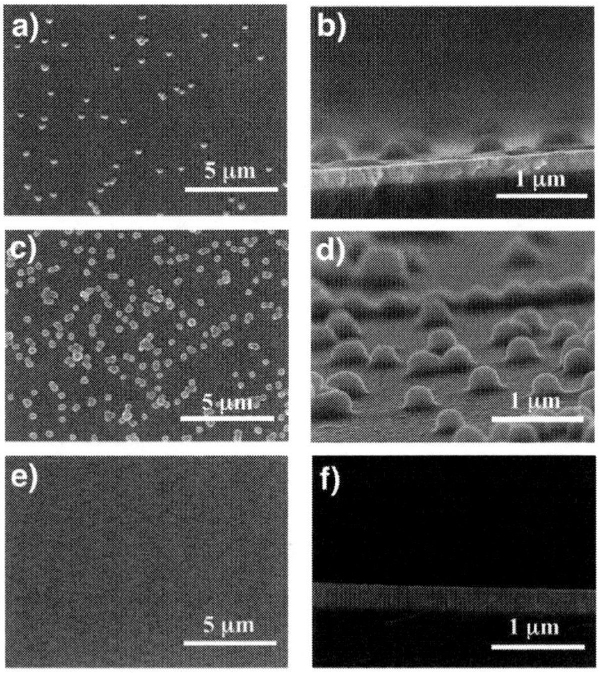

Figure 2. SEM images of modified electrodes: (a) (C_{60}P(3)+P3DT)/ITO, (b) cross-sectional image of (C_{60}P(3)+P3DT)/ITO, (c) (C_{60}P(10)+P3DT)/ITO, (d) cross-sectional image of (C_{60}P(10)+P3DT)/ITO, (e) P3DT/ITO, and (f) cross-sectional image of P3DT/ITO.

The absorption spectra of the modified electrodes are shown in Figure 3(a). The peak intensities of (C_{60}P(3)+P3DT)/ITO, (C_{60}P(10)+P3DT)/ITO, and P3DT/ITO are not very different from each other. In particular, the profiles of the spectra of (C_{60}P(3)+P3DT)/ITO and P3DT/ITO are almost the same, indicating that in both these modified electrodes, similar amounts of P3DT are adsorbed . A significant increase in the baseline was observed in (C_{60}P(10)+P3DT)/ITO, which may be attributed to light scattering by C_{60}Ps.

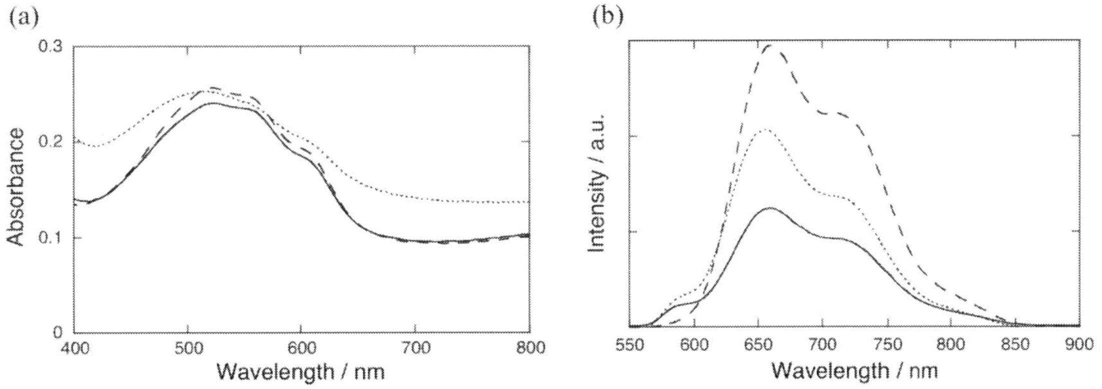

Figure 3. (a) Absorption spectra and (b) fluorescence spectra of (C_{60}P(3)+P3DT)/ITO (solid line), (C_{60}P(10)+P3DT)/ITO (dotted line), and P3DT/ITO (dashed line).

Figure 3(b) shows the fluorescence spectra of the modified electrodes excited at 520 nm. The fluorescence from polythiophene in (C_{60}P(3 or 10)+P3DT)/ITO was significantly quenched as compared to that in P3DT/ITO, indicating the occurrence of electron transfer quenching from the photoexcited state of polythiophene by C_{60}P. The fluorescence intensity of (C_{60}P(10)+P3DT)/ITO is larger than that of (C_{60}P(3)+P3DT)/ITO, which also may be attributed to light scattering by C_{60}Ps.

All modified electrodes generated stable cathodic photocurrents in the presence of methylvilogen as a sacrificial electron acceptor. Figure 4(a) shows photocurrent action spectra of the modified electrodes (C_{60}P(3 or 10)+P3DT)/ITO and P3DT/ITO (E = 0V vs. Ag/AgCl (sat. KCl)). All action spectra show quite similar profiles and broad peaks around the 550–600 nm region. These broad peaks correspond to the absorption band of the P3DT film. It is interesting that the photocurrent generated by (C_{60}P(10)+P3DT)/ITO is larger that of generated by (C_{60}P(3)+P3DT)/ITO in the 400-520 nm region. On the contrary, the photocurrent generated by (C_{60}P(10)+P3DT)/ITO is smaller that of generated by (C_{60}P(3)+P3DT)/ITO in longer wavelength region than 530 nm. These changing of photocurrent intensities are not may due to the effect of incorporation amount of C_{60}P in polythiophene. Possible reasons for these changing of photocurrents are changing of refractive index or light scattering properties of the C_{60}P+P3HT mixed films. Detail optical measurements of the mixed films may clear the reason for these changing of photocurrent.

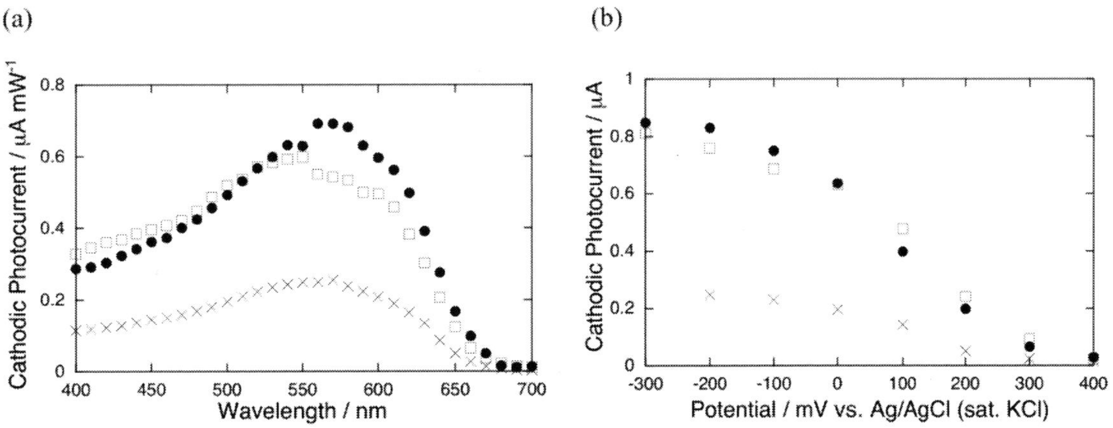

Figure 4. (a) Photocurrent action spectra and (b) applied potential dependence of photocurrent of (C_{60}P(3)+P3DT/ITO (●), (C_{60}P(10)+P3DT)/ITO (□), and P3DT/ITO (×), at E = 0 V vs. Ag/AgCl (sat. KCl aq.).

Figure 4(b) shows the dependences of photocurrents of the modified electrodes on applied potential (λ_{ex} = 520 nm). In all cases, the photocurrent increased with positively increasing applied potential. The photocurrent intensities of (C_{60}P(3 or 10)+P3DT)/ITO are similar and considerably larger than that of P3DT/ITO. It is noteworthy that the photocurrents generated by (C_{60}P(3)+P3DT)/ITO are roughly three times enhanced than those generated by P3DT/ITO, even one order smaller coverage of C_{60}Ps in this study than the C_{60}P deposited P3DT film in our previous case [18]. This photocurrent enhancement appears to be due to the efficient electron transfer between C_{60}P and P3DT in the mixture film, which is caused by the buried structure of the C_{60}P in the P3DT film.

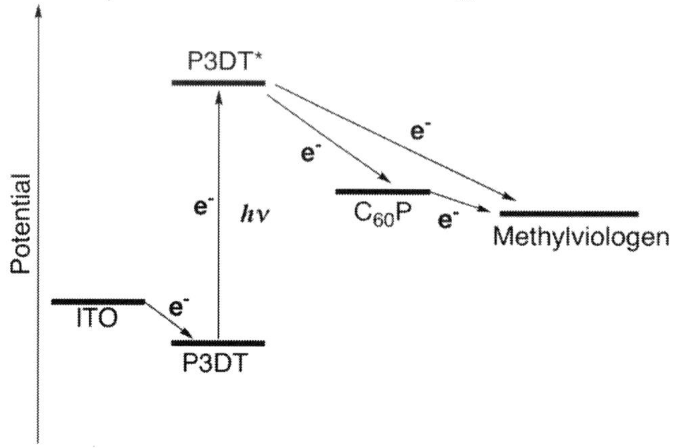

Figure 5. Mechanism of photocurrent generation.

The possible photocurrent generation mechanism is shown in Figure 5, which is a combination of (1) photoinduced electron transfer from photoexcited P3DT to C_{60}P or methylviologen and (2) subsequent electron transfer from the reduced C_{60}P to methylviologen in the electrolyte solution.

In this study, the incorporation of C_{60}Ps into the P3DT film enhanced the photocurrent signal considerably as compared to the presence of P3DT alone. Accordingly, such a photocurrent enhancement must be attributed to the electron accepting property of C_{60}P.

4. Conclusion

We demonstrated that C_{60}P causes an enhancement of photocurrent generated by a polythiophene-modified electrode. We also verified that the contact condition between C_{60}P and the P3DT film is crucial for photocurrent enhancement. Though we succeeded in incorporating a different amount of C_{60}P into the P3DT film, we could not elucidate the photocurrent enhancement effect. Further studies are now in progress to address this issue.

Acknowledgement

T. A is grateful for Grants-in-Aid for Scientific Research for Priority Area "Super-Hierarchical Structures" (Nos. 18039030 and 19022028) from the Ministry of Education, Culture, Sports, Science and Technology, Japan. This research was also partially supported by a Grant-in-Aid for Scientific Research (C) (No. 22550123) from Japan Society for the Promotion of Science (JSPS).

References

[1] Akiyama T, Yamada S, 2001 *Trends Photochem. Photobiol.* **8** 67
[2] Akiyama T, Fukuyama T, Sugawa K, Yoneda H, Yamada S, 2009 *Synt. Met.* **159** 965

[3] Akiyama T, Yoneda H, Fukuyama T, Sugawa K, Yamada S, Takechi K, Shiga T, Motohiro T, Nakayama H, Kohama K, 2009 *Jpn. J. Appl. Phys.* **48** 04C172
[4] Yu G, Gao J, Hummelen J C, Wudl F, Heeger A J, 1995 *Science* **270** 1789
[5] Shaheen S E, Brabec C J, Sariciftci N S, Padinger F, Fromherz T, Hummelen J C, 2001 *Appl. Phys. Lett.* **78** 841
[6] Hoppe H, Sariciftci N S, 2004 *J. Mater. Res.* **19** 1924
[7] Kobayashi S, Takenobu T, S. M, Fujiwara A, Iwasa T, 2003 *Appl. Phys. Lett.* **82** 4581
[8] Geckeler K E, Hirsch A, 1993 *J. Am. Chem. Soc.* **115** 3850
[9] Chen K, Caldwell W B, Mirkin C A, 1993 *J. Am. Chem. Soc.* **115** 1193
[10] Caldwell W B, Chen K, Mirkin C A, Babinec S J, 1993 *Langmuir* **9** 1945
[11] Matsuoka K-i, Akiyama T, Yamada S, 2008 *J. Phys. Chem. C* **112** 7015
[12] Wang Y-M, Kamat P V, Patterson L K, 1993 *J. Phys. Chem.* **97** 8793
[13] Kamat P V, Barazzouk S, Hotchandani S, Thomas G, K., 2000 *Chem. Eur. J.* **6** 3914
[14] Kamat P V, Barazzouk S, Thomas G, K., Hotchandani S, 2000 *J. Phys. Chem. B* **104** 4014
[15] Hasobe T, Imahori H, Kamat P V, Fukuzumi S, 2003 *J. Am. Chem. Soc.* **125** 14962
[16] Hwu J R, Kuo T Y, Chang T M, Patel H V, Yong K T, 1996 *Fullerene Sci. Technol.* **4** 407
[17] Ulüg A, Mete A, Ulüg B, 1997 *Fullerene Sci. Technol.* **5** 1651
[18] Matsuoka K-i, Seo H, Akiyama T, Yamada S, 2007 *Chem. Lett.* **36** 934
[19] Matsuoka K-i, Matsumura S, Akiyama T, Yamada S, 2008 *Chem. Lett.* **37** 932

Effect of Proton Irradiation on 2DEG in AlGaN/GaN Heterostructures

A. Abderrahmane[1], S. Koide[1], T. Tahara[1], S. Sato[3], T. Ohshima[3] , H.Okada[1, 2], and A. Sandhu[1,2]

[1]Department of Electrical and Electronic Information Engineering, Toyohashi University of Technology, 1-1 Hibarigaoka, Tempaku-cho, Toyohashi, Aichi 441-8580, Japan

[2]Electronics-Inspired Interdisciplinary Research Institute (EIIRIS), Toyohashi University of Technology, 1-1 Hibarigaoka, Tempaku-cho, Toyohashi, Aichi 441-8580, Japan

[3]Quantum Beam Science Directorate, Japan Atomic Energy Agency (JAEA), 1233 Watanuki-cho, Takasaki, Gunma 370-1292, Japan

E-mail: Abderrahmane@eiiris.tut.ac.jp

Abstract. Low temperature Hall effect measurements were carried on AlGaN/GaN micro-Hall effect sensors before and after irradiation with 380 keV and fluence of 10^{14} protons/cm^2 protons. The sheet electron density after irradiation did not show significant changes but there was a dramatic decrease in the electron mobility of the heterostructures. Prior to irradiation, the observation of well-defined Landau plateaus in the Hall resistance and Shubnikov-de Haas oscillations (SdH) at 4.5 T was indicative of the high quality the heterojunction confining the two-dimensional electron gas (2DEG) at the AlGaN/GaN interface of micro-Hall effect sensors. In contrast, the Landau plateaus disappeared after irradiation and the threshold magnetic field required for the observation of the SdH increased, which was accompanied by a decrease of the electron mobility. Temperature dependent magnetoresistance measurements were used to deduce the effective mass and the quantum scattering time before irradiation. A negative magnetoresistance was observed at low magnetic fields which is related to weak localization and parabolic negative magnetoresistance attributed to electron-electron interaction in both samples.

1. Introduction

Hall Effect magnetic sensors based on III-nitrides are promising for applications in space technology and other such harsh environments. Such applications necessitate device operation at high temperatures and under harmful radiation. Specifically, AlGaN/GaN Hall effect sensors are excellent candidates for measuring magnetic fields in such environments [1].

At low temperatures, three kinds scattering in the AlGaN/GaN heterostructure are dominant: interface roughness, alloy disorder [2] and impurity scattering [3]. Reports show a constant sheet density from room temperature to low temperature, and temperature independent mobility at very low temperature [4-6].

Redwing *et al.* have reported quantum Hall effect AlGaN/GaN heterostructures to exhibit clear Landau plateaus at a mobility of 7500 (cm^2/Vs) [5] and Wang *et al.* for mobilities exceeding 10^4 (cm^2/Vs) [7]. Shubnikov-de Haas oscillations are used to determine the effective mass and quantum scattering times, which give valuable insights into the dominant scattering mechanisms in the two-dimensional electron gas 2DEG [8-10].

The effect of proton irradiation on the quantum Hall effect in AlGaN/GaN micro-Hall sensor is not clear inspite of being an important area of research for 'hard-electronics' for devices used in space and other such environments. Here, we describe the results of a systematic study on the magnetotransport and quantum Hall effect of AlGaN/GaN micro-Hall sensors before irradiation and after proton irradiation. This study showed the existence of the 2DEG layer even after irradiation and stability of the sheet electron density but significant degradation of the mobility after irradiation.

2. Experimental.

The AlGaN/GaN micro-Hall effect sensors were grown by MOCVD on sapphire substrates. The structures consisted of a 2μm GaN layer, a 25 nm unintentionally doped $Al_{0.25}Ga_{0.75}N$ layer and Ti/Al/Ni/Au Ohmic contacts. Samples were irradiated with 380 keV protons at a fluence of 10^{14} protons/cm^2 at the Takasaki Ion Accelerators for Advanced Radiation Application. Van der Pauw measurements were carried out from 5K to room temperature, with a100μA drive current for non-irradiated samples and 30μA for irradiated ones. The magnetoresistance measurements were performed in a cryogenic liquid helium cryostat from 1.6 to 300°K with magnetic fields of upto 10 T produced by a superconductor magnet.

The room temperature electron mobilities were 2324 cm^2/Vs and 1627 cm^2/Vs for non-irradiated and irradiated samples, respectively.

3. Results and Discussion

As shown in Fig. 1, the sheet electron density before and after irradiation was stable over all the temperatures studied with a slight increase near room temperature (RT). The increase of the sheet electron density at RT maybe due to the thermal activation of bulk carriers. Since the sheet electron density is inversely proportional to the absolute sensitivity according to the equation $S_A = I/qN_S$ we conclude that the absolute sensitivity is stable over this range of temperature after irradiation. The sheet resistance shown in Fig. 2, increased with increasing temperature for both samples, which is related to the decrease of the mobility as shown in Fig. 3.

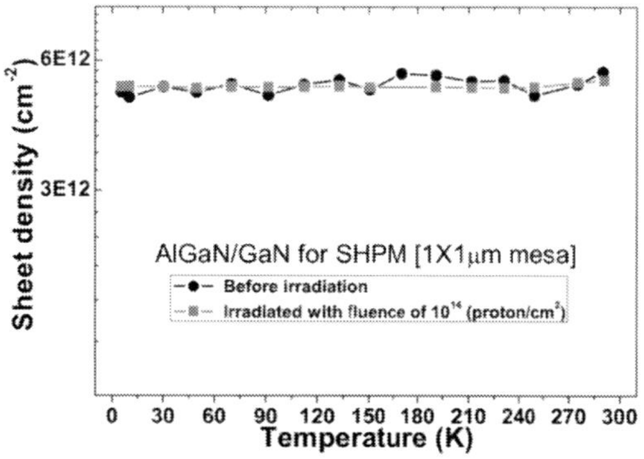

Figure 1. Temperature dependence of the sheet density of the AlGaN/GaN before and after irradiation.

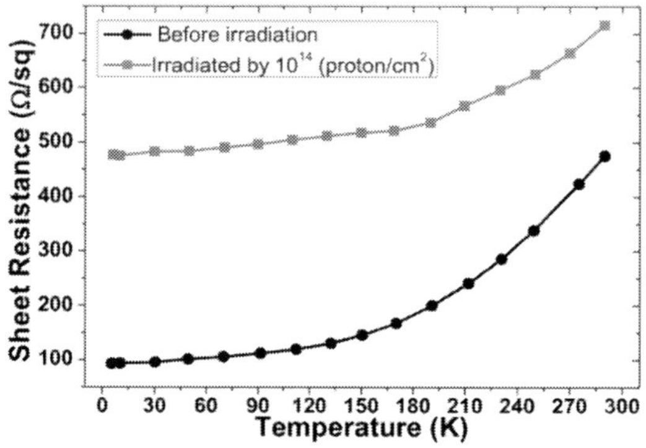

Figure 2. Temperature dependence of the sheet resistance of the AlGaN/GaN before and after irradiation.

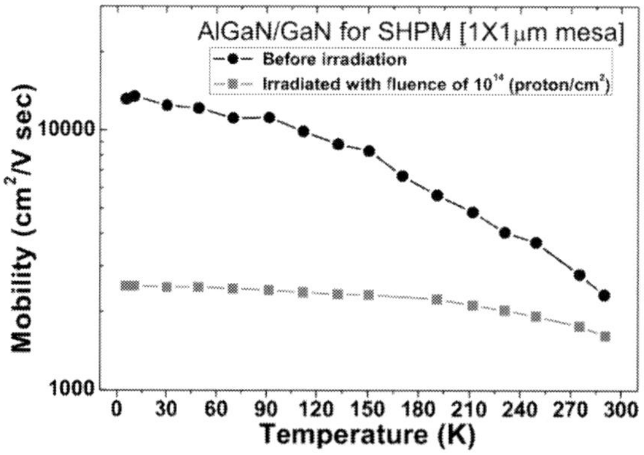

Figure 3. Temperature dependence of the mobility of the AlGaN/GaN before and after irradiation.

Rate of change of mobility can be divided in three regions: (1) lower than 90 °K the mobility is almost constant in this region, with the three probable scattering mechanisms being interface roughness, alloy disorder, and impurities scatterings. Ling *et al.* report on the observation of a change in the surface roughness of a GaN layer after proton irradiation due to impurities or point defects [11]. Increases in interface roughness and/or impurities near the AlGaN/GaN interface can lead to a dramatic decrease in the mobility after irradiation. (2) An intermediate region where the aforementioned scattering are less pronounced and acoustic phonon scattering begins to dominate. (3) At room temperature, where interface roughness and impurity scatterings can be neglected and optical phonon scattering dominates, which explains the decrease of the rate of change of the mobility near room temperature.

Figure 4 shows the quantum Hall resistance of the micro-Hall sensor before and after irradiation. The sample before irradiation showed clear Landau plateaus, which started to disappear at 14°K . After irradiation, the Landau plateaus disappear and the Hall resistance becomes linear but this result does not necessarily mean the absence of the 2DEG.

The origin of the Landau levels is due to the 2DEG edge transport at low temperature, electrons can move freely along the interface without scattering which give constant Hall resistance and the magnetoresistance tends to zero. But increases in electron scattering at the interface can deflect electrons to the bulk and this effect explains the disappearance of the Landau levels and increases of the magnetoresistance. This is the reason why the magnetoresistance increased after irradiation as shown Fig. 5.

The decrease of the minima in the oscillations indicates the absence of parallel conduction. The oscillations are clear from magnetic fields of about 4.5 T before irradiation and from about 8 T after irradiation. This increase in the threshold magnetic field is due to the reduction of the mobility after irradiation.

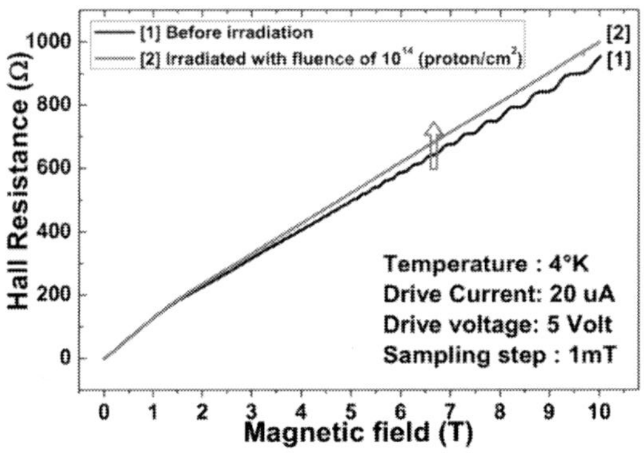

Figure 4. Quantum Hall resistance as a function of magnetic field for AlGaN/GaN micro-Hall sensor before and after irradiation.

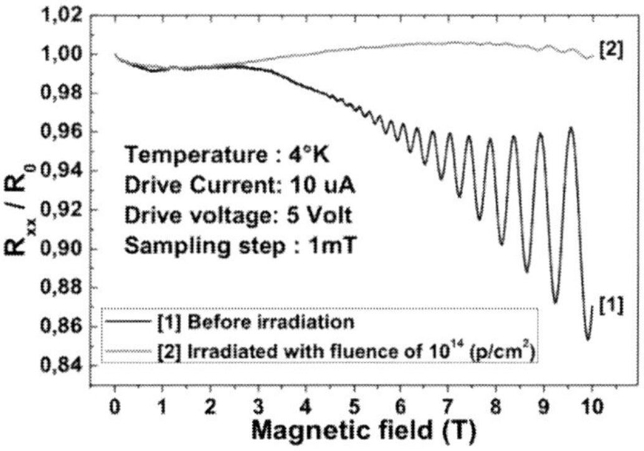

Figure 5. The magnetoresistance as a function of magnetic field for AlGaN/GaN micro-Hall sensor before and after irradiation.

The Landau levels also disappeared with increasing temperature and drive currents. The temperature dependence of the quantum Hall resistance is shown in Fig. 6 and the current dependence of the Hall resistance in Fig. 7, where the disappearance of the Landau plateaus in this case is due to electron heating phenomenon [12].

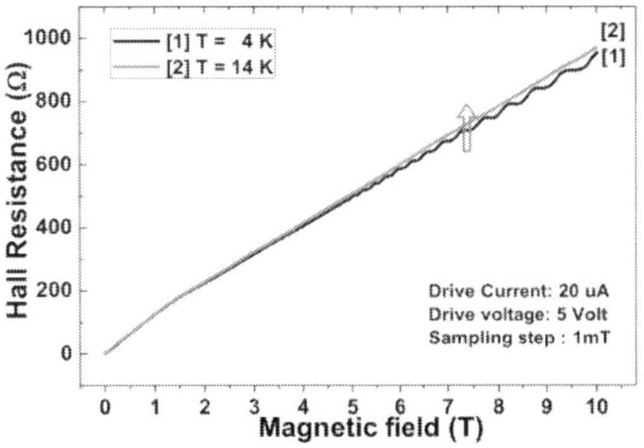

Figure 6. Quantum Hall resistance as a function of magnetic field for AlGaN/GaN micro-Hall sensor using 20uA drive current for two value of temperature.

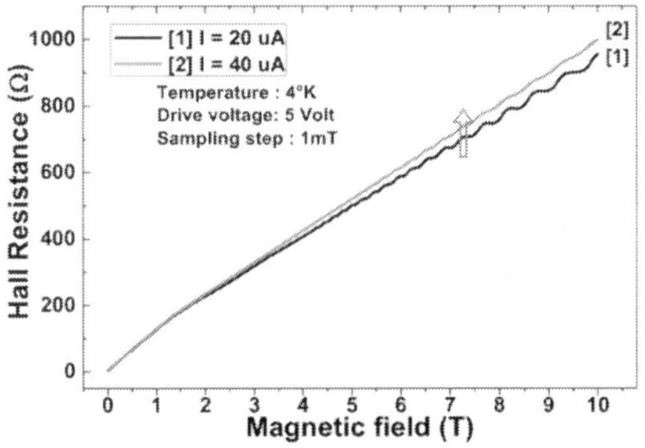

Figure 7. Quantum Hall resistance as a function of magnetic field for AlGaN/GaN micro-Hall sensor at 4°K for two value of current.

Both the non-irradiated and irradiated showed weak localization for magentic fields less than 1T, and a linear dependence of the magnetoresistance as function of square of the magnetic field, as shown in Fig. 8 for a sample before irradiation, which is related to electron-electron interaction according to the equation (1):

$$\rho_{xx} \approx \rho_0 - \rho_0^2(1 - \mu^2 B^2)\, \delta\sigma_{xx}^{ee}(T) \tag{1}$$

where ρ_{xx} represent the magnetoresistivity, ρ_0 is the resistivity at zero magnetic field, μ is the mobility and $\delta\sigma^{ee}(T)$ is the correction term due to electron-electron interaction at different temperatures.

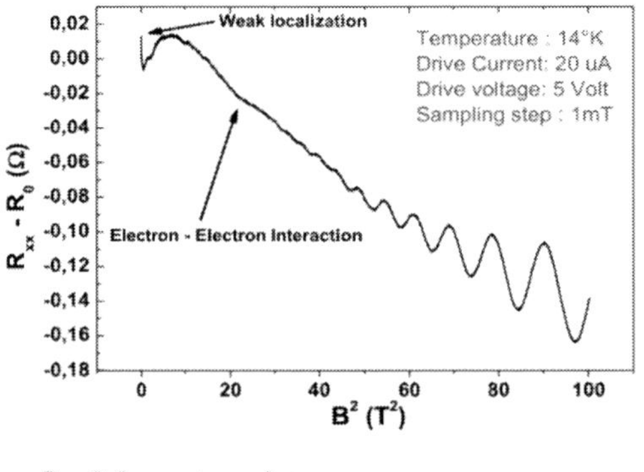

Figure 8. Magnetoresistance versus square of the magnetic field for sample before irradiation.

Temperature-dependent SdH oscillations are shown in the Fig. 9 for a non-irradiated sample. The oscillations became more pronounced at higher magnetic fields and tended to damp with increasing the temperature. The oscillating portion of the magnetoresistance can be expressed as:

$$\frac{1}{2}\frac{\Delta R_{xx}}{R_0} = 2\frac{\chi}{\sinh(\chi)}\exp\left(\frac{-\pi}{\omega_c\tau_q}\right)\cos\left(\frac{2\pi\Delta E}{\hbar\omega_c} - \pi\right) \quad (2)$$

where $\omega_c = eB/m^*$ is the cyclotron frequency, m^* the effective mass at the Fermi level, τ_q the quantum scattering time, $\chi = 2\pi^2 k_B T/\hbar\omega_c$.

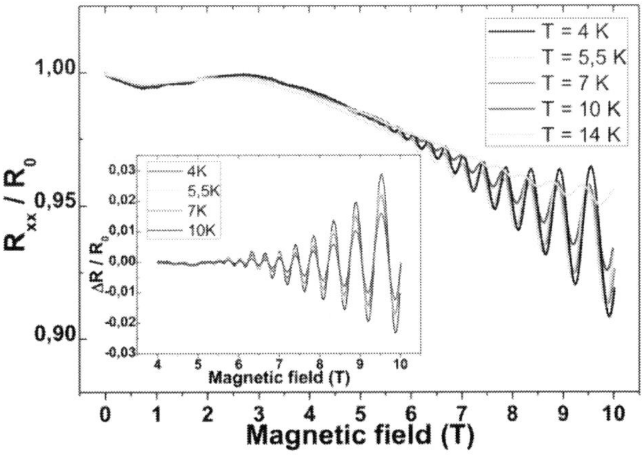

Figure 9. Shubnikov de Haas oscillations at different temperature values.
The inset shows oscillating component of the magnetoresistance

We determined the effective mass from the temperature dependence of the oscillating component amplitude, shown in the inset of Fig. 9 at a fixed magnetic field. The amplitude A of the SdH can be given by:

$$\ln\left(\frac{A}{T}\right) = C - \frac{2\pi^2 k_B m^*}{e\hbar B}T \quad (3)$$

where C is a temperature independent term, by plotting ln(A/T) versus T we deduce directly the effective mass from the slope which is equal to $2\pi^2 k_B m^*/e\hbar B$.

And in order to obtain the quantum scattering we plot the equation

$$\ln\left(\frac{1}{4}\frac{\Delta R\sinh(\chi)}{R_0\chi}\right) = C - \frac{\pi m^*}{e\tau_q}\frac{1}{B} \quad (4)$$

The effective mass of the sample before irradiation is approximately $0.20m_e$ at 6.3 Tesla. And the quantum scattering time equal to 63.8fs a value close to those reported before [6,13]. The classical scattering time τ_c is experimentaly determined from the mobility using the equation $\mu = e\tau_c/m^*$ and it is approximately equal to 1,38ps. The ratio τ_c/τ_q can give us an idea about the scattering dominant in our device.

Hsu and Walukiewicz [14] propose that only a ratio value between 1.5 and 9 allows dominant scattering in the 2DEG AlGaN/GaN and in this case short range scattering mechanism such as interface roughness scattering dominate. In our case the ratio is equal to about 21 a value reported before [6]. It remains to be confirm the ratio τ_c/τ_q and then the scattering dominant in the dominant scattering in the irradiated sample.

4. Conclusion

We investigated the effect of high energy and high fluence proton irradiation on magnetoelectric properties of AlGaN/GaN micro-Hall sensors from 5.4°K to room temperature. The sensors show good resistance versus the irradiation translated by the stability of the sheet density therefore the stability of the absolute sensitivity of the sensor. However, the proton irradiation damaged the electrical properties of the sensor indicated by the dramatically decrease of the mobility at low temperature by rate of about 81% at 5,4°K. The existing of the 2DEG system either after irradiation with high energy was confirmed by investigation the magnetotransport measurements at low temperature and which show Shubnikov de Haas oscillations at high magnetic field. Damping of the Shubnikov de Haas oscillations and disappearance of Landau plateaus after irradiation were related to the degradation in the mobility causing by increasing the scattering at the interface.

5. References

[1] H. Okada, A. Abderrahmane, S. Koide, H. Takahashi, S. Sato, T. Ohshima and A. Sandhu; Journal of Physics: Conference Series **352**, 01 (2012) 012010.

[2] S. B. Lisesivdin, S. Acar, M. Kasap, S. Ozcelik, S. Gokden and E. Ozbay; Semiconductor science and Technology **22** (2007) 543.

[3] A. Biswas, A. Ghosal, Hasanujjaman, S. Khan; International Journal of Scientific & Engineering Research **2** (2011) 2229-5518 9.

[4] S. B. Lisesivdin, S. Demirezen, M. D. Caliskan, A. Yildiz, M. Kasap, S. Ozcelik1 and E. Ozbay; Semiconductor science and Technology **23** (2008) 095008.

[5] J. M. Redwing, M. A. Tischler, J. S. Flynn, S. Elhamri, M. Ahoujja, R. S. Newrock and W. C. Mitchel; Applied Physics Letters **69** (1996) 963.

[6] S. Elhamri, W. C. Mitchel, W. D. Mitchell, R. Berney, M. Ahoujja, J. C. Roberts, P. Rajagopal, T. Gehrke, E. L. Piner, K. J. Linthicum; Journal of Electronic Materials **34**, 4 (2005) 444-449.

[7] T. Wang, Y. Ohno, M. Lachab, D. Nakagawa, T. Shirahama, S. Sakai, and H. Ohno; Applied Physics Letters **74** (1999) 3531.

[8] M. Ahoujja, W. C. Mitchel, S. Elhamri, R. S. Newrock, D. B. Mast, J. M. Redwing, M. A. Tischler, J. S. Flynn; Journal of Electronic Materials **27**,4 (1998) 210-214.

[9] S. Elhamri, R. Berney, W. C. Mitchel, W. D. Mitchell, J. C. Roberts, P. Rajagopal, T. Gehrke, E. L. Piner and K. J. Linthicum; Journal of Applied Physics **95** (2004) 7982.

[10] D. R. Hang, C.-T. Liang, C. F. Huang, Y. H. Chang, Y. F. Chen, H. X. Jiang, and J. Y. Lin; Applied Physics Letters **79** (2001) 66.

[11] Ling Lü, Yue Hao, XueFeng Zheng, JinCheng Zhang, ShengRui Xu, ZhiYu Lin, Shan Ai and FanNa Meng; Science China Technological Sciences **55**, 9 (2012) 2432-2435.

[12] Lin Li-Hung, Chen Kui-Ming, Han Shiou-Shian, C. T. Liang, Hsueh Wen-Chang, Kuang Yao Chen, Sun Zhi-Hao, P. H. Chang, N. C. Chen, Change Chin-An; Physica E, **40**, 2 (2007) 343-346.

[13] S.V. Danylyuk, S.A. Vitusevich, B. Podor, A.E. Belyaev, A.Yu. Avksentyev, V. Tilak, J. Smart, A. Vertiatchikh, L.F. Eastman; Microelectronics Journal **34** (2003) 575–577.

[14] L. Hsu and W. Walukiewicz; Applied Physics Letters **80** (2002) 2508.

Proximity and Force Characteristics of CMC Touch Sensor with Square/Dome-shaped Sensor Elements

T Kawamura[1], N Inaguma[1], Y Kakizaki[1], H Yamada[1] and K Tani[1]

[1] Dept. of Human and Information Systems, Gifu University, 1-1 Yanagido, Gifu 501-1193, Japan

E-mail: tkz-kawa12@gifu-u.ac.jp

Abstract. A tactile sensor called Carbon Micro Coil (CMC) touch sensor was developed by CMC Technology Development Co., Ltd. The sensor's elements used in the experiments of this paper are made of silicon rubber containing CMCs several micrometers in diameter. One of the elements is molded into a square 30 mm on a side and 3 mm thick; the other is a dome 16 mm in diameter and 2 mm height. CMCs in the sensor element contribute to the electrical conductivity and the sensor element is considered to constitute an LCR circuit. When an object approaches to the sensor element or the sensor element is deformed mechanically, the impedance changes, and the CMC sensor detects the impedance changes by measuring the modulation of amplitude and phase of an input excitation signal to the sensor element. The CMC sensor also creates voltage signals of the R- and LC-components separately according to the amplitude and phase modulation. In this paper, the characteristics of the CMC sensor with respect to its proximity and force senses are investigated. First, the output of the CMC sensor with the square-shaped sensor element is measured when an object approaches to the sensor element. Next, the output of the CMC sensor with the dome-shaped sensor element is measured when fine deformations of 1 to 5 μm are applied to the sensor element under variable compression force. The results suggest that the CMC sensor can measure the force variance applied to the sensor element as well as the distance between the sensor element and an object.

1. Introduction

A tactile sensor called Carbon Micro Coil (CMC) touch sensor [1, 2] was developed by CMC Technology Development Co., Ltd. A CMC is a coil of micrometer size made of carbon and the sensor element is made of silicon rubber containing CMCs. The CMC sensor is able to respond to stimulation of micrometer order that causes fine deformation of the sensor element. On the other hand, human tactile sense is very sensitive so that humans can detect force variance applied to a fingertip on the order of several grams and easily grip and operate an object with their fingers. Therefore, it is expected that a sensitive tactile sensor such as a human tactile sense will be developed and enable robots to support human activities at home or office as humans do.

Several researchers have already developed tactile sensors to measure force and/or deformation applied to a sensor element. The finger-mounted tactile sensor (Kikuuwe et al. [3]) is made of a sheet of polyvinylidene fluoride (PVDF) film and detects voltage changes in the film caused by vibrations resulting from an object's surface roughness such as textile fabrics. Also, the slip-detecting tactile sensor (Teshigawara et al. [4, 5]) consists of thin pressure-conductive rubber sandwiched between two conductive film sheets and detects the slip of an object on the conductive rubber. In addition, the optical three-axis tactile sensor (Ohka and Yussof et al. [6, 7]) is comprised of several columnar feelers made of silicon rubber. The haptic sensor developed using GelForce technology (Kamiyama and Sato et al. [8, 9]) utilizes a transparent rubber as a sensor element. These sensors measure the magnitude and direction of a compression force by capturing the deformation of their sensor elements.

Meanwhile, authors of this paper are developing a tactile sensor to measure the distance between a robot fingertip and an object before the object touches the fingertip, as well as compression force applied to the fingertip while gripping the object. Such tactile sensor can be a very useful sensor for a robot to grip and operate an object because it enables a robot to realize an object's approach to the robot finger before touching an object.

In this paper, the characteristics of the CMC sensor with square- and dome-shaped sensor elements with respect to its proximity and force senses are investigated. One of the elements is molded into a square 30 mm on a side and 3 mm thick; the other is a dome 16 mm in diameter and 2 mm height. First, to clarify the proximity characteristics of the CMC sensor with the square-shaped sensor element, the CMC sensor's output is measured when an object approaches to the sensor element. Next, to measure the force characteristics of the CMC sensor with the dome-shaped sensor element, the sensor's output is measured when fine deformations of 1 to 5 μm are applied to the sensor element under variable compression force.

2. Fundamental principles of CMC touch sensor

Figure 1 shows CMC sensor elements used in this study. The sensor elements are made of silicone rubber and each sensor element contains 10 percent CMCs by weight. One of the elements is a square 30 mm on a side and 3 mm thick; the other is a dome 16 mm in diameter and 2 mm height. CMCs in a sensor element contribute to the electrical conductivity and the sensor element is considered to constitute an LCR circuit. When an object approaches to the sensor element or touches and deforms the sensor element mechanically, the impedance in the LCR circuit changes. The CMC sensor signal processing device (CMC Technology Development Co., Ltd.) detects the impedance changes by measuring the modulation of amplitude and phase of an input excitation signal to the sensor element. The CMC sensor also creates voltage signals of the R- and LC-components separately according to the amplitude and phase modulation.

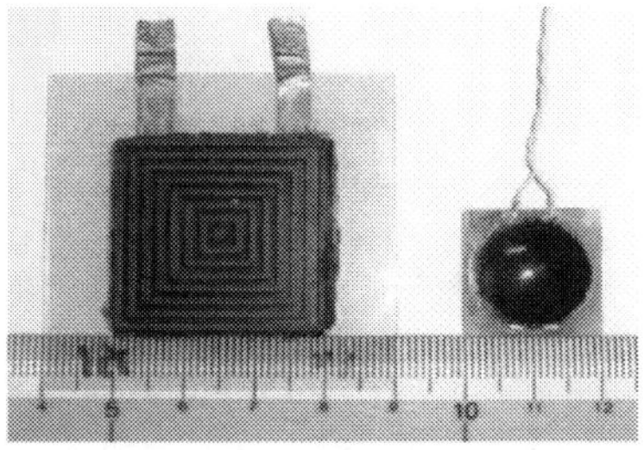

Figure 1. Square- and dome-shaped CMC sensor elements.

3. Proximity characteristics of CMC touch sensor with a square-shaped sensor element

A measurement system (Fig. 2) was developed to measure the proximity characteristics of the CMC touch sensor with a square-shaped sensor element. This measurement system consists of the CMC touch sensor and an X-stage driven by a stepping motor (Kohzu Precision Co., Ltd.). An acrylic plate is installed on a steel block as an object. In the experiment, the object approaches to the square-shaped CMC sensor element by the X-stage.

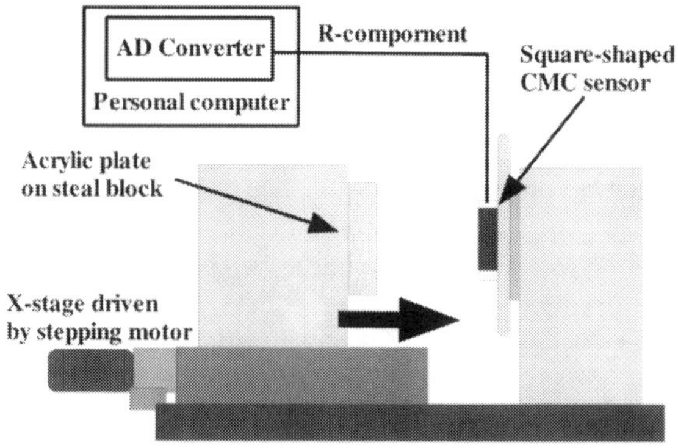

Figure 2. Measurement system of the proximity characteristics of the CMC touch sensor with a square-shaped sensor element.

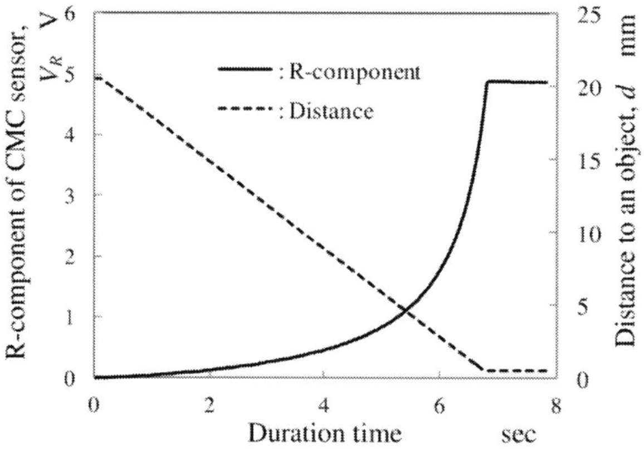

Figure 3. Relationship between the output of CMC sensor's R-component and the distance between the sensor element and the object when the object approaches to the CMC sensor.

Figure 3 shows the output of the CMC sensor when the object approaches to the CMC sensor element. The horizontal axis expresses the duration time while the left-side vertical axis expresses the R-component of CMC sensor V_R [V] and the right-side vertical axis expresses the distance d [mm] from the CMC sensor element to the object. The object was first located at the distance of 20 mm from the sensor element and the CMC sensor's output was set to 0 V. As the figure shows, V_R increases from 0 to approx. 5 V and d decreases from 20 to 0 mm as the object approaches to the sensor element. Therefore, it is considered that the CMC sensor can detect an object's approach to the sensor element.

Next, to investigate the influence of the velocity of object's approach to the sensor element, the outputs of CMC sensor's R-component were measured at the distances of 0.5, 1, 2, 3, 4, 5, 10, and 15 mm from the CMC sensor element to the object when the object was approached at different velocities within the range of 0.5 to 3.0 mm/s. Figure 4 shows the output of CMC sensor's R-component. The values of the CMC sensor's output measured at each distance almost keep constant for the different velocities. Consequently, it is found that the CMC sensor can measure the distance between the sensor element and the object regardless of the velocity of object's approach.

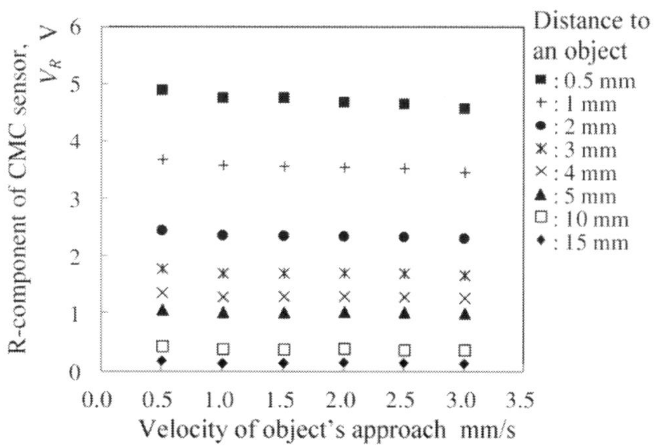

Figure 4. Output of CMC sensor's R-component measured at the distances from the CMC sensor element to the object for the different velocities of object's approach to the CMC sensor.

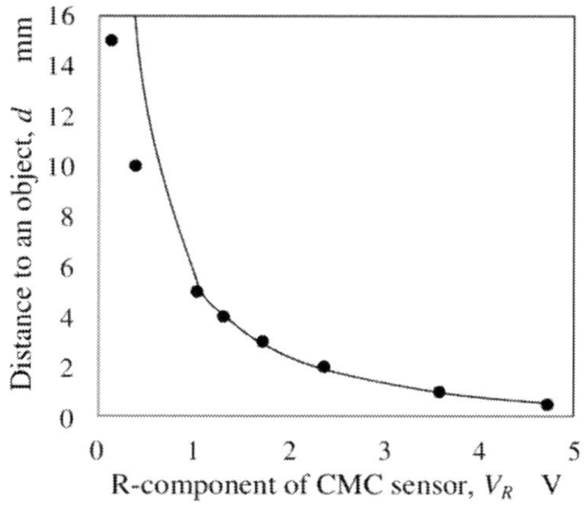

Figure 5. Relationship between the output of CMC sensor's R-component and the distance between the CMC sensor element and the object.

Figure 5 shows the relationship between the output of CMC sensor's R-component and the distance between the sensor element and the object. The horizontal axis is the R-component of CMC sensor V_R [V] while the vertical axis is the distance d [mm] from the sensor element to the object. The equation is determined as follows:

$$d = \frac{6.377}{V_R} - 0.831. \quad (1)$$

The result shows that the CMC sensor can measure the distance between the sensor element and the object on the order of millimetres.

4. Force characteristics of CMC touch sensor with a dome-shaped sensor element

A measurement system (Fig. 6) was developed that uses a CMC touch sensor and a force sensor to measure the force characteristics of the CMC touch sensor with a dome-shaped sensor element. This measurement system consists of the CMC touch sensor, a force sensor (BL AUTOTEC), and a wedge-type Z stage driven by a stepping motor (Kohzu Precision Co., Ltd). A dome-shaped CMC sensor element is placed on the force sensor. The Z stage pushes the CMC element against a fixed acrylic plate perpendicularly to apply compression force to the sensor element. In the experiment, compression forces of 5, 50, 100, 150, and 200 gf and fine deformations of 1, 2, 3... and 10 μm were applied to the sensor element and the CMC sensor and the force sensor sampled the difference in the CMC sensor's output and the difference in the force sensor's output simultaneously before and after each fine deformation was repeatedly applied to the sensor element in addition to a constant compression force.

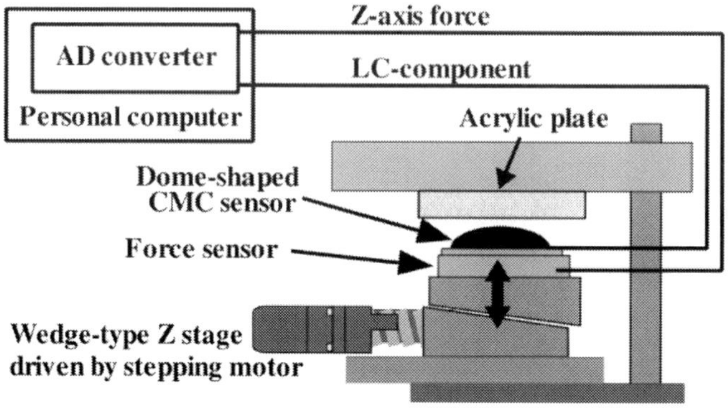

Figure 6. Measurement system of the force characteristics of the CMC touch sensor with a dome-shaped sensor element.

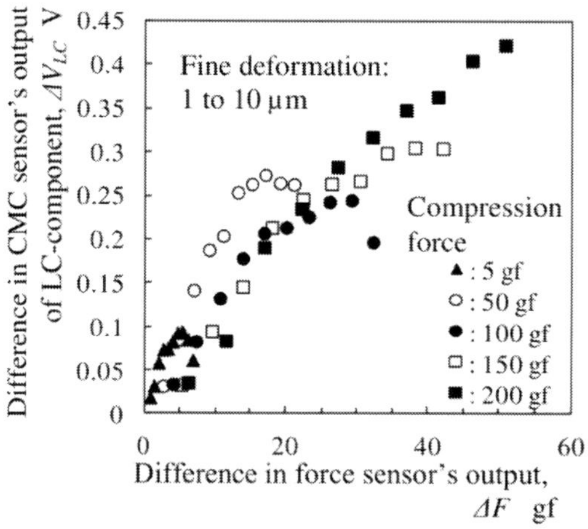

Figure 7. Relationship between the output of force sensor and the output of CMC sensor's LC-component when fine deformations of 1 to 10 μm are given to the CMC sensor element under variable compression force within the range of 5 to 200 gf.

Figure 7 shows the relationship between the difference in the CMC sensor's LC-component and the difference in the force sensor's output (force variance). Each mark in the figure represents the relationship between the difference in the CMC sensor's output and the difference in the force sensor's output when the magnitude of fine deformation increases from 1 to 10 μm under a constant compression force. From the results, it can be noted that when the compression force is in the range of 50 to 200 gf and the magnitude of force variance is within the range of 0 and approx. 20 gf, the difference in the force sensor's output and the difference in the CMC sensor's LC-component increase linearly as the fine deformation increases. Also, it is found that the tendency in the CMC sensor for the compression force of 50 gf is different from the tendency for the compression force of 100 to 200 gf and that is dependent on the compression force.

Also, Fig. 8 shows the relationship between the force sensor's output and the CMC sensor's output for fine deformation within the range of 1 to 5 μm and for compression force within the range of 100 to 200 gf. Those values are extracted from Fig. 7. The horizontal axis expresses the difference in the CMC sensor's LC-component, ΔV_{LC} [V] while the vertical axis expresses the difference in the force sensor, ΔF [gf]. The straight line is calculated using the least squares method and the equation that estimates force variance of CMC sensor element is as follows:

$$\Delta F = 90.63 \times \Delta V_{LC} \ (R^2 = 0.93), \tag{2}$$

where R^2 is the coefficient of determination. The magnitude of force variance of CMC sensor element can be estimated from the CMC sensor's output using the equation.

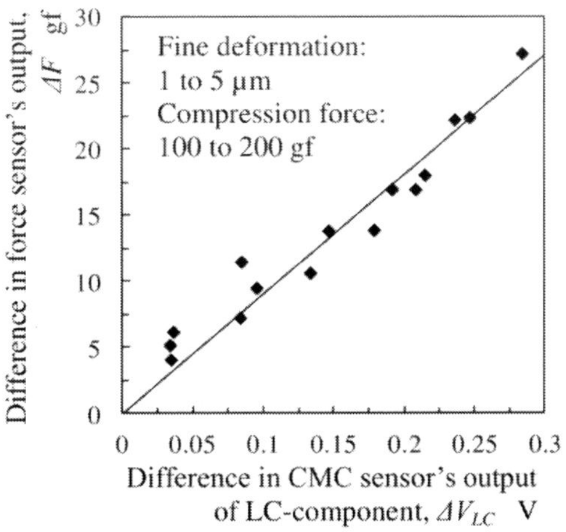

Figure 8. Relationship between the difference in the output of CMC sensor's LC-component and the difference in the output of force sensor for fine deformations of 1 to 5 μm under compression force within the range of 100 to 200 gf.

Next, an experiment was conducted to estimate the force variance of the dome-type CMC sensor element from CMC sensor's output. In the experiment, using the measurement system (Fig. 6) as an evaluation system, the CMC sensor's output and the force sensor's output were sampled when the CMC sensor element was compressed with compression forces of 100, 150, and 200 gf and fine deformations of 1, 2, 3, 4, and 5 μm were applied to the sensor element. Each fine deformation was applied twenty times repeatedly to the sensor element in addition to a constant compression force.

Figure 9 shows the force variance estimated from the output of CMC sensor's LC-component using Eq. (2). The horizontal axis expresses the force variance measured by the force sensor, which is the difference in the force sensor's output, while the vertical axis expresses the force variance estimated from the CMC sensor's output. The dotted line drawn in the figure means the two values of force variance are equal. We can see that the marks within the range of approx. 7 to 25 gf seem to be distributed near the dotted line. Therefore, it is found that the CMC sensor with the dome-type element can measure the magnitude of force variance of the sensor element.

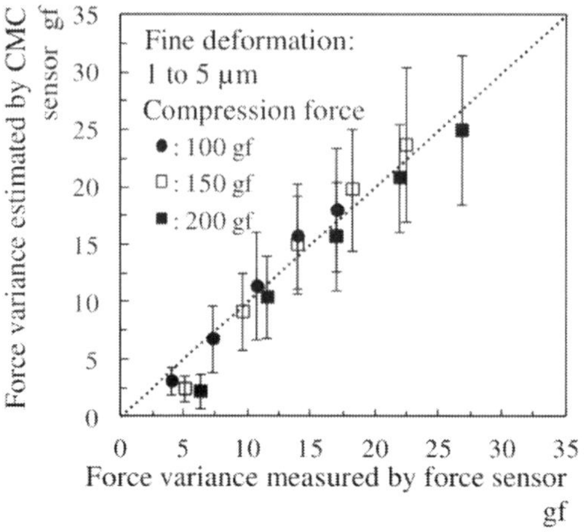

Figure 9. Estimated force variance using CMC sensor's output when fine deformations of 1 to 5 μm are applied to the sensor element under compression force within the range of 100 to 200 gf.

5. Conclusion

In this study, to develop a tactile sensor for a robot that can measure the distance from the fingertip to an object before touching the object as well as the force variance applied to the fingertip while gripping the object, the characteristics of the CMC touch sensor with respect to its proximity and force senses were investigated. First, the relationship between the output of the CMC sensor with the square-shaped sensor element and the distance from the sensor element to an object was determined. Next, the relationship between the output of the CMC sensor with the dome-shaped sensor element and the force variance applied to the sensor element was clarified.

In future work, the authors will develop a tactile sensor that has the senses of force and proximity by merging the characteristics of square- and dome-shaped sensor elements into one sensor element.

Acknowledgements

The authors would like to thank Kenji Kawabe (President of CMC Technology Development Co., Ltd.) for enabling the use of the CMC touch sensor.

References

[1] Kuzuya C, Ueda A and Kawabe K 2004 Application of CMC to tactile sensor *Materials Integration,* **17(8)** 9–16 (in Japanese)

[2] Homma M, Morita H, Maeno T, Konyo M and Motojima S 2006 Electromechanical conversion mechanism of a tactile sensor using carbon micro coil inside an elastic material *J. Robotics and Mechatronics* **18(3)** 235–41

[3] Kikuuwe R, Nakamura K and Yamamoto M 2012 Finger-mounted tactile sensor for evaluating surfaces *J. Robotics and Mechatronics* **24(3)** 430–40

[4] Teshigawara S, Tadakuma K, Ming A, Ishikawa M and Shimojo M 2009 Development of high-sensitivity slip sensor using special characteristics of pressure conductive rubber. *Proc. IEEE Int. Conf. on Robotics and Automation* 3289–94

[5] Teshigawara S, Tadakuma K, Ming A, Ishikawa M and Shimojo M 2009 High speed and high sensitivity slip sensor utilizing characteristics of conductive rubber –relationship between shear deformation of conductive rubber and resistance change– *J. Robotics and Mechatronics* **21(2)** 200–08

[6] Ohka M, Mitsuya Y, Matsunaga Y and Takeuchi S 2004 Sensing characteristics of an optical three-axis tactile sensor under combined loading *Robotica* **22** 213–21

[7] Yussof H, Morisawa N, Wada J and Ohka M 2009 Handling capabilities of two robot hands

equipped with optical three-axis tactile sensor. *Proc. 18th IEEE Int. Symp. on Robot and Human Interactive Communication* 165–70

[8] Kamiyama K, Vlack K, Kajimoto H, Kawakami N and Tachi S 2005 Vision-based sensor for real-time measuring of surface traction fields *IEEE Computer Graphics and Applications* **25(1)** 68–75

[9] Kamiyama K, Kawakami N and Tachi S 2010 Finger-shaped GelForce sensor for measuring surface traction fields for robotic hand *IEEE Transactions on Haptics* **3(1)** 37–47

Journal of Physics: Conference Series **433** (2013) 012013 doi:10.1088/1742-6596/433/1/012013

Visualization of flowing current in braided carbon fiber reinforced plastics using SQUID gradiometer for nondestructive evaluation

Y Hatsukade[1], K Yoshida[1], T Kage[1], Y Takai[2], M S Aly-Hassan[3], A Nakai[4], H Hamada[3], and S Tanaka[1]
[1]Department of Environmental & Life Sciences. Toyohashi University of Technology, 1-1 Hibarigaoka, Tenpaku-cho, Toyohashi, Aichi 441-8580, Japan
[2]Osaka Sangyo University, 3-1-1 Nakagaito, Daito, Osaka 574-8530, Japan
[3]Advanced Fibro-Science Division, Kyoto Institute of Technology, Matsugasaki, Sakyo-ku, Kyoto 606-8585, Japan
[4]Gifu University, 1-1 Yanagidao, Gif City 501-1193, Japan

E-mail: hatukade@ens.tut.ac.jp

Abstract. In this paper, visualization of flowing current in various braided carbon fiber reinforced plastics (CFRPs) was demonstrated using high-temperature superconductor (HTS) superconducting quantum interference device (SQUID) gradiometer, in order to study electrical properties and integrity of the braided CFRP samples. Step-by-step tensile loading was also applied to the samples, in order to study their mechanical properties and destructive mechanism. Experimental results indicated that the addition of carbon nano fibers and middle-end carbon fiber bundles attributed to modify not only the mechanical properties, but also the electrical properties of the samples. Combining the results by the both methods, a scenario of the destructive mechanism of one sample was estimated.

1. Introduction

Recently, braided carbon fiber reinforced plastics (CFRPs), in which carbon fiber bundles are interlaced in a braided textile (see Figure 1(a)), have been developed and studied to achieve superior properties compared to conventional woven CFRPs, because of the continuity of the carbon fiber bundles [1 - 3]. Furthermore, it is expected that the addition of carbon nanofillers in the matrix of the braided CFRPs (Figure 1(b)) and insertion of middle-end carbon fiber bundles in the braided textile (Figure 1(c)) would improve some braided CFRP's properties [4, 5]. However, nondestructive evaluation (NDE) technique for the braided CFRPs is yet established, because the fine diameter of the carbon fibers and the complexity of the braided textile require a novel NDE technique having high sensitivity and spatial resolution. Since tensile loading applied to the braided CFRPs is mainly maintained by the continuous carbon-fiber bundles, it is important for the NDE technique to detect breakages of the carbon fiber bundles, whose electric conductivity can change significantly when they are broken. Thus, we have proposed a NDE technique using a high-temperature superconductor (HTS) superconducting quantum interference device (SQUID) gradiometer having extremely high sensitivity to electric current, which generates magnetic field, and sufficiently high spatial resolution for the braided CFRPs [6]. So far, we demonstrated that the discontinuity of the braided carbon-fiber bundles in tubular-braided CFRPs could be well detected using the SQUID-NDE technique by visualizing electric current distribution in the CFRP samples while injecting ac currents into the samples.

In this study, we prepared various braided CFRP samples based on flat-braided CFRP panels. The samples included carbon nanofibers (CNFs) or the middle-end carbon fiber bundles. The mechanical properties and destructive mechanism of the samples were studied with observation and measurements of stress-strain curves, while applying step-by-step tensile loading to the samples. The SQUID-NDE technique to visualize flowing currents in the samples was tested for evaluating the integrity of the carbon fiber bundles in the samples, before, during, and after the tensile loading.

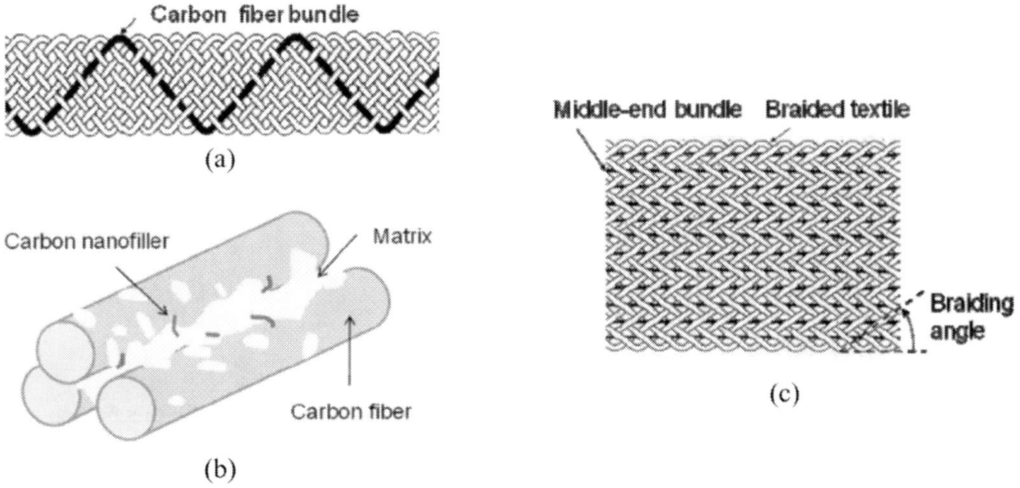

Figure 1. (a) Braided textile with continuous carbon fiber bundles in flat-braided CFRPs. In actual samples, neighboring bundles are touching with each other. (b) Dispersion of carbon nanofillers in matrix and carbon fibers. (c) Inserted middle-end bundles in braided textile.

2. Braided CFRP samples and step-by-step tensile tests

2.1 Samples

For this study, ±45° flat-braided CFRP panels with and without CNFs were fabricated. These panels contain 25 carbon-fiber bundles (HTS40, Toho Tenax) braided together at angles of ±45° into flat textiles, and epoxy resin (JER828, Japan Epoxy Resins Co. Ltd.). Each carbon-fiber bundle was composed of 12000 carbon fibers with the average diameter and specific electrical resistivity of 7 μm and 1.6×10^{-3} Ω·cm. The average tensile strength, tensile modulus, and elongation of the bundles were 4200 MPa, 240 GPa, and 1.8 %, respectively. Vapor grown CNFs (VGCF, Showa Denko K.K.) were dispersed throughout one panel, which was labeled as 45HTS-NF. The average diameter, length, and specific resistivity of the CNFs were about 150 nm, 10 μm, and 10^{-4} Ω·cm, respectively. The dimensions of the sample 45HTS-NF were 151 mm × 41 mm × 1.0 mm. From observation of a sample of the same kind as 45HTS-NF, using scanning electron microscopy, it was demonstrated that the CNFs were distributed not only in the epoxy resin matrix, but also in between the carbon-fiber bundles as schematically shown in Figure 1b [5]. For comparison, the CFRP panel without the CNFs was prepared and labeled as 45HTS. The dimensions of 45HTS were almost same as 45HTS-NF.

Three ±25° flat-braided CFRP panels, with different middle-end carbon fiber bundles were also fabricated. These CFRP panels contained 25 carbon fiber bundles (UTS50, Toho Tenax) braided at angles of ±25° into flat textiles, and the same epoxy resin as 45HTS. The carbon-fiber bundle UTS50 was composed of 12000 carbon fibers with the average diameter and specific electrical resistivity of 6.9 μm and 1.6×10^{-3} Ω·cm. The tensile strength, tensile modulus, and elongation of the bundles were 4900 MPa, 240 GPa, and 2%, respectively. In these panels, three different middle-end carbon fiber bundles were inserted: ten UTS50 bundles, ten XN05 (Nippon Graphite Fiber Corp.) bundles, and ten XN60 (Nippon Graphite Fiber Corp.) bundles, respectively. The parameters of the used carbon fiber bundles were summarized in Table 1. These three panels were labeled as 25UT-UT50, 25UT-XN05, and 25UT-XN60, respectively. We note that the electric resistivity of XN05 is highest (2.8×10^{-3} Ω·cm), while that of XN60 is lowest (0.7×10^{-3} Ω·cm). The dimensions of the samples 25UT-UT50, 25UT-XN05, and 25UT-XN60, were 250 mm × 28 mm × 1.5-1.7 mm.

Table 1. Parameters of carbon fiber bundles used for flat-braided textiles (HTS40 and UTS50) and middle-end bundles (UTS50, XN05, and XN60).

Grade	Filament count	Tensile strength [MPa]	Tensile modulus [GPa]	Elongation [%]	Filament diameter [μm]	Electrical resistivity [Ω·cm]
HTS40	12000	4200	240	1.8	7	1.6×10^{-3}
UTS50	12000	4900	240	2	6.9	1.6×10^{-3}
XN05	6000	1100	54	2	10	2.8×10^{-3}
XN60	6000	3430	620	0.6	10	0.7×10^{-3}

2.2 Step-by-step tensile tests

The CFRP samples were tested by step-by-step tensile loading to clarify their mechanical properties and destructive mechanisms. Tensile loading was applied to each sample in the longitudinal direction at a crosshead speed of 1 mm/min., while one surface of each sample was monitored and the stress-strain curve was measured using strain gauges. The surfaces of the samples were painted at times with white watercolor to observe surface defects. Once the initial tensile loading was applied to each sample, the loading was increased gradually. When an obvious decrease in mechanical strength was occurred, or when any defect was identified, the loading was stopped, and then released to apply below-described SQUID-NDE technique to the sample, in order to evaluate the integrity of their carbon-fiber bundles and electrical properties. We note that the stress-strain curves of the most samples did not return to the starting point after release of the loading. Therefore, after the application of the SQUID-NDE technique, we applied loading to the samples again until the measured strain reached to the maximum value at the previous loading, and thereafter, incremental quantities of strain and stress were measured, and then added to the initial stress-strain curves, for simplicity of plotting. By this process, a continuous stress-strain curve was obtained for each sample.

Figure 2 shows the stress-strain curves of the samples 45HTS-NF and 45HTS, with photographs of sample surfaces at certain damage stages. In the case of 45HTS-NF, the stress increased nonlinearly with increasing strain from the virgin stage, indicated by "a". When the stress and strain reached 340 MPa and 0.055, cracks occurred on the right side of the sample (indicated by the dotted oval), although no corresponding decrease in stress was measured ("b"). This means that the cracks were only matrix cracks on the matrix-rich side, and not accompanied by carbon-fiber fracture or delamination. Just before the strain reached about 0.06, the final fracture suddenly occurred at the maximum measured stress of 350 MPa ("c").

In the case of 45HTS, the stress also increased nonlinearly with the strain, similarly to 45HTS-NF, from the virgin stage ("d"). When the strain reached 0.056, at the maximum stress of 275 MPa, the stress dropped with the occurrence of a crack in the right-lower area with an angle of about 30° ("e"). After this first drop in stress, slight increases and drops repeated with propagation of new zigzag cracks along the first crack. This repeated increase and drop indicated that delamination between the carbon-fiber bundles and relocation of the bundles, occurred in turn. The final fracture occurred with the carbon-fiber fracture when the strain reached over 0.072 ("f"). The results demonstrated that the addition of CNFs could give greater mechanical tensile strength to the braided CFRP, although the measured maximum strain was smaller than that of the braided CFRP without CNFs.

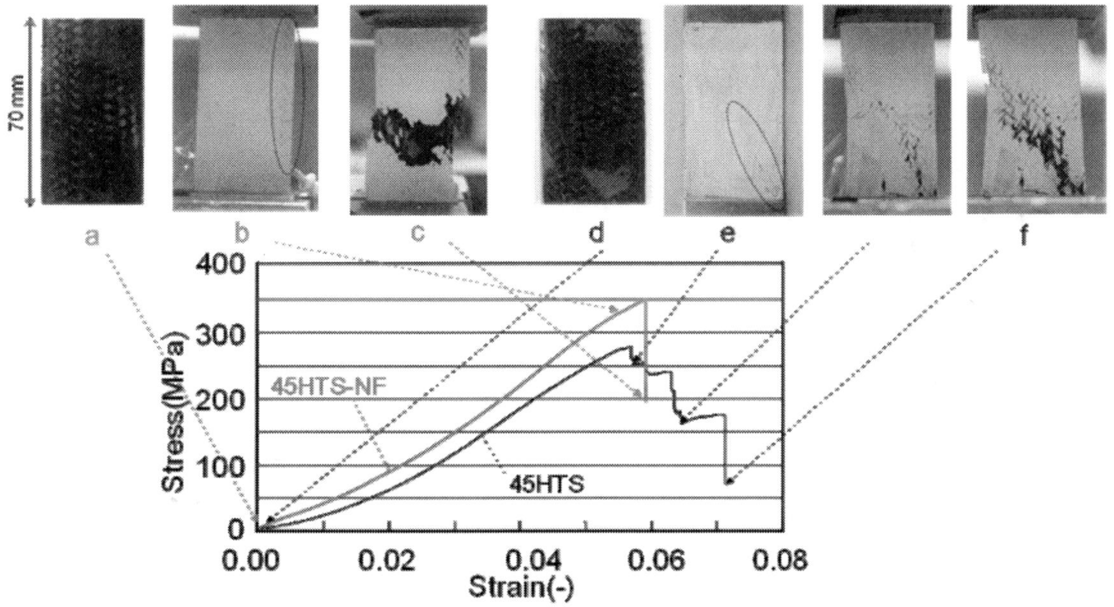

Figure 2. Stress-strain curves of samples 45HTS-NF and 45HTS. Photographs of the samples at certain damage stages are shown together. Some cracks are indicated with dotted ovals. Photographs at "a" and "d" were taken before they were painted by white watercolor.

Figure 3 shows the stress-strain curves of the samples 25UT-UT50, 25UT-XN05, and 25UT-XN60, with photographs of sample surfaces at certain damage stages. In the case of 25UT-UT50, the stress increased linearly with the strain, with the initial tensile modulus of about 59 GPa. When the stress and strain reached 1055 MPa and 0.0178, the final fracture suddenly occurred near the center of the sample ("o"). We note that any cracks did not occur until the final fracture. It was interesting that slanting zigzag cracks at an angle of about 45° was observed on the lower part of the sample. These cracks were thought to be caused by reaction of the final fracture. Vertical zigzag cracks were observed along some middle-end bundles.

In the case of 25UT-XN05, the stress increased almost linearly with the strain, with the initial tensile modulus of about 58 GPa, similarly to 25UT-UT50. When the stress and strain reached 700 MPa and 0.012, small horizontal cracks occurred ("l"). After the occurrence of the cracks, the tensile modulus slightly decreased, and when the stress and strain reached 800 MPa and 0.0145, the final fracture occurred in the upper part, concomitantly with occurrence of slanting zigzag cracks at an angle of about 45° in the bottom part, similarly to 25UT-UT50 ("m"). However, vertical zigzag cracks, which occurred in 25UT-UT50, were not observed.

In the case of 25UT-XN60, the stress increased linearly with the strain, with the higher initial tensile modulus of about 104 GPa, than those of 25UT-UT50 and 25UT-XN05. When the stress and strain reached 255 MPa and 0.0025, small horizontal cracks occurred ("h"). After the occurrence of the cracks, the tensile modulus slightly decreased. Occurrence of new cracks and slight decrease in the modulus repeated a few times such as indicated by "i", and when the stress and strain reached 550 MPa and 0.009, the final fracture occurred in the center part of the sample ("j"). Slanting zigzag cracks at an angle of about 25° occurred in both the top and bottom parts. Vertical zigzag cracks, which occurred in 25UT-UT50, were not observed, but peeling-off of the watercolor paint was observed in relatively wide parts on 25UT-XN60.

From those results, it was shown that 25UT-UT50, in which the same carbon fiber bundles were used in the textile and middle-end bundle, had the greatest destructive strength and strain among the samples. 25UT-XN05, in which XN05 with much smaller tensile modulus and the same elongation was used as the middle-end bundles, had the same initial tensile modulus, although the destructive strength and strain were smaller than those of 25UT-UT50. This should be because of the 4 times smaller tensile strength of the XN05 than that of UTS50. 25UT-XN60, in which XN60 with higher tensile modulus and smaller elongation was used as the middle-end bundles, had the highest initial tensile modulus among the samples. The destructive strength and strain of 25UT-XN60 were smallest probably because of the smallest elongation of XN60.

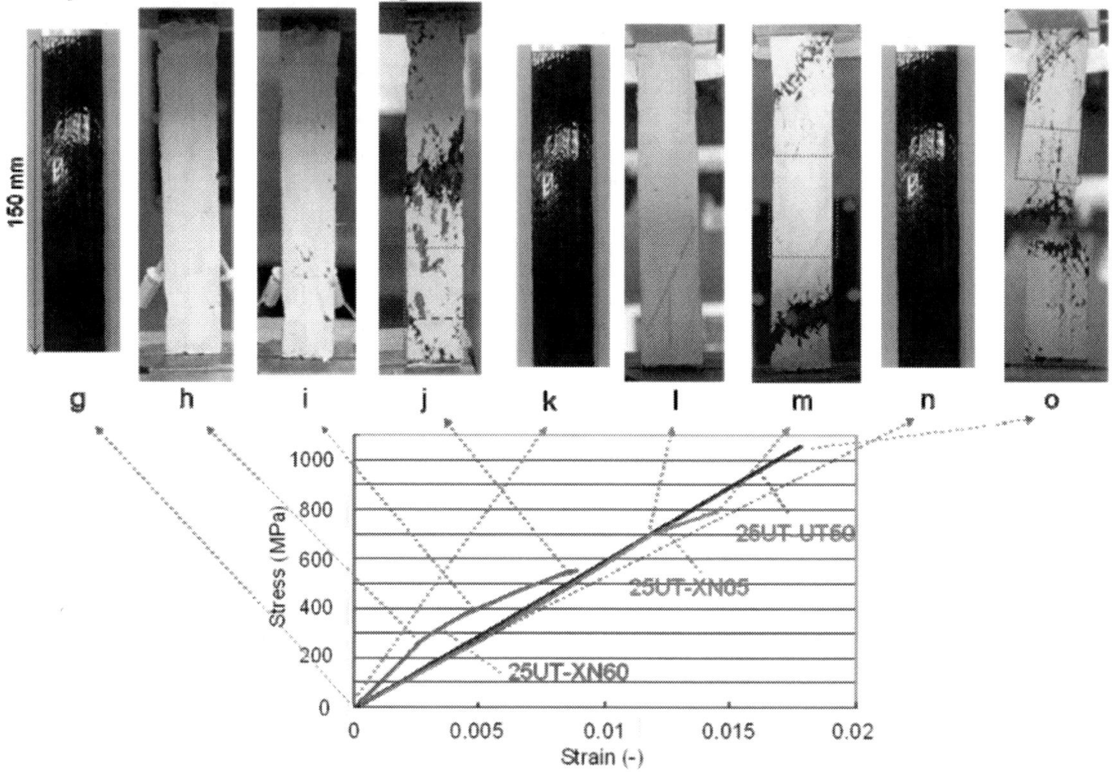

Figure 3. Stress-strain curves of samples 25UT-UT50, 25UT-XN05, and 25UT-NX60. Photographs of the samples at certain damage stages are shown together. Photographs at respective virgin stages were taken before they were painted by white watercolor.

3. Visualization of flowing current using HTS-SQUID gradiometer

3.1 Visualization method

At the certain damage stages before, during, and after the tensile loading, ex-situ SQUID-NDE technique was applied to the samples, using a HTS-SQUID gradiometer with current injection. A planar first-order HTS-SQUID gradiometer, with small differential pickup coil and a small baseline, functions as a current detector with high spatial resolution when a liftoff between the gradiometer and a conductive panel sample is short enough such that the liftoff is comparable to the baseline, as described elsewhere [7]. In this study, we put electrical terminals on both ends of each sample to induce an ac current in the longitudinal direction, while applying an ac voltage across each sample. In such case, distribution of the resultant flowing current in the sample depended on distribution of electrical resistivity in the sample. The distribution of the electrical resistivity was determined by the geometry of the braided textile, and contacting resistivity between the bundles in the braided CFRPs. The CNFs existing in between the bundles, and the middle-end carbon-fiber bundles interlaced in the braided textile would also attribute to the current distribution. When such a sample, in which an ac current is being induced, is set on the x-y plane, the induced current densities J_x and J_y in the sample have similar distributions as the magnetic field gradients dB_z/dy and $-dB_z/dx$ in an x-y plane near the sample. These relationships $\mu_0 J_x \approx dB_z/dy$, and $\mu_0 J_y \approx -dB_z/dx$, between the currents and the field gradients, can be derived from the Maxwell's equation, $\nabla \times \mathbf{H} = \mathbf{J}$ [7].

Based on this principle, we employed a SQUID-NDE system using a small HTS-SQUID gradiometer and an automated xy-stage, whose details were described elsewhere [6]. Schematic diagram of the NDE system is shown in Figure 4. An HTS-SQUID gradiometer with a differential pickup coil, composed of two square coils with size of 1 mm x 1 mm, was used. DC bias current of about 170 μA was applied to the SQUID to drive the gradiometer. We set each sample on the xy-stage, and induced an ac current of 7 mA at 800 Hz in the longitudinal directions in the cases of 45HTS-NF and 45HTS. In the cases of 25UT-UT50, 25UT-XN05, and 25UT-XN60, the induced currents were 2.6 mA, 4.2 mA, and 10.8 mA at 800 Hz, respectively, which depended on the electrical resistivity of each sample. The distributions of the field gradients dB_z/dy and dB_z/dx above each sample were scanned with a liftoff of 2–3 mm, while changing the direction of the gradiometer to measure dB_z/dy at first, and then dB_z/dx next. The scanning intervals in the x and y directions were 1 mm. The measured area above the samples was the area shown in the photographs in Figures 2 and 3. We converted the field gradient distributions of dB_z/dy and $-dB_z/dx$ into the distributions of current components J_x and J_y based on the field gradient to current conversion scheme [7]. In this paper, the units of the converted current distributions were shown with those of the measured field gradients dB_z/dy and $-dB_z/dx$.

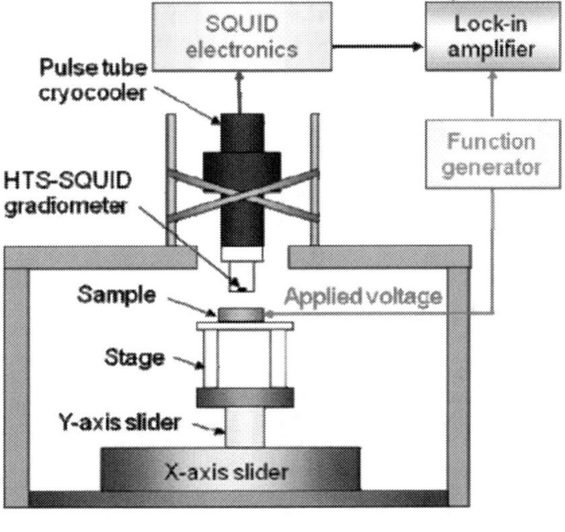

Figure 4. Schematic diagram of NDE system using HTS-SQUID gradiometer.

3.2 Measurement results of 45HTS-NF and 45HTS

Figures 5 and 6 show the measurement results of the converted J_x and J_y distributions in 45HTS-NF and 45HTS. The left-upper and left-lower current maps in Figure 5 show the J_x and J_y distributions at the virgin stage "a" with a liftoff of 2 mm. In the current map of J_x, two peaks were measured along both sides, whereas the current was weaker in the middle area of the sample. Conversely, a stripe pattern along the braided bundles was clearly visible in the current map of J_y. The same stripe pattern can be hardly seen in the J_x map because the peaks along the sides were dominant. From the stripe patterns, it could be seen that the currents flowed mainly along the ±45° bundles continuously. The patterns of the same kind were also observed in the case of 45HTS at the virgin stage "d", as shown in Figure 6. However, the peak along the edges observed in 45HTS was weaker and narrower, and the

Journal of Physics: Conference Series **433** (2013) 012013 doi:10.1088/1742-6596/433/1/012013

stripe pattern was less uniform than those of 45HTS-NF. It is inferred that these differences were attributed to the addition of CNFs in 45HTS-NF, because it is thought that CNFs in between the bundles should decrease the contacting electrical resistivity between the bundles, thereby averaging (or smoothing) the current density among the bundles and increasing the J_x component, which should flow in the x direction through neighboring bundles owning to the assistance of the CNFs [4]. The current distributions shown in the middle-upper and middle-lower maps of Figure 5 (at "b") did not quite differ from those at "a", with the exception of less spatial resolution owning to the longer liftoff of 3 mm. In the tensile test, 45HTS-NF broke suddenly, and most of the bundles were fractured. This damage mechanism was quite different from that of 45HTS and was likely due to the presence of the CNFs, which increased the maximum stress of the sample, as shown in Figure 2. Because of the CNFs, not only the strength of the matrix, but also the strength of the mechanical connection between the bundles should be increased. Thus, delamination between the bundles and relocation of the bundles, which occurred in 45HTS, did not occur in 45HTS-NF. This led to the firm transmission of stress between neighboring bundles fixed with the strengthened matrix, resulting in the sudden final fracture. At the final stage "c", only several bundles were not broken in 45HTS-NF, as shown in the right-upper and right-lower maps of Figure 5, because several strong current appeared.

In the case of 45HTS, it appears that the first crack did not cause a clear change in the current distribution, as shown in the middle-upper and middle-lower maps of Figure 6 ("e"). As mentioned, it was estimated that the crack, which caused the drop in stress, should cause the delamination and relocation of the bundles. However, because the mechanical and electrical connections between the neighbouring bundles in 45HTS should be weaker than those of 45HTS-NF, the delamination and relocation did not seriously affect to the current distribution. After the final fracture, it was estimated that most bundles in the right-side area of the first crack line, which is indicated by the dotted line in Figure 6, were broken, whereas those in the left-side area of the line were almost not damaged, from observation of the current distribution. We note that there was a low current density area on the first crack line even at the virgin stage "d", which is indicated by the dotted line. This may indicate a possibility to predict the location of the final fracture by this NDE technique.

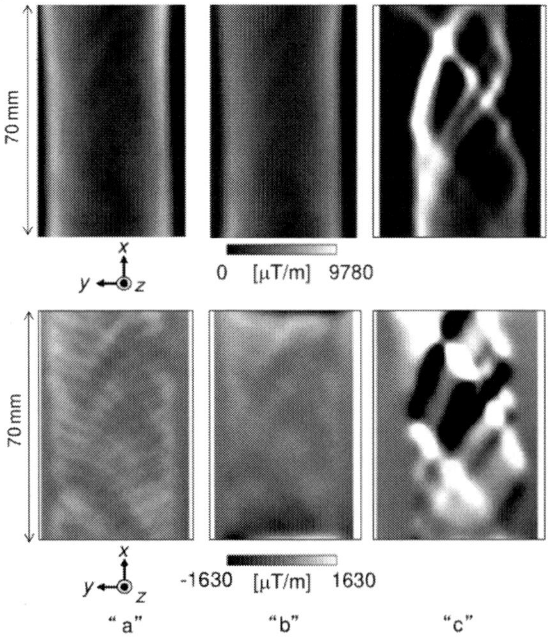

Figure 5. Measured J_x (upper) and J_y (lower) maps of 45HTS-NF at damage stages "a", "b", and "c".

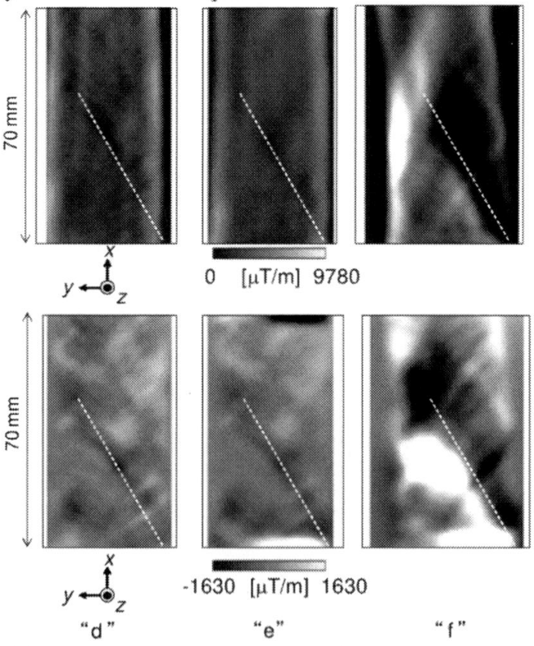

Figure 6. Measured J_x (upper) and J_y (lower) maps of 45HTS at damage stages "d", "e", and "f".

3.3 Measurement results of 25UT-UT50, 25UT-XN05, and 25UT-XN60

Figure 7 shows the measurement results of the converted J_x and J_y distributions in 25UT-UT50 at "n" and "o" in Figure 3, 25UT-XN05 at "k" to "m", and 25UT-XN60 at "g" to "j", respectively. The scales of the current maps in Figure 7 were normalized according to the currents induced in the respective samples. In the case of J_x distributions, white region means large current toward the $+x$ direction. Black regions above both the samples' edges in the figures are the edge effects [7]. In the case of J_y distributions, white region means current toward the $+y$ direction, while black region means current toward the $-y$ direction. At the respective virgin stages ("g", "k", and "n"), the current distributions were quite different. In the case of 25UT-XN60 with the middle-end bundles of the lowest resistivity, the current in the x direction was dominant compared to the other results, while the current in the $\pm y$ directions, which appeared along the braided textile, was weakest relative to the other results. These indicate that the induced current flowed mainly in the middle-end bundles XN60, and partially in the braided textile made of UTS50. In the rest of the samples, the current in the x direction in 25UT-UT50 was more dominant relative to that in 25UT-XN05. These results well agreed with the fact that the resistivity of UTS50 was lower than that of XN05. From the results shown in Figure 7, it was shown that the observed cracks before the final fractures, which occurred in 25UT-XN60 at "h" and "i", and in 25UT-XN05 at "l", caused the reduction of the samples' tensile modules, but did not change the current distributions significantly. It indicates that the surface cracks occurred in 25UT-XN05 and 25UT-XN60 were likely due to rupture of small number of the middle-end bundles in the respective samples, and/or debonding between the middle-end bundles and the braided textile. It is thought that the energy of the rupture or the debonding was transmitted to the surface, resulting in the observed surface cracks.

We further investigated the samples after their final fractures to elucidate the destructive mechanisms of the samples. We cut out the samples such that cut-out parts did not have visible damage, to study the inside of the broken samples. Only in the case of 25UT-XN60, the cut-out sample had a visible damage, which occurred at the final fracture. The measured areas are shown on the photographs in Figure 3 by dotted lines. The field gradient distributions above the cut-out parts were measured by the same visualization method, and they were converted into the current distributions. The measurement results are also shown in Figure 7. In the most results, the current distributions in the cut-out parts look similar to those in the samples before the final fractures, especially concerning the flowing current in the braided textile as seen in the J_y distributions. In the case of 25UT-TU50 at "o", the edge effects are clearly seen in the both ends of the cut-out sample because the sample was short. Concerning the flowing currents in the middle-end bundles, the J_x current distributions in 25UT-UT50 and 25UT-XN60 "j" have some areas with lower current density, which are indicated by dotted circles. Especially, in the case of 25UT-XN60 at "j" with less influence of the edge effect, the current decrease along the middle-end bundles are more evident. Figure 8 shows the line scan results of J_x at "g", "i" and "j" in Figure 7. The scanned lines are indicated with broken lines in the figure. As shown in Figure 8, the current density around 10 mm in the y-axis decreased with the propagation of the damage stages. The current density decrease in the area from 16 – 27 mm in the y-axis after the final fracture must be originated in the visible slant zigzag crack near the bottom.

Considering that the distribution of the current J_y flowing in the braided textile of 25UT-XN60 changed little between before and after the final fracture, the destructive mechanism of 25UT-XN60 is thought to be following: first, debonding between the braided fabric and the middle-end yarns should occur at the damage stages such as "b", and "c" in Figure 2, because of the difference in the elongation characteristics between UTS50 and XN60. From the current visualization results, it is indicated that the base textile was not damaged even at the final fracture except for the final fracture part. At the final fracture, not only at the fracture part but also in some parts of the sample, the middle-end bundles broke inside the sample. Taking account of this scenario, it can be said that the debonding between the base textile and the middle-end bundles is the key factor in the destructive mechanism.

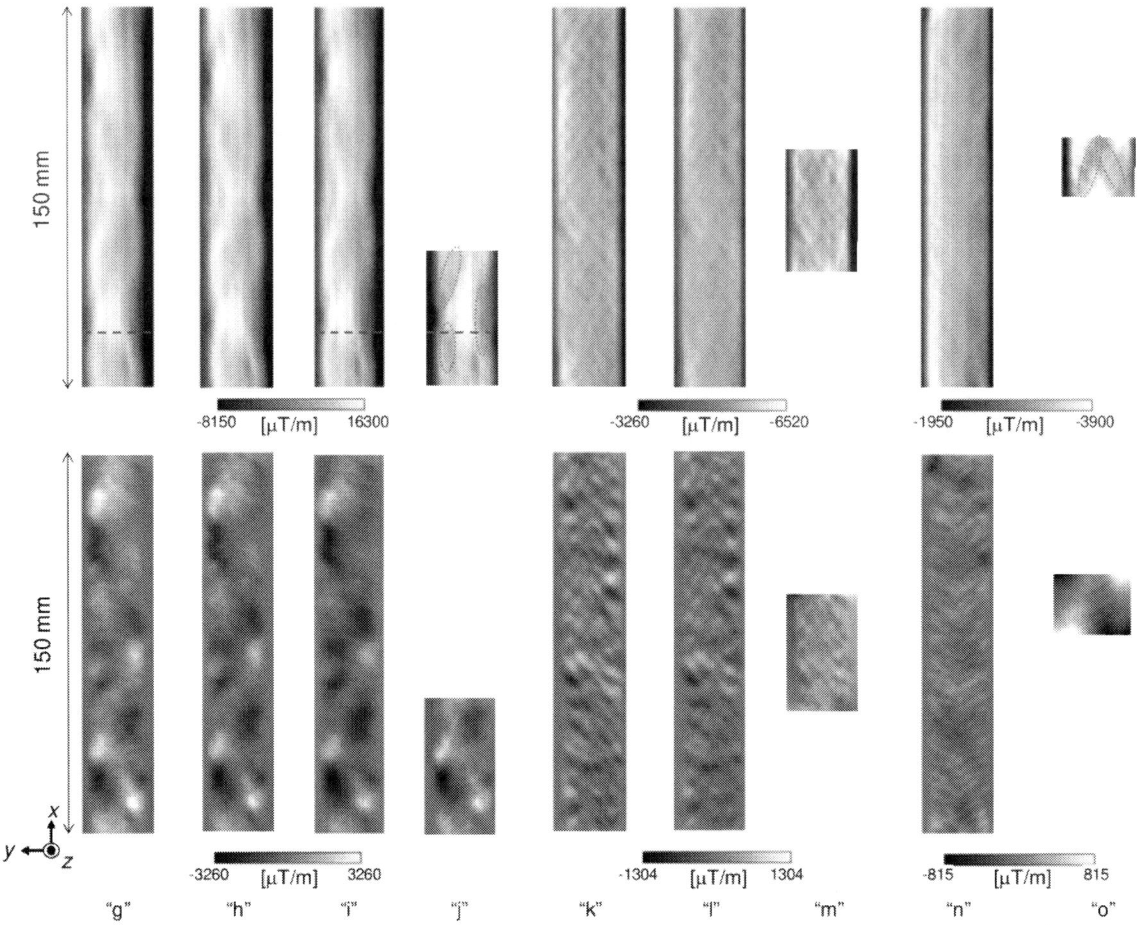

Figure 7. Measured J_x (upper) and J_y (lower) maps of 25UT-XN60 at "g" to "j", 25UT-XN05 at "k" and "m", and 25UT-UT50 at "n" and "o".

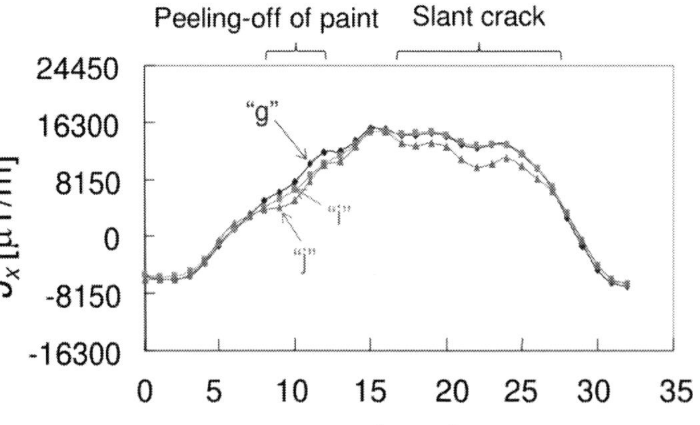

Figure 8. Line scan results of J_x distribution on 25UT-XN60 at "g", "i", and "j" in Figure 7.

4. Conclusions

Destructive test of the step-by-step tensile loading and current visualization method using HTS-SQUID gradiometer were applied to the braided CFRP samples with the CNFs and the different middle-end fibers. The experimental results indicated that the addition of the CNFs and the different middle-end bundles attributed to modify not only the mechanical properties, but also the electrical properties of the samples. Combining the results by the both methods, a scenario of the destructive mechanism of 25UT-XN60 was also estimated. It is expected that the current visualization method using HTS-SQUID gradiometer will be a useful tool to assist for study of destructive mechanism of new advanced composites like braided CFPR. For practical applications, in-situ NDE method without contact to a sample is desirable. We are studying an eddy-current-based NDE method for the composites by improving the SQUID-NDE system.

References

[1] Chou T -W, Ko F K 1989 Textile Structural Composites. in *Composite Materials Series Vol. 3* (Elsevier Science Publishers BV, Amsterdam), pp. 1-26
[2] Falzon P J, Herszberg I 1998 *Composites Science and Technology* **58** 253
[3] Nakai A, Ohki T, Takeda N, Hamada H 2001 *Composite Structures* **52** 315
[4] Gojny F H, Wichmann M H G, Fiedler B, Kinloch I A, Bauhofer W, Windle A H, Schulte K 2006 *Polymer* **47** 2036
[5] Takai Y 2011 Design for advanced carbon fiber braided composites (Ph. D. Thesis, Kyoto Institute of Technology, Kyoto)
[6] Shinyama Y, Hatsukade Y, Takai Y, Aly-Hassan M S, Nakai A, Hamada H, Tanaka S 2011 *Physica C* **471** 1242
[7] Hatsukade Y, Aly-Hassan M S, Kasai N, Takashima H, Hatta H, Ishiyama A 2003 *IEEE Trans. Appl. Supercond.* **13** 207

Journal of Physics: Conference Series **433** (2013) 012014 doi:10.1088/1742-6596/433/1/012014

A Microfluidic Flow-switching Device Powered by *Vorticella* Stalk

M Nagai[1*], K Tanizaki[1], Y Hayasaka[1], T Kawashima[1] and T Shibata[1]

[1]Department of Mechanical Engineering, Toyohashi University of Technology, 1-1 Hibarigaoka, Tenpaku-cho, Toyohashi, Aichi 441-8580, Japan

*E-mail: nagai@me.tut.ac.jp

Abstract. Bioactuators are an attractive alternative for mechanical components of MEMS devices. We propose a flow-switching device active to calcium ion based on bioactuator of *Vorticella*. We develop a fundamental procedure for immobilization of *Vorticella* in a microfluidic chamber and control of contraction and extension of stalks. Cells were trapped in microfluidic chambers and allowed to adhere. After treatment of cells, stalks were contracted and extended by injecting solutions. Flow speed changed during the motion. Our developed method presents a strategy for application of bioactuator.

1. Introduction

Microfluidic device is widely used for chemical and bioanalysis applications [1]. Whereas application of active elements is preferred for controllable systems [2], traditional active elements enlarge the systems because a driving circuit is required. Use of cellular actuators (bioactuators) can eliminates an external circuit and enables us to achieve miniaturization and functionalization of the systems, since energy conversion cycle is contained in a cell body. Bioactutors are an attractive alternative of active elements. We have employed a contractile stalk from *Vorticella* [3]. Stalk of *Vorticella* is several tens of μm in length and composed of filamentous spasmoneme. After permeabilization of stalk, it contracts and extends in Ca^{2+} solution at the concentration range from 10^{-8} to 10^{-5} M [4]. In this paper, we propose a flow-switching device active to Ca^{2+} based on *Vorticella* stalk. Figure 1 shows a flow-switching mechanism powered by the motion of *Vorticella*. We develop a method to immobilize this highly motile organism in a targeted microchamber. We demonstrate control of its reversible linear actuation by changing concentration of Ca^{2+} and measure change in flow speed.

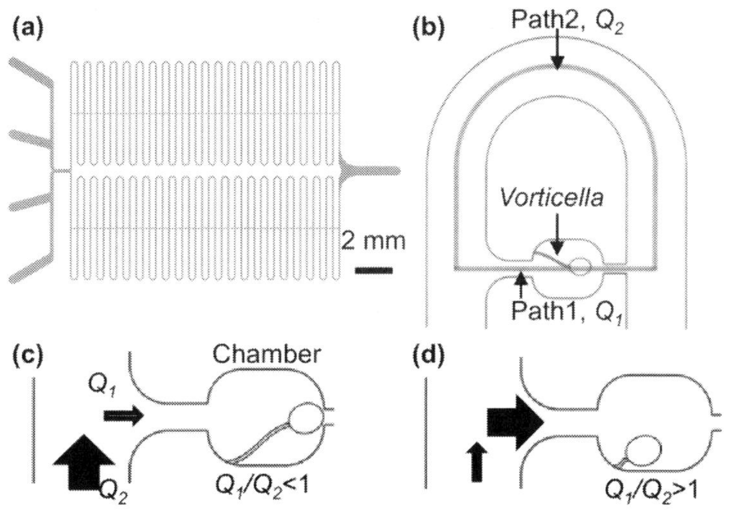

Figure 1. Working principle of a microfluidic flow-switching device. Flow ratio of two streams is switched by contraction and extension of *Vorticella*.

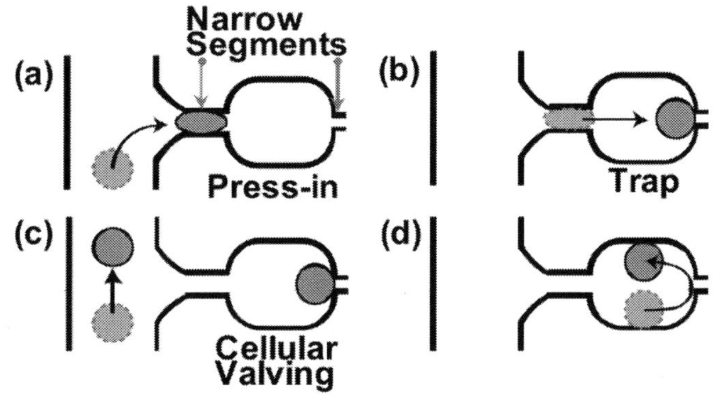

Figure 2. Schematic illustration of trapping of *Vorticella*.

2. Experimental Methods

2.1. Design and fabrication of a fluidic device for cell immobilization and motion control

We design and fabricate a flow channel device (Figure 1a) for immobilization and motion control of *Vorticella*. We use the approach developed by W. Tan *et al.* [5] and modify the design to trap a highly motile cell in a microchamber. Figure 2 shows hydrodynamic trapping of *Vorticella* into a microchamber with two narrow segments. A cell is pressed into the entry narrow segment, trapped in a target microfluidic chamber (Figure 2a, b). Ratio of flow through trapping channel (Q_1 through path1) to by-pass channel (Q_2 through path2) was designed to satisfy $Q_1/Q_2>1$ to trap *Vorticella* in a chamber. Q_1/Q_2 is expressed as follows,

$$\frac{Q_1}{Q_2} = \left(\frac{C_2(\alpha_2)}{C_1(\alpha_1)}\right)\left(\frac{L_2}{L_1}\right)\left(\frac{W_2+H}{W_1+H}\right)^2\left(\frac{W_1}{W_2}\right)^3 \tag{1}$$

where L is channel length, W is channel width, H is channel height, and α is aspect ratio. $C(\alpha)$ denotes a constant that is a function of α. Q_1/Q_2 is designed to be 5.23. A channel was fabricated in PDMS (Sylgard 184) by standard soft-lithography [6]. We used SU-8 3050 as a mold of flow channel. SU-8 was spin-coated on silicon wafer at 3,000 rpm, baked, and exposed. The dimensions of the channel are listed in Table 1.

Table 1. Dimensions of microfluidic channels to trap a cell.

	Path 1		Path 2		
	Width, W_1 µm	Length, L_1µm	Width, W_2 µm	Length, L_2 µm	Height, H µm
Design	16	40	100	5600	30
Fabricated *(n=5)*	18.3± 1.3	40.9± 1.0	104.0± 0.6	5517± 3	34.9 ±0.6

2.2. Cell preparation

Vorticella for injection was prepared according to the method reported by Matsumoto [7]. We stored the prepared cells at a concentration of 4×10^3 cells/µL in a reservoir and injected them into a flow channel by gravitational force. Flow speed was adjusted to 50 µm/sec at which *Vorticella* was trapped.

Vorticella was treated and controlled with three solutions shown in Table 2. Cells adhered to the surface was permeabilized by treating with surfactant solution containing saponin to control the stalk length by Ca^{2+}. EGTA solution collects Ca^{2+} from a stalk and extends the stalk. Concentration of free Ca^{2+} was adjusted to 1 µM under equilibrium condition with EGTA (ethylene glycol tetraacetic acid) and Ca^{2+}.

Table 2. Composition of solutions for treatment of *Vorticella*.

Surfactant solution	0.1%(w/w) Saponin, 0.1 M KCl, 4 mM EGTA, 50 mM Tris-malate buffer pH 7.0
Extension solution	0.1 M KCl, 4 mM EGTA, 50 mM Tris-maleate buffer pH 7.0
Contraction solution	0.1 M KCl, 3 mM $CaCl_2$, 4 mM EGTA, 50 mM Tris-maleate buffer pH 7.0

2.3. Observation

Cells were observed under the inverted microscope (Eclipse TE2000-U, Nikon Instruments Inc.) equipped with objective lens ×10 or ×20. Movies were recorded with a CCD camera at 5 fps. Stalk length was analyzed with image analysis software (Image J). Flow speed was measured by tracking fluorescent particles (φ0.5µm).

3. Results and discussion

3.1. Trapping and immobilizing Vorticella

Figure 3 shows a single cell trap in a chamber. When *Vorticella* reached an entry site, the cell was introduced into a chamber via Path1 and immobilized on a trapping site (Figure 3a). After the cell was trapped, the next coming cell reaching the entry site flowed via Path2 (Figure 3b-d).

After *Vorticella* was trapped, flow speed was reduced. Although *Vorticella* freely swum in the chamber (Figure 4a-c), the cell was unable to move outside the chamber for itself. This is because the width of an entry site (W_1, 18.3 μm) is smaller than *Vorticella* with a diameter of 30 μm. 16h after introduction, most of the cells adhered in the chamber and grew their stalks as shown in Figure 4d.

Figure 5 is cell number distribution of immobilized *Vorticella* in microchambers. 46 cells were trapped in 39 of 68 chambers. 16 h after the trap, 26 cells adhered in the chambers. Stalks were treated with surfactant solution before motion control. Following surfactant treatment, total 12 cells remained in the chambers. Cells trapped in the chambers decreased from 34 to four. The number of chambers with two to three cells decreased from five to three.

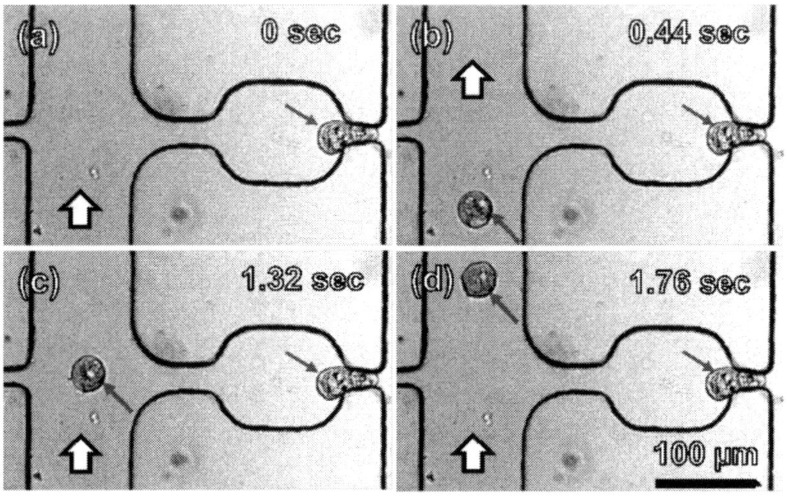

Figure 3. Trapping of a single *Vorticella* in a chamber.

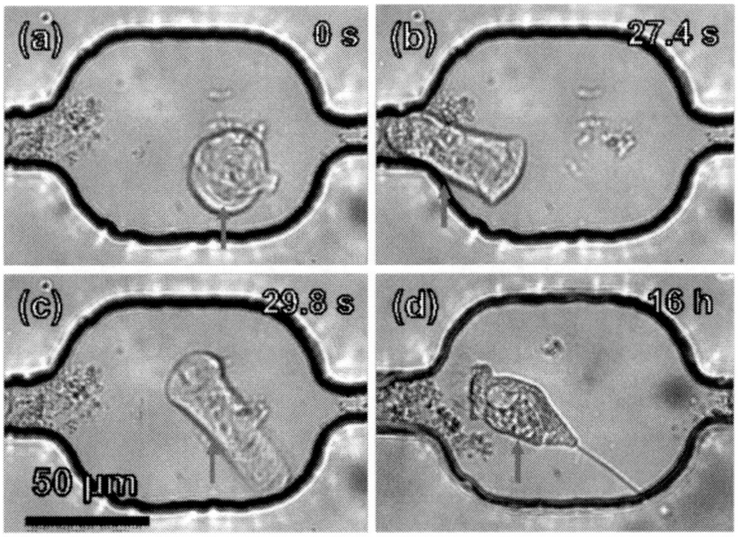

Figure 4. (a)-(c) Motion of pressed-in *Vorticella* in a microchamber. (d) *Vorticella* adhered in a chamber.

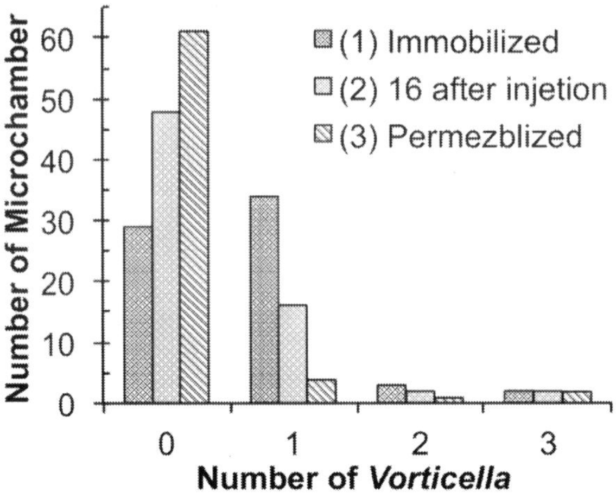

Figure 5. Histogram of number of *Vorticella* in a single chamber during trap and cell treatment.

3.2. Motion control of stalk

Figure 6 and 7 shows control of stalk contraction and extension in a microchamber by changing the concentration of Ca^{2+}. By alternately injecting chelator solution (containing EGTA for removing Ca^{2+}) and contraction solution, we controlled stalk length reversibly. The lengths of stretched and contracted stalk were 35.4 ± 9.13 μm and 9.37 ± 3.29 μm (n=6), respectively. Transition time from extension to contraction and from contraction to extension was 9.11 ± 0.63 s (n=3) and 14.9 ± 1.28 s (n=3), respectively. Change in flow speed was measured by tracking fluorescent particles. Flow speed was changed accompanied by the motion of *Vorticella*. A flow speed through path1 was 345 ± 99 μm/s (n=4) in contracted state, while the speed was 138 ± 36 μm/s (n=4) in extended state. This decrease may be attributed to the increase of fluidic resistance. Change of flow speeds shows the possibility of flow switching.

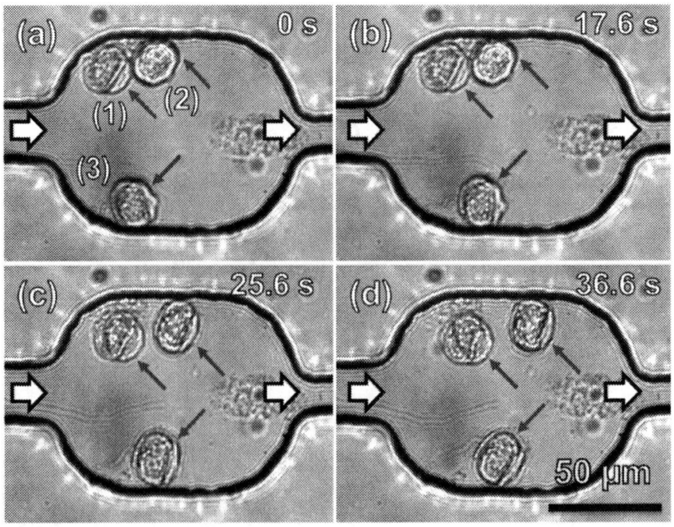

Figure 6. Micrographs of three cells in a single chamber. Time series extension of the three stalks.

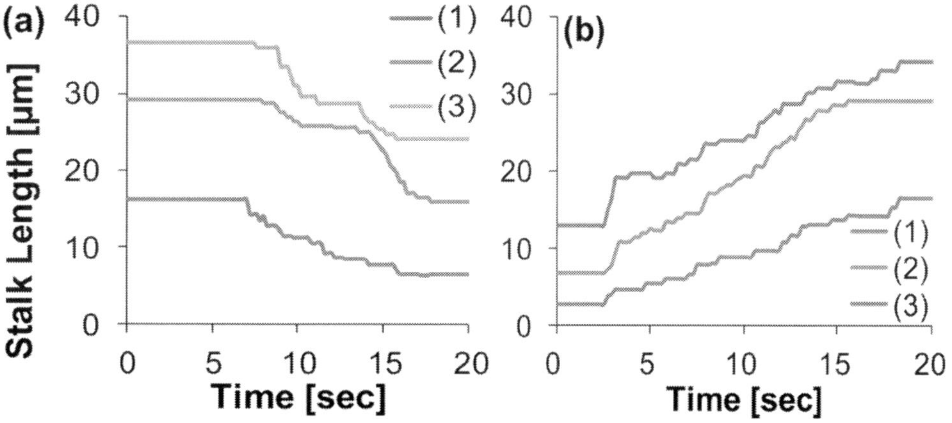

Figure 7. Reversible control of three stalks by changing the concentration of Ca^{2+}. (a) Contraction of the stalks by injection of Ca^{2+} solution. (b) Extension of the stalks by injection of chelator solution.

4. Conclusions

In this research, we proposed a flow-switching device active to Ca^{2+} based on *Vorticella*. We demonstrated immobilization of *Vorticella* in a microchamber. Reversible actuation of *Vorticella* was achieved in a microchamber by exchanging solutions. Change of flow speeds shows the possibility of flow switching. Our developed method has the potential to accelerate the application of bioactuators.

References

[1] Ohno K, Tachikawa K and Manz A 2008 Microfluidics: applications for analytical purposes in chemistry and biochemistry *Electrophoresis* **29** 4443

[2] Bell D, Lu T, Fleck N and Spearing S 2005 MEMS actuators and sensors: observations on their performance and selection for purpose *Journal of Micromechanics and Microengineering* **15** S153

[3] Nagai M, Ryu S, Thorsen T, Matsudaira P and Fujita H 2010 Chemical control of Vorticella bioactuator using microfluidics *Lab on a Chip* **10** 1574

[4] Katoh K and Naitoh Y 1994 Control of cellular contraction by calcium in Vorticella *Journal of experimental biology* **189** 163

[5] Tan W H and Takeuchi S 2007 A trap-and-release integrated microfluidic system for dynamic microarray applications *Proceedings of the National Academy of Sciences* **104** 1146

[6] Anderson J R, Chiu D T, Wu H, Schueller O J A and Whitesides G M 2000 Fabrication of microfluidic systems in poly (dimethylsiloxane) *Electrophoresis* **21** 27

[7] Matsumoto N, Nagai M, Hayasaka Y, Oguri M, Kawashima T and Shibata T 2012 Flow Regulator Powered by the Stalk of Vorticella convarallia *Journal of Physics: Conference Series* **352** 012030

The Weak and Strong Oblique Shock Waves Appeared on the Carbon Dioxide Two-phase Flow in the Ejector Refrigeration Cycle

Y Kawamura[1], **M Nakagawa**[1]
[1]Department of Mechanical Engineering, Toyohashi University of Technology, 1-1 Hibarigaoka, Tenpaku-cho, Toyohashi-city, Aichi, 441-8580, Japan

E-mail: kawamura@nak.me.tut.ac.jp

Abstract. After the Great East Japan Earthquake, the saving energy became the one of the most important issues. Especially, saving the electrical power for air-conditioning at summer time is a problem of great urgency. We have been developing the ejector refrigeration system which can improve coefficient of performance of refrigeration systems by converting exhausted expansion energy into useful pressure energy. The performance of the ejector is greatly affected by the pressure recovery at the mixing section in the ejector. It is elucidated by our recent research on the carbon dioxide refrigeration cycle that the pressure recovery at the ejector is performed by oblique shock waves appeared in the mixing section. The purpose of this research is to clarify the characteristics of the two-phase flow oblique shock waves in the supersonic carbon dioxide two-phase flow. It is shown by the theoretical analyses that the two types of oblique shock waves occur on the supersonic two-phase flow. One is the week shock wave behind of which the flow is still a subsonic state. The other is the strong oblique shock wave which has a large pressure recovery. And the experimental research also carried out by using the carbon dioxide two-phase flow channel. The theoretical results are compared with the experiment.

1. Introduction

The speed of sound in high-speed two-phase flow, which has been used extensively in areas such as total flow systems for geothermal power plants and in refrigeration cycles, is extremely low compared to that of single-phase flow [1]; moreover, a characteristic of the two-phase flow is its easy acceleration into a supersonic state. In general, both shock and expansion waves occur with rapid changes in the flow path of the supersonic state, and the effect of these waves on the flow field cannot be neglected. In addition, the shock and expansion waves that occur in the two-phase flow differ from those in a single-phase gas flow, and the pressure field tends to disperse. Therefore, we can expect a combination of more complex phenomena than those in a single-phase gas flow. We have studied the ejector-refrigeration cycle [2], which can improve the refrigeration cycle by converting the expansion energy that has so far been wastefully discarded into useful compression energy. The performance of the ejector [3] is considerably affected by pressure recovery in the mixing section inside the ejector. In past studies, because the speed of sound for the two-phase flow is low compared to that for the single-

[1] To whom any correspondence should be addressed.

phase flow, an oblique shock wave was generated at the nozzle exit of the ejector interior that affected the flow field of the supersonic two-phase flow inside the mixing section. Furthermore, it was found that most of the pressure recovery in the ejector depends on the kinetic energy of the two-phase flow in the mixing section. Theoretical studies have shown that in supersonic flow, there are two types of oblique shock waves with two-phase flow. One exists at the mach angle and is a weak oblique shock wave, which remains supersonic behind the shock wave. The other is a strong oblique shock wave, which is subsonic behind the shock wave, whose pressure is considerably recovered similar to that in case of a normal shock wave. The conditions under which the strong and weak oblique shock waves are generated were not discussed in these studies, thereby clarifying that the characteristics of these oblique shock waves is important for elucidating the pressure-recovery phenomenon of the ejector. Accordingly, in this study, both strong and weak oblique shock waves were generated on an analytical semi-infinite plane to determine the conditions under which they arise and their characteristics, with the aim of comparing them to experimental results.

2. Basic equations for compressible two-phase flow

In our analysis, transport phenomena such as momentum, heat, and mass transports were considered between the gas and liquid phase. Two-phase flow is assumed to consist of a gas phase in the continuous phase and droplets in the dispersal phase by using the following assumptions:

- Phase changes occur in the flow field according to the heat-transfer limit.;
- The effect of friction other than interphase interactions is neglected.
- All droplets are spheres with diameter d, and any breakup or coalescence due to collisions among droplets is not considered.
- Both gas and liquid phases are assumed to refer to real gases.

Now, we provide the respective equations for the conservation of mass for gas and liquid phases:

$$\frac{\partial}{\partial t}\left(\alpha\rho_g\right)+\frac{\partial}{\partial x_j}\left(\alpha\rho_g v_{gj}\right)=\dot{m}, \tag{1}$$

$$\frac{\partial}{\partial t}\{(1-\alpha)\rho_l\}+\frac{\partial}{\partial x_j}\{(1-\alpha)\rho_l v_{lj}\}=-\dot{m}. \tag{2}$$

Here, α expresses the void fraction represented by the volume fraction of the gas phase in the two-phase flow, ρ is density, $\dot{m}$ is the rate of evaporation, and v is flow velocity with the respective suffixes g and l denoting gas and liquid phases, and i represents each of the directions in two dimensions.

Similarly, the momentum-conservation equations for each of the gas and liquid phases are as follows:

$$\frac{\partial}{\partial t}\left(\alpha\rho_g v_{gi}\right)+\frac{\partial}{\partial x_j}\left(\alpha\rho_g v_{gj}v_{gi}\right)=\dot{m}v_{ki}-(1-\alpha)\rho_l\frac{v_{gi}-v_{li}}{\tau_V}-\alpha\frac{\partial p}{\partial x_i}, \tag{3}$$

$$\frac{\partial}{\partial t}\{(1-\alpha)\rho_l v_{li}\}+\frac{\partial}{\partial x_j}\{(1-\alpha)\rho_l v_{lj}v_{li}\}=-\dot{m}v_{ki}+(1-\alpha)\rho_l\frac{v_{gi}-v_{li}}{\tau_V}-(1-\alpha)\frac{\partial p}{\partial x_i}. \tag{4}$$

The velocity of evaporation is $v_{ki}=(v_{li}\,;\dot{m}\geq 0\,, v_{gi}\,;\dot{m}<0)$, and the momentum relaxation time is $\tau_V=\rho d^2/(18\mu_g)$.

The total-energy equation is:

$$\frac{\partial}{\partial t}\left\{\alpha\rho_g\left(e_g+\frac{v_{gj}^2}{2}\right)+(1-\alpha)\rho_l\left(e_l+\frac{v_{lj}^2}{2}\right)\right\}+\frac{\partial}{\partial x_j}\left\{\alpha\rho_g v_{gj}\left(h_g+\frac{v_{gj}^2}{2}\right)+(1-\alpha)\rho_l v_{lj}\left(h_l+\frac{v_{lj}^2}{2}\right)\right\}=0. \tag{5}$$

Here, e represents internal energy, and h represents enthalpy. This equation does not contain a term for transport between phases.

From the heat-transfer equation for each phase, we can write the following equations considering entropy:

$$T_g \frac{D_g s_g}{D_g t} = C_{pg} \frac{1-\alpha}{\alpha} \frac{T_s - T_g}{\tau_{Tg}}, \tag{6}$$

$$T_l \frac{D_l s_l}{D_l t} = C_{pl} \frac{T_s - T_l}{\tau_{Tl}}. \tag{7}$$

Here, C_{pg} and C_{pl} represent the specific heat of gas and liquid phases, respectively, and then, the thermal-relaxation times are $\tau_{Tg} = \rho_g C_{pg} d/(6h_{Tg})$ and $\tau_{Tl} = \rho_l C_{pl} d/(6h_{Tl})$. Please refer to the reference section for details.

We used the NIST REFPROP v.8 software for calculating the thermophysical properties.

3. Relational expression for oblique shock waves in extreme states

Though momentum, heat, and mass transports are considered in this study, because phase changes occur according to the heat-transfer limit, only momentum and heat transports are important. Extreme states exist depending on the velocity at which these transport phenomena progress, and for extremely high transport velocity, they are referred to as equilibrium states. On the other hand, for low or negligible transport velocity they are referred to as frozen states. In this section, we illustrate the method to determine the relational expressions for the shock waves in each of these extreme states.

3.1. Jump equation for the normal shock wave in momentum and thermal-equilibrium states

Interphase velocity is equal to temperature in the momentum and thermal-equilibrium states. If we denote the forward state for the normal shock wave with suffix 1 and the rear state with suffix 2, the conservation equations for mass, momentum, and energy for a uniform steady flow are given by integrating Eqs. (1) + (2), (3) + (4), and (5):

- mass-conservation equation

$$\{\alpha_1 \rho_{g1} + (1-\alpha_1)\rho_{l1}\} w_{n1} = \{\alpha_2 \rho_{g2} + (1-\alpha_2)\rho_{l2}\} w_{n2}, \tag{8}$$

- momentum-conservation equation

$$\{\alpha_1 \rho_{g1} + (1-\alpha_1)\rho_{l1}\} w_{n1}^2 + p_1 = \{\alpha_2 \rho_{g2} + (1-\alpha_2)\rho_{l2}\} w_{n2}^2 + p_2, \tag{9}$$

- energy-conservation equation

$$\alpha_1 \rho_{g1} w_{n1}\left(\frac{w_{n1}^2}{2} + h_{g1}\right) + (1-\alpha_1)\rho_{l1} w_{n1}\left(\frac{w_{n1}^2}{2} + h_{l1}\right) = \alpha_2 \rho_{g2} w_{n2}\left(\frac{w_{n2}^2}{2} + h_{g2}\right) + (1-\alpha_2)\rho_{l2} w_{n2}\left(\frac{w_{n2}^2}{2} + h_{l2}\right). \tag{10}$$

Because the physical quantities for the forward shock wave are known, the unknowns terms in Eqs. (8), (9), and (10) are $\alpha_2, \rho_{g2}, \rho_{l2}, w_{n2}, p_2, h_{g2}$, and h_{l2}. However, $h_{g2}, h_{l2}, \rho_{g2}$, and ρ_{l2} can be calculated from the pressure corresponding to the saturation state of thermal equilibrium in the two-phase flow. Therefore, the three unknown terms α_2, w_{n2}, and p_2 can be obtained using Eqs. (8), (9), and (10).

3.2. Jump equation for the normal shock wave in momentum equilibrium and thermal frozen states

For the thermal frozen state in which only a small amount of heat is transferred between gas and liquid phases, because the thermodynamic variable is a function of pressure and entropy, when adiabatic conditions are applied to Eqs. (8) and (9), we get Eq. (11) in addition to Eqs. (8), (9), and (10).

$$s_{g1} = s_{g2}, \quad s_{l1} = s_{l2}. \tag{11}$$

From the tabulated thermodynamic properties, $\rho_{g2}, \rho_{l2}, h_{g2}$, and h_{l2} can be estimated using the pressure, p_2, and entropy, s_2, for mutual phases. Therefore, the five unknown terms $\alpha_2, w_{n2}, p_2, s_{g2}$, and s_{l2} can be found by solving Eqs. (8), (9), (10), and (11).

3.3. Jump equation for the normal shock wave in the momentum frozen and thermal equilibrium states

If the mutual forces acting on the phases are reduced to approximately zero, the mutual phases flow independently. Therefore, the momentum equation is required for both gas and the liquid phases.

- mass-conservation equation

$$\alpha_1\rho_{g1}w_{gn1} + (1-\alpha_1)\rho_{l1}w_{ln1} = \alpha_2\rho_{g2}w_{gn2} + (1-\alpha_2)\rho_{l2}w_{ln2}, \tag{12}$$

- total momentum-conservation equation

$$\alpha_1\rho_{g1}w_{gn1}^2 + (1-\alpha_1)\rho_{l1}w_{ln1}^2 + p_1 = \alpha_2\rho_{g2}w_{gn2}^2 + (1-\alpha_2)\rho_{l2}w_{ln2}^2 + p_2, \tag{13}$$

- droplet momentum-conservation equation

$$\frac{1}{2}w_{ln2}^2 - \frac{1}{2}w_{ln1}^2 + \int_1^2 \frac{1}{\rho_l}\frac{dp}{dT}dT = 0, \tag{14}$$

- energy-conservation equation

$$\alpha_1\rho_{g1}w_{gn1}\left(\frac{w_{gn1}^2}{2} + h_{g1}\right) + (1-\alpha_1)\rho_{l1}w_{ln1}\left(\frac{w_{ln1}^2}{2} + h_{l1}\right) = \alpha_2\rho_{g2}w_{gn2}\left(\frac{w_{gn2}^2}{2} + h_{g2}\right) + (1-\alpha_2)\rho_{l2}w_{ln2}\left(\frac{w_{ln2}^2}{2} + h_{l2}\right). \tag{15}$$

Similar to the momentum and thermal-equilibrium states, $\rho_{g2}, \rho_{l2}, h_{g2}$, and h_{l2} can be calculated from the pressure corresponding to the saturation state of themal equilibrium in the two-phase flow. Therefore, the four unknown terms $\alpha_2, w_{gn2}, w_{ln2}$, and p_2 can then be found from Eqs. (12), (13), (14), and (15).

3.4. Jump equation for the normal shock wave in the momentum and thermal frozen states

By assuming that all transport phenomena can be neglected, mass, momentum, and energy for each phase are conserved.

- mass-conservation equation

$$\alpha_1\rho_{g1}w_{gn1} = \alpha_2\rho_{g2}w_{gn2}, \tag{16}$$

$$(1-\alpha_1)\rho_{l1}w_{ln1} = (1-\alpha_2)\rho_{l2}w_{ln2}, \tag{17}$$

- momentum-conservation equation

$$\alpha_1\rho_{g1}w_{gn1}^2 + p_1 = \alpha_2\rho_{g2}w_{gn2}^2 + p_2, \tag{18}$$

$$(1-\alpha_1)\rho_{l1}w_{ln1}^2 + p_1 = (1-\alpha_2)\rho_{l2}w_{ln2}^2 + p_2, \tag{19}$$

- energy-conservation equation

$$\alpha_1\rho_{g1}w_{gn1}\left(\frac{w_{gn1}^2}{2} + h_{g1}\right) = \alpha_2\rho_{g2}w_{gn2}\left(\frac{w_{gn2}^2}{2} + h_{g2}\right), \tag{20}$$

$$(1-\alpha_1)\rho_{l1}w_{ln1}\left(\frac{w_{ln1}^2}{2} + h_{l1}\right) = (1-\alpha_2)\rho_{l2}w_{ln2}\left(\frac{w_{ln2}^2}{2} + h_{l2}\right). \tag{21}$$

The terms h_{g2} and h_{l2} can be estimated using the mutual phase pressure p_2 and the densities ρ_{g2} and ρ_{l2}. Therefore, the six unknown terms $\alpha_2, w_{gn2}, w_{ln2}, \rho_{g2}, \rho_{l2}$, and p_2 can be estimated using Eqs. (16)–(21).

3.5. Relationship between normal and oblique shock waves

The equation for the oblique shock wave can be calculated from its relationship with the normal shock wave. As shown in figure 1, because the velocity in the tangential direction does not change for the oblique shock wave, Eq. (22) is obtained:

$$\frac{\tan(\phi-\theta)}{\tan\phi}=\frac{w_{n2}}{w_{n1}} \quad i.e. \quad \theta=\phi-\tan^{-1}\left(\frac{w_{n2}}{w_{n1}}\tan\phi\right). \tag{22}$$

Here, θ is the turning angle of the inclined wall, ϕ is the shock-wave angle for the oblique shock wave, and w_n is the normal velocity. The shock-wave angle ϕ can be calculated from Eq. (22) using the turning angle of the inclined wall, θ, and the ratio of normal velocities, w_{n2}/w_{n1}. The ratio of the normal velocities is similar to that for the normal shock wave. Therefore, if the turning angle of the inclined wall is given, the oblique shock-wave angle is determined using the relational expression for the normal shock wave.

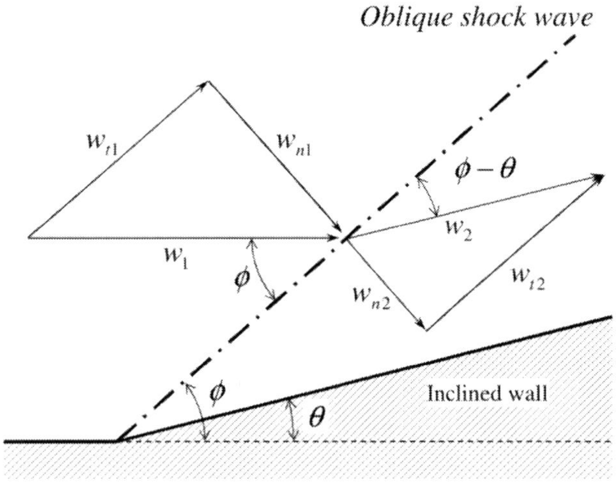

Figure 1. Velocity vector diagram for the oblique shock wave.

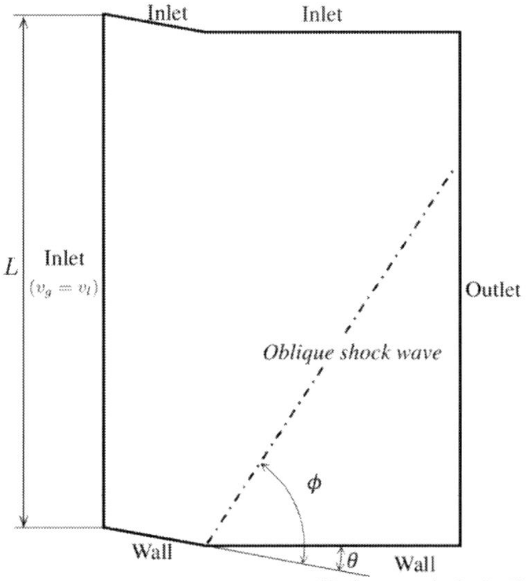

Figure 2. Calculation domain for the weak oblique shock wave.

4. Relationship between strong and weak oblique shock waves

4.1. Weak oblique shock wave

4.1.1. Analysis method for the weak oblique shock wave. Weak oblique shock waves are generated by mach waves, and in supersonic flow, entropy production is small. In addition, with only slight disturbance to the flow field of weak oblique shock waves, the back of the shock wave is supersonic at the inlet.

In past studies [4][5], weak oblique shock waves were simulated using numerical analysis. In numerical analysis, Constrained Interpolation Profile Scheme method is applied to the basic equations of the compressible two-phase flow, which considers momentum, heat, and mass transports. Non-steady calculations are performed, and a steady-state solution is obtained at which a constant value is reached. The calculation domain in figure 2 has the same shape as that of the inclined wall so that oblique shock waves are generated over a semi-infinite plane. The inclined wall has an angle of θ. For the domain, we make 70 divisions in the direction horizontal to the inclined wall and 100 divisions in the direction perpendicular to the wall. In the calculation domain, we define a typical length of L for the inlet-section height and a typical velocity of u for the gas-flow rate in the inlet section. The two-phase flow enters into the domain from the left side and flows out from the top and right sides. On the bottom side of the wall surface, the component of the gas velocity in the direction perpendicular to the wall is 0, and the horizontal component is symmetrical with the wall. It has been assumed that, after collision with the wall, the liquid flows as a liquid film with no reflection. Because the volume of such a liquid film is small, the volume of the liquid-film flow and the boundary layer were neglected in the analysis. Therefore, as for the outlet, the boundary conditions of the free outflow for the liquid are given. The inflow conditions were that constant pressure and velocity were maintained, and for an equilibrium state, the velocities of both gas and liquid phases at the inlet were assumed equal.

This analysis was performed by simulating the four extreme states of the transport phenomena for the two-phase flow, total-equilibrium state, thermal frozen state, momentum frozen state, and total frozen state. The momentum relaxation time, τ_V, and the thermal relaxation times, τ_{Tg} and τ_{Tl}, are important in simulating these extreme states. The relaxation times are functions of the droplet size. Therefore, the analysis of weak oblique shock waves proceeds by assuming each non-dimensional relaxation time as shown in table 1.

Table 1. Supposition of the non-dimensional relaxation time.

Extream state	Momentum relaxation time		Thermal relaxation time
	Droplet size d	τ_V'	τ_{Tg}' , τ_{Tl}'
[a]m:eq , t:eq	0.1 μm	1.42×10^{-4}	1.42×10^{-4}
[b]m:eq , t:fr	0.1 μm	1.42×10^{-4}	1.42×10^{4}
[c]m:fr , t:eq	1000 μm	1.42×10^{4}	1.42×10^{-4}
[d]m:fr , t:fr	1000 μm	1.42×10^{4}	1.42×10^{4}

[a] Momentum and thermal equilibrium state.
[b] Momentum equilibrium and thermal frozen state.
[c] Momentum frozen and thermal equilibrium state.
[d] Momentum and thermal frozen state.

4.1.2. Characteristics of weak oblique shock waves. Figure 3 shows the results of the numerical analysis for weak oblique shock waves and the differences in the oblique shock waves by assuming an inflow equilibrium Mach number $M_{e1} = 2.17$ for the three extreme states: (a) the total equilibrium state, (b) the thermal frozen state, and (c) the momentum frozen state. In figure 3, according to the differences between these extreme states, differences appear in the generated shock-wave angle of the oblique shock wave. To compare the results of the numerical analysis with the theoretical values for the oblique shock waves in the extreme states, the relationship between the shock-wave angle ϕ and the inflow equilibrium Mach number M_{e1} is shown in figure 4, and the relationship between the pressure ratio $p_2/p_1 = p_{21}$ for the front and back of the shock wave and the inflow Mach number, M_{e1}, is shown in figure 5. Figures 4 and 5 are the results for a constant turning angle of the inclined wall, $\theta = 10°$, and for the calculations in the previous section, which uses the relational expression for the oblique shock waves in the extreme states. The curves in the diagram are shown as a red solid line for the momentum and thermal equilibrium state, blue dashed line for the momentum equilibrium and thermal frozen state, green dashed–dotted line for the momentum frozen and thermal equilibrium state, and purple dashed-two dotted line for the momentum and thermal frozen state. The domain for the weak oblique shock wave is the region ranging from the point of inflection in figure 4 with a smaller shock-wave angle ϕ, but in figure 5, it is the region ranging from the point of inflection for which the pressure ratio, p_{21}, tends to increase by a smaller amount with increasing inflow equilibrium Mach number M_{e1}. In addition, the domain for the strong oblique shock wave, which will be discussed later, is the region in figure 4 ranging from the point of inflection with the larger shock-wave angle ϕ, but in figure 5 is the region ranging from the point of inflection for which the pressure ratio p_{21} tends to increase by a larger amount with increasing inflow equilibrium Mach number, M_{e1}.

The angles of the oblique shock waves obtained from the numerical analysis are shown in figure 4 as circles. From figure 4, because the numerical-analysis results match approximately with the theoretical curves for the oblique shock waves in the extreme states, the weak oblique shock waves may be predicted using this type of analysis.

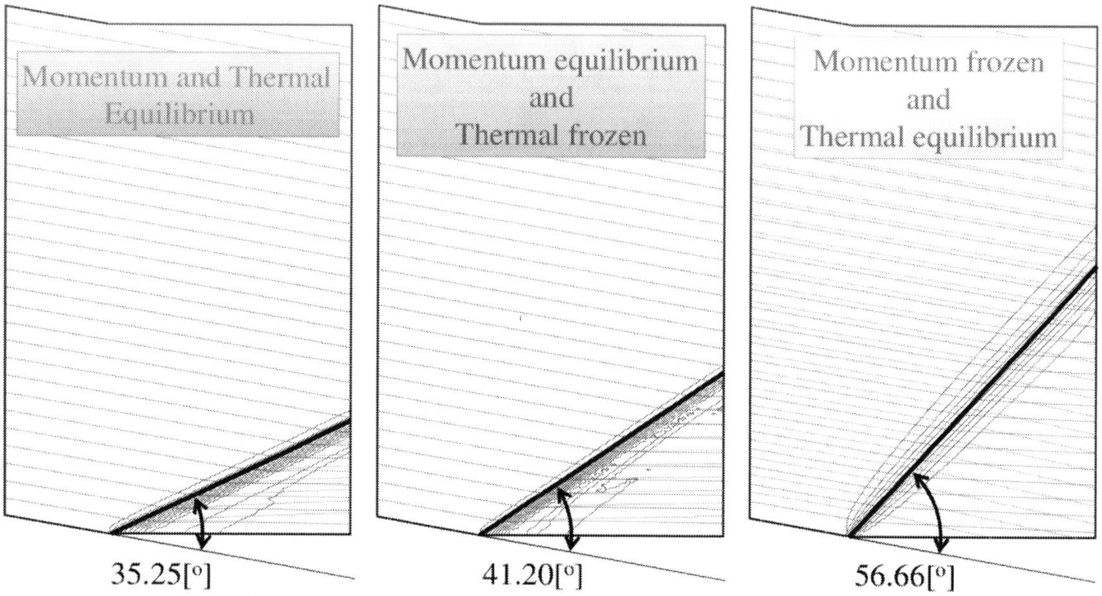

(a) (b) (c)

Figure 3. Weak oblique shock waves for $M_{e1} = 2.17$.

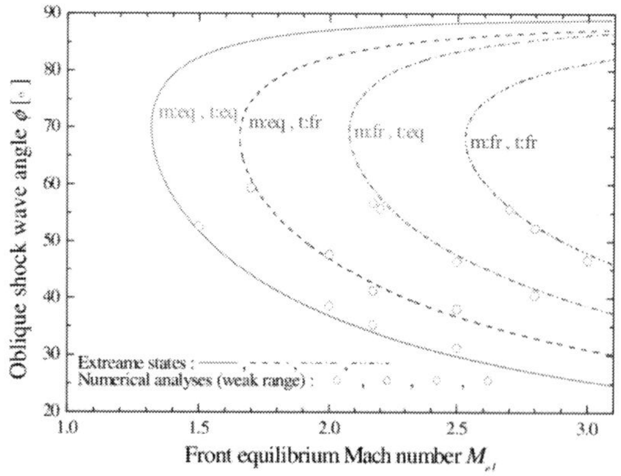

Figure 4. Shock wave angles versus front Mach number for extreame states.

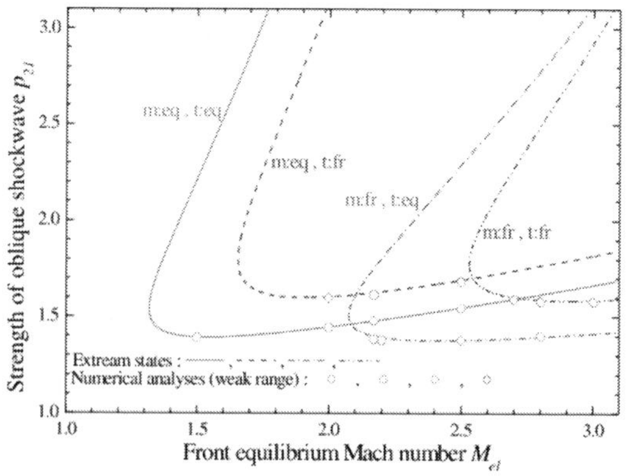

Figure 5. Strength of oblique shock wave versus front Mach number for extreme states.

4.2. Strong oblique shock waves

4.2.1. Analysis method for strong oblique shock waves. Another solution for oblique shock waves are strong oblique shock waves, as is the case for normal shock waves, with distinctive large changes in the physical quantities in the front and back of the shock wave and within its interior. Phenomena such as viscous stress and heat transfer occur in a non-equilibrium state. In addition, an increase in the changes in the physical quantities means that the flow changes from supersonic to subsonic. The calculation domain created to predict the strong oblique shock waves on a semi-infinite plane is shown in figure 6. For the calculation domain, we defined the number of divisions in the direction horizontal to the inclined wall as 120 and the divisions in the direction perpendicular to the inclined wall as 100. Moreover, in this domain, we defined a typical length, L, for the height of the inlet section and a typical velocity, u, for the gas-flow rate in the inlet section. The boundary and inlet conditions are defined similarly for the weak oblique shock waves.

The flow at the back of the shock wave, in case of strong oblique shock waves, is subsonic. In general, within the supersonic flow, because the flow exceeds maximum propagation velocity, the effects on the downstream cannot be propagated forward. However, within the subsonic flow, the effects on the downstream can be propagated forward. Therefore, not only the inflow conditions were given but also the initial distribution as well as the physical quantities derived from the relational expression for the oblique shock wave in extreme states are given for the back of the shock wave shown in the gray region in figure 6. The non-dimensional relaxation times used in the analysis of the strong oblique shock wave in the extreme states were assumed to be the same as those shown in table 1.

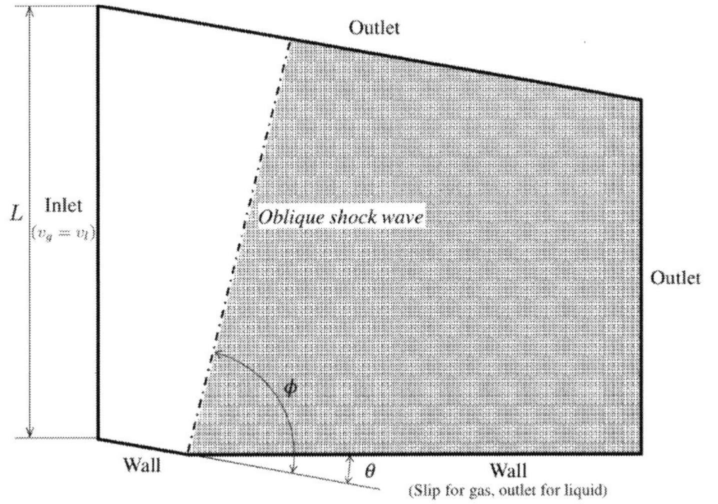

Figure 6. Calculation domain for strong oblique shock wave.

4.2.2. Establishing the initial distribution for the calculation domain. The physical quantities that are important in the initial distribution for the calculation domain are the pressure ratio, p_{21}, for the front and back of the shock wave and the shock-wave angle, ϕ. Both values are computed from the previously mentioned shock-wave relational expression. However, if the analysis is performed using these two values, an extremely unstable solution is obtained for given arbitrary values. The analysis results are given below. First, figure 7 shows the results for the case in which the pressure ratio is large for initial conditions, rather than just the required pressure ratio, to sustain an oblique shock wave at a given shock-wave angle. In figure 7, we can observe that the fundamental oblique shock wave moves in front of the inclined wall and is generated with an angle approximately equal to the normal shock wave. On the other hand, figure 8, which is the opposite case of figure 7, shows the results for a small pressure ratio. In figure 8, from the turning starting point of the inclined wall, a weak oblique shock wave is generated in the backward direction with a small shock-wave angle. Then, at some arbitrary position, a shock wave that is close to a normal shock wave is generated from the inclined wall, and the oblique shock waves that are weak in the forward direction overlap to generate a comparatively strong oblique shock wave. This configuration conforms to the λ-shaped shock wave in gas dynamics.

If the time-step progresses further, the results in figure 7 proceed forward, and those in figure 8 proceed backward. Because the domain for the strong oblique shock wave is an extremely unstable region, these waves are sensitive to very slight differences in the initial values and tend to migrate toward a stable solution.

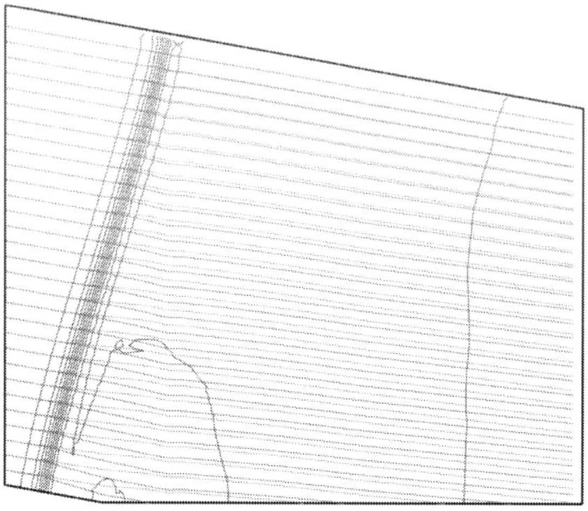

Figure 7. Case of large pressure ratio.

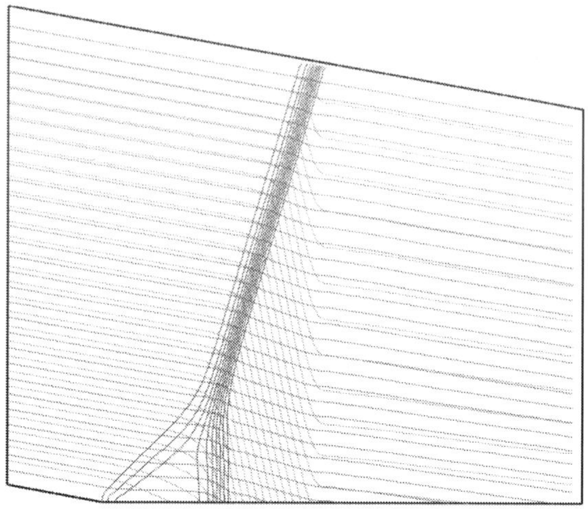

Figure 8. Case of small pressure ratio.

4.2.3. Characteristics of strong oblique shock waves. The numerical-analysis results for strong oblique shock waves are shown in figure 9. In figure 9, extreme states are shown for the total equilibrium state with $(M_{e1} = 1.5)$ in (a), the thermal frozen state with $(M_{e1} = 1.7)$ in (b), and the total frozen state with $(M_{e1} = 2.6)$ in (c). From figure 9, it can be observed that, even if the inflow Mach number for each of the extreme states is different, there is no considerable difference in the shock-wave angle for the generated oblique shock wave. By plotting the shock-wave angle obtained by this analysis as triangles on the theoretical curves, as shown in figure 4, we get figure 10. From figure 10, given that the numerical-analysis results are an approximate match with the theoretical curves for the oblique shock waves in extreme states, this demonstrates the possibility of predicting strong oblique shock waves by using this type of analysis.

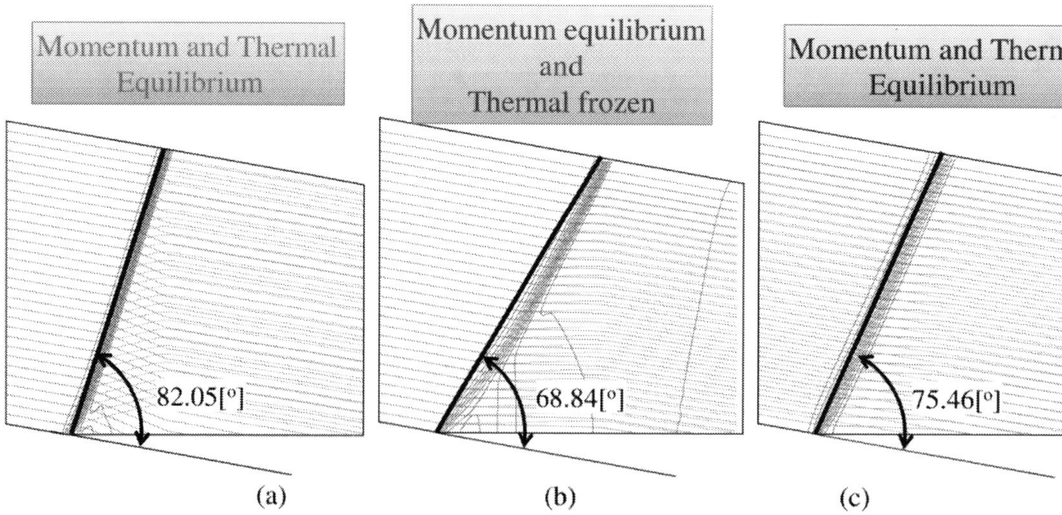

Figure 9. Strong oblique shock waves from numerical analysis.

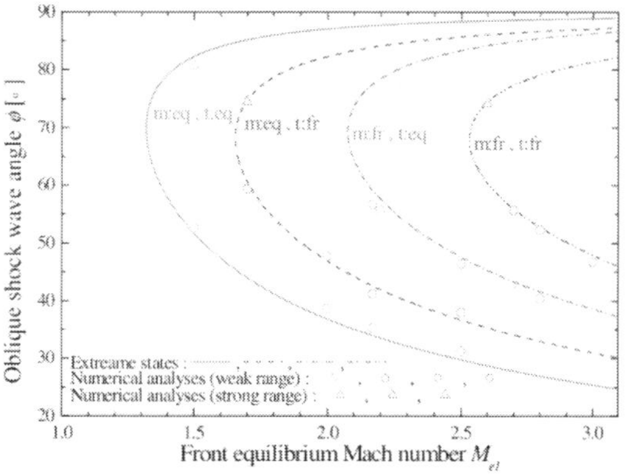

Figure 10. Oblique shock wave angles for the strong shock wave range.

4.3. Comparison with the oblique shock waves obtained by experiments

In past studies [5], the authors have experimentally measured the oblique shock waves generated in a convergent-divergent (Laval) nozzle using carbon dioxide gas as a refrigerant. The static-pressure distribution inside the nozzle in the experiments is shown in figure 11. For the measured oblique shock waves, when a third-order approximation was made for the pressure distribution around the points at which the pressure started rising and a cross-correlation was performed, the average value was found to be 67°. This result demonstrates that for this shock wave angle, which was shown theoretically for carbon dioxide gas, the experimentally generated oblique shock waves exhibit a relatively strong pressure ratio of the front and back of the shock wave. Accordingly, if we plot the angles for the strong oblique shock waves obtained by numerical analysis and the shock wave obtained from experiments on the shock-wave-angle-relationship diagram, we get figure 12. We can observe from figure 12 that the shock-wave angle for the oblique shock wave obtained from experiments is approximately equal to the strong oblique shock-wave region.

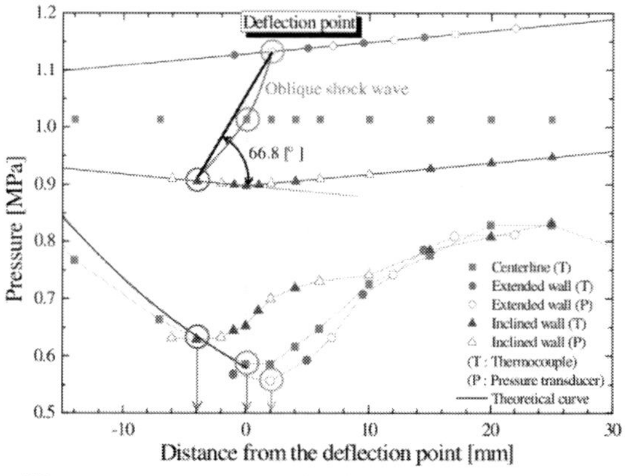

Figure 11. Pressure distribution at the nozzle outlet from the experiment.

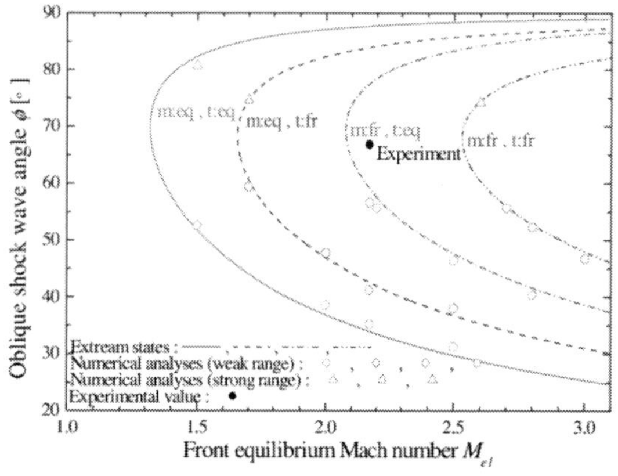

Figure 12. Comparison between experiment and analysis.

5. Conclusion

We analytically investigated the weak and strong oblique shock waves that are generated in the supersonic two-phase flow of carbon-dioxide gas and found the following wave characteristics:

- For each respective extreme state, a stable solution for weak oblique shock waves was found with outflow boundary conditions.

- The angles for the weak oblique shock waves varied considerably with differences in the transport processes, but the solution matches with the theoretical solution and demonstrates that this analysis was correct.
- Because the back of the strong oblique shock waves is subsonic, it is necessary to define the pressure at the back of the wave. For a small change in the pressure at the back of the wave, a solution was obtained in which either the oblique shock wave becomes a normal shock wave that progresses forward or becomes a wave associated with a λ-shaped weak oblique shock wave that retreats backward.
- Strong oblique shock waves are moderately unstable, but steady-state solutions were obtained with this analysis method for each respective extreme state.
- We revealed that the oblique shock waves obtained in the experiments within the two-phase flow nozzle for carbon dioxide are strong waves.

6. Reference

[1] L.D. Landau E.M. Lifshitz 1952 Fluid Mechanics Pergamon Press 422

[2] Nakagawa, M., Takeuchi, H. and Nakajima, K. 1998 Performance of Two Phase Ejector in Refrigeration Cycle *Transactions of the Japan Society of Mechanical Engineers* Series B Vol 64 No 625 pp 3060-3067

[3] Nakagawa, M. 2004 Refrigeration Cycle with Two-Phase Ejector *Refrigeration* Vol 79 No 925 pp 856-861

[4] Harada, A. and Nakagawa, M. 2010 Theoretical Analysis of the Two-Phase Oblique Shock Waves in an Ejector with Momentum and Temperature Relaxation *Transactions of the Japan Society of Refrigerating and Air Conditioning Engineers* Vol 27 No 3 pp 13-21

[5] Kawamura, Y. and Nakagawa, M. 2010 Investigation on Oblique Shock Waves Occurred in the Supersonic Carbon Dioxide Two-phase Flow *9th International Conference on Heat Transfer, Fluid Mechanics and Thermodynamics* pp 939-945

Optimizing Structure of LED Light Bulb for Heat Transfer

T. Kobayashi[*1], D. Itami[1], R. Hashimoto[2], T. Takashina[3], H. Kanematsu[4], K. Mizuta[5] and Y. Utsumi[6]

[1] Department of Electronics & Control Engineering, Tsuyama National College of Technology, Okayama, Japan
[2] Collaborative Research Center, Hiroshima University, Hiroshima, Japan
[3] International Industry Academia Collaboration Division, Hiroshima University, Hiroshima, Japan
[4] Department of Materials Science & Engineering, Suzuka National College of Technology, Mie, Japan
[5] Hiroshima R&D Center, Mitsubishi Heavy Industries, Ltd., Hiroshima, Japan
[6] Laboratory of Advanced Science and Technology for Industry, University of Hyogo, Hyogo, Japan

E-mail: t-koba@tsuyama-ct.ac.jp

Abstract. In this paper, in order to optimize the heat transfer structure of LED light bulb, the effects of various parameters on the temperature of the LED device were systematically analyzed, and a design guideline was shown. Although LED device has become popular due to its high-efficiency and long life, the design issues on the heat transfer structure of LED light bulbs has still remained. Because the original efficiency and life of the LED device can not be obtained due to the local temperature rise of LED element and the surrounding polymer molding material. Therefore, heat transfer analysis by finite element method was conducted systematically by changing parameters such as the shape, number and thickness of the radiating fin of the LED. As a result, advantage of open type structure was shown, and the proper design guidance for the structure of the fin shape was obtained.

1. Introduction

In recent years, LEDs have been used for traffic signals, lighting in road tunnel and lamp for home due to their long life and high efficiency, further the applications to the head lamp and tail lamp of automobile has been started as well [1]-[2]. However, the LEDs are point heat sources, thus the potential efficiency cannot be obtained, because the local temperature of the polymer molding materials [3], peripheral devices and LED element itself increases when LEDs are assembled to LED bulb, namely the heat transfer design issues have been preserved. For example, it is known that the luminous efficiency may decrease 5-8% when the temperature rise 10°C, life is halved for every for every 10°C rise in temperature, furthermore, the specific issues of LED have been reported, such as thermal degradation of the material of the feeding unit and the mounting part of LED element, and the increase of failure rate due to insulation failure and fatigue failure resulting from thermal stress [2], [4], [5].

It is considered, in general as well as electronic equipment, that the heat transfer problems in the structures can be qualitatively solved by reducing the ambient temperature and the increase in the thermal conductivity and the target radiation area. However, the compression of development period and cost is difficult by the conventional processes, such as prototyping and performance evaluation test, since there are many parameters. Therefore, it is considered to be effective to utilize Computer Aided Engineering (CAE), which has been frequently used in automotive industry, in the development of the LED lighting. However, there is little report on the thermal design of the LED lamp using the simulation. A group of the anthers has reported that the effect of heat dissipation area is greater than that of cross-sectional area of heat conduction as for the heat dissipation structure of LED bulb, and that an open structure will be effective to prevent the heat from, being trapped inside the bulb [6].

In the present work, focusing mainly on the shape of heat radiation fin, the number of the LED elements, the thickness of parts, the presence or absence of the outer cylinder, a thermal analysis by finite element method was performed, and the guidelines for the design of the structure was obtained.

2. Analysis methods and conditions.

2.1. Analysis method

Eight parameters were allocated to 18 cases using Design of experiments (DOE), then three dimensional steady-state heat transfer analysis using the finite element method [7]. From the obtained temperature distribution, multiple regression analysis on the effects of the parameters on the maximum temperature was carried out; the influence of each parameter was quantified by creating a regression equation. First, as shown in Table1, more specifically, three levels of condition were set in each of the eight parameters, the presence or absence of the outer cylinder, the number of heat radiation fin, the height of the heat radiation fin, the thickness of the fin, the thickness of the upper plate, the material, the internal temperature of the outer cylinder, and the number of LED tip. Then, as shown in the table of condition in Fig.2, 18 conditions were allocated using DOE. Based on these cases, axially symmetrical part models for analysis were created using 3 dimensional CAD. An example of the entire shape model is shown in Fig.1. Fig.2 shows an example of the axially symmetrical part model.

2.2. Analysis conditions

The heat generation of the LED bulb was fixed to 6[W] as a whole, that is 2[W] per element when the number of LED elements are three, and 1[W] per element for six LED elements, the influence of the number of LED elements on the maximum temperature was investigate. Assuming a resin molding, the shape of the LED element has a disk shape with a height of 2mm and a diameter of 4mm. The diameter of the mounting substrate of LED element was fixed at 60mm. In addition, the ambient temperature was set at 20[℃], and the heat transfer coefficient was applied to 5[W/m^{2}K] to be used in calm condition of a typical convection. Further, assuming the heat build-up in the outer cylinder, the temperature in the outer cylinder was varied in the range from 20 to 45[℃]. As the material of the radiation fin, Aluminium alloy, which can be easily formed by die-casting or stamping, and has reasonable cost, was selected and compared with copper and stainless steel. Table 3 shows the material properties used. With respect to the heat dissipation structure, the number of fins, the height of fin, the thickness of fin, were varied in the practical range, and also examined the effect of the presence or absence of the outer cylinder, that is, open type and closed type.

Table1 Calculation condition for the analysis

Parameter	Level		
	1	2	3
with/without cylinder	with	without	
the number of fin	3	12	24
the height of fin [mm]	5	10	15
the thickness of fin [mm]	1	2	3
the thickness of substrate [mm]	1	2	3
material	Type 304	Al	Cu
inner temperature [℃]	20	30	45
the numberof LED	1	3	6

Table2 18L orthgonal table

No.	with/without cylinder		the number of fin	the height of fin[mm]	the thickness of fin[mm]	the thickness of substrate[mm]	material		inner temperature[℃]	the number of LED
1	with	1	3	10.5	1	1	Type 304	1	22	1
2	with	1	3	15.5	2	2	Al	2	30	3
3	with	1	3	20.5	3	3	Cu	3	45	6
4	with	1	12	10.5	1	2	Al	2	45	6
5	with	1	12	15.5	2	3	Cu	3	22	1
6	with	1	12	20.5	3	1	Type 304	1	30	3
7	with	1	24	10.5	1	1	Cu	3	30	6
8	with	1	24	15.5	2	2	Type 304	1	45	1
9	with	1	24	20.5	3	3	Al	2	22	3
10	with/without	0	3	10.5	1	3	Al	2	30	1
11	with/without	0	3	15.5	2	1	Cu	3	45	3
12	with/without	0	3	20.5	3	2	Type 304	1	22	6
13	with/without	0	12	10.5	1	3	Type 304	1	45	3
14	with/without	0	12	15.5	2	1	Al	2	22	6
15	with/without	0	12	20.5	3	2	Cu	3	30	1
16	with/without	0	24	10.5	1	2	Cu	3	22	3
17	with/without	0	24	15.5	2	3	Type 304	1	30	6
18	with/without	0	24	20.5	3	1	Al	2	45	1

Table3 Property of materials

material	Type 304	Cu	Al
density[kg/m^3]	7930	8880	2693
efficiency of heat transfer[W/℃]	16.7	398	235
spescific heat[J/kg・℃]	590	380	860

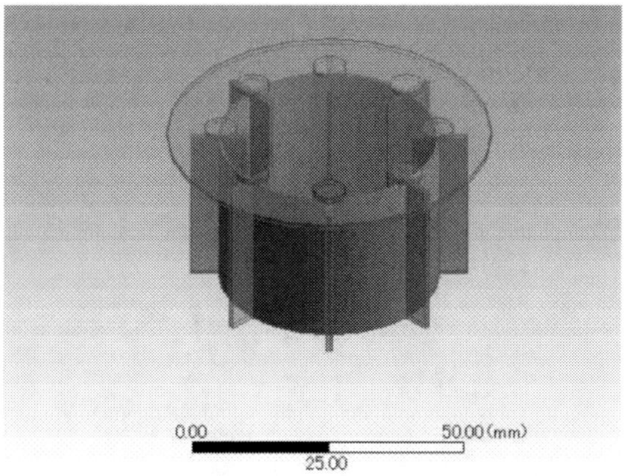

Fig.1 Basic model to use for analysis

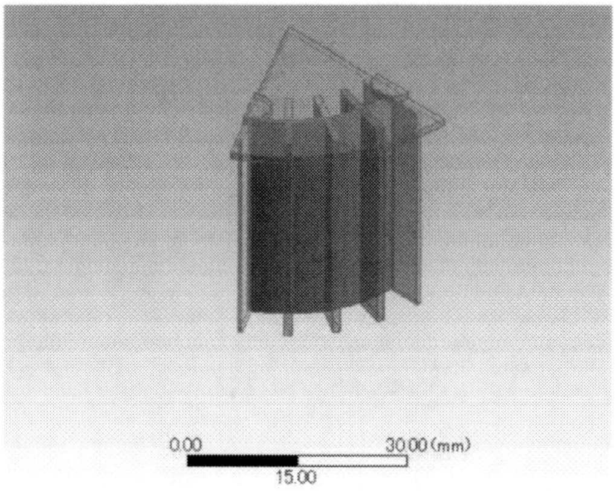

Fig.2 Model example after parameter change

3. Analysis of results and Discussion

The 18 conditions allocated into orthogonal table were analyzed, and then multiple regression analysis on the effects of the parameters on the maximum temperature was conducted using the temperature distribution obtained.
The obtained regression equation is shown below.

T_{max}=190.08+3.27Qa-1.95Qb-2.78Qc+0Qd-9.18Qe-15.39Qf+0.617Qg-6.17Qh

where, T_{max} : maximum temperature, Q_a : with or without cylinder, Q_b : the number of fin, Q_c : the height of fin, Q_d : the thickness of fin, Q_e : the thickness of substrate, Q_f : material, Q_g : inner temperature, Q_h : the number of LED chips .

In Fig3, impact factor of each parameter on T_{max} in the range of present condition, namely the value multiplied the coefficient of the above equation Q_a by the range of each parameter (the range between level 1 to level 3) Δx_i , is shown.

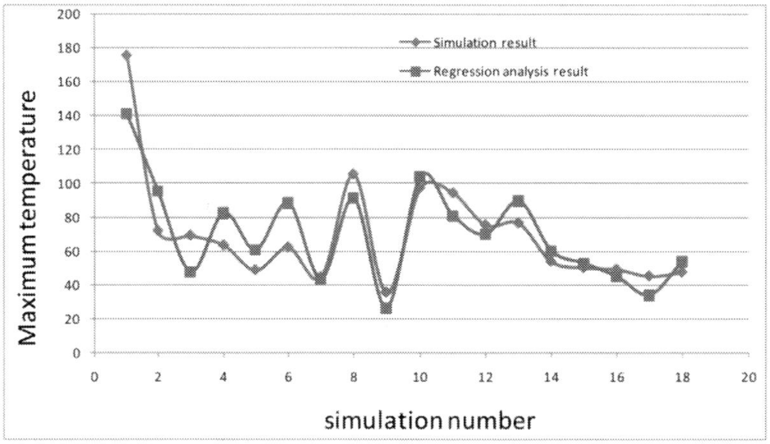

Fig.3 Comparison between equation of regression and result of the finite element method

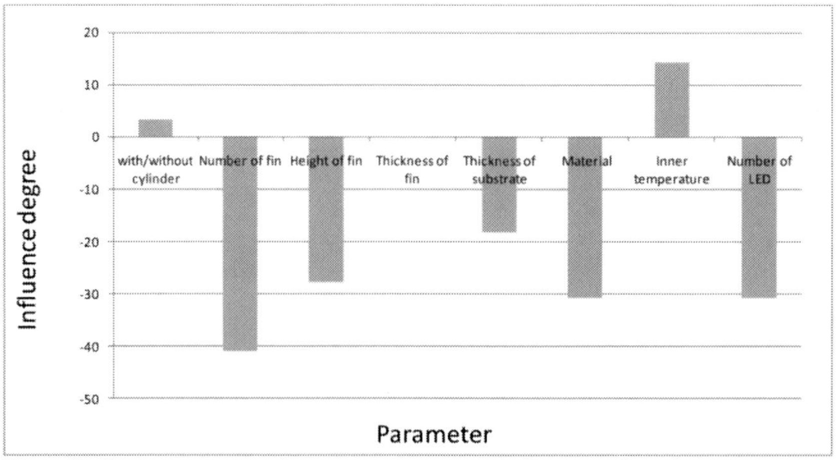

Fig.4 Influence on maximum temperature degree of each parameter

The impact factor on the maximum temperature T_{max} is greater in the order of the number of fin, the number of LED chips, the material of fin, the height of fin and the thickness of the substrate, that is the order of cooling effect.

Using the regression equation, the best model showing the lowest T_{max}, the worst model showing the highest T_{max} and the reasonable model considering manufacturing cost and ease of maintenance explained later, were selected.

The parameters used are shown in Table4 and the created models are shown in Fig.5, Fig.6 and Fig.7.

Regarding the best model, in which the material of fin is copper with the highest heat conductivity, and six LED chips are placed on the substrate, and the 24 fins have the height of 15mm. Further the structure has no cylinder, that is the open type which prevents the increase of inner temperature.

The calculated result by FEM showed a maximum temperature T_{max} of 38℃. The worst model shown in Fig.6 has smaller area of fin, and has only on LED chip placed at the center of the substrate, made of austenitic steel with lower heat conductivity. The maximum temperature T_{max} in the case was 186℃.

Consequently, the difference between the best and worst models was found to be greater than 120℃

The above mentioned best model is appropriate only for cooling effect, in practical application, manufacturing method, production cost, ease of maintenance and design properties.

The reasonable model similar to the best model is shown in Fig.7.In this case the maximum temperature the is 43℃. The model has a little simpler fin structure and has the open type as well as the best model. It is suggested that the practicality of the model is high although the maximum temperature is a little higher than that of the best model.

Table 4 Parameter of each model

parameter	best	reasonable	worst
with/without cylinder	without	without	with
the number of fin	24	24	3
the height of fin[mm]	20.5	15.5	10.5
the thickness of fin[mm]	1	1	1
the thickness of substrate[mm]	3	3	1
material	Cu	al	sus304
inner temperature	22	22	45
the number of LED chips	6	3	1
maximum temperature[℃]	38	43	186
minimum temperature[℃]	37	40	51

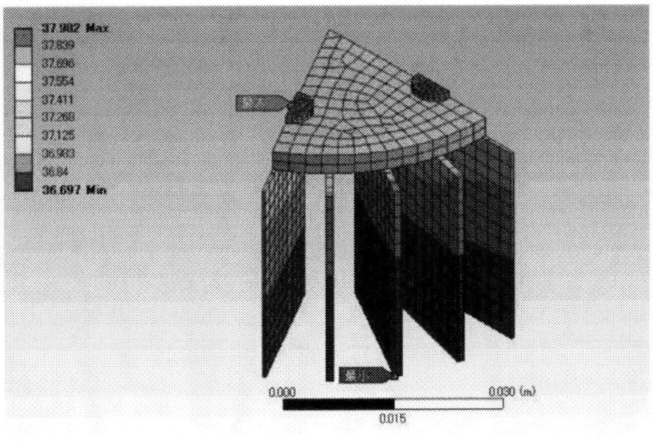

Fig.5 Best model

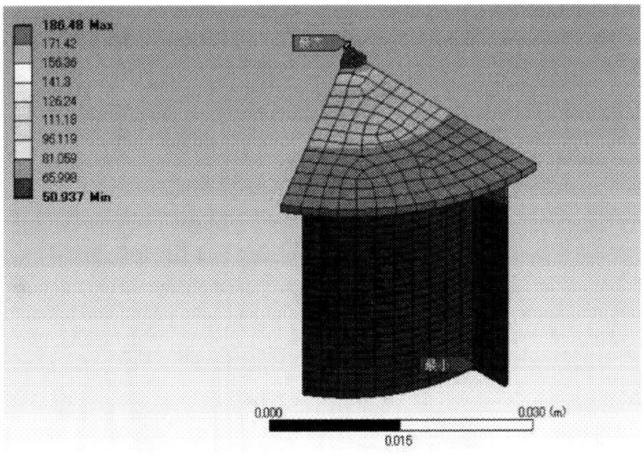

Fig.6 Worst model

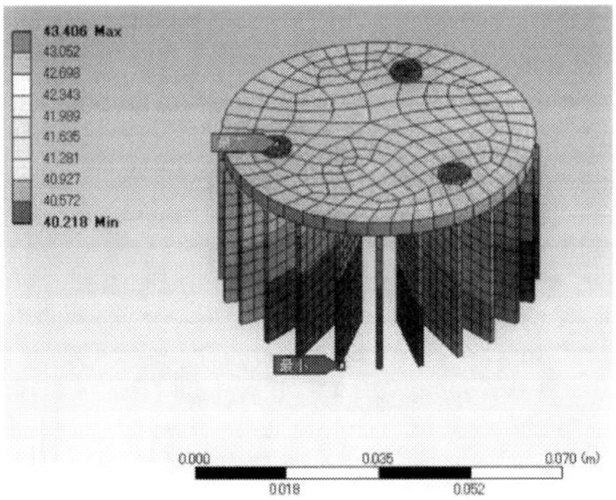

Fig.7 Reasonable model

4. Conclusion

With regard to a LED bulb with an diameter of 60mm and a power of 6 watt, the effect of the heat dissipation structure, the presence or absence of outer cylinder, the number of fins, and the number of LED tips, and so forth on the maximum temperature of the LED chips was analysed by the FEM method in order to minimize the maximum temperature which affect electric efficiency and life, and the following results were obtained.

(1) It was found that the impact factor on the maximum temperature of the LED chips is greater in the order of the number
(2) It was confirmed that the open type structure without outer cylinder is effective to prevent the increase of inner temperature.
(3) Within the studied condition, the worst structure showed a lowest max. temperature of 186℃, however the best model can decrease the max. temperature to 38℃.
(4) The calculated result with the FEM method was analyzed the regression analysis, and then the regression equation was formed to estimate the maximum temperature. Using the equation, without the change of CAD diagram and the FEM analysis

References

[1] Nozawa T 2008 *Nikkei Electronics* Nov.3rd pp 87-97
[2] Office of Energy Efficiency and Renewable Energy, U.S. Department of Energy, July 2011, *Solid-state Lighting Research and Development: Manufacturing Roadmap*
[3] Qin Y Y and Ron Hui S Y 2010 Comparative Study on the Structural Designs of LED Devices and Systems Based on the General Photo-Electro-Thermal Theory *IEEE Trans. Power Electronics* **25** pp507-513
[4] Chou H and Yang T 2007 Development of Junction Temperature and Driving Current upon Emission Spectra of LEDs *Proc. 1st Int. Conf. White LEDs and Solid State Lighting* pp202-
[5] Osawa S, Izumi M and Sakamoto S 2010 *Toshiba Technical Review* **65** 7 pp 8-11
[6] Kobayashi T, Sakate Y, Hashimoto R, Takashina T, Kanematsu H, Mizuta K and Utsumi Y 2012 Research on optimization of cooling structure of LED element *The-Asia-Pacific Interdisciplinary Research Conference* Irago,Japan
[7] Cybernet systems co., ltd., 2010, ANSYS, Retrieved from http://www.cybernet.co.jp/ansys/

(Na, Bi)TiO_3 based lead-free ferroelectric thin films on Si substrate for pyroelectric infrared sensors

D Akai[1], R Yoshita[2] and M Ishida[1,2]

[1] Electronics-Inspired Interdisciplinary Research Institute (EIIRIS), Toyohashi University of Technology, 1-1 Hibarigaoka, Tempaku-cho, Toyohashi 441-8580, Japan
[2] Department of Electrical and Electronic Information Engineering, Toyohashi University of Technology, 1-1 Hibarigaoka, Tempaku-cho, Toyohashi 441-8580, Japan

E-mail: akai@vbl.tut.ac.jp

Abstract. In this study, we report ferroelectric and pyroelectric properties of $(Na_{0.5}Bi_{0.5})TiO_3$-$BaTiO_3$ (NBT-BT) thin films on Si substrates using chemical solution deposition for the first time. The NBT-BT thin films deposited on Pt/Ti/SiO_2/Si substrates have exhibited a typical hysteresis loop with remnant polarization of 5 $\mu C/cm^2$ and coercive field of 80 kV/cm. Furthermore NBT-BT films showed pyroelectricity with pyroelectric coefficient of 0.6×10^{-8} C/cm^2K. Monolithic-integration of Si electronics and lead-free ferroelectric NBT thin films has been archived using SiN passivation layer. It was previously believed that LSI processes could not incorporate any sodium-containing material which would cause characteristic degradation, such as threshold voltage shift. In this work, no threshold voltage shift in MOS characteristics was observed using this SiN layer. The SiN layer not only blocked diffusion from NBT chemistry, but also from crystallized NBT films during NBT formation process.

1. Introduction

In recent years, ferroelectric thin films have become important because of their numerous useful and attractive applications, such as ferroelectric random access memories (FeRAMs), micro actuators, piezoelectric micromachined ultrasonic transducers and pyroelectric infrared (IR) sensors. To realize such devices, thin film fabrication of ferroelectric materials and the integration of ferroelectric thin films on Si substrates are required. Previously, we reported highly sensitive pyroelectric IR image sensor arrays using epitaxial $Pb(Zr,Ti)O_3$ (PZT) thin films deposited on epitaxial γ-Al_2O_3(100)/Si(100) substrates [1]. PZT thin films are, today, widely used due to their good ferroelectric properties and high compatibility with LSI processes. However, lead-based ferroelectric materials, such as PZT, are not environmentally friendly because of their lead toxicity. Therefore lead-free ferroelectric materials with excellent properties are needed. $Na_{0.5}Bi_{0.5}TiO_3$ (NBT) and NBT based solid systems, such as $Na_{0.5}Bi_{0.5}TiO_3$–$BaTiO_3$ (NBT-BT) [2,3] and $Na_{0.5}Bi_{0.5}TiO_3$-$K_{0.5}Bi_{0.5}TiO_3$ (NBT-KBT) [4-6] , are promising lead-free materials due to their excellent ferroelectric properties. NBT has a perovskite-type structure, as typified by PZT, and several studies have reported a relatively large remnant polarization (Pr) of 38 $\mu C/cm^2$ at room temperature, high curie temperature (Tc) of 320 °C [7] and pyroelectric coefficient (p) of 2.5×10^{-8} C/cm^2K [8]. However, it is difficult to introduce NBT into an

LSI process because NBT contains sodium which causes degraded characteristics of Si devices, such as threshold voltage shifts.

Usually, Si circuitry is passivated by a SiO_2 film. The Diffusion coefficient of sodium in SiO_2 is about 10^{-6} cm^2/s. This is an extremely higher value than for other impurities in Si devices, such as Arsenic (As), Antimony (Sb) and Boron (B). According to rough calculations, sodium could pass through SiO_2 films about 10 μm in thickness. Therefore sodium can easily reach circuitry, passing through the SiO_2 passivation layer during thermal processes such as NBT deposition. Several studies have reported that SiN films were able to block diffusion of alkali ions including sodium. [9]

In this study, first NBT based thin films were deposited on the SiO_2/Si substrates by chemical solution deposition (CSD), and crystallographic orientations, ferroelectric properties and pyroelectric properties were investigated. Then, we investigated the feasibility of employing a SiN passivation layer in the integration processes of NBT films with Si electronics to create Si based lead-free ferroelectric devices. SiN is a conventional passivation material in LSI processes to block mobile ions, such as sodium and potassium.

2. Experimental

2.1. NBT-based ferroelectric thin film

50-nm-thick SiO_2 films were formed on p-Si(100) substrates by thermal oxidation. Pt/Ti stacked layer were deposited by sputtering on the SiO_2/Si substrate for bottom electrodes. NBT-BT films were prepared as NBT-based thin films. The precursor solution was prepared using sodium nitrate, bismuth nitrate, titanium isopropoxide, and barium acetate as starting materials.

The preparation procedure of precursor solution for NBT-BT is schematically described in figure.1. NBT-BT precursor solution with a nominal composition of 94 mol% of NBT and 6 mol% of BT was synthesized.

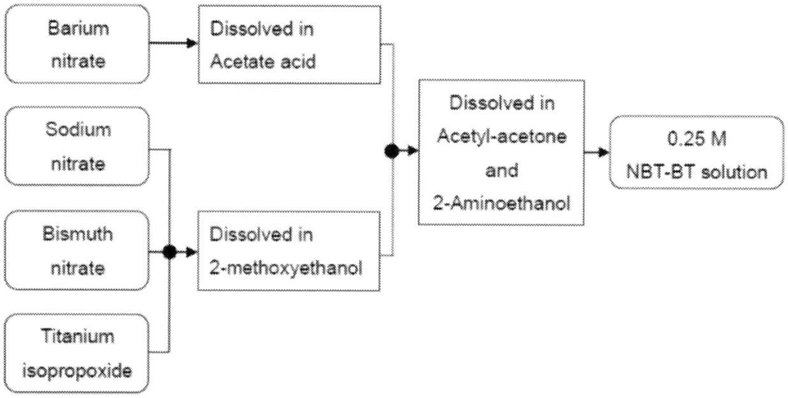

Figure 1. Preparation flow of the NBT-BT precursor solution.

The precursor solution was spin-coated on Pt/Ti/SiO_2/Si substrates at 3500 rpm for 30 s. The drying condition was 150 °C for 5 min, and the pre-fired condition was 450 °C for 5 min in air with a hotplate. The final annealing condition was 650 °C for 5 min in oxygen atmosphere with rapid thermal annealing system. The coating process was repeated three times to obtain the about 120 nm thick NBT-BT films.

The crystallographic orientations of NBT-BT films were analyzed by XRD. The ferroelectric properties were measured using a standardized ferroelectric test system (Toyo Corp. FCE) with Pt/NBT-BT/Pt capacitors by making circular Pt top electrodes 500 μm in diameter using RF magnetron sputtering through a shadow mask. The pyroelectric properties were measured using a Keithley 6517 electrometer and a fine temperature control system (Toyo Corp. FCEOPT/PY1).

2.2. Sodium blocking test

To confirm the sodium ion blocking by SiN films, a simple MOS capacitor has been used. Figure 2 shows the sample preparation processes. First, 50-nm-thick SiO_2 films were formed on p-Si(100) substrates by thermal oxidation to act as oxide of an MOS capacitor. SiN films were deposited on the oxidized substrate by plasma enhanced chemical vapor deposition (PECVD) as the alkali ion blocking layer. Thicknesses of the SiN film were 100 nm or 200 nm. 150-nm-thick NBT films were coated on the SiN layer by a same method described above. After crystallization of the NBT films, the NBT and SiN films were removed by BHF solution and hot phosphoric acid, respectively. An indium electrode was formed on the back of the substrates. Capacitance-Voltage (C-V) characteristics of the MOS capacitors were measured at a frequency of 1 kHz using mercury probe system (CVmap92A, Four Dimensions, Inc) and an impedance analyser (4285A, Agilent).

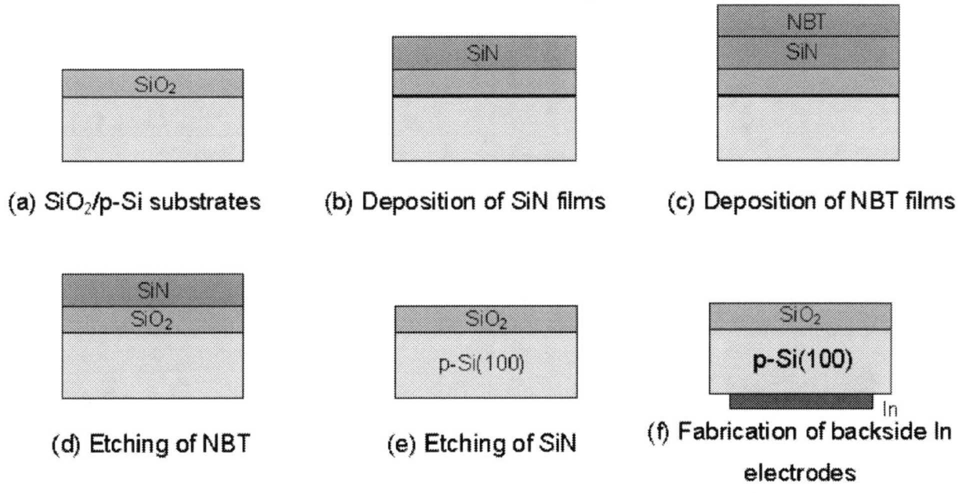

Figure 2. Fabrication processes of MOS capacitor for sodium blocking test.

2.3. Integration of pyroelectric infrared sensor using NBT based thin films and Si electronics

An integrated pyroelectric sensor, comprising NBT thin films, N-MOSFETs and N-Junction FETs (N-JFETs), was fabricated on an epitaxial γ-Al_2O_3 (100)/Si(100) substrate. The fabrication process flow is described below. (a) Epitaxial γ-Al_2O_3 films, 50 nm in thickness, were deposited on Si(100) substrates by metal organic CVD [10]. A partial area of γ-Al_2O_3 films was etched to create an LSI fabrication area. (b) N-MOSFETs and N-JFETs were fabricated by standard LSI process on the LSI fabrication area. (c) Passivation SiN/SiO_2 layers were deposited by PECVD. The passivation layer on the Al_2O_3 film was removed to form a sensor structure. (d) The Pt/Ti/TiN layer was formed by RF sputtering. 150-nm-thick NBT films were coated by CSD described above. 100-nm-thick SRO films were deposited by RF sputtering. (e) The SRO/NBT/Pt/Ti/TiN structures were patterned by a conventional photolithography technique and dry etching processes. (f) A 500-nm-thick SiN film as an inter-layer dielectric was deposited by PECVD and the Al-Si wiring was formed by DC sputtering. V_{GS}-I_D characteristics of the N-MOSFETs and the N-JFETs were measured with a semiconductor parameter analyser (4155B, Agilent).

3. Results and Discussion

3.1. NBT-based ferroelectric thin film

Figure.3 shows the XRD pattern of NBT-BT thin films deposited on $Pt/Ti/SiO_2/Si(100)$ substrates with various annealing temperature. Only the peaks from NBT-BT were observed at annealing temperature of 650 °C.

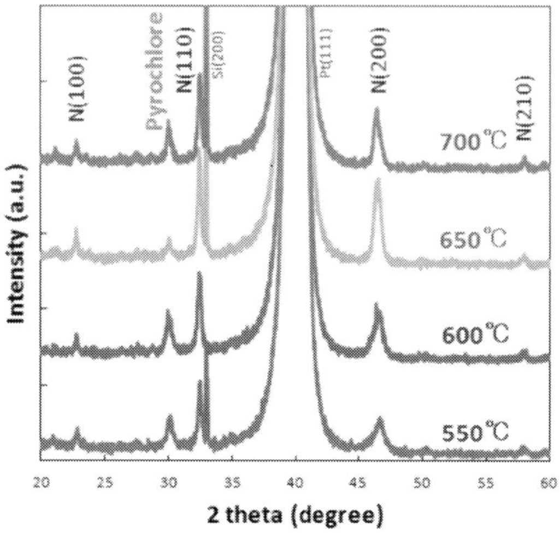

Figure 3. XRD patterns of NBT-BT films on $Pt/Ti/SiO_2/Si$ substrate annealing at various temperature.

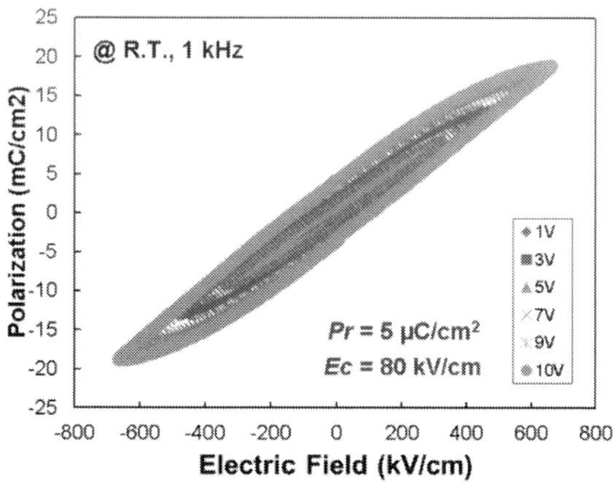

Figure 4. Typical ferroelectric hysteresis loop of NBT-BT films on $Pt/Ti/SiO_2/Si$ substrate with various sweeping voltage

Figure.4 shows ferroelectric properties measured at a frequency of 1 kHz for the Pt/NBT-BT/Pt capacitor structure. The obtained NBT films on Pt/Ti bottom electrode showed a hysteresis loop with remnant polarization (*Pr*) of 5 $\mu C/cm^2$ and coercive filed (*Ec*) of 80 kV/cm at sweeping voltage of ±10 V. These values are in good agreement with the reported value deposited on $Pt/Ti/SiO_2/Si$ substrates by sol-gel method [11].

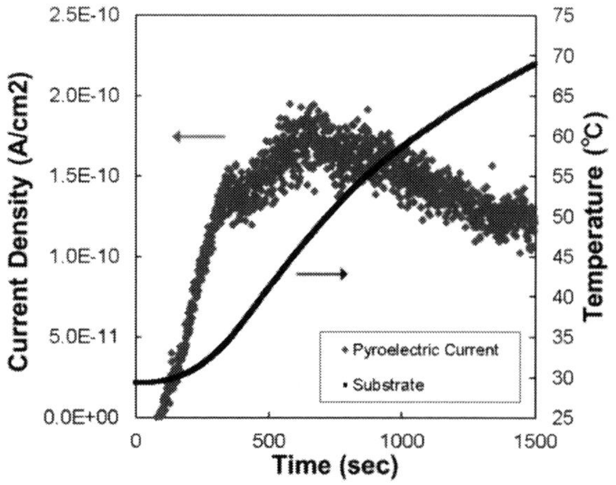

Figure 5. Pyroelectric current of the NBT-BT thin film.

Figure.5 shows the pyroelectric current of NBT films. A pyroelectric coefficient was defined as the following Eq. (1).

$$I_p = pA\frac{dT}{dt} \quad (1)$$

where *Ip* is the pyroelectric current, *p* is the pyroelectric coefficient, *A* is the area of the top electrode and d*T*/d*t* is the temperature gradient of NBT-BT films. A pyroelectric coefficient was calculated using the Eq. (1). The average pyroelectric coefficient of NBT-BT films is 0.6×10^{-8} C/cm^{2}K. This is the first time report of pyroelecricity of the NBT-BT thin film on Si substrates.

3.2. Sodium Blocking
Figures 6 (a) and (b) show the CV characteristics of MOS capacitors using various thicknesses of SiN passivation layer and as-formed MOS capacitors (SiN films and NBT films were not deposited). Without a SiN passivation layer, CV characteristics exhibited an anomalous capacitance value and hysteresis, as indicated by arrows shown in Fig. 6(a). It is suggested that mobile ions contaminate the SiO_2 film during NBT film deposition. On the other hand, Fig. 6(b) shows that no hysteresis is observed in MOS capacitors using a SiN passivation layer. Compared with the as-formed MOS capacitor, small threshold voltage shifts of 0.1 V in the positive direction were observed in an NBT formed MOS capacitor. This is not diffusion of sodium ions. If positive sodium ions exist in the SiO_2, the flat band voltage shifts in the negative direction. These results indicate sodium diffusion during deposition of NBT films could be prevented by the SiN passivation layer.

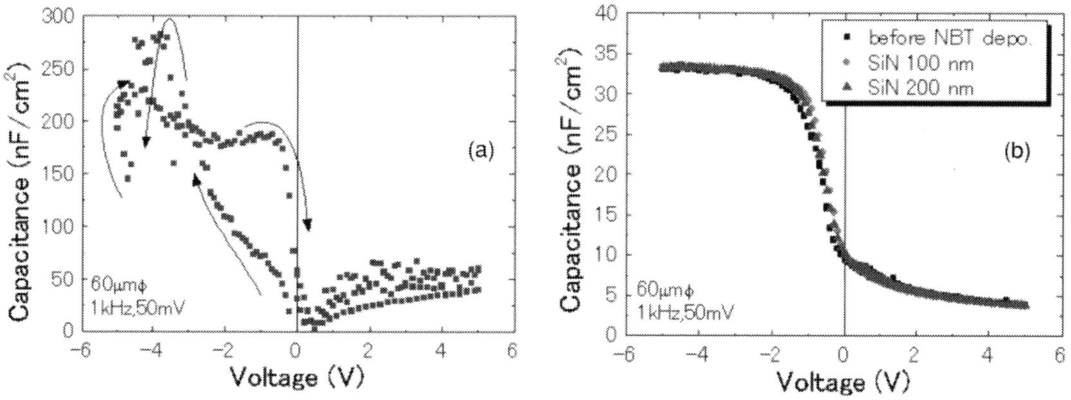

Figure 6. C-V characteristics of MOS capacitors after NBT thin film deposition. (a) without a SiN layer and (b) with a SiN layer under the NBT film.

3.3. Integration of pyroelectric infrared sensor using NBT based thin films and Si electronics
The V_{GS}-I_D characteristics of the circuitry on the fabricated device are shown in Fig. 7. Both the N-MOSFET and the N-JFET operated properly, without threshold voltage shifts. This result also shows that the SiN passivation layer blocked sodium diffusion during the NBT deposition and patterning processes.

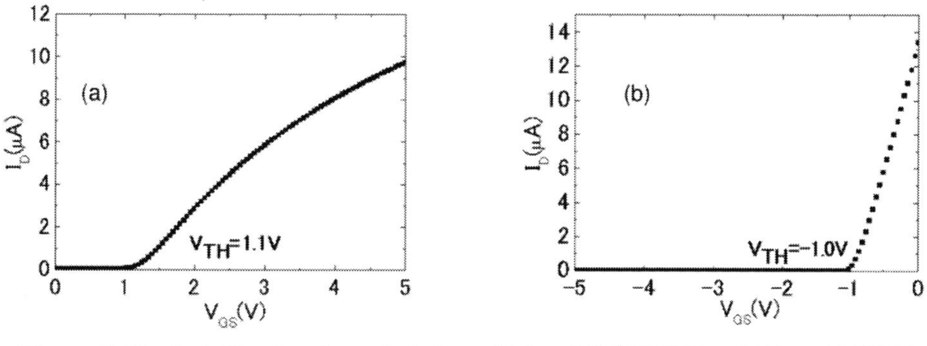

Figure 7. Typical V_{GS}-I_{DS} characteristics of (a) an NMOSFET and (b) an N-JFET. Both FETs operated well.

4. Conclusion

First, NBT-BT thin films have been deposited onto Pt/Ti/SiO_2/Si substrates by CSD. NBT-BT films showed a typical hysteresis loop with remnant polarization of 5 $\mu C/cm^2$ and a coercive field of 80 kV/cm. Pyroelectricity was obtained with pyroelectric coefficient of 0.6×10^{-8} C/cm^2K

Then, we demonstrated that a SiN film could be used as a passivation layer between Si electronics and an NBT layer to prevent diffusion of sodium ions or atoms into the Si electronics region. No threshold voltage shift in MOS C-V characteristics was observed when using a SiN passivation layer. The SiN layer not only blocked diffusion from the NBT solution, but also from crystallized NBT films during the NBT formation process. Also we have achieved monolithic integration of MOSFETs and lead-free ferroelectric material, which includes potassium, on Si substrates. These results can be applied to future integrated sensors using lead-free ferroelectric materials.

References

[1] Akai D, Kawazu N, Sugai T, Sawada K and Ishida M 2009 *Proceedings of SPIE* **7419** 741905
[2] Takenaka T, Maruyama K and Sakata K 1991 *Jpn. J. Appl. Phys.* **30** 2236.
[3] Chu B J, Chen D R, Li G R and Yin Q R 2002 *J. Eur. Ceram. Soc.* **22** 2115
[4] Sasaki A, Chiba T, Mamiya Y and Otsuki E 1999 *Jpn. J. Appl. Phys.* **38** 5564.
[5] Wang T B, Wang L E, Lu Y K and Zhou D P 1998 *J. Chin. Ceram.* **14** 14.
[6] Li Y, Chen W, Zhou J, Xu Q, Sun H, Liao M 2005 *Ceram. Int.* **31** 139
[7] Smolensky G A, Isupov V A, Agranovskaya A I and Krainik N N 1961 *Soc. Phys. Solid State* **2** 2651
[8] Hagiyev M S, Ismaizade L H and Abiyev A K 1984 *Ferroelectrics* **56** 215
[9] Sze S M 1981 *Physics of semiconductor devices 2nd ed* (New York: Wiley)
[10] Ishida M, Katakabe I and Nakamura T 1988 *Appl. Phys. Lett.* **52** 1326
[11] Yu T, Kwok K W and Chan H L W 2007 *Thin Solid Films* **515** 3563

Two-phase Flow Ejector as Water Refrigerant by Using Waste Heat

H Yamanaka[1], M Nakagawa[1]

[1]Department of Mechanical Engineering, Toyohashi University of Technology, 1-1 Hibarigaoka Tenpaku-tyo, Toyohashi-city, Aichi, 441-8122, Japan

E-mail: yamanaka@nak.me.tut.ac.jp

Abstract. Energy saving and the use of clean energy sources have recently become significant issues. It is expected that clean energy sources such as solar panels and fuel cells will be installed in many private dwellings. However, when electrical power is generated, exhaust heat is simultaneously produced. Especially for the summer season, the development of refrigeration systems that can use this waste heat is highly desirable. One approach is an ejector that can reduce the mechanical compression work required in a normal refrigeration cycle.

We focus on the use of water as a refrigerant, since this can be safely implemented in private dwellings. Although the energy conversion efficiency is low, it is promising because it can use heat that would otherwise be discarded. However, a steam ejector refrigeration cycle requires a large amount of energy to change saturated water into vapour. Thus, we propose a more efficient two-phase flow ejector cycle.

Experiments were carried out in which the quality of the two-phase flow from a tank was varied, and the efficiency of the ejector and nozzle was determined. The results show that a vacuum state can be achieved and suction exerted with a two-phase flow state at the ejector nozzle inlet.

1. Introduction

Widespread interest in environmental preservation has grown in recent years. At the same time, concern has also grown about fossil fuel depletion due to rising energy demand, and energy conservation has become a major issue worldwide. In the wake of the Great East Japan Earthquake, moreover, strong doubts have arisen particularly in Japan in regard to the continued use of nuclear energy. For all of these reasons, efforts to develop and use new energy sources and natural energy are expected to expand.

Solar power and solid oxide and other types of fuel cells are expected to proliferate for household use, as new sources of energy. At present, however, their power generation is accompanied by waste heat production, and no related means of efficient energy conversion has as yet been achieved. It is important to recover and use the energy of waste heat, which is currently simply discarded, and thereby increase the total efficiency of household power systems. In winter, this waste heat can be used for home and water heating, but little progress has been made toward its summertime utilization even though its use in air conditioners, which are the largest power consumers in the summer season, would bring a substantial increase in total system efficiency.

In one proposed concept, an ejector[1-2] would be mounted on the air conditioner and utilize the low-temperature waste heat to carry out compression, in place of a dedicated compressor. Water is a safe, shared resource with a small environmental burden, and would represent an ideal refrigerant for home air conditioners.

In existing air conditioners, the refrigerant is compressed by an electrically powered compressor. Ejector systems, which convert heat energy to kinetic energy, could further convert this to compression energy and thus take over the burden of compression work in an air conditioner. Some progress has been made in research on ejector systems[3-5], but the energy conversion efficiency with water as a refrigerant is generally regarded as unacceptably low. Even with a low efficiency, however, its effective utilization of heat that is otherwise simply discarded would constitute a considerable improvement in the overall efficiency of electric power systems.

One way to increase the performance of such an ejector air conditioning system would be to use supersonic two-phase flow in place of vapour. To drive the vapour cycle requires only the addition of heat energy equal to the latent heat, in order to produce vapour. Figure 1 shows the refrigeration cycle of a water refrigerant ejector in a Mollier (p-h) diagram, which represents the conversion of the heat energy of high-temperature high-pressure water to kinetic energy and its return to a low-temperature low-pressure state. If the refrigeration cycle in the two-phase flow state can be accomplished, the required heat quantity for the cycle can be reduced and, in comparison with the vapour ejector cycle, it can serve as an air conditioner with very high efficiency.

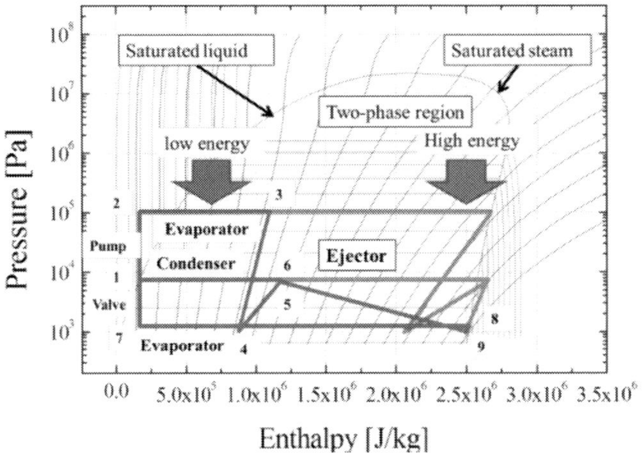

Figure 1. P-h diagram of water ejector refrigeration cycle.

2. Objective

Various studies have been performed on the vapour ejector air conditioning cycle with water refrigerants, but the quantity of energy required for a phase change of water from liquid to vapour has been inordinately large. If a two-phase flow ejector cycle can be used instead of a vapour ejector cycle, however, the cycle itself would accomplish the energy saving.

The authors have performed studies on nozzles as a form of basic research on water refrigerant two-phase flow ejectors[6], and have shown that the energy conversion efficiency of the nozzle plays an extremely important role in ejectors using water as a refrigerant. In the present study, the objectives were to verify the achievement of a vacuum state at the outlet of a two-phase flow nozzle, to accomplish input of water refrigerant to the ejector in the two-phase flow state, and to investigate the ejector pressure rise characteristics and suction capability under these circumstances.

3. Theory

3.1. Nomenclature

w	[W/kg]	Specific work needed for compression
h	[J/kg]	Specific enthalpy
s	[J/(kgK)]	Entropy
q	[W/kg]	Specific heat
p	[Pa]	Pressure
u	[m/s]	Velocity
A	[m2]	Area
G	[kg/s]	Mass flow rate
z	[m]	Coordinate along flow from the throat
x	[-]	Quality
g	[-]	Mass flow ratio
COP	[-]	Coefficient of performance
Special characters		
ρ	[kg/m^3]	Density
Subscripts		
s		Saturated state (for pressure)
d		Driving flow
s		Suction flow
c		Critical condition
ex		Experimental result
$mainth$		Isentropic Homogeneous Equilibrium model

3.2. Coefficient of performance

Figure 2 is a schematic of the air conditioning cycle of the water refrigerant ejector investigated in this study. The change of state in pressure and enthalpy is depicted in the Molliere diagram in Figure 1. The numbers shown in this cycle correspond to those shown in the Molliere diagram. The cycle comprises two component cycles. One is the conventional Rankin cycle, moving in turn from the condenser to the pump to the high-temperature evaporator and back to the condenser via the ejector. The other is the vapour-compression refrigeration cycle, moving from the condenser to the expansion valve, the low-temperature evaporator, the ejector, and back to the condenser. In this scheme, the ejector serves the role of a heat engine turbine in the former component cycle and an air conditioner compressor in the latter component cycle. The high-temperature evaporator is heated by heat that would otherwise be discarded as waste heat, and the low-temperature evaporator acts to cool the designated space on the low-temperature side in the refrigeration cycle.

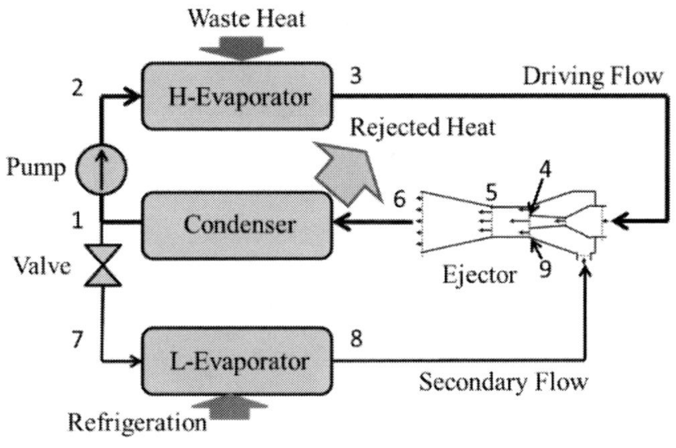

Figure 2. Ejector refrigeration system in this study.

To assess the performance of this water refrigerant ejector refrigeration cycle, the coefficient of performance (COP) was calculated with the following assumptions.

Friction loss in the various components of the system was disregarded and the pump, evaporators, and condensers were thus regarded as ideal devices. With subscripts 1 and 2 referring to the states at the pump inlet and outlet, respectively, the pump work w_{21} per unit mass flow is obtained from the change in enthalpy h as

$$w_{21} = h_2 - h_1, \quad s_2 = s_1 \tag{1}$$

where s represents entropy.

With the outlet on the high-temperature side designated as 3, and the evaporator inlet as 2, the heat energy absorption quantity q_{32} is similarly expressed in terms of an increase in enthalpy as

$$q_{32} = h_3 - h_2, \quad p_3 = p_2 \tag{2}$$

Ordinarily, the pressure p_2 at the high-temperature evaporator inlet is taken as being lower than the waste heat saturated vapour pressure $p_s(100^\circ\mathrm{C})$, but here we assume an ideal heat exchanger, so that $p_2 = p_s(100^\circ\mathrm{C})$. The saturated vapour pressure $p_s(T)$ is a function of the temperature T. Similarly, for the low-temperature side evaporator, we have

$$q_{87} = h_8 - h_7, \quad p_8 = p_7 \tag{3}$$

The state of the inlet of the evaporator on the low-temperature side is designated by the subscript 7, and at the outlet by the subscript 8. p_7 is then the saturated vapour pressure $p_s(10^\circ\mathrm{C})$ of the air conditioner. The condenser inlet section and the ejector diffuser outlet section are represented by the subscript 6, and the quantity of heat q_{61} released to the atmosphere is

$$q_{61} = h_6 - h_1, \quad p_6 = p_1 \tag{4}$$

where p_6 is based on the assumption of high summer temperatures and is thus taken to be $p_s(40^\circ\mathrm{C})$.

As in the conventional treatment, the process at the expansion valve is taken as an isenthalpic change and thus

$$h_7 = h_1 \tag{5}$$

The ejector consists of four distinct components: the driving flow nozzle, the suction nozzle, the mixer and the diffuser. We assume that the steps in each component are ideal heat exchange processes. The driving flow, suction flow nozzle, and mixer states are indicated by the subscripts 4, 5, and 9, respectively.

$$\frac{u_4^2}{2} = (h_3 - h_4), \quad s_3 = s_4 \tag{6}$$

$$\frac{u_9^2}{2} = (h_9 - h_8), \quad s_9 = s_8 \tag{7}$$

$$h_6 = h_5 + \frac{u_5^2}{2}, \quad s_6 = s_5 \tag{8}$$

Since the cross sections of the channels in the ejector inlet and outlet are large in comparison with the ejector's internal channel cross section, the kinetic energy at the inlet and outlet can be disregarded.

The laws of conservation of mass, momentum, and energy in the mixer, where the driving flow and suction flow are mixed, are thus expressed as

$$\rho_4 u_4 A_4 + \rho_9 u_9 A_9 = \rho_5 u_5 A_5 \tag{9}$$

$$\rho_4 u_4^2 A_4 + \rho_9 u_9^2 A_9 + A_4 p_4 + A_9 p_9 = \rho_5 u_5^2 A_5 + A_5 p_5 \tag{10}$$

$$\frac{u_4^2}{2} + h_4 + \frac{u_9^2}{2} + h_9 = \frac{u_5^2}{2} + h_5 \tag{11}$$

and, since the driving flow nozzle and the suction flow nozzle join at the mixer inlet, the following relation applies.

$$p_9 = p_4, \qquad A_4 + A_9 = A_5 \tag{12}$$

The driving flow G_d and suction flow G_s are given by

$$G_d = \rho_4 u_4 A_4, \qquad G_d + \rho_9 u_9 A_9 \qquad where\, g = G_s / G_d \tag{13}$$

All of the thermal parameters in this system are thus determined by the flow ratio g and the nozzle outlet pressure p_4. Using ρ_4, for example, which is determined by p_4, the nozzle outlet cross-sectional area A_4 is determined from the continuity equation and thus depends on p_4.

The COP, which represents the thermal efficiency of this ejector cycle, is defined as the ratio of the heat absorption quantities of the high-temperature and low-temperature evaporators as

$$COP = \frac{G_s}{G_d}\frac{(h_8 - h_7)}{(h_3 - h_2)} \tag{14}$$

The flow ratio g was obtained so that the ejector outlet pressure p_0 would match the condenser pressure and thus the saturated vapour pressure of the atmosphere. p_4 was varied to determine the value at which the COP of this system became maximal.

Based on these assumptions, calculations were performed to obtain the relation between the ejector inlet flow quality and the COP of the water-refrigerant two-phase flow ejector. The results are depicted in Figure 3. As noted above, these results were calculated with the assumption of summer use, a waste-heat temperature of 100°C, an atmospheric temperature of 40°C, and a cooling temperature of 10°C. As shown in Figure 3, the COP was high when the inlet flow was low in quality and maximal at an inlet flow quantity of 0.3.

This indicates that the two-phase flow ejector system is more effective than a vapour ejector system as an air conditioner utilizing waste heat.

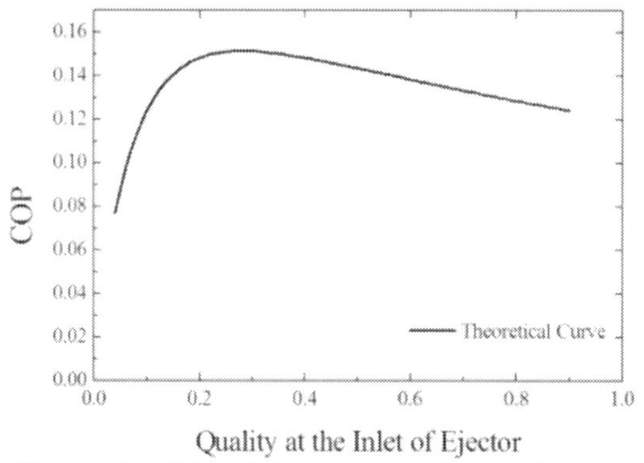

Figure 3. COP of the ejector refrigeration system.

4. Experimental apparatus

The experimental apparatus is shown schematically in Figure 4. It essentially consists of a high-pressure tank, a heater, a saturated vapour and a saturated liquid mixer, a two-phase flow ejector, a condenser, and a flowmeter. As the purpose of the experiments was to investigate the performance of the ejector, it was not necessary to run the air conditioning cycle. To simplify the apparatus, we assumed a simulated cycle in which the ejector outlet was at atmospheric pressure, and implemented the cycle in an open loop with the outlet open to the atmosphere. The tank was wrapped in thick insulation and the piping and apparatus were wrapped in neoprene rubber, thus insulating the flow channels. The water was heated in the tank by a heater and maintained at a constant temperature with a temperature control unit. The experiment was performed with an internal tank temperature of 150°C, a saturated vapour flow from the top part of the tank and a saturated liquid from the bottom part. The phase flow rates were controlled using flow regulating valves, and the quality of the two-phase flow was arbitrarily controlled. The saturated steam flow was measured using a capillary flowmeter, and the total flow was measured using a coriolis flow meter mounted at the outlet of the apparatus. As it is difficult to use a capillary flow meter with saturated liquids, their flow rates were calculated from the total flow and the saturated vapour flow. It was then possible to obtain the flow quality at the ejector inlet.

As an additional measure to simplify the apparatus, the suction performance of the ejector suction flow was assessed for air suction rather than water vapour refrigerant suction.

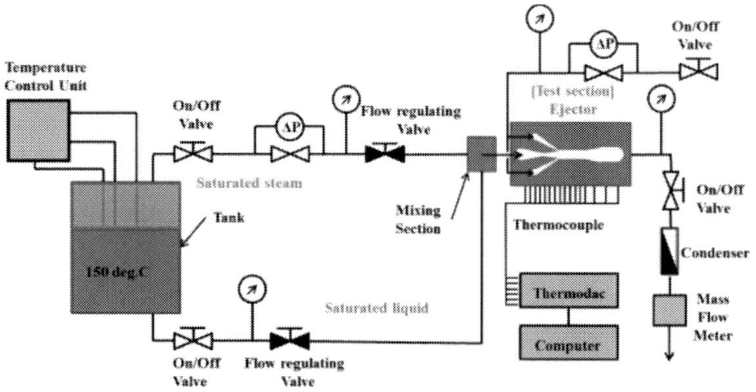

Figure 4. Experimental apparatus.

Figure 5. Structure of two-phase ejector.

As the specific volume of the saturated vapour was nearly 1,000 times that of the saturated liquid, the diameter of the inlet on the liquid side was made small in comparison with that for the vapour inlet so that the flow rate of the phases would be of the same order during the mixing. To obtain a large heat transfer on the droplet surfaces in both phases, it was deemed important to make the droplets small.

The experimentally used ejector was designed for a vapour flow of 1.0 g/s with an inlet flow quality of 1.0. Figure 5 shows a schematic of the ejector assembly. The heated water flows through the flow channel formed by plates of PES, which has a low thermal conductivity, and stainless steel. In the ejector flow channels, pressure taps were included at 7 points in the nozzle section and 7 points in the mixer and the diffuser. K type thermocouples were inserted into the pressure taps, the ejector internal pressure was measured, and the pressure distribution in the flow direction was obtained.

5. Experimental results

In the experiments, the vapour quality at the ejector inlet was varied for experimental verification of whether or not two-phase flow to the ejector actually yields higher performance than vapour flow as indicated by the theoretical calculation. In each experiment, we measured the flow rate of each phase from the tank, the internal pressure distribution of the ejector in the flow direction, the rate of suction flow into the ejector, and the ejector inlet and outlet pressures. Our previous studies had clearly shown that the nozzle's energy conversion efficiency greatly affected the ejector performance, and in the present study, therefore, the nozzle performance and ejector performance were obtained for various flow qualities at ejector inlet.

5.1. Verification of nozzle performance

5.1.1. Shock wave in the nozzle. The pressure distribution in the nozzle in the direction of flow was measured for various inlet flow qualities. Figure 6 shows the results at qualities of 0.20 and 0.75. The x-axis represents the distance from the inlet throat in the direction of the flow, with the throat at 0 mm and the nozzle outlet at 10 mm. The distributions shown were obtained with no suction flow for various ejector outlet pressures p_{out}, and with suction flow. When the ejector outlet pressure was increased, a moderate or sharp pressure recovery was observed in the nozzle divergence region. The results showed that changes in the outlet pressure and the presence or absence of suction had no effect on the convergent nozzle, thus indicating that the flow in the nozzle was supersonic, with the pressure rise in the ejector representing the shock wave generated by this flow. The gentle pressure rise at low flow quality in comparison with the sharper rise at high vapour quality is considered to be a reflection of the experimental results shown in Figure 6.

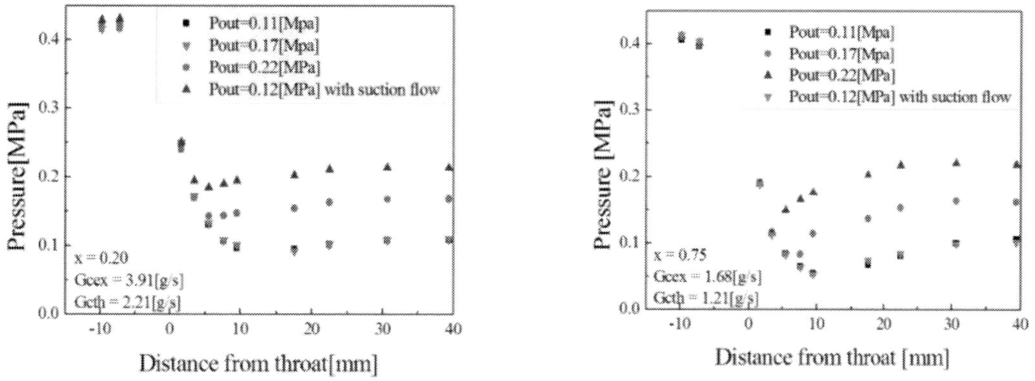

Figure 6. Pressure distribution (x=0.20 and 0.75).

5.1.2. omogeneous flow. As described in the previous section, the state in the convergent nozzle is constant regardless of the state in the nozzle outlet. As the flow rate and the vapour quality in the throat are constant, the flow rate is not changed by the backpressure. This is a critical state, and the flow rate in this state is referred to as the critical flow rate.

In these experiments, as noted above, saturated vapour and saturated liquid were mixed to form the two-phase flow upstream from the ejector. It was therefore difficult to assess the phase distribution and gas-liquid slip. One method that can be applied for this purpose has been described by Isbin et al., Starkman et al., and Akagawa[7] in their assessments of flow conditions, by comparing critical flow rate corrections with homogeneous flow models.

Figure 7 shows the relation between the flow quality at the ejector inlet and the error correction, with the theoretical critical flow rate at the nozzle throat designated as G_{cth} and the measured flow rate as G_{cex}. The error tends to increase at low flow qualities. The results are in general agreement with those of other researchers, and the two-phase flow mixed upstream from the ejector may therefore be treated as a homogeneous flow.

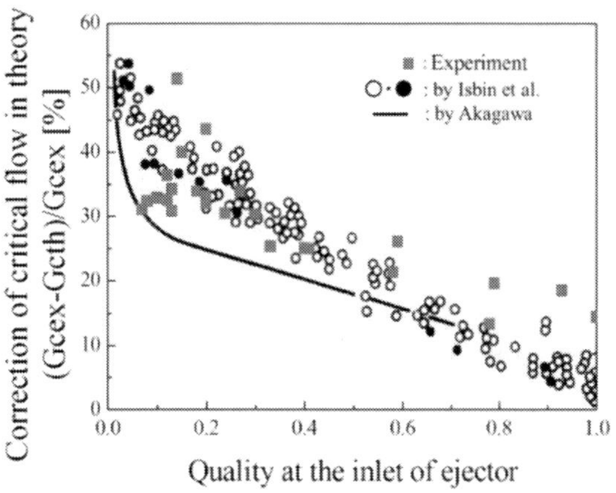

Figure 7. Critical flow correction with homogeneous flow model.

5.1.3. Nozzle efficiency. Figure 8 shows the pressure distribution in the nozzle for different inlet flow qualities, using colour-coded symbols. The corresponding curves represent the pressure distribution as calculated using the theoretical IHE model from the inlet conditions for the different flow qualities. Since the experimental distributions are higher than those given by the IHE model, expansion in the nozzle was insufficient. Thus, as described in the previous section for the critical flow correction, this tendency is marked at low flow qualities and indicates the occurrence of larger nonequilibrium phenomena at low qualities.

The two-phase flow velocity at the nozzle outlet can be calculated from the pressure distribution in the nozzle, using the law of conservation of momentum. Figure 9 shows the energy efficiency of the nozzle as derived from the outlet flow velocity and the adiabatic enthalpy drop. The nozzle efficiency declines with decreasing flow quality from a value of about 80% in the vapour state to about 60% for a flow quality of 0.2.

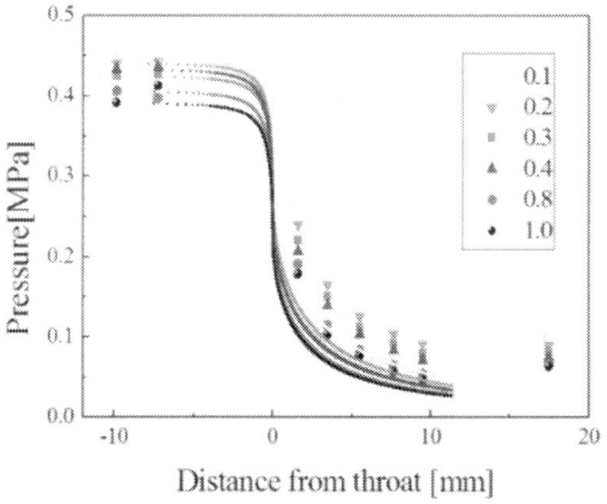

Figure 8. Pressure distribution for different inlet qualities.

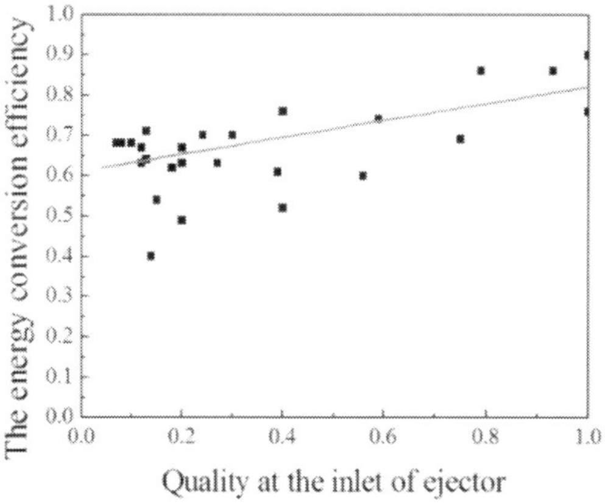

Figure 9. Nozzle energy conversion efficiency.

5.2. Assessment of ejector performance

5.2.1. Ejector suction force. The ejector suction performance was investigated experimentally by gradually opening the valve on the suction side of the ejector for suction flow to the ejector mixer and measuring the suction force. The suction flow rate was measured using a capillary flowmeter. Figure 10 shows the relation between the suction flow rate and the suction stream pressure rise *dp* at the ejector with the driving-flow flow inlet conditions held constant. The pressure rise *dp* represents the ejector suction force, and the results show that the ejector used in these experiments induced a maximum suction pressure of 0.08 MPa. The pressure difference between the ejector outlet and the suction flow inlet tended to decrease with increasing suction flow rate, and the ejector thus performs the same function as a constant-rotation compressor. The maximum suction flow rate shown in the right of the figure is an important indicator of performance of the ejector. Figure 11 shows the relation between the inlet flow quality and the achieved suction flow rate.

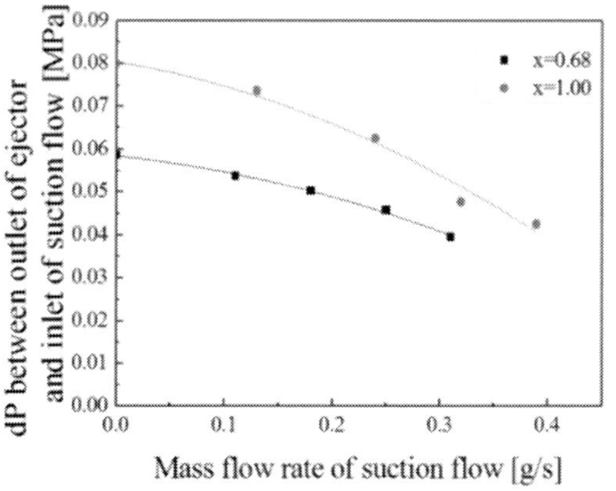

Figure 10. Pressure recovery of ejector.

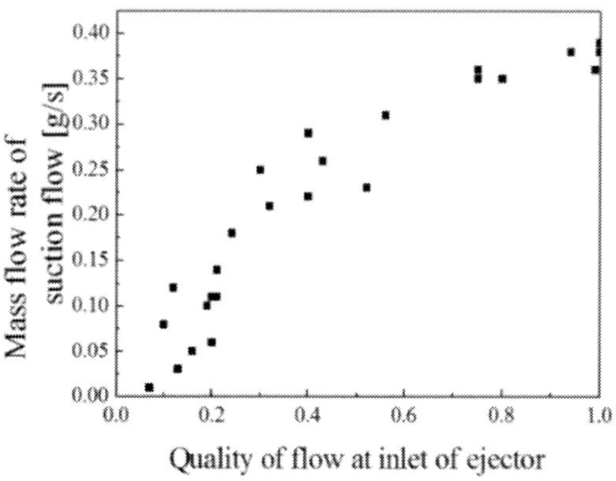

Figure 11. Mass flow rate of suction flow.

The suction flow rate is maximal when the flow quality is high. To induce air suction at the nozzle outlet, it is necessary for the fluid to accelerate at the driving flow nozzle and expand sufficiently to drop below atmospheric pressure. Comparison with Figure 8 clearly indicates that the nozzle outlet enters a higher vacuum state and the suction flow rate becomes maximal in the high flow-quality region. Substantial acceleration is therefore essential to induce suction flow at the nozzle by the ejector, and a high-efficiency nozzle is therefore necessary.

5.2.2. Calculation of ejector efficiency. The functions of the ejector are to convert the high-temperature high-pressure heat energy of the driving flow to kinetic energy and thereby induce suction flow and conversion to pressure work. Here we define the ejector efficiency η_{eje}, which is an indicator of the performance in terms of the actual pressure rise of the suction flow induced by the driving flow, as

$$\eta_{eje} = \frac{G_s \dfrac{p_6 - p_8}{\rho_s}}{G_d dh_n} \tag{14}$$

with the provision that dh_n is the adiabatic enthalpy drop from the ejector driving flow inlet to the ejector outlet, $p_6 - p_8$ is the pressure rise from the ejector suction flow inlet to the ejector outlet, and ρ_s is the air density and equals to $(\partial p/\partial h)_s$.

Typical results are shown in Figure 12. The higher ejector efficiencies were obtained for the case of high in quality at inlet rather than case of low inlet quality. It was also found that there were differences in the ability to raise the pressure of the suction fluid at a given suction flow rate with different inlet flow qualities. Furthermore, it was clear that the efficiency rises toward the right and thus shows that the ejector efficiency increases with suction flow and also, only with high inlet quality, the efficiency had the maximum value.

Figure 13 shows the values at which the ejector efficiency became maximal in the experiments on inlet flow quality. The tendency shown in this figure is similar to that found in the graph for maximal suction flow in Figure 11, which indicates that the ejector efficiency becomes high for high flow quality accompanying high suction flow, and becomes low for low flow quality.

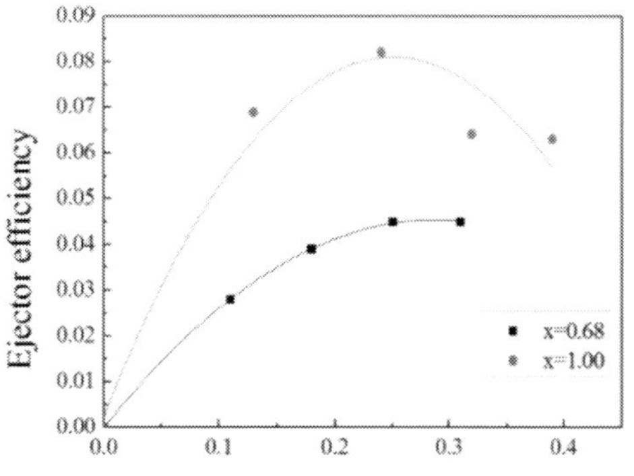

Figure 12. Ejector efficiency.

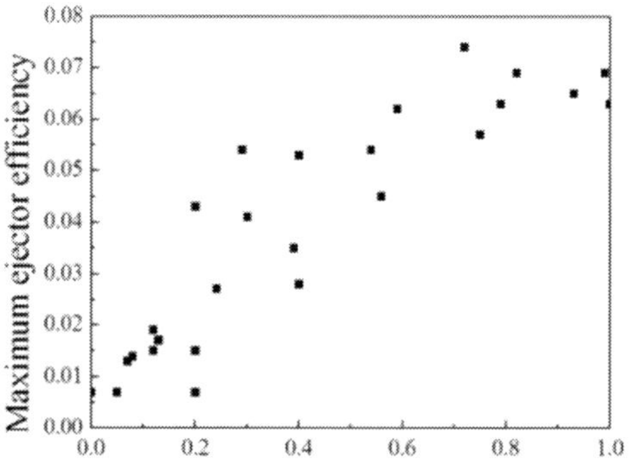

Figure 13. Maximum ejector efficiency.

Although the ejector efficiency values obtained in these experiments may seem rather low, they nevertheless demonstrate that these values can be achieved with no more than a laboratory-scale ejector. This refrigeration cycle is driven by waste heat and thus has considerable potential even at low efficiencies through its capability for utilization of heat energy that would otherwise simply be discarded. Further studies will be necessary to develop an ejector design that will yield higher efficiencies.

6. Conclusion

In the present study, a two-phase flow ejector was fabricated with the ultimate objective of realizing a two-phase flow ejector air conditioning system that could achieve the performance level shown by theoretical calculations. Its performance was assessed in experimental operation while inputting a water refrigerant in a two-phase flow state, and the following results were obtained:

1. It was found to be possible to obtain a vacuum state at the nozzle outlet under a two-phase flow condition at the ejector driving flow inlet and to induce a suction flow.
2. The energy conversion efficiency of the two-phase flow nozzle of the ejector was 80% at an inlet flow quality x=1.0, and declined with decreasing inlet flow quality to a value of 60% at x=0.2.
3. The efficiency of the ejector was 0.08 at a nozzle inlet flow quality of x=1.0 and declined to 0.04 at x=0.3.
4. Although the experimental results showed rather low ejector efficiency, the two-phase flow ejector air conditioning system may nonetheless be expected to provide a useful cooling capacity through its utilization of waste heat that would otherwise be discarded.

References

[1] Nakagawa M 2004 Refrigeration Cycle with Two-Phase Ejector *Refrigeration* **79**, No. 925 856-861

[2] Nakagawa M and Harada A 2010 Two-phase Ejector for Refrigeration Cycle and Shock Waves Appearing in the Supersonic Two-phase Flow Nozzle *proceeding of Japanese J. Multiphase Flow* **24** *No.1* .21-28

[3] Nakagawa M, Sato C, 2010 Performance of the ejector for the steam refrigeration cycle, *proceeding of the conference of Japanese Institution of Mechanical Engineers in Tokai area*

[4] G.K. Alexis and E.D. Rogdakis, 2003 A vertification study of steam-ejector refrigeration model, *Proceedings of Applied Thermal Engineering* **23** 29-36

[5] Chunnanond K and Aphornratana S 2003 An experimental investigation of a steam ejector refrigerator: the analysis of the pressure profile along the ejector,

[6] Yamanaka H and Nakagawa M 2012 Supersonic Nozzle Flow in the Two-Phase Ejector as Water Refrigeration System by Using Waste Heat *9th International Conference on Heat Transefer, Fluid Mechanics and Thermodynamics* 933-938

[7] Japanese Institution of Mechanical Engineers 2006 The handbook about the technique of gas-liquid two-phase flow *published by Corona* 430-435

Synthesis and study of magnetic properties of Co-doped anatase TiO_2 nanoparticles

S Yano, A Kurokawa, H Takeuchi, T Yanoh, K Onuma, T Kondo, K Miike, T Miyasaka and Y Ichiyanagi*

Department of Physics, Graduate School of Engineering, Yokohama National University, 79-5 Tokiwadai, Hodogaya, Yokohama, Kanagawa 240-8501, Japan

*E-mail: yuko@ynu.ac.jp; +81-(0)45-339-4185

Abstract. Anatase phase $Ti_{0.97}Co_{0.03}O_2$ nanoparticles were synthesized using two different preparation methods. The prepared samples were investigated by X-ray diffraction (XRD), X-ray fluorescence (XRF) analysis, and magnetization measurements. The particle diameters were controlled by adjusting the annealing temperature, and they ranged from 4.2–7.0 nm for $Ti_{0.97}Co_{0.03}O_2$ encapsulated in amorphous SiO_2, and from 14–61 nm for $Ti_{0.97}Co_{0.03}O_2$ that was not encapsulated in amorphous SiO_2. Magnetization measurements were performed on the 7.0-nm sample from the former sample set, and on the 14-nm sample from the latter sample set, and both these samples exhibited an identical crystal structure and an equal number of magnetic ions, as confirmed by XRF and XRD measurements. The magnetization curves for the two samples indicated their ferromagnetic behavior at room temperature. However, the magnetic parameters of saturation magnetization M_S and coercive force H_C were clearly different for these two samples. We attribute this difference in magnetic properties between these samples to magnetic impurities present due to the different degrees of solubility of Co ions in the TiO_2 lattice.

1. Introduction

Diluted magnetic semiconductors (DMSs) have been studied because of their potential application in spintronics devices [1]. In this light, Co-doped anatase TiO_2 thin films, which are a class of DMSs, are reported to exhibit Curie temperatures (T_C) that are above room temperature [2]. Subsequently, extensive research has been carried out on Co-doped anatase TiO_2 thin films [2-6]. However, in contrast to thin films, few studies have focused on nanoparticles. It is noteworthy that current methods for preparing Co-doped anatase TiO_2 thin films are complicated, and they require the use of special techniques such as molecular beam epitaxy, pulsed laser deposition, chemical vapor deposition, and sputtering. In addition, certain preparation methods require special conditions such as vacuum conditions and a reductive atmosphere in order to achieve room-temperature ferromagnetism [7-10]. Moreover, it is known that the magnetic properties of DMSs depend on the preparation method and conditions [10].

In our previous studies, we prepared magnetic nanoparticles, including those of 3*d* transition metals such as metal oxides and hydroxides, and investigated their magnetic [11, 12] and thermal properties [13, 14] along with their structural analyses [15, 16]. In this study, we prepared Co-doped anatase TiO_2 nanoparticles using two simple and novel preparation methods. We investigated the magnetic properties of these materials, and we observed their room-temperature ferromagnetic behavior.

2. Experimental

2.1. Sample preparation

Co-doped anatase TiO_2 ($Ti_{0.97}Co_{0.03}O_2$) nanoparticles were synthesized using two different methods.

Method 1: Aqueous solutions of $TiCl_4$ and $CoCl_2 \cdot 6H_2O$ were mixed, and the pH of the mixture was adjusted to about 6.8 using aqueous ammonia to obtain precipitates. The molar ratio of the prepared reagent was Ti:Co = 0.97:0.03. The obtained precipitates were washed several times with distilled water and dried at 330 K in a thermostat. The as-prepared samples were subjected to heat treatment in a furnace in air at annealing temperatures between 723 K and 873 K for 10 hours.

Method 2: Aqueous solutions of $TiCl_4$, $CoCl_2 \cdot 6H_2O$, and $Na_2SiO_3 \cdot 9H_2O$ were mixed, and the pH of the mixture was adjusted to about 6.8 using aqueous ammonia to obtain precipitates. The molar ratio of the prepared reagent was Ti:Co:Si = 0.97:0.03:1.00. Subsequent to mixing the solutions, the obtained precipitates were washed and dried in the same manner as in method 1. The as-prepared samples were subjected to heat treatment in a furnace in air at annealing temperatures between 873 K and 1073 K for 10 hours.

2.2. Property analysis

Each sample was examined using CuKα X-ray diffraction (XRD, λ = 0.154 nm) and X-ray fluorescence analysis (XRF). The average sample diameters were analyzed by fitting the theoretical 101 diffraction peak of the anatase phase to the measured diffraction peak of each sample. The theoretical diffraction patterns were calculated using the Stokes & Wilson method [16] by means of crystallite size distribution analysis software (Ver. 1.3, Rigaku Corporation, Japan). Magnetization measurements were carried out at 300 K under a magnetic field of ±50 kOe using a SQUID magnetometer.

Chemical analysis was performed in order to confirm the molar ratio of the ions in each of the obtained samples. The X-ray fluorescence analysis patterns of the samples prepared via the two methods are shown in figure 1. Subsequently, the actual Ti and Co content was determined by using fundamental parameter method. The molar ratio of the sample obtained via method 1 was Ti:Co = 0.969:0.031, and that of the sample obtained via method 2 was Ti:Co = 0.967:0.033. We confirmed that the Ti and Co ions present in each sample as charged quantities contained no impurity ions. Actual Ti and Co content is summarized in Table 1.

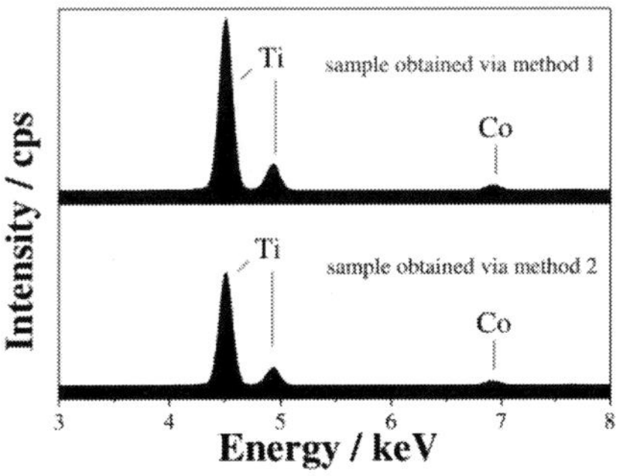

Figure 1. X-ray fluorescence analysis patterns of samples prepared by two different methods.

Table 1. Actual Ti and Co content, as determined by XRF.

	Ti content (%)	Co content (%)
Sample obtained via method 1	96.9	3.1
Sample obtained via method 2	96.7	3.3

3. Results and Discussion

3.1. X-ray diffraction

The X-ray diffraction patterns of samples obtained via two methods are shown in figures 2 and 3. Figure 2 shows the XRD patterns of samples prepared via method 1 (sample set 1). In the case of the sample that was annealed at 723 K, all the detectable peaks can be indexed as those corresponding to the anatase phase. The peaks corresponding to the rutile phase and $CoTiO_3$ were clearly detected in addition to the peaks corresponding to the anatase phase in the samples annealed at 773 and 873 K. The average diameters were estimated to be 14, 31, and 61 nm for the samples annealed at 723, 773, and 873 K, respectively. Figure 3 shows the XRD patterns of samples prepared via method 2 (sample set 2). In the case of the samples annealed at 873 K, 923 K, and 973 K, all the detectable peaks can be indexed as those corresponding to the anatase phase. The peaks corresponding to the rutile phase appeared in the sample annealed at 1023 K, and all the detectable peaks can be indexed as those corresponding to the rutile phase and the SiO_2 crystal in the sample annealed at 1073 K. The average diameters of the samples obtained via method 2 were estimated to be 4.2, 6.0, 7.0, and 13 nm at annealing temperatures of 873, 923, 973, and 1023 K, respectively. Figure 4 shows the relationship between the particle diameter and the annealing temperature for each sample set. From these results, we determined that the average diameter of $Ti_{0.97}Co_{0.03}O_2$ nanoparticles could be controlled by adjusting the temperature of the annealing process. The results of the XRD analysis indicated the following:

(1) Samples prepared via method 2 exhibit a higher anatase-to-rutile-phase transition temperature when compared with that for samples prepared via method 1.

(2) The particle sizes of the samples in set 2 were smaller than those of the samples in set 1 when they were annealed at the same temperature, and the particle sizes in the case of the set 2 samples grew more slowly than those of the set 1 samples.

(3) The $CoTiO_3$ peak was not detected in the set 2 samples regardless of the annealing temperature.

It has been speculated that $Ti_{0.97}Co_{0.03}O_2$ nanoparticles prepared by method 2 are encapsulated in amorphous SiO_2 [15]. It has been previously reported that TiO_2 encapsulated in SiO_2 leads to the suppression of the anatase-to-rutile transition [17]. Even in our study, anatase-to-rutile-phase transition was suppressed, and as a result, a higher anatase-to-rutile-phase transition temperature was observed. In addition, the presence of amorphous SiO_2 contributed towards nanometer-level control of particle sizes of the samples in set 2. Furthermore, the XRD results show that the $CoTiO_3$ peak was detected in sample set 1. We speculate that this could be due to the fact that Co ions may not be completely soluble in the TiO_2 lattice.

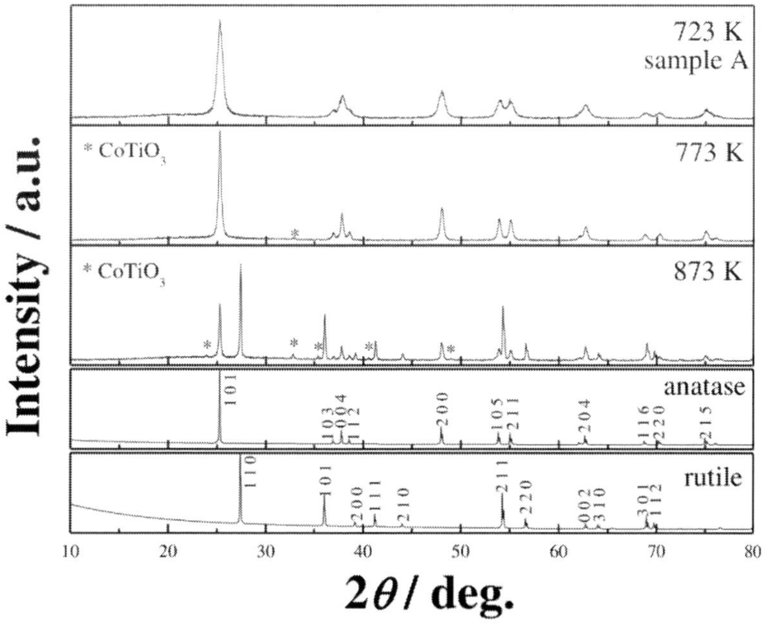

Figure 2. X-ray powder diffraction patterns of samples prepared via method 1 and annealed at 723, 773, and 873 K in an air atmosphere.

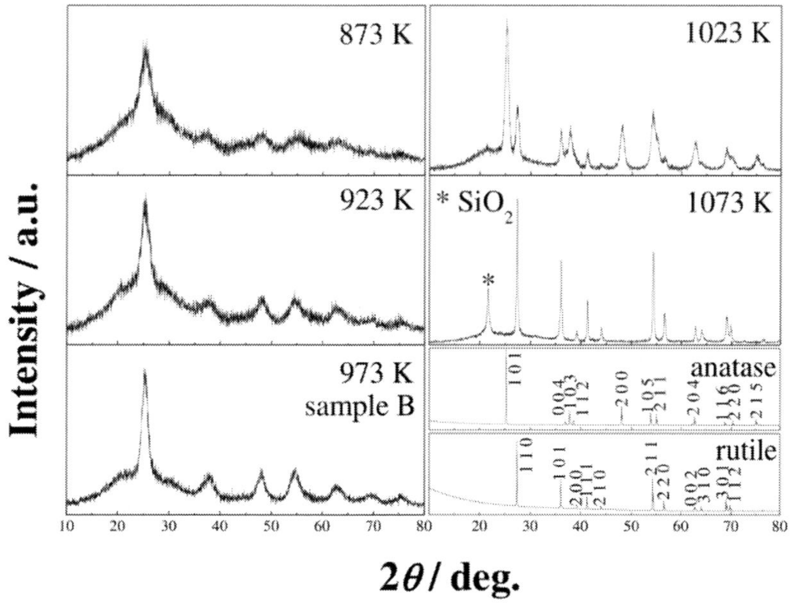

Figure 3. X-ray powder diffraction patterns of samples prepared via method 2 and annealed at 873, 923, 973, 1023, and 1073 K in an air atmosphere.

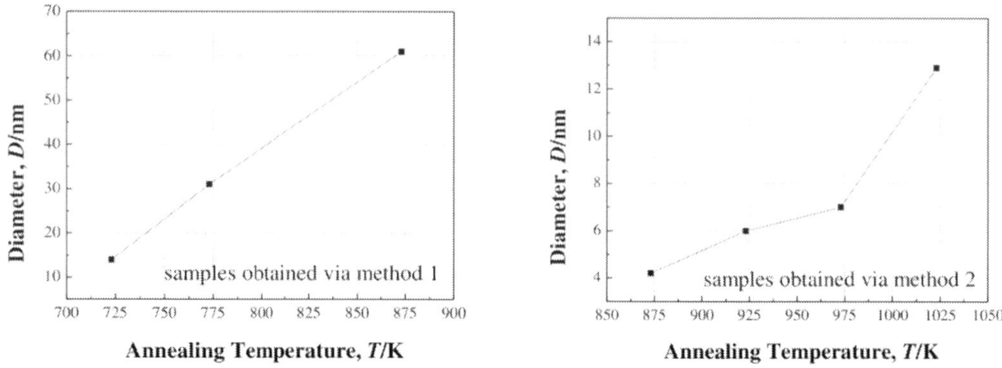

Figure 4. Average diameters estimated from the 101 X-ray diffraction peaks of samples obtained via methods 1 and 2.

3.2. Magnetization measurements

At this stage of the study, the two types of anatase structures prepared via the two above-mentioned methods were compared in terms of magnetization measurements. Figure 5 shows the magnetization curves at room temperature (300 K) of the sample annealed at 723 K via method 1 (named "sample A") and the sample annealed at 973 K via method 2 (named "sample B"). Both samples exhibit an identical crystal structure and an equal number of magnetic ions, as confirmed by XRF and XRD measurements. It was found that both samples exhibited ferromagnetism at room temperature. For sample A, the saturation magnetization per Co ion, M_S, was $0.8 \times 10^{-3}\mu_B$, and the magnetic coercive force, H_C, was 32.5 Oe. In the case of sample B, the saturation magnetization per Co ion, M_S, was $2.5 \times 10^{-3}\mu_B$, and the corresponding H_C value was 71 Oe. Remarkably, both the samples exhibited different magnetic parameter values regardless of the fact that they possessed the same crystal structure and the same number of magnetic ions. In the case of sample A (method 1), we speculate that Co ions might not be completely soluble in the TiO_2 lattice; a $CoTiO_3$ peak was detected for samples annealed at high temperatures. Co ions that did not dissolve in the TiO_2 lattice may have led to the formation of oxidative products such as CoO, Co_3O_4, and $CoTiO_3$, which exhibit Néel temperatures of $T_N = 291$, 33, 38 K, respectively. These impurities might contribute to the decrease in the magnetic moment of Co per ion in this sample.

The results of the magnetization measurements at room temperature (300 K) in the case of samples obtained via method 2 (two different particle sizes) are shown in figure 6. In the figure, the 7.0-nm-diameter sample is the same as the sample for which the magnetization curve is shown in figure 5. The saturation magnetization per Co ion, M_S, for the 4.2-nm-diameter sample was $1.0 \times 10^{-3}\mu_B$, and the magnetic coercive force H_C was 44 Oe. We confirmed that 4.2-nm sample exhibited a decreased ferromagnetic interaction when compared with the 7.0-nm sample. We attribute this variation in magnetic parameters between the 4.2- and 7.0-nm samples to the different degrees of solubility of Co ions in the TiO_2 lattice as well as the difference in the magnetization curves of the two samples (figure 5). In order to confirm the impurities existing in above-mentioned samples, XAFS and XPS measurements will be performed in the near future [18].

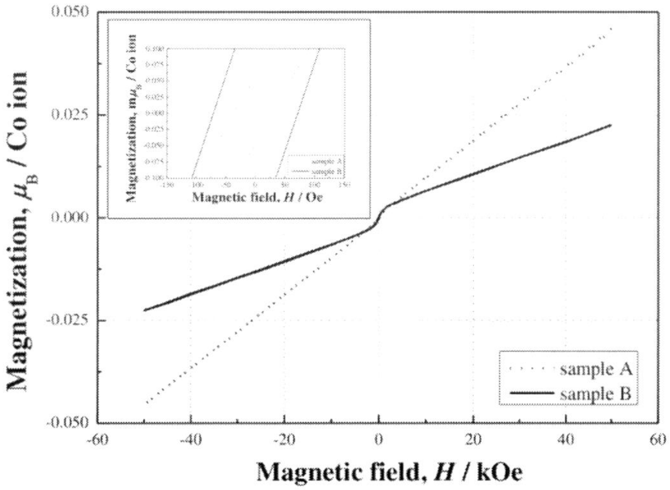

Figure 5. Magnetization curves for samples A and B at 300 K.

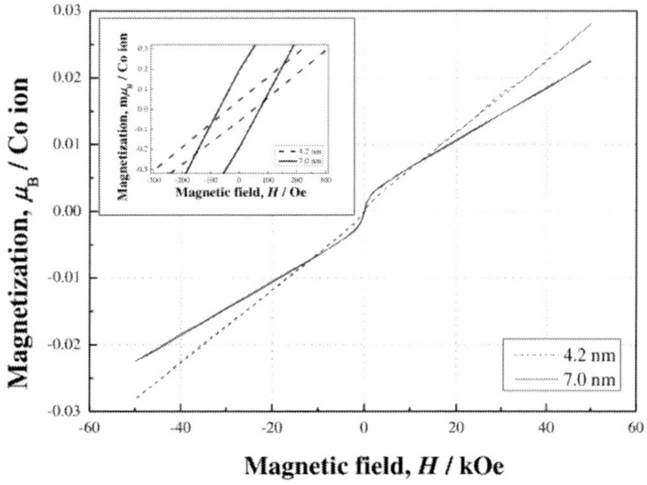

Figure 6. Magnetization curves at 300 K for samples obtained via method 2 for particle diameters of 4.2 and 7.0 nm.

4. Conclusion

We successfully prepared Co-doped anatase TiO_2 nanoparticles that exhibited ferromagnetism at room temperature via two simple and novel preparation methods. The XRD results showed that the crystal structures of both sample A (obtained via method 1; annealed at 723 K) and sample B (obtained via method 2; annealed at 973 K) are anatase single-phase structures. The magnetic properties of each sample were examined; the magnetic properties of the two sample sets prepared by these methods were dissimilar. Nevertheless, each representative sample exhibited an identical crystal structure and an equal number of magnetic ions, as confirmed by XRF and XRD measurements. We speculate that the solubility of Co ions in the TiO_2 lattice depends on the sample preparation method, and that magnetic impurities interfere with the ferromagnetism of Co-doped anatase TiO_2.

Acknowledgments
This study was partially supported by a grant provided by the Precursory Research for Embryonic Science and Technology program of the Japan Science and Technology Agency to Y. I.

References

[1] Ohno H 1998 *Science* **281** 951–956
[2] Matsumoto Y, Murakami M, Shono T, Hasegawa T, Fukumura T, Kawasaki M, Ahmet P, Chikyow T, Koshihara S and H. Koinuma 2001 *Science* **291** 854–856
[3] Ohtsuki T, Chainani A, Eguchi R, Matsunami M, Takata Y, Taguchi M, Nishino Y, Tamasaku K, Yabashi M, Ishikawa T, Oura M, Senba Y, Ohashi H and Shin S 2011 *Phys. Rev. Lett.* **106** 047602
[4] Manivannan A, Seehra M S, Majumder S B and Katiyar R S 2003 *Appl. Phys. Lett.* **83** 111–113
[5] Matsumoto Y, Takahashi R, Murakami M, Koida T, Fan X J, Hasegawa T, Fukumura T, Kawasaki M, Koshihara S and Kounuma H 2001 *Jpn. J. Appl. Phys.* **40** L1204–L1206
[6] Fukumura T, Toyosaki H, Ueno K, Nakano M and Kawasaki M 2008 *New J. Phys.* **10** 055018
[7] Manivannan A, Glaspell G and Seehra M S 2003 *J. Appl. Phys.* **94** 6994–6996
[8] Bryan J D, Heald S M, Chambers S A and Gamelin D R 2004 *J. Am. Chem. Soc.* **126** 11640–11647
[9] Choudhury B, Choudhury A, Maidul Islam A K M, Alagarsamy P and Mukherjee M 2011 *J. Magnetism Magnetic Mater.* **323** 440–446
[10] Li J-G, Büchel R, Isobe M, Mori T and Ishigaki T 2009 *J. Phys. Chem. C* **113** 8009–8015
[11] Ichiyanagi Y, Kubota M, Moritake S, Kanazawa Y, Yamada T and Uehashi T 2007 *J. Magnetism Magnetic Mater.* **310** 2378–2380
[12] Ichiyanagi Y, Uehashi T and Yamada S 2004 *Physica Status Solidi (c)* **12** 3485–3488
[13] Kubota M, Kanazawa Y, Nasu K, Moritake S, Kawaji H, Atake T and Ichiyanagi Y 2008 *J. Therm. Anal. Calorim.* **92** 461–463
[14] Kimura S, Mashino T, Hiroki T, Shigeoka D, Sakai N, Zhu L and Ichiyanagi Y 2012 *Thermochimica Acta* **532** 119–122
[15] Moro Y, Katayanagi H, Kimura S, Shigeoka D, Hiroki T, Mashino T and Ichiyanagi Y 2010 *Surf. Interface Anal.* **42** 1655–1658
[16] Ida T, Shimazaki S, Hibino H and Toraya H 2004 *J. Appl. Crystallogr.* **36** 1107–1115
[17] Suyama Y and Kato A 1978 *J. Ceram. Soc. Jpn.* **86** 119–125
[18] Wei H, Yao T, Pan Z, Mai C, Sun Z, Wu Z, Hu F, Jiang Y and Yan W 2009 *J. Appl. Phys.* **105** 043903

Observation of the dynamics of magnetically induced chains of sub-micron superparamagnetic beads in aqueous solutions by laser light scattering

Y Tanizawa[1,*], T Tashiro[1], P J Ko[2], and A Sandhu[1,2]
[1] Department of Electrical and Electronic Information Engineering, Toyohashi University of Technology, 1-1 Hibarigaoka, Tempaku-cho, Toyohashi, Aichi 441-8580, Japan
[2] Electronics-Inspired Interdisciplinary Research Institute (EIIRIS), Toyohashi University of Technology, 1-1 Hibarigaoka, Tempaku-cho, Toyohashi, Aichi 441-8580, Japan

E-mail: tanizawa-y@eiiris.tut.ac.jp

Abstract. Optical monitoring the behaviour of magnetically induced self-assembled chains of superparamagnetic beads (SPBs) are of interest for biomedical applications such as biosensors. However, it is difficult to directly monitor magnetically induced self-assembly of sub-micron nano-beads with conventional optical microscopes. Here, we describe the optical observation of the dynamics of magnetically induced self-assembled rotating chains of 130 nm SPBs in aqueous solutions by laser light scattering. Magnetic fields of ~1 kOe were applied to control the self-assembly chains of SPBs and their behaviour analyzed by monitoring the intensity of laser light scattered from the chain structures. We compared the light scattering from chains that were formed only by the application of external fields with chains formed by beads functionalized by EDC, where chemical reactions lead to the bonding of individual beads to form chains. The EDC experiments are a precursor to experiments on molecular recognition applications for biomedical diagnostics.

1. Introduction

Superparamagnetic beads (SPBs) are used for a wide range of biomedical applications including drug delivery, magnetic separation, magnetic resonance imaging (MRI), and enzyme-linked immunosorbent assay [1, 2]. The size of SPBs can be controlled to be comparable to actual biological materials such as proteins (10 –100 nm), viruses (20 –450 nm), and bacteria (0.2 μm –5 μm). Notably, SPBs offer the following unique properties for medical diagnostic applications: SPBs can be manipulated by external magnetic field gradients; external magnetic fields applied to the SPBs are not screened by biological materials or liquids; and it is easy to functionalize SPBs with antibodies, enzymes, proteins, and ligands [2]. To-date, research on the dynamics of SPBs in aqueous solutions has been limited to optical observations of micrometer sized SPBs with conventional white light microscopes [1, 3]. However, the spatial resolution of such methods is limited to the micrometer range, thus optical microscopes are not applicable for monitoring the behavior of sub-micrometer magnetic beads.

Here, we describe the use of laser light scattering for monitoring COOH-functionalized 130 nm sized SPBs (nanobeads), where changes in the amplitude of magneto-optical scattering (MS) from magnetic beads in an aqueous solution enabled the quantitative determination of the concentration of

1-ethyl-3-(3-dimethylaminopropyl) carbodiimide hydrochloride (EDC) in the solution. Furthermore, we tracked single SPBs in solution to monitor the interaction of small numbers of SPBs under external magnetic fields. These experiments enabled the determination of the concentration of EDC by monitoring rotating SPBs functionalized with carboxylic acid (COOH) in less than five minutes.

2. Methods and Experimental

The experiments were conducted using a modified laser light scattering microscope with integrated electromagnetic coils for producing rotating magnetic fields (Figure 1). The laser light scattering microscope enabled the real-time tracking nano-SPBs dispersed in aqueous solutions. The 635 nm line of the laser was guided to pass through the sample of liquid containing the magnetic nanobeads. The laser beam is refracts at the interface between the liquid and the optical element through which the light passes, with the light travelling parallel to the glass-sample interface. Nano-SPBs in the path of the laser beam, which is approximately 100 μm wide and 15 μm deep, are visualized by x20 microscope objective lens of an optical microscope aligned at 90 degrees to the laser beam axis. The microscope is fitted with a CCD camera, which collects the light scattered by particles in the field of view [4, 5]. During the experiments, we collected the intensity of scattered light and monitored the structures of the diffraction patterns, frame-by-frame during the measurements. The measurements were made using in-plane rotating fields upto ~1100 Oe with rotation frequencies of 0–5 Hz.

The experiments were conducted using 130 nm diameter superparamagnetic beads (Nanomag-D, Micromod) with concentrations of 15.8×10^{8} and 6.10×10^{8} particles/mL. The SPBs were functionalized with carboxyl groups.

After dilution and ultrasonic cleaning the nano-SPBs suspensions were slowly introduced into the observation chamber. Next, the field was applied, the focus adjusted, and images of 100 μm × 80 μm area recorded with a 640 × 480 pixel (8 bit) CCD camera operating 30 frames per second. The images were processed using ImageJ software [6] to determine the scattering area and intensity of light scattered by each SPB and chains formed by the external field.

In the 1-ethyl-3-(3-dimethylaminopropyl) carbodiimide hydrochloride (EDC) experiments, the nano-SPBs were dispersed in deionized (DI) water at 2000-times dilution, where the concentrations of the EDC aqueous solution were 0.001, 0.01 and 0.2 M in DI water. The EDC aqueous solution was added to the nano-SPBs solution until the ratio was SPBs : EDC = 10 : 1. The mixture was shaken for five times and introduced into measurement chamber of the laser scattering microscope focused on the nano-SPBs in solution. Next, the external magnetic field applied and images recorded. The external magnetic field was rotated at $f = 0.1$ Hz, $H = 400$ Oe and was turned off after recording for 3, 5, and 7 min after mixing the solutions. The video data were converted to the scattering area data by processing the images with ImageJ software. We compared the results of the scattering area data with and without EDC.

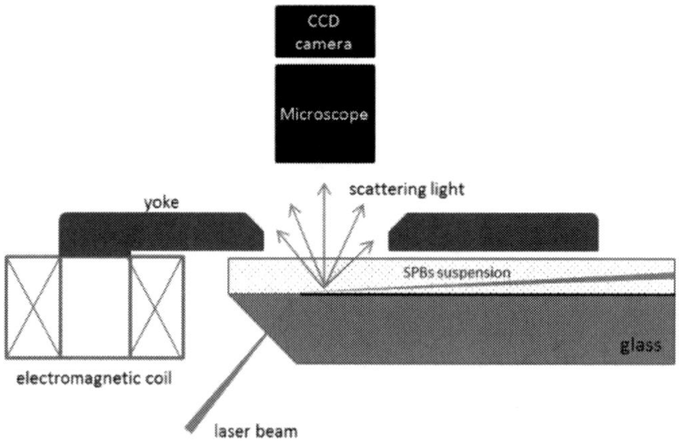

Figure 1. Schematic of the laser light scattering microscope integrated with four electromagnetic coils.

3. Results and Discussion

3.1. Observation of two nano-SPBs

Figure 2 (a) shows tracking of two nano-SPBs in solution without an external rotating magnetic field for sevral video frames. Frame 4 shows two 130 nm SPBs (A and B), getting close to each other. In frame 5, the center to center distance approaches the sum of the individual diameters of SPBs A and and B, resulting in the aggregation into structure C. In frame 8 the structure C separates back into the single SPBs A and B. The movement of SPBs in the liquid suspension is due to Brownian motion. The variation of scattered light intensity with function of time is shown in Figure 2 (b).

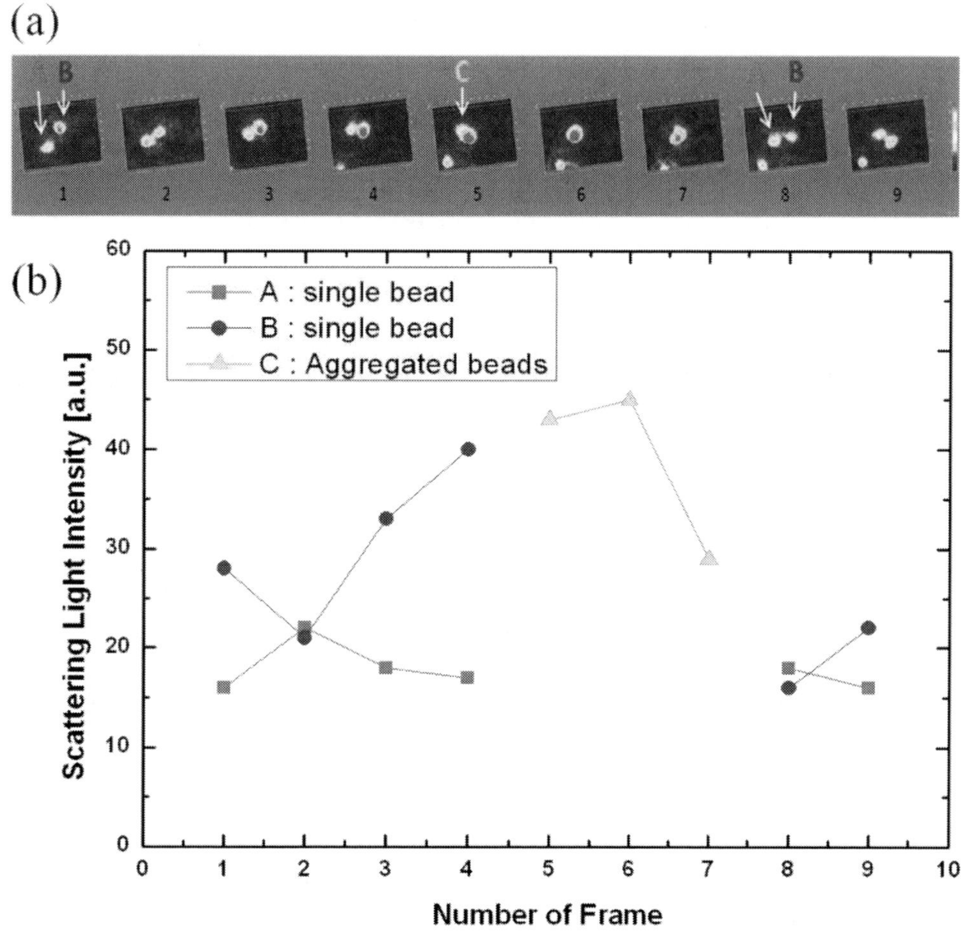

Figure 2. Tracking of two nano-SPBs in solution without an external rotating magnetic field. (a) Image of tracking of two SPBs at frame-by-frame, (b) Intensity of scattered light from the two SPBs being tracked as function of number of frames.

3.2. Comparison of with and without EDC under an external rotating magnetic field

Figure 3 shows intensity of light scattered by chains formed by 130 nm SPBs under an external rotating magnetic field of 1000 Oe at 0.1 Hz with respect to angle (θ). The maximum (Smax) and minimum (Smin) optical scattering for a given external rotating magnetic field occurred along the lowest (θ = 0, 180, 360....) and highest (θ = 90, 270....) scattering intensity with respect to the incident laser light, respectively. The light scattering by SPBs chains depends on the angle (θ) between the direction of SPBs chain and the incident laser light. Here, we define the magneto-optical scattering (MS) effect in terms of $\Delta S = (Smax - Smin) / Smax \times 100$, where the Smin and Smax are the optical scattering at $\theta = 0°$ and $90°$, respectively. The video frame images used to monitor the scattering of SPB chains under rotating fields at $\theta = 0°$ and $90°$ are shown in Figure 3 (b).

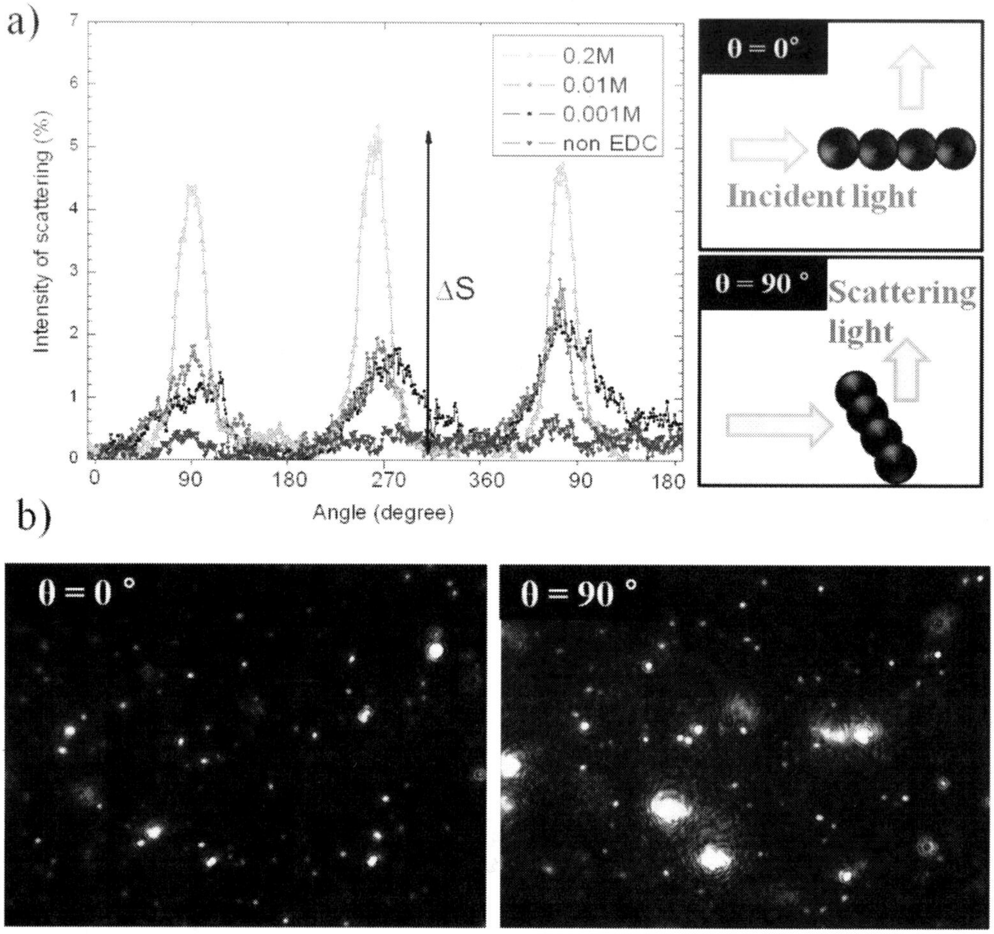

Figure 3. (a) Optical scattering of 130 nm SPBs with respect to the angle (θ) between the external magnetic field and the incident laser light at $\theta = 0°$ and $\theta = 90°$. (b) Video frames of the optical scattering of SPBs at $\theta = 0°$ and $\theta = 90°$, respectively.

Figure 4 shows ΔS of SPBs functionalized with carboxylic acid (COOH) containing EDC (0, 0.001, 0.01 and 0.2 M) as a function of the time after applying the external rotating field. The ΔS of non-EDC increased due to Brownian motion in magnetically-induced chains due to chain-chain interactions and the formation of thicker, column-like chains. Notably, with increasing EDC concentration, ΔS increased significantly due to forces resulting from COOH-EDC conjugation [7] contributing significantly to enhancing the bead to bead bonds in the rotating SPB chains, which leads to an increase in ΔS. Thus, biomedical applications are possible by monitoring changes in the ΔS corresponding to COOH-EDC conjugation due to the addition of EDC into the colloidal solution.

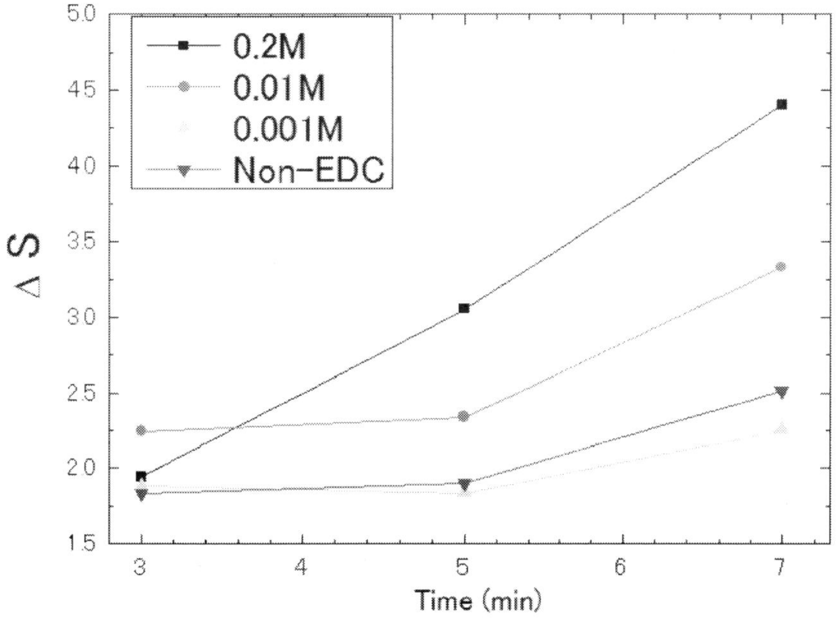

Figure 4. ΔS of SPBs functionalized with carboxylic acid (COOH) containing EDC (0, 0.001, 0.01 and 0.2 M) as a function of time after applying the external rotating field.

4. Conclusions

We described the scattering of laser light from individual and chains of 130 nm SPBs in aqueous solutions under rotating magnetic fields. Coalescence of individual nano-SPBs was clearly observed. Furthermore, we observed a clear relationship between ΔS and the mean length of chains of 130 nm SPBs in terms of the laser light scattering intensity. The addition of EDC to solutions containing functionalized 130 nm SPBs led to clear increases in ΔS, due to enhanced forces from COOH-EDC conjugation between bead to bead bonds in the rotating SPB chains. Finally, these results may find applications in biomedical diagnostics.

Acknowledgments

This work was supported in part by a grant from the Ministry of Education, Culture, Sports, Science, and Technology, Japan in the form of the Special Coordination Funds for Promoting Science and Technology for the Integrated Research Institute.

References

[1] Park S Y, Handa H and Sandhu A 2010 *IEEE Trans. Nano Lett.* **10** 446-451
[2] Marszałł M P 2011 *Pharm. Res.* **28** 480–483
[3] Anik K V, Antonio A G and Mark A H 2003 *Langmuir* **19** 8646-8653
[4] Carr B, Hole P, Malloy A, Nelson P, Wright M and Smith J 2009 *European Journal of Parenteral & Pharmaceutical Sciences* **14(2)** 45-50
[5] Carr R, Smith J, Nelson P, Hole P, Malloy A, Weld A and Warren J 2008 *J. Nanopart. Res.*
[6] Rasband, W.S., U. S. National Institutes of Health, Bethesda, Maryland, USA, http://imagej.nih.gov/ij/, 1997-2012.
[7] Eric V and Mark B 2009 *Chem. Soc. Rev.* **38** 606–631

Study on performance simulation of polymer electrolyte fuel cell for preventing degradation

T Kobayashi[1*] T Fukuda[2] M Doi[1] R Hashimoto[3] H Kanematsu[4] and Y Utsumi[5]
[1] Dep. of Electronics & Control Eng., Tsuyama National College of Technology, Okayama, Japan
[2] Dep. of Mechanical Eng., Nagoya Institute of Technology, Nagoya, Japan
[3]Collaborative Research Center, Hiroshima University, Hiroshima, Japan
[4]Dep. of Materials Sci. & Eng., Suzuka National College of Technology, Mie, Japan
[5]Laboratory of Advanced Science and Technology for Industry, University of Hyogo, Hyogo, Japan

*E-mail: t-koba@tsuyama-ct.ac.jp;+81-(0)868-24-8215

Abstract. In the present study, the distribution of water content in the membrane of PEFC was analyzed by using a numerical simulation as well as understanding the behavior of internal moisture of PEFC. Eight parameters were selected for the simulation then 18 combinations of the parameters were allocated by design of experiments, thus the data obtained were analyzed by multiple regression analysis to understand the influence factor of operating conditions quantitatively. As a result, the influence of the operating parameters on the dryness of the membrane for the anode side and the cathode side of PEFC was quantitatively shown by using the method of the multiple regression analysis. Further it was found that the area where cerium carbonate ought to be coated for preventing the degradation without decreasing performance.

1. Introduction

Polymer electrolyte fuel cell (PEFC) is expected as one of the efficient use of fossil fuels. However, there have been still remaining obstacles to be overcome for commercial use. Electrolytic membrane in the PEFC is degraded with decreasing water content in the membrane [1]. It has been reported that low humidity degrades the polymer membrane, resulting in a decrease in cell voltage, a hydrogen cross leak from the anode to the cathode, and finally, stopping power generation [2]. The degraded polymer membrane became much thinner. This damage might be due to radicals formed in the MEA [2]; therefore, a study examined the effect of a radical capturing layer and showed that an MEA with the radical capturing layer exhibits greater endurance—10 to 20 times that of a conventional MEA [3]. However, excessive amount of coating of the radical capturing layer causes performance drop of PEFC [3]-[4].

In the present study, the distribution of water content in the membrane of PEFC was analyzed by using a numerical simulation as well as understanding the behavior of internal moisture of PEFC. The simulation data obtained were analyzed by multiple regression analysis to understand the degree of influence of operating conditions quantitatively. Furthermore, the calculations using the diffusion coefficients of water applied some researchers were conducted to clarify the influence of the properties of the membrane on the dryness [5].

2. Performance Simulation

A two dimensional model considering the transfer of electric charge and mass was constructed. In the simulation model, the direction of gas flow channel can be divided into n elements, the direction of the membrane thickness is consist of anode gas channel, membrane electrode assembly(MEA), cathode gas diffusion layer and anode gas flow channel. In the present paper, the direction of gas flow channel was divided into 10 elements, and iterative calculation was conducted. Moisture to pass through the polymer membrane is transferred by the following three factors. 1) moisture migration by the reaction. 2) moisture migration by electro-osmosis. 3) moisture migration by diffusion. The above factors are expressed as the following equation which is the sum of each moisture migration.

$$N(x) = \frac{I(x)}{2\mathrm{F}} + n_d \frac{I(x)}{\mathrm{F}} + D_w \frac{dc_w}{dy} \tag{2.1}$$

Where, $N(x)$: Molar flux of water in the membrane[mol/s/cm^2], $I(x)$: current density[A/cm^2], F : Faraday constant[C/mol], n_d : Electro-osmosis coefficient [-], Dw : Diffusion coefficient [cm^2/s], c_w : Water content in the membrane [mol/cm^3] , Accordingly, using the equation , the behavior of water can be understand. In addition, the following equation was used for the water content in the membrane obtained in the experiment by Springer et al.[6]

$$c_w = \frac{\rho_{\mathrm{m,dry}}}{\mathrm{M_{m,dry}}} (0.043 + 17.8a + 39.8a^2 + 36.0a^3) \qquad (a \leq 1) \tag{2.2}$$

where, $\rho_{\mathrm{m,dry}}$: Density of the film dried [g/cm^3], $\mathrm{M_{m,dry}}$: Equivalent weight of the membrane dried [g/mol], a : Water activity [-]. The simulation was performed assuming a case of using pure hydrogen as the fuel gas, air as the oxidizing gas (N_2:79%, O_2:21%).Physical properties required for the simulation, are the same as those in the simulation used by Onda et al.[5]

2.1 Equation for the characteristics of the membrane

Springer et al[6] expressed as a function of percentage of the water content of the membrane by measuring to the diffusion coefficient and electro-osmosis coefficient in a experimental of PEFC. Here,λ water content ratio defined by water activity, and the relationship between water activity and moisture content is expressed in the following equation.

$$\lambda_k = 0.043 + 17.81a_k - 39.85a_k^2 + 36.0a_k^3 \quad (0 < a_k \leq 1) \tag{2.3}$$

However, a_k is the water activity of the k side (anode or cathode).Theoretical water content is a ration of the number of sulfone group which works as a carrier of proton in the membrane, to the number of water molecule in the membrane. The water content is calculated with the equation. In this paper, this value is referred to as the water content.

$$cH_2O,k = \frac{\rho_{\mathrm{m,dry}}}{\mathrm{M_{m,dry}}} \lambda_k = \frac{\rho_{\mathrm{m,dry}}}{\mathrm{M_{m,dry}}} (0.043 + 17.81a_k + 39.85{a_k}^2 + 36.0{a_k}^3) \quad (0 < a_k \leq 1) \tag{2.4}$$

where, $\rho_{\mathrm{m,dry}}$: Density of the film dried[g/cm^3], $\mathrm{M_{m,dry}}$: Equivalent weight of the film dried[g/mol] The equations of the electro-osmosis coefficient n_d and the diffusion coefficient D_{H20} are shown in equation (4) and (5).

$$n_d = 2.522\lambda_a = 0.0049 + 2.02a_a - 4.53a_a^2 + 4.09a_a^3 \qquad (0 < a_a \leq 1) \tag{2.5}$$

$$D_{H_2O} = (0.0049 + 2.02a_a - 4.53{a_a}^2 + 4.90{a_a}^3)D_0 \exp\{1268\left(\frac{1}{303} - \frac{1}{T + 273.15}\right)\}$$

$$(0 < a_a \leq 1) \qquad (2.6)$$

where, D_o : the effective water diffusion coefficient in the membrane in the standard condition[cm^2/s], T : Cell temperature [℃]. The electric conductivity of membrane, which was expressed as a function of the water content in the anode side, is shown below.

$$\sigma_m = (0.00514\lambda_a - 0.00326)D_0 \exp\left\{1268\left(\frac{1}{303} - \frac{1}{T + 273.15}\right)\right\} \qquad (2.7)$$

Here, the activity of water in the anode side is

$$a_a = \frac{x_{H_2O,a}P}{P^{sat}_{H_2O,a}} = \left(\frac{M^v_{H_2O,a}}{M^v_{H_2O,a} + M_{H_2}}\right)\frac{P}{P^{sat}_{H_2O,a}} \qquad (2.8)$$

and the activity of water in the cathode side is written in the next formula

$$a_c = \frac{x_{H_2O,c}P}{P^{sat}_{H_2O,c}} = \left(\frac{M^v_{H_2O,c}}{M^v_{H_2O,c} + M_{O_2} + M_{N_2}}\right)\frac{P}{P^{sat}_{H_2O,c}} \qquad (2.9)$$

where, x_{H2O} is molar fraction. The amount of saturated water vapor required to determine the activity of water is expressed in the following equation.

$$\log 10(P^{sat}_{H_2O,k}) = 2.95 \times 10 - 2T_k - 9.18 \times 10 - 5T_k^2 + 1.44 \times 10 - 7T_k^3 - 2.18 \qquad (2.10)$$

3. The simulation conditions

In the simulation, a single cell of one flow path is used as shown in Figure 1, and operating temperature and temperature distribution are kept constant. The initial conditions of the simulation are current density [A/cm^2],operating temperature [℃], H_2 humidification temperature [℃], air humidification temperature [℃], H_2 utilization [%], and the air utilization [%], and the values shown in Table 1 are used.

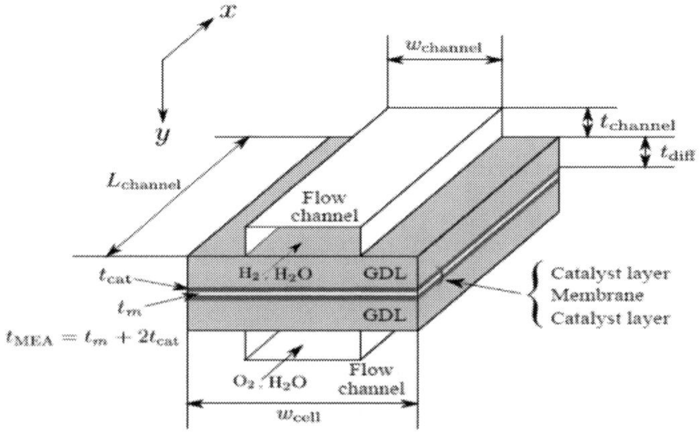

Figure 1 Dimensions of the simulation

Table 1 Eight parameters

	First standard	Second standard	Third standard	Range
H_2 entrance flow quantity	1	2	3	1
Cell temperature	50	55	60	10
H_2 humidification temperature	50	55	60	10
Air humidification temperature	50	55	60	10
H_2 utilization	30	50	80	50
Air utilization	15	40	60	45
Current density	0.2	0.5	1	0.8
Air entrance flow quantity	1	2	3	2

4. Simulation result

The distribution of water content in Polymer Membrane was calculated in tar 18 conditions using 9 parameters in Table 1. Some examples of the results are shown in Figure 2. It was found that the water content of anode side was higher than that of cathode side probably due to the produced water by power generation. In addition, the water content of cathode side tends to increases from the inlet to outlet direction, on the contrary the trend reverses in the anode side. The reason was suggested that the cathode gas has low humidity due to the lower water saturation temperature in than operation temperature in cathode, and that the water molecules in anode transfer toward cathode by the osmic effect flow. Firstly, dryness and the range of dryness were defined shown in Figure 2. In Figure 2.1 and Figure 2.3, the water contents saturate at around 0.025[mol/cm^3]. The saturated value was defined as the dryness of zero, For example, the estimation method of the dryness was shown in Figure 2.1. The average value c of max and min of water content was determined, then the ratio of *a* to *b* is defined as the dryness.

$$(b\text{-}c)/b \quad = a/b \quad \times \quad 100 \quad = \quad \text{dryness} \quad [\%] \qquad (4.1)$$

The range of the dryness was defined as the region not saturated in water content, shown in Figure 2.2 and Figure 2.4. Eight parameters were allocated to 18 conditions using Design of Experiment (DOE), then the dryness and the range of dryness were calculated by the above mentioned simulation.

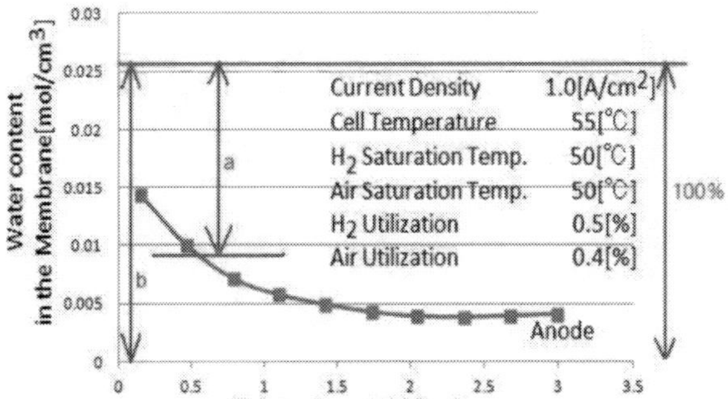

Figure 2.1 Dryness of membrane (anode side)

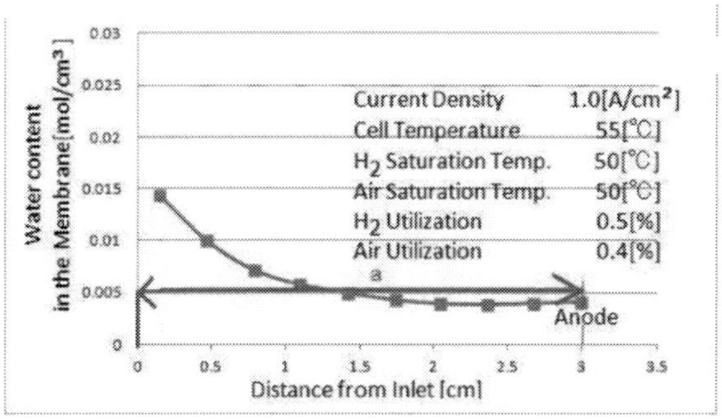

Figure 2.2 The range of dryness (anode side)

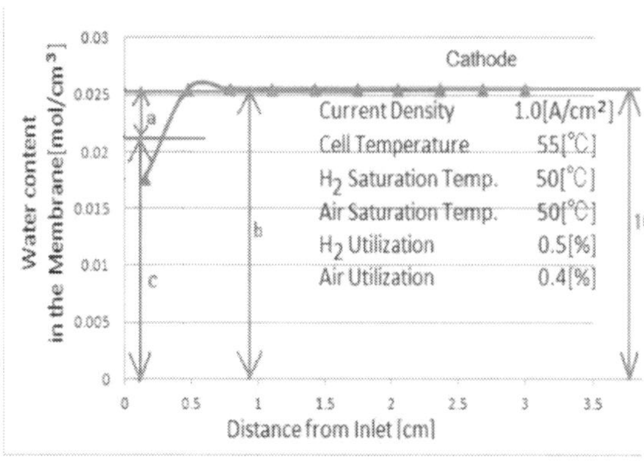

Figure 2.3The range of dryness (cathode side)

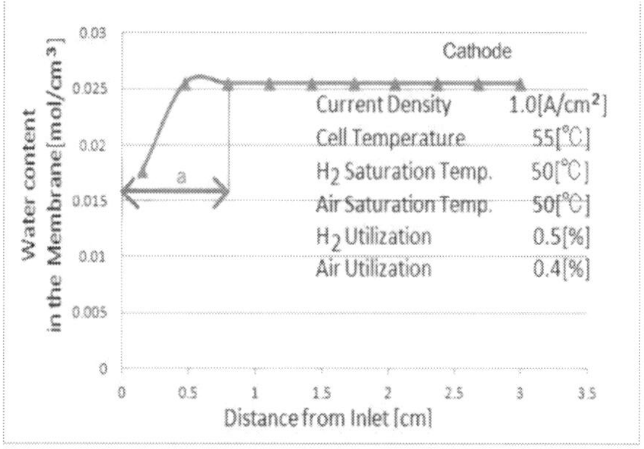

Figure 2.4 Dryness of membrane (cathode side)

Figure 2 Dryness of membrane

From the obtained data, multiple regression analysis on the effects of the parameters on the dryness and the range of the dryness were carried out, and regression equations were created. The obtained coefficients and the equations are shown in the lower part. The influence factor of each parameter is the value multiplied the coefficient of the equation as by the range of each parameter (the range between level 1 to level 13) Δx_i.

Table 2 Part of the multiple regression analysis result

	Coefficient
Intercept	-121.24
x_1	11.96122
x_2	5.081
x_3	0
x_4	-2.78408
x_5	-0.41412
x_6	-0.23742
x_7	40.84643
x_8	9.107994

Anode side-dryness

$$Y_1=-121.24+11.961x_1+5.081x_2+0x_3-2.784x_4-0.414x_5-0.237x_6+40.846x_7+9.107x_8 \quad (4.2)$$

Anode side- range of dryness

$$Y_2=-5.063+0.549x_1+0.199x_2+0x_3-0.095x_4-0.027x_5-0.011x_6+1.97x_7+0.434x_8 \quad (4.3)$$

Cathode side-dryness

$$Y_1=-13.863+2.032x_1+1.643x_2+0x_3-1.467x_4-0.002x_5-0.042x_6+3.111x_7+4.572x_8 \quad (4.4)$$

Cathode side- range of dryness

$$Y_2=-1.915+0.239x_1+0.094x_2+0x_3-0.067x_4+0.008x_5-0.014x_6+0.487x_7+0.159x_8 \quad (4.5)$$

Where, Y_1:dryness of membrane[-], Y_2:the range of dryness[-], x_1: H_2 entrance flowing quantity[cc/min], x_2: cell temperature[℃], x_3: H_2 humidify temperature[℃], x_4: air humidify temperature[℃], x_5: H_2 utilization[%], x_6:air utilization[%], x_7:current density[A/cm^2], x_8:air entrance flowing quantity[cc/min].

Figure 3 shows the influence factors of the eight parameters. It was found that in anode side, the influential parameters on the dryness art cell temperature can suppress the drying of the polymer membrane. The parameters influenced on the range of dryness were same as those for the dryness. In cathode side, cell temperature and current density influenced heavily on the dryness, and H_2 utilization ratios and air humidify temperature can suppress the dryness. The parameters influenced on the range of dryness in cathode side are same as those of the dryness, however the parameters suppress drying are air humidify temperature and air utilization ratio.

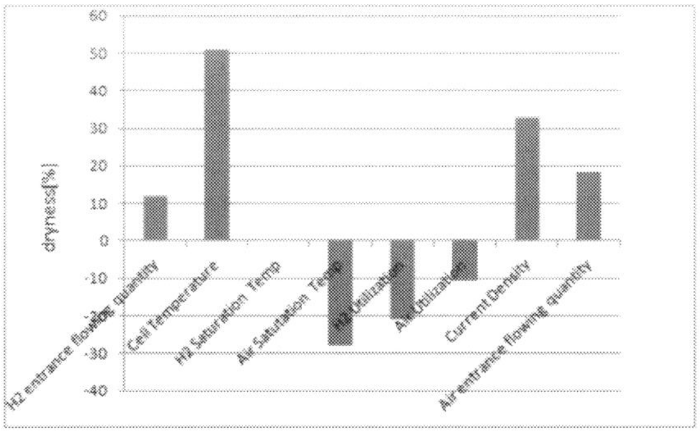

Figure 3.1 Dryness of membrane (anode side)

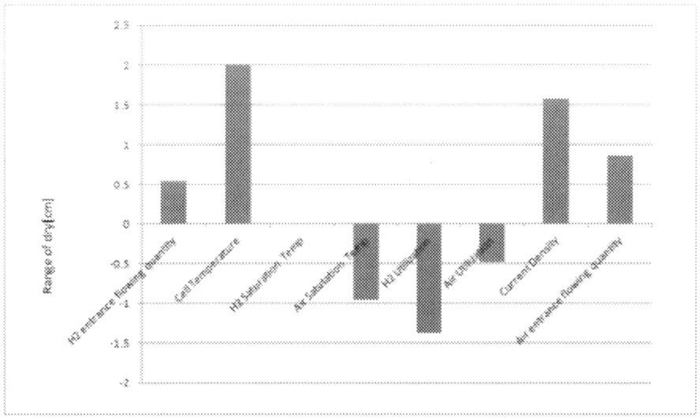

Figure 3.2 The range of dryness (anode side)

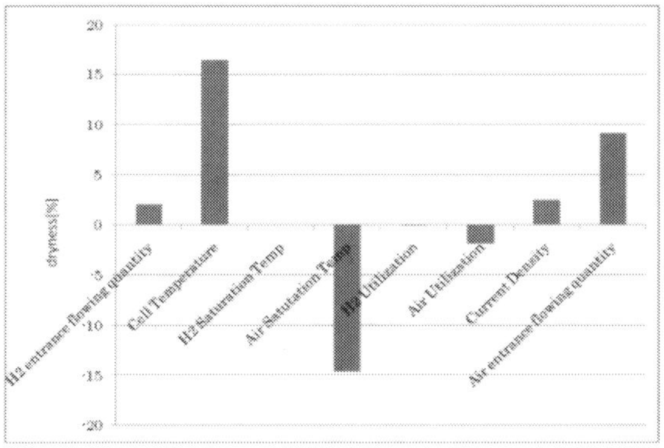

Figure 3.3 Dryness of membrane (cathode side)

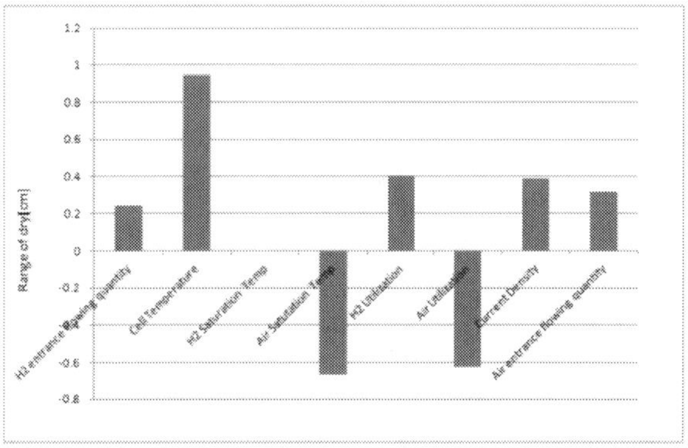

Figure 3.4 The range of dryness (cathode side)

Figure 3 The influence factors of the eight parameters

5. Influence of physical properties

In the above calculation, the physical property values used by Springer and Nguyen were used. To investigate the effect of the physical property values used in the equations, calculation was carried out using the values Zawodzinski were used. In the calculation, current density and H_2 utilization which effect the dryness and the range of dryness respectively were varied keeping the other parameters constant. The diffusion coefficients of water in the polymer membrane compared were shown in Figure 4, where the dependence of relative humidity on the diffusion coefficient is significantly different.

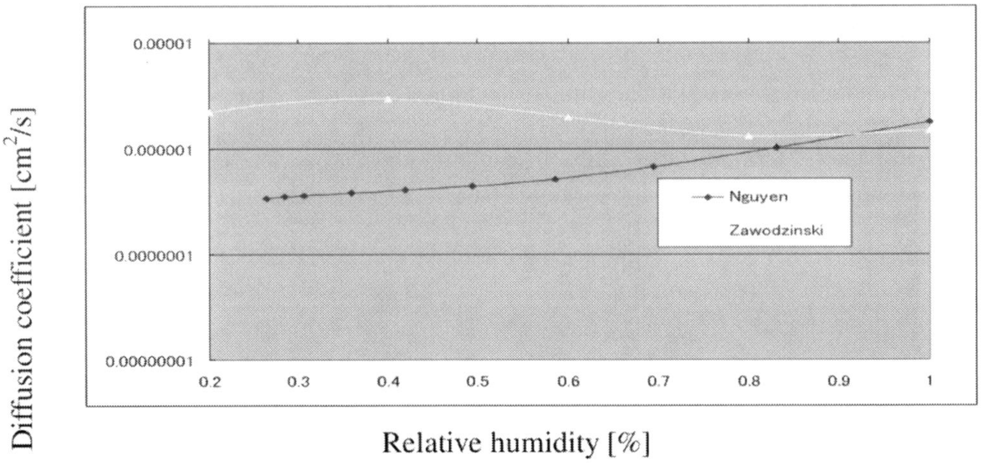

Figure 4 Distribution of the diffusion coefficient

It was found that the difference in dryness was only 5.8% in average, regardless of the significant difference shown in Figure 5 and Figure 6, thus the diffusion coefficient does not influence on the simulation significantly.

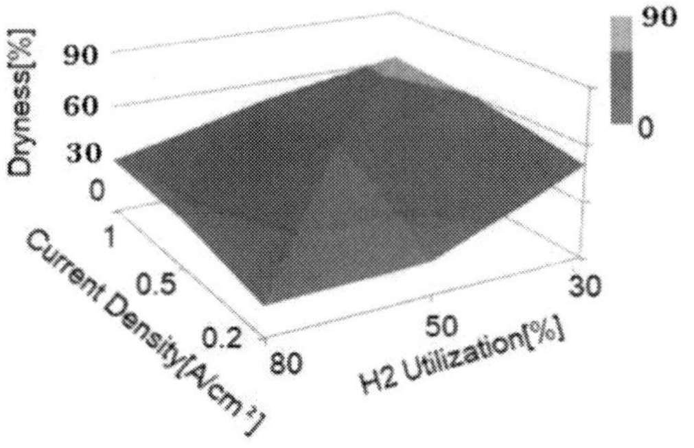

Figure 5 Dryness influence rate

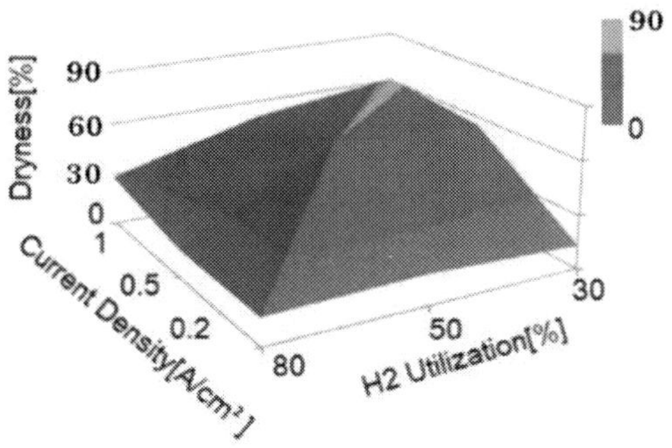

Figure 6 Dryness influence rate (other researchers)

6. Coating region of radical capturing layer

In section 4, the multiple regression equations were made for the dryness and the range of dryness in anode and cathode side. Using the equations, the map for coating region of the radical capturing layer was created in order to prevent the polymer membrane from degradation due to radical.

Two parameters having greater influence were chosen in the calculation. The range of dryness in anode side is shown in Figure 8. In this case, cell temperature and air humidity temperature were selected as the two parameters. For the cathode side , current density and air utilization ratio were varied to show the coating region of the radical capturing layer , shown in Figure 7. The vertical axis of Figure 7 and Figure 8 is the distance from inlet of gas flow channel to the region where the coating of capturing layer is required. Regarding the cathode side shown in Figure 7 the region required the coating of the capturing layer was one third of the length of gas flow channel under the condition of high air utilization ratio and low current density. In the case of high air utilization ratio and high

current density, the coating region was around a half of the length of gas flow channel. When the air utilization ratio is low , the coating region becomes wider.
Therefore, it is recommended that an air utilization of around 40% or more and the coating of one third of gas flow channel are applied in the operation condition. There is a report where the degradation occurs in anode side. Therefore, the region required the coating in anode side is shown in Figure 8. The region required coating was only one sixth of the gas flow channel in a condition of high cell temperature and air humidity temperature.

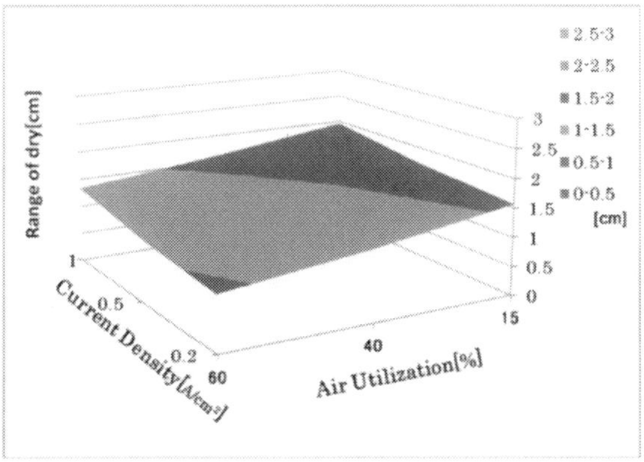

Figure 7The range of dryness (cathode side)

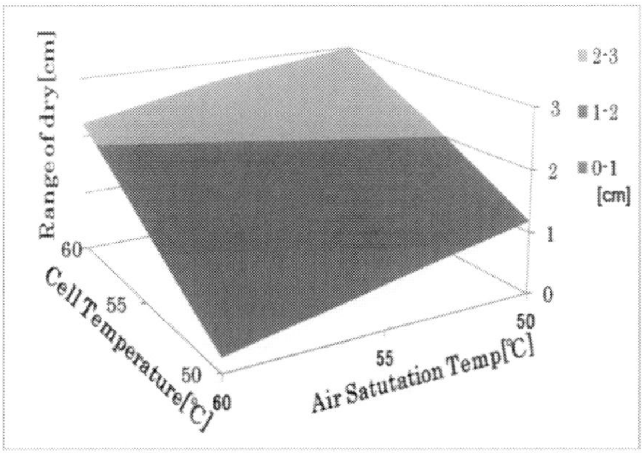

Figure 8 The range of dryness(anode side)

7. Conclusion

The distribution of water content in the membrane of PEFC was analyzed by using a numerical simulation as well as understanding the behavior of interned moisture of PEFC , and the flowing results are obtained.

1) The influence of the operating parameters on the dryness of the membrane for the anode side and the cathode side of PEFC was quantitatively shown by using the method of the multiple regression analysis.
2) Further it was found that the area where cerium carbonate ought to be coated for preventing the degradation without decreasing performance.
   2.1) Regarding the cathode side shown in Figure 7 the region required the coating of the capturing layer is one third of the length of gas flow channel under the condition of high air utilization ratio and low current density. In the case of high air utilization ratio and high current density, the coating region is around a half of the length of gas flow channel.
   2.2) Anode side region required coating is only one sixth of the gas flow channel in a condition of high cell temperature and air humidity temperature.

8. Reference

[1] Hori M, Yu J, Kobayashi K and Kato M 2005, Control of Degradation by Water Management *Proc of the 12th FCDIC Fuel Cell Symp.* pp. 89–92.
[2] Itou H, Tsurumaki S, Moriga T, Yamada A, Inoue G, Matsukuma Y and Minemoto M 2009, Study on Deterioration Mechanism of Polymer Electrolyte Fuel Cell *J. Chem. Eng. Japan* **35** pp.184–190.
[3] Itou H, Tsurumaki S, Moriga T, Yamada A, Nojima S, Inoue G, Matsukuma Y and Minemoto M 2009, Prevention of Degradation of a Polymer Electrolyte Fuel Cell *J. Chem. Eng. Japan* **35**

pp.304–311.

[4] Kobayashi T, Hirai E, Itoh H, Moriga T, 2011 Development of Production Technology for Membrane-Electrode Assemblies with Radical Capturing Layer *ASME 5th International Conference on Energy Sustainability & 9th Fuel Cell Science Engineering and Technology Conference ESFuelCell2011* Washington DC USA.

[5] Onda K, Araki T, Taniuchi T, Sunakawa D, Wakahara K and Nagahama M 2007 *J Electrochem. Soc* **154** No. 2 B247-B257

[6] Springer T E, Zawodzinski T A and Gottesfeld S 1991 Polymer Electrolyte Fuel Cell Model *J. Electrochem. Soc* **138** No.8 2334.

[7] Nguyen T V and White R E 1993 A water and Heat Management Model for Proton-Exchange-Membrane Fuel Cell *J. Electrochem Soc* **140** No.8 2178.

Performance comparison of microbial fuel cells equipped with different membrane electrode assemblies

O Rubaba[1], Y Araki[1], S Yamamoto[1], K Suzuki[1], H Sakamoto[2], A Matsuda[2] and H Futamata[1]

[1] Department of Material Sciences and Chemical Engineering, Shizuoka University, 3-5-1 Jyohoku, Naka-ku, Hamamatsu 432-8561, Japan
[2] Department of Electrical and Electronic Information Engineering, Toyohashi University of Technology, 1-1 Hibarigaoka, Tempaku, Toyohashi, Aichi, 441-8580, Japan

owenrubaba@yahoo.co.uk

Abstract. It is important for practical use of microbial fuel cells (MFCs) to not only develop new materials including electrodes and proton exchange membranes but also to understand the bacterial community structure related to electricity generation. Here, four kinds of novel membrane electrode assemblies (MEAs) were made. Four lactate fed MFCs equipped with the membranes were characterized by electrochemical, molecular-dependent and molecular-independent methods. MFC1 equipped with Nafion 117-type MEA (18 μm thickness) exhibited the highest performance. Although the other MEAs with different configurations of three kinds of polymers; poly (diallyldimethylammonium chloride), polyallylamine hydrochloride and poly (2-acrylamino-2-methyl -1-propanesulfonic acid) had thicknesses of about 0.3 μm (MEA 2 and 3) and 1.0 μm (MEA4), their power densities were lower. Denaturing gradient gel electrophoresis (DGGE) and phylogenetic analyses showed that anaerobic bacteria dominated in anode biofilms of MFC1. A bacterium completely corresponding to nucleotide sequence of one of the DGGE bands was isolated from the anode biofilm in MFC1. Interestingly, BLAST search indicated that the bacterium (named strain RO1) belonged to the genus of gram positive bacterium, *Propioniferax*. It was confirmed that strain RO1 was capable of producing electricity and constructing biofilm on the anode surface in pure culture MFC. These results suggested that the property of MEA affects significantly the bacterial community structure, thereby influencing the MFC-performance.

1. Introduction

Microbial fuel cells (MFCs) are devices that exploit microorganisms as "biocatalysts" for generating electric power from organic matter, which have been studied about energy recovery from waste as electrical power [1, 2, 3 ,4, 5] and generating power from marine aquatic sediments [6, 7], rice paddy field soil [8, 9], wheat straw biomass [10] or cellulose [11].

Although power densities have been improving steadily from a few mW m^{-2} to several hundreds of mW m^{-2}, there is need for optimization for improved power densities and development of large-scale devises [2]. Increasing the power density is one of the greatest challenges for the practical application of MFCs. Potential loses include the anodic reaction limited by microbial activity, cathodic reaction limited by the abiotic electron-accepting reaction, and oxygen intrusion through the cathode and proton exchange membrane [8, 4]. Furthermore, it is generally accepted that thinner membranes are ideal in that they exhibit lower internal resistance which is one of most important parameters [5]. It is important for accomplishment of this challenge not only to develop the materials of electrodes and proton exchange membranes but also to analyze electricity-producing bacteria.

While several membranes have been tested in chemical fuel cells, not much work on membranes has been reported with MFCs. Hence, the aims of this study were to evaluate novel ultrathin membrane electrode assemblies (MEAs) by analyzing the electrochemical properties and bacterial community structures. Four kinds of MEAs were used; Nafion 117-type MEA with a thickness of 18 μm, the other three had different configurations of three kinds of polymers; poly (diallyldimethylammonium chloride) (PDDA), polyallylamine hydrochloride (PAH) and poly (2-acrylamino-2-methyl -1-propanesulfonic acid) (PAMPS). A bacterium with 100% similarity to nucleotide sequence of one of the DGGE bands was isolated from the anode biofilm of MFC1. The electricity-producing properties and phylogenetic position were also analyzed.

2. Materials and Methods

2.1. Membrane electrode assembly

Four kinds of membrane electrode assemblies (MEAs) were made by using a casting method or a layer-by-layer method. For the casting method, 0.4 mL solution of Nafion 117 [5% (w/w) in lower aliphatic alcohols and water, Aldrich] was applied on the surface of carbon paper electrode electroplated with platinum (TGP-H-060, CHEMIX Co., Ltd., Japan. 0.5 mg Pt cm^{-2}) and then dried up. Five mL of Nafion 117 solution [5% (w/w) in lower aliphatic alcohols and water, Aldrich] and 5 mL of dH_2O were mixed and then concentrated to around 3 mL at 80°C by evaporating the alcohols in the Nafion solution. 0.6 mL of the concentrated Nafion solution was then coated onto the carbon paper and then dried up at 90°C (denoted Nafion-type MEA). The Nafion 117-type MEA (MEA1) was used on MFC1.

The other MEAs (MEA2, 3 and 4) were made by the layer-by-layer method as described previously [12, 13]. These MEAs had different configurations of three kinds of polymers, PDDA, PAH and PAMPS. MEA2 consisted of five layers of $(PDDA/PAMPS)/(PAH/PAMPS)_4$ stacked on a carbon paper electrode electroplated with platinum (0.5 mg Pt cm^{-2}) and was used on MFC2. MEA3 consisted of five layers of $(PDDA/PAMPS)/ (PAH/PAMPS)_4$ stacked on a carbon paper electrode electroplated with platinum (2.0 mg Pt cm^{-2}) and was used on MFC3. MEA4 consisted of ten layers of $(PDDA/PAMPS)/(PAH/PAMPS)_9$ stacked on a carbon paper electrode electroplated with platinum (2.0 mg Pt cm^{-2}) and was used on MFC4. The thicknesses of MEA1, 2, 3, and 4 were approximately 18 μm, 0.3 μm, 0.3 μm, and 1.0 μm, respectively.

2.2. MFC configuration and operation

Mediator less air-cathode MFCs were constructed to evaluate novel MEAs by electrochemical and microbiological analyses. Each MEA was used as both proton exchange membrane on one side and cathode electrode on the other side, providing a total projected cathode surface area (on one side) of 16 cm^2. 135 pieces of cubic (125 mm^3) graphite felts (SOHGOH-C Co., Ltd. Yokohama, Japan) were packed into the anode chamber (36 mL in capacity) and the total area of anode surface was 0.02425 m^2. 20 of these pieces were directly connected to platinum wires (0.3 mm; AlfaAesar).

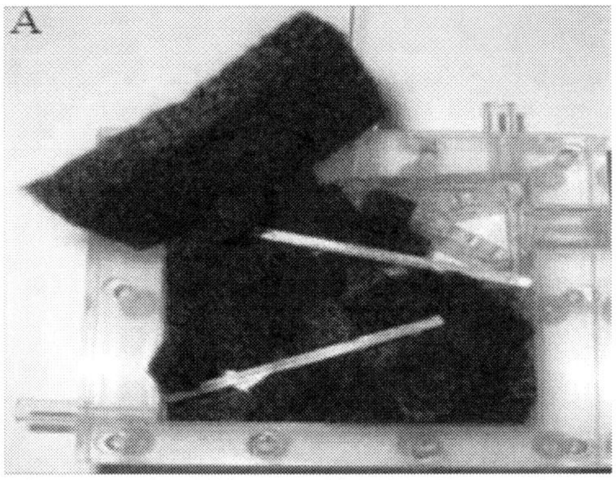

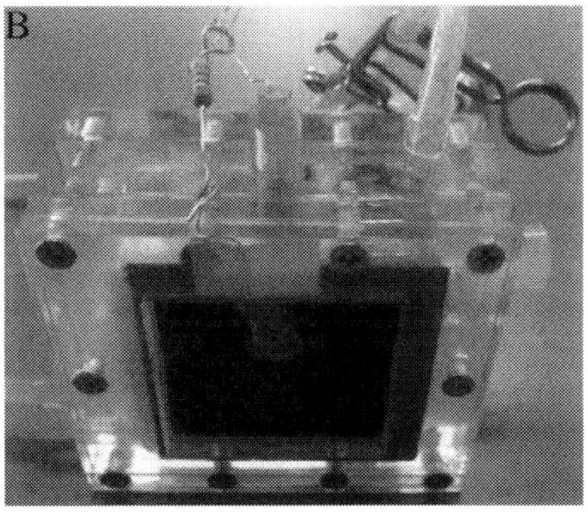

Figure 1. Graphite packing in the anode (A) and the MFC design used in this study (B)

A chemically defined anoxic medium designated BE was used for cultivation. The BE medium was modified from previously reported media [14, 15, 16], and contained 0.5 g of KH_2PO_4, 0.20 g of $MgSO_4{\cdot}7H_2O$, 0.15 g of $CaCl_2{\cdot}2H_2O$, 0.5 g of NH_4Cl, 2.5 g of $NaHCO_3$, 1.0 mL of Se/W solution [17], 1.0 mL of trace elements SL8 solution [18], and 1.0 mL of vitamin solution PV1 [19] per liter. 0.4 g of paddy field soil was inoculated into each MFC containing 20 mM lactate (DL-lactic acid, Sigma) and the electrodes were then connected with an external resistor (10 Ω) 5 days after inoculation. The MFCs were enriched by recirculation of BE medium at a rate of 25 mL h^{-1}. Lactate was added to be 20 mM in the anode chamber whenever the cell voltage dropped to baseline levels. Another MFC that was also constructed using the same materials was run under open circuit conditions as a control (denoted control MFC).

2.3. Electrochemical Measurements and operation

The voltage (V) across a 10 Ω resistor was measured every 5 min using a data loggar GL200A (GRAPHTEC, co. Ltd., Yokohama, Japan) with a data acquisition system connected to a personal computer. Power (P) was calculated according to $P = VI$, and maximum power densities were calculated using the maximum power values that remained constant for approximately 4 hours (corresponding to each substrate feeding) divided by the total apparent surface area of the electrode (0.02025 m^2). Polarization curves were obtained by imposing a linear potential decrease of 1 mV s^{-1} from the open circuit voltage potential to a cell potential of 0 mV followed by a linear voltage increase of 1 mV s^{-1} of the cell potential by a potentiostat HSV-110 (Hokuto Denko, Tokyo, Japan). Coulombic efficiency was obtained by calculating the ratio of total recovered coulombs by integrating the current over time to the theoretical amount of coulombs that can be produced from lactate [20].

2.4. HPLC and organic acids

Organic acids were measured using a high-pressure liquid chromatograph (GL-7400 series, GL Science Inc., Shinjuku-ku, Tokyo, Japan) equipped with a Shodex RSpak KC-811 column (300 × 8.0 mm) (SHOWA DENKO Co. Ltd., Kanagawa, Japan) and UV detector. Column oven was set at 50°C. Samples were eluted with 0.1% H_3PO_4 solution and elutes were monitored at 210 nm. Formate, pyruvate, lactate, butyrate and acetate were identified according to the retention time and the concentration was determined by comparing the peak area with that of its respective standard sample.

2.5. Microbial analysis

2.5.1. *DNA extraction.* Aliquots of anolytic culture (1 mL) from MFCs were centrifuged for 10 min at 4°C and 20,000 ×g, and cells were collected. Anode samples (2 pieces) from MFCs were taken out and washed thoroughly in 10 mM sodium phosphate buffer (pH 7.0) and were crushed using a sterile pipette tip. DNA was extracted according to the conventional method described previously [14].

2.5.2. *PCR-DGGE.* To analyze the total bacterial community structure, the variable V3 region of bacterial 16S rRNA genes (corresponding to positions 341 to 534 in the *Escherichia coli* sequence) was amplified using primers P2 and P3 (containing a 40-bp GC clump) [21] and a thermal cycler (ASTEC PC320, Astec Co,. Ltd., Fukuoka, Japan) as described previously [21]. A DCode DGGE system (Bio-Rad, CA, USA) was used for electrophoresis following the manufacturer's instructions. A total of 10 μL of a PCR-amplified mixture was subjected to electrophoresis in a 10% (wt/vol) polyacrylamide gel at 200 V for 3.5 h. The gel gradients used for separation, which were applied in parallel to the electrophoresis direction, were 40 to 60%. After electrophoresis, the gel was stained with SYBR Green I (Molecular Probes) for 30 min according to the manufacturer's instructions. The nucleotide sequences of DGGE bands were determined as described previously [22].

2.5.3. *Isolation of bacteria.* BE medium was used for isolation of anaerobic bacteria. Lactate (final concentration 20 mM) was used as an electron donor and gellan gum was added to a final concentration of 2.0% w/v. Either Na_2SO_4 or Fe(III) EDTA was used as a sole electron acceptor at a final concentration of 20 mM and 5 mM, respectively. The diluted culture suspension was directly spread onto a solid plate containing either medium under anaerobic conditions. The plate was placed in a nylon bag (Mitsubishi Gas Chemical Company, Tokyo, Japan) with a catalyst sachet (AnaeroPouch-Anaero; Mitsubishi Gas Chemical Company) and was then quickly sealed with a clip. One-tenth strength Trypticase soy broth (TSB) medium (1/10TSB) was also used for isolation of aerobic bacteria. After the plates were incubated at 25°C for three weeks, colonies formed were picked and spread onto the same medium to confirm their purity.

2.5.4. *Sequencing of 16S rRNA gene.* Purified bacterium, named strain RO1, was incubated in an anaerobic BE medium containing lactate (20 mM) and Na_2SO_4 (20 mM) at 25°C for 10 days. The DNA was extracted from cells using a method described previously [14]. The quality and quantity of the extracted DNA was checked by measuring the absorbance at 260 and 280 nm. Almost-full-length fragment of 16S rRNA gene was amplified using the 27f (5'-AGAGTTTGATCCTGGC TCAG-3', *E. coli* 16S rRNA gene positions 8 to 27) and 1525r (5'-AAGGAGGTGATC CAGCC-3', positions 1525 to 1542). Amplification was performed with the PC320 thermal cycler (ASTEC) by using a 50 μL reaction volume as described previously [14].

2.5.5. *Scanning electron microscopy observation.* Anode electrode particles were taken out and were washed three times with 10 mM phosphate buffer. The anodes were fixed in a 2.5% solution of glutaraldehyde and processed through an ethanol dehydration series (i.e.10, 15, 25, 50, 75, 100% v/v ethanol. Each treatment was carried out for 15 min and repeated twice) as described previously [23]. The samples were then coated with gold using a coater device (SC-701AT Quick Auto Coater, Sanyo, Tokyo, Japan) set at 50Å and observed using a scanning electron microscope (S3000N, HITACH, Japan) at 7 kV.

2.5.6. *Accession numbers.* The nucleotide sequence data reported here have been deposited in the DDBJ under accession numbers AB693055 to AB693099.

3. Results and Discussion

3.1. Characterization of current generation

The electricity producing capabilities were different from each other. Unexpectedly, MFC1 equipped with Nafion 117-type MEA had the highest powert density although it had the thickest membrane (Figure 2). MFC3 and MFC4 produced comparatively very low power production.

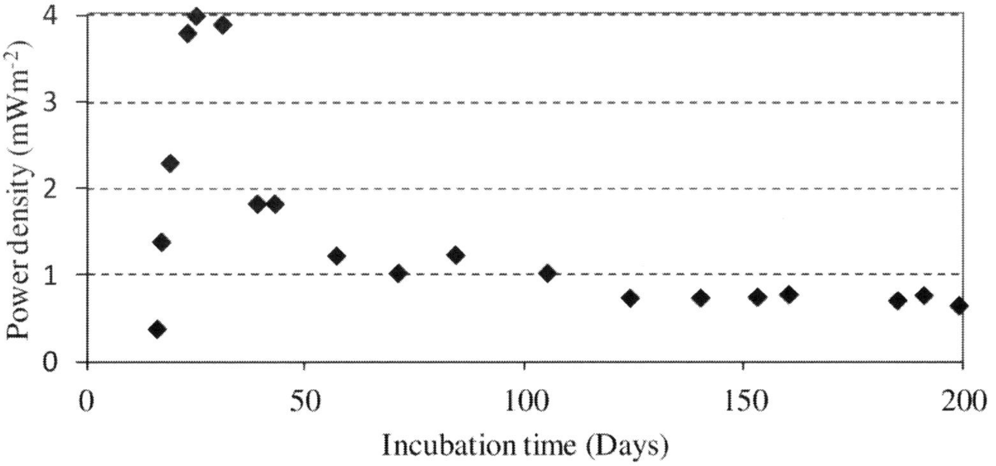

Figure 2. MFC1 power density

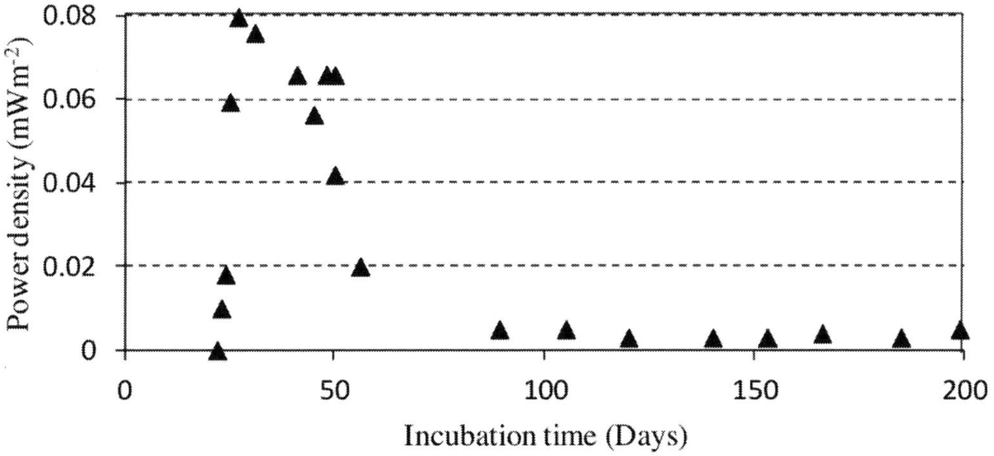

Figure 3. MFC2 power density

MFC1 and MFC2 had a similar trend of an initially high power production followed by a lower, stable and reproducible phase (Figure 2 and 3). Power production occurred in cycles depending on degradation of lactate and concentration of resultant organic acids. The power generation of MFC1 started on day 12 and the power density was about 4 mW m^{-2} during day 20 to day 40. The power density slightly decreased later, and was stable at around 1 mW m^{-2}. Correspondingly, the coulombic efficiency of MFC1 changed from 35 ± 4% (during day 20 to day 40) to 16 ± 2% (after day 50). This result indicated that the electron flow changed dramatically in the anode of MFC1. The power density of MFC2 equipped with MEA2 was as high as 0.08 mW m^{-2} during day 25 to day 50, and finally became stable at less than 0.02 mW m^{-2}. The coulombic efficiencies of MFC2 were slightly less than

5%. These results suggested that bacterial community succession occurred and their community structures were different from each other.
Polarization curve analyses showed that the electricity-producing properties of MFC1 changed (Table 1). Polarization curve analyses of MFC2, 3 and 4 could not be performed due to their low power production. In the initial stage until day 38, MFC1 maximum power and current densities were 5.8 mW m^{-2} and 95 mA m^{-2}, respectively. However these values decreased to 1.2 mW m^{-2} and 37 mA m^{-2} in the final stage from day 158 to day 197.

Table 1. MFC1 Linear Sweep Voltammetry showing electricity producing properties

Day	P_{max} (mW/m^2)	I_{max} (mA/m^2)	R_{int} (ohms)	OCV (mV)
18	6.2	90	90	205
38	5.3	100	80	198
158	1.0	34.5	187	124
170	1.0	34.3	175	126
197	1.3	39.8	146	131

Along with these changes, internal resistance values increased from 90 Ω to 170 Ω. Since the MEAs were completely covered by a biofilm when the experiment was terminated, it was inferred that the fouling of MEA caused the decrease of proton permeability, resulting in changes in electrochemical properties. It has also been reported that the fouling of the cathode results in a reduction of coulombic efficiency due to both utilization of organic acids and a decrease in the active cathode surface area by the cathode biofilm [24].

3.2. Bacterial community structure

Bacterial community structures of MFCs were analyzed by DGGE technique and were then analyzed phylogenetically (Table 2). Since the power densities of MFC1 and MFC2 were significantly different from each other, it was predicted that their bacterial community structures would be also different from each other. Interestingly, dominant anolyte-bacteria and biofilm-bacteria from MFC1 and MFC2 did not make distinct clusters based on MFC of origin. The bacteria from both MFCs were present in almost all clusters, suggesting that members of these bacterial communities were almost similar (Table 2). It therefore appears relative abundances of the members or other factors like oxygen intrusion into the MFCs may be responsible for the differences in power production.
Almost half of DGGE bands sequenced from MFC1 and MFC2, which were mostly from anode biofilm, belonged to *Firmicutes* containing the genera of *Pelosinus* and *Psychrosinus*. *Firmicutes* are generally known as fermenters and it has been reported that *Pelosinus fermentans* strains are capable of fermenting lactate and coupling its oxidation to Fe(III) and metal reduction [25, 26]. *Psychrosinus fermentans* FCF9 is reported as an obligately anaerobic lactate fermenter [27]. These results suggest that these bacterial community structures adapted to transfer of electrons from lactate to anode electrode.

Table 2. DGGE band sequence analysis

MFC	Band	Nearest neighbor	Accession number	Similarity
MFC1	1-1	*Sphingobacterium multivorum*	HE800586	95%
	1-4	*Oscillatoria* sp. INT-06	AB046728	100%
	1-6	*Veillonellaceae bacterium* FCF9B	DQ767882	97%
	1-8	*Delftia* sp. BCA19	HE716888	95%
	1-9	*Leclercia adecarboxylata* strain MN11SW4	JX501721	100%
	1-11	*Enterobacter hormaechei* strain skg0062	HQ322394	97%
	1-12	*Pelosinus* sp. UFO1	EU215386	97%
	1-14	*Dechlorosoma* sp. PCC	AY126453	100%
	1-15	*Propionispora hippei*	AJ508928	99%
	1-16	*Psychrosinus fermentans* strain FCF9	DQ767881	92%
	1-17	*Sphingobacterium multivorum*	HE800586	100%
	1-19	*Pseudomonas indica* strain NR4_17	JX843378	100%
	1-20	*Veillonellaceae bacterium* FCF9B	DQ767882	97%
	1-21	*Oscillatoria* sp. INT-06	AB046728	100%
MFC2	2-1	*Azospirillum* sp. C10	JX842779	100%
	2-2	*Azospirillum* sp. C10	JX842779	100%
	2-3	*Dechlorospirillum* sp. DB	AY530551	100%
	2-4	*Pseudomonas stutzeri*	FR871674	97%
Control	C-1	*Sphingobacterium multivorum*	HE800586	100%
	C-2	*Chryseobacterium* sp. JC145	HE962139	97%
	C-3	*Delftia* sp. aer2	JX501738	99%
	C-4	*Citrobacter freundii* strain HME8594	JX426059	99%
	C-5	*Sporotalea propionica*	AM258975	97%
	C-6	*Acinetobacter* sp. A1060	JX266319	100%
	C-7	*Klebsiella* sp. IARI-ALC-11	JX512192	100%
	C-8	*Klebsiella* sp. IARI-ALC-11	JX512192	100%
	C-9	*Pelosinus* sp. UFO1	EU215386	96%
	C-10	*Delftia* sp. aer2	JX501738	99%
	C-11	*Anaerovibrio burkinabensis* DSM 6283(T)	AJ010961	96%
Biofilm	A1-1	*Pseudomonas* sp. ISTDF	GQ202116	92%
	A1-2	*Clostridium* sp. SW001	HM755724	95%
	A1-3	*Sporotalea propionica*	AM258975	97%
	A1-4	*Propioniferax* sp. P7	EU109728	100%
	A1-5	*Clostridium quercicolum*	AJ010962	97%
	A1-6	*Desulfovibrio inopinatus*	AF177276	91%
	A1-7	*Pelobacter propionicus* DSM 2379	CP000482	93%
	A1-8	*Dechlorosoma* sp. PCC	AY126453	100%
	A1-9	*Veillonellaceae bacterium* FCF9B	DQ767882	96%
	A2-10	*Macellibacteroides fermentans* strain LIND7H	HQ020488	99%
	A2-11	*Clostridium quercicolum*	AJ010962	97%
	A2-12	*Sporotalea propionica*	AM258975	98%
	A2-13	*Desulfovibrio desulfuricans* strain E062	JX267091	99%
	AC-14	*Macellibacteroides fermentans* strain LIND7H	HQ020488	99%
	AC-15	*Klebsiella* sp. IARI-ALC-11	JX512192	100%

Other dominant biofilm-bacteria of MFC1 and MFC2 belonged to several classes, α-, β-, γ-, δ-*Proteobacteria*, *Bacteriodetes*, and *Actinobacteria* (Table 2). A bacterium with 100% similarity to the DGGE band A1-4 was isolated in this study. The strain was named *Propioniferax* sp. strain RO1 due to its 98.2% similarity to Gram-positive *Propioniferax* sp. P7 (Table 2). To investigate whether the stain RO1 was capable of producing electricity, the strain RO1 was inoculated into an air-cathode MFC equipped with normal Nafion 117 membrane (170 μm thickness) and lactate as an electron donor. The electrochemical analysis showed that the strain RO1 was capable of producing electricity (Table 3). This is the first report of electricity production for the genus of *Propioniferax*. We do not know the exact mechanism for electron transfer by this bacterium.

Table 3. *Propioniferax* strain RO1 pure culture MFC Linear Sweep Voltammetry showing electricity producing properties

Day	P_{max} (mW/m^2)	I_{max} (mA/m^2)	R_{int} (ohms)	OCV (mV)
13	1.24	13.8	17586	584
29	1.23	21.28	5750	359
40	1.73	34.82	3762	349

SEM observation showed presence of a biofilm of strain RO1 on anode without filamentous structures (Figure. 4), raising the possibility of direct electron transfer. Direct electron transfer to an insoluble electron acceptor was once considered incompatible with a Gram-positive bacterial envelope [28]. However, it was demonstrated that Gram-positive *Thermincola potens* strain JR has extracellular electron transfer mechanism using *c*-type cytochromes [29].

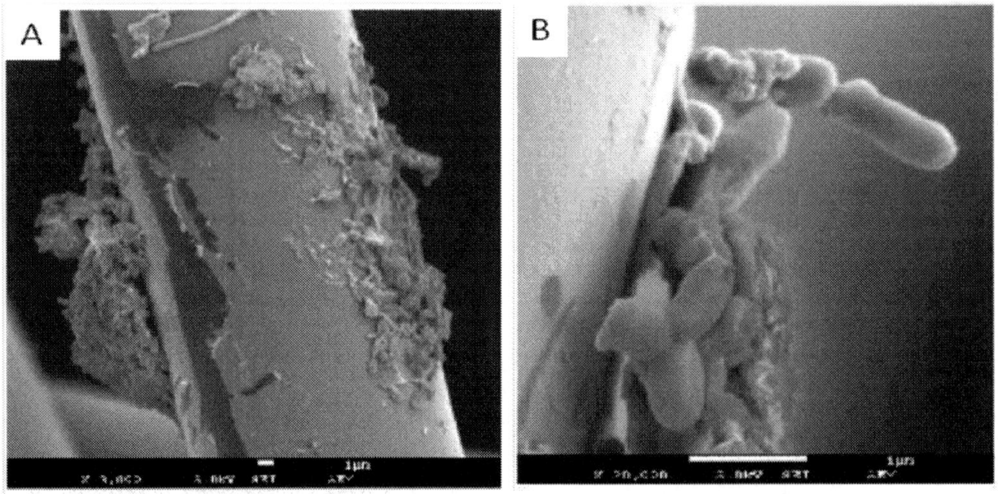

Figure 4. SEM images of *Propioniferax* strain RO1 pure culture attached to graphite anode (A) and a close up image (B)

Part of our future work will be to investigate how the strain RO1 produces electricity, i.e., analyses of *c*-type cytochromes or mediator-like compounds.

Most of DGGE bands from anolytic cultures of MFC1 and MFC2 were closely related to genera of *Sphingobacterium*, *Pseudomonas*, *Enterobacter*, and *Azospirillum* (Table 2), which are facultative anaerobes. These results suggested that these bacteria were not only lactate-fermenters but also oxygen-scavengers. Since some strains belonging to these genera are known as exoelectrogens [30, 11], it was suggested that anolytic bacteria also play a role in electricity production. SEM observation revealed that there were chasms and pores in MEA2, MEA3, and MEA4 (data not shown), suggesting oxygen-intrusion into the anode chamber. MEA1 had no chasms and pores, resulting in higher electricity production by MFC1 compared to the other MFCs. It was demonstrated that temporary oxygen intrusion into anode chamber is not necessarily irreparable damage to MFC performance, because bacterial communites adapted swiftly from semi-aerobic to anaerobic conditions [31, 32]. However, it was indicated that MFC performance is damaged seriously by continuous oxygen intrusion. Thicker membranes have been shown to give a lower rate crossover but tend to have higher resistances, reducing the cell performance. Therefore, there exists an optimal thickness for the membranes, which gives the best performance [33]

Table 4. Cell densities on the anode surface on day 112

Electron acceptor	MFC 1 (cfu/cm^2 anode)	MFC2 (cfu/cm^2 anode)	Control (cfu/cm^2 anode)
FE(III) EDTA	$3.9 \pm 2.4 \times 10^5$	$7.3 \pm 0.2 \times 10^4$	$5.5 \pm 3.7 \times 10^4$
Na_2SO_4	$1.7 \pm 5.0 \times 10^5$	$2.5 \pm 4.2 \times 10^5$	$6.2 \pm 3.1 \times 10^4$
O_2	$4.5 \pm 0.9 \times 10^7$	$1.7 \pm 0.1 \times 10^7$	$5.0 \pm 0.1 \times 10^5$

The differences of oxygen-permeability affected the population densities. Significantly, Fe (III)-reducing bacterial population of MFC1 was approximately 5-fold higher than that of MFC2 (Table 4), which may be one of the reason why the current density of MFC1 was higher than that of MFC2.

4. Conclusion

It was demonstrated that bacterial community structure is significantly affected by the device feature. The gram-positive *Propioniferax* sp. RO1 was isolated from the anode of MFC1, and may have played a role in electricity generation in the MFC. Since the layer-by-layer method is capable of making a very thin layer, an effective MEA might be made using an extremely flat cathode. It is important to evaluate MFC performance equipped with new PEM by electrochemical and microbiological analyses. This subject is currently under investigation.

Acknowledgements

This research was supported in a part by grant 22310045 funded by the Ministry of Education, Sports, Culture, Science and Technology (Japan), and also partially founded by ALCA project, Japan Science and Technology Agency.

References

[1] Feng Y, Wang X, Logan B E and Lee H 2008 *Appl. Microbiol. Biotechnol.* **78** 873-880
[2] Logan B E, Hamelers B, Rozendal R, Schroder U, Keller J, Freguia S, Aelterman P, Verstraete

W and Rabaey, K 2006 *Environ. Sci. Technol.* **40** 5181-519.
[3] Logan B E and Regan J M 2006 *Trends. Microbiol.* **14** 512-518
[4] Rabaey K and Verstraete W 2005 *Trends Biotechnol.* **23** 291-298
[5] Shimoyama T, Komukai S, Yamazawa A, Ueno Y, Logan B E and Watanabe K 2008 *Appl. Microbial. Biotechnol.* **80** 325-330
[6] Lovley D R 2006 *Curr. Opin. Biotechnol.* **17** 327-332
[7] Tender L M Reimers C E, Stecher III H A, Holmes D E, Bond D R, Lowy D A, Pilobello K, Fertig S J and Lovley D R 2002 *Nat. Biotechnol.* **20** 821-825
[8] Ishii S, Shimoyama T, Hotta Y and Watanabe K 2008 *BMC Microbiol.* 8, 6 2008, 8:6 doi:10.1186/1471-2180-8-6 (This article is available from: http://www.biomedcentral.com/1471-2180/8/6)
[9] Kaku N, Yonezawa N, Kodama Y and Watanabe K 2008 *Appl. Microbial. Biotechnol.* **79** 43-49
[10] Zhang Y, Min B, Huang L and Angelidaki I 2009 *Appl. Environ. Microbiol.* **75** 3389-3395
[11] Rezaei F, Xing D, Wagner R and Regan J M 2009 *Appl. Environ. Microbiol.* **75** 3673-3678
[12] Sakamoto H, Daiko Y, Katagiri K, Muto H, Sakai M and Matsuda A 2010 *Solid State Ionics*.**181** 210-214
[13] Sakamoto H, Daiko Y, Muto H, Sakai M and Matsuda A 2012 *Soild State Ionics*. doi:10.1016/j.ssi.2012.01.025
[14] Futamata H, Yoshida N, Kurog, T, Kaiya S and Hiraishi A 2007 *The ISME J.* **1** 471-479
[15] Futamata H, Kaiya S, Sugawara M and Hiraishi A 2009 *Microbes Environ.* **24** 330-337
[16] Ishii S, Watanabe K, Yabuki S, Logan B E and Sekiguchi Y 2008b *Appl. Envrion. Microbiol.* **74** 7348-7355
[17] Löffler F E, Sanfold R A and Tiedje J M 1996 *Appl. Environ. Microbiol.* **62** 3809-3818
[18] Biebl H and Pfennig N 1978 *Arch. Microbiol.* **117** 9-16
[19] Hiraishi A, Yonemitsu Y, Matsushita M, Shin Y K, Kuraishi H and Kawahara K 2002 *Arch. Microbiol.* **178** 45-52
[20] Liu H and Logan B E 2004 *Envion. Sci. Technol.* **38** 4040-4046
[21] Müyzer G, de Waal E G and Uitterlinden A G 1993 *Appl. Environ. Microbiol.* **59** 695-700
[22] Futamata H, Nagano Y, Watanabe K and Hiraishi A 2005 *Appl. Environ. Microbiol.* **71** 904-911
[23] Gorby Y A *et al* 2006 *Proc. Natl. Acad. Sci. USA.* **103** 11358-11363
[24] Chung K, Fujiki I and Okabe S 2011 *Bioresource Technol.* **102** 355-360
[25] Shelobolina E S *et al* 2007 *Int. J. Syst. Evol. Microbiol.* **57** 126-135
[26] Mosher J J *et al* 2012 *Appl. Environ. Microbiol.* **78** 2082-2091
[27] Matthew S M, Jung D O, Madigan M T 2008 *FEMS Microbiol. Lett.* **287** 121-127
[28] Pham T H *et al* 2008 *Appl. Microbiol. Biotechnol.* **77** 1119-1129
[29] Wrighton K C *et al* 2011 *Appl. Envrion. Microbiol.* **77** 7633-7639
[30] Rabaey K, Boon N, Sicilliano S D, Verhaege M, Verstraete and W 2004. *Appl. Environ. Microbiol.* **70** 5373-5382
[31] Futamata H, Bretschger O, Cheung A, Kan J, Rubaba O and Nealson K H. 2012 *J. Biosci. Bioeng.* http://dx.doi.org/10.1016/j.jbiosc.2012.07.016
[32] Oh S E, Kim J R, Joo J H and Logan B E. 2009 *Water Sci. Technol.* **60** 1311-1317.
[33] Liu JG, Zhao T S, Liang Z X and Chen R 2006 *J. Power Sources.* **153** 61-67

Journal of Physics: Conference Series **433** (2013) 012023 doi:10.1088/1742-6596/433/1/012023

Fabrication and characterization of polymer tandem solar cells

Haruto Maruhashi*, Atsushi Suzuki, Tsuyoshi Akiyama and Takeo Oku

Department of Materials Science, The University of Shiga Prefecture
2500 Hassaka, Hikone, Shiga 522-8533, Japan

*E-mail: zn21hmaruhashi@ec.usp.ac.jp, suzuki@mat.usp.ac.jp

Abstract. Polymer tandem solar cells were fabricated and characterized. Copolymer poly(3-hexylthiophene-2,5-diyl) (P3HT) and poly((4,4-octyldithieno(3,2-b:20,30-d)silole)-2,6-diyl-alt-(2,1,3-benzothiadiazole)-4,7-diyl) (PSBTBT) were used as p-type semiconductors, and [6,6]-phenyl-C_{61}-butyric acid methyl ester (PCBM) was used as an n-type semiconductor. The open circuit voltage was improved up to 0.69 V by using the tandem structure of P3HT:PCBM and PSBTBT:PCBM layers as compared with those of P3HT:PCBM and PSBTBT:PCBM single cell of 0.53 V and 0.10 V, respectively. Absorption of the tandem cell was in the range of 400 to 800 nm, which corresponded to a sum of those in P3HT:PCBM and PSBTBT:PCBM single cells.

1. Introduction

Organic solar cells have advantage to develop as flexible, low-cost and light weight products. Tandem solar cells have an effective way to harvest a high performance of solar radiation by combining two or more solar cells with different energy bands. In fabricating tandem solar cells, active layer materials based on thiophene block-copolymer:[6,6]-phenyl-C_{61}-butyric acid methyl ester (PCBM) and poly(3-hexylthiophene-2,5-diyl) (P3HT):PCBM as back and front cells have been used for utilizing narrow band gap polymers in a wide range of wavelength, and provided improvement of power conversion efficiency [1]. The copolymer poly((4,4-octyldithieno(3,2-b:20,30-d)silole)-2,6-diyl-alt-(2,1,3-benzothiadiazole)-4,7-diyl) (PSBTBT) is silole-containing polymer. Molecular structure of PSBTBT is shown in Fig. 1. PSBTBT is one of gathering attention low band gap polymers. PSBTBT has advanced features with a wide wavelength and carrier mobility (3×10^{-3} cm^2 V^{-2} s^{-1}) and advantages for narrow band-gap polymer solar cells [2, 3, 11].

The purpose of the present work is fabrication and characterization of polymer tandem solar cells with an inverted structure using PSBTBT [2, 3] as narrow band gap copolymer. Narrow band gap polymer with long absorption has advantage to develop as organic solar cell with a high performance of photovoltaic properties. That feature makes that tandem organic solar has absorption with broad wavelength. Tandem solar cell has a great potential to utilize sunlight effectively with increasing open circle voltage. Photovoltaic mechanism and optical properties were characterized for optimization of thickness and annealing condition by thermal treatment.

2. Experimental

As device fabrication, the device structure of the inverted polymer tandem solar cell is shown in Fig. 1. Photovoltaic cells were fabricated on indium tin oxide (ITO) coated glass substrates. The TiO_2 (Solaronix, HT-LSC) layer was spin-coated at 3000 rpm for 60 s and thermally annealed at 120 °C for 5 min. The P3HT:PCBM at a 5:4 weight ratio in *o*-dichlorobenzene solution (25 mg of P3HT in 1 mL of solvent) was spin-coated at 700 rpm for 60 sec on top of a layer of TiO_2. The polyethylene-dioxythiophen doped with polystyrene-sulfonic acid (PEDOT:PSS) (Sigma Aldrich) was spin-coated at 2000 rpm for 60 sec on top of layer of P3HT:PCBM and thermally annealed at 150 °C for 5 min. MoO_3 layer was thermally evaporated on top of the bottom cell. After evaporating, the TiO_2 (Solaronix, HT-LSC) layer was spin-coated at 3000 rpm for 60 sec and thermally annealed at 120 °C for 5 min on top of a layer of MoO_3. The second active layer PSBTBT:PCBM at a 5:4 weight ratio in *o*-dichlorobenzene solution (25 mg of PSBTBT in 1 mL of solvent) was spin-coated at 700 rpm for 60 sec. The PEDOT:PSS was spin-coated at 2000 rpm for 60 sec on top of layer of PSBTBT:PCBM and thermally annealed at 150 °C for 5 min. Finally, Au as the anode were evaporated and annealed at 150 °C for 5 min. The other tandem cell was prepared by two active layers of the P3HT:PCBM as front cell and P3HT:PCBM as back cell. The current density-voltage (J-V) characteristics (Hokuto Denko, HSV-110) of the solar cells were measured under illumination at 100 mW/cm^{-2} by using an AM 1.5 solar simulator (San-ei Electric, XES-301S). The solar cells were illuminated through the side of the ITO substrates and the illuminated area was 0.16 cm^2. Optical absorption of the solar cells was investigated by UV-vis spectroscopy (JASCO V-670). Incident photon to current conversion efficiency (IPCE) were measured by PECCELL PEC-S20. Surface morphology was observed by atomic force microscope (AFM, SII SPA-400) using contact mode.

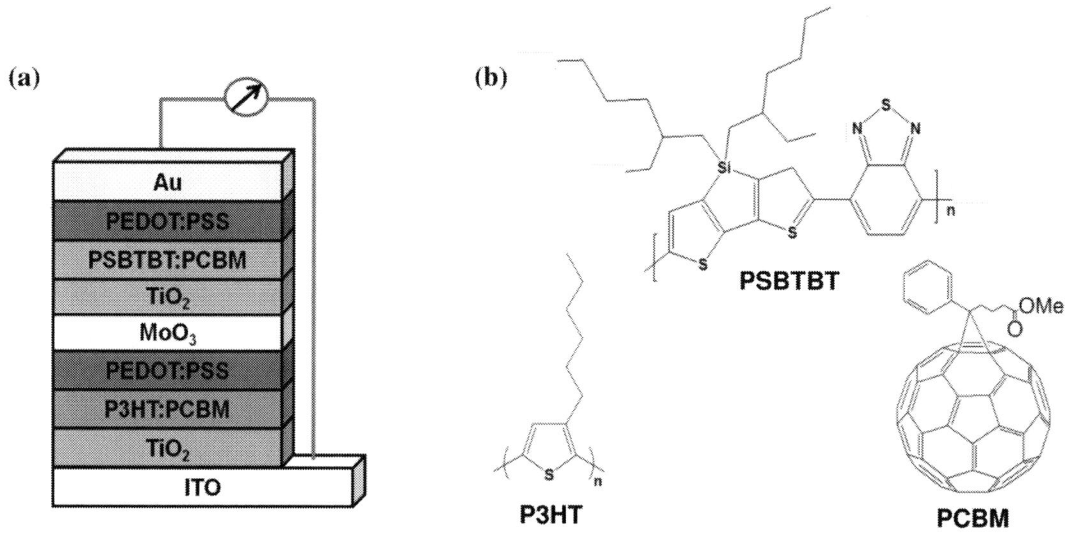

Figure 1. (a) Structure of P3HT/PSBTBT tandem solar cell. (b) Molecular structures of PSBTBT, P3HT and PCBM.

3. Results and Discussion

Photo-induced currents density depending on voltage (J-V) characterizations of the heterojunction solar cells in tandem and single cell were measured. The light induced J-V curves displayed an increase of current for increasing bias voltage with measured parameters including open circuit voltage (V_{oc}), short circuit current density (J_{sc}) and conversion efficiency (η). The current density-voltage characteristics in tandem and single solar cells showed a photovoltaic characteristic behavior with an increase of short circuit current density with increasing voltage. The performance had improvement of open circuit voltage with sum of single cells. The measured parameters of P3HT:PCBM + PSBTBT:PCBM tandem solar cell were obtained to be $J_{SC} = 2.1\times10^{-2}$ mA cm^{-2}, V_{OC} = 0.69 V, FF = 0.26, $\eta = 3.6\times10^{-3}$, and those of P3HT:PCBM + P3HT:PCBM tandem cell were J_{SC} = 2.2×10^{-2} mA cm^{-2}, V_{OC} = 0.58 V, FF = 0.34, $\eta = 3.8\times10^{-2}$. In the case of single cell using P3HT:PCBM the V_{OC} and J_{SC} were obtained to be 3.23 mA cm^{-2}, 0.53 V, and those of single cell using PSBTBT:PCBM 2.2×10^{-2} mA cm^{-2}, 0.10 V respectively. Comparison between the parameters in tandem and single cells was investigated by the experimental results. The photovoltaic performance in the tandem solar cell revealed improvement of the measured parameters with a slight rising of V_{oc} with lowing value of J_{sc}, FF and η. These results in the photovoltaic performance were originated in light-induced separation, charge transfer and carrier diffusion in the microstructure near interface between the active layer of P3HT:PCBM and PSBTBT:PCBM, P3HT:PCBM and P3HT:PCBM through intermediate layer with a broad wavelength.

Table 1. Measured parameters of the present tandem and single solar cells.

Sample	Structure	Jsc (mA cm^{-2})	Voc (V)	FF	η (%)
P3HT:PCBM + PSBTBT:PCBM	tandem	0.021	0.69	0.26	3.6×10^{-3}
P3HT:PCBM + P3HT:PCBM	tandem	0.19	0.58	0.34	3.8×10^{-2}
P3HT:PCBM	single	3.23	0.53	0.49	0.84
PSBTBT:PCBM	single	0.022	0.10	0.26	5.5×10^{-4}

Figure 2 (a) shows UV-vis spectra and (b) IPCE of the heterojunction solar cells in tandem and single solar cells. The heterojunction solar cell in tandem solar cell had a strong optical absorption at 320 nm, which was converted to energy level at 3.6 eV. The optical absorption corresponded identification of PCBM at 320 nm as energy band and in the range of 400 to 800 nm as energy band for P3HT and PSBTBT. There existed a wide excited energy of P3HT and PSBTBT in the photosensitive layer. The absorption of the tandem cell corresponded to a sum of those in P3HT:PCBM and PSBTBT:PCBM single cells in the range of 400 to 650 nm and 450 to 800 nm. IPCE was estimated to be from 0.16 - 0.02 in the range of 300 to 800 nm in wavelength.

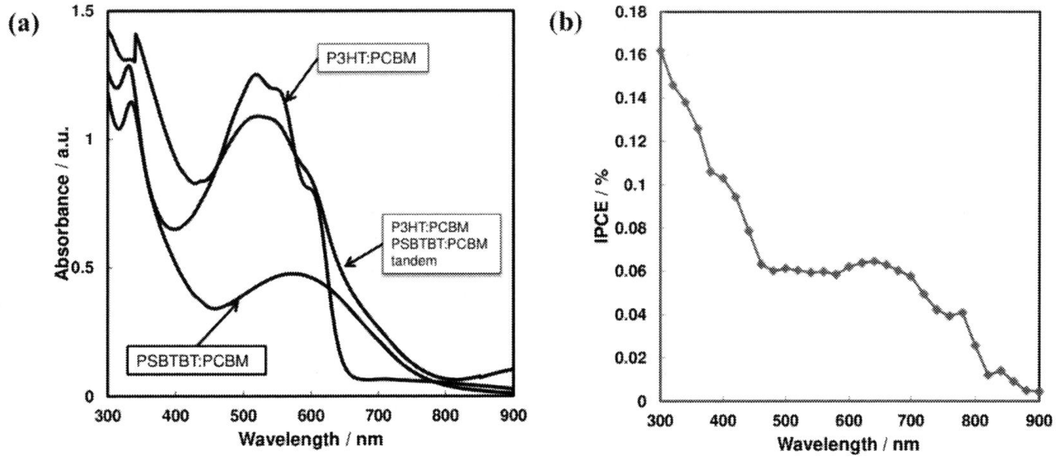

Figure 2. (a) UV-vis absorption spectra and (b) IPCE of P3HT:PCBM + PSBTBT:PCBM tandem cell.

Figure 3(a) and 3(b) show AFM images of P3HT:PCBM and PSBTBT:PCBM thin films after annealing at 150 °C, respectively. The thin film of P3HT:PCBM has rough and irregular surface in continuous phase with phase separation as shown in Fig. 4, yielding a low mobility of carriers across interface between PSBTBT:PCBM phase [7-9]. In contrast, the surface interface in PSBTBT:PCBM thin film showed flat structure with several holes. These holes were from 10 to 60 nm in depth. P3HT:PCBM and PSBTBT:PCBM films were both about 300 nm thick.

(a) (b)

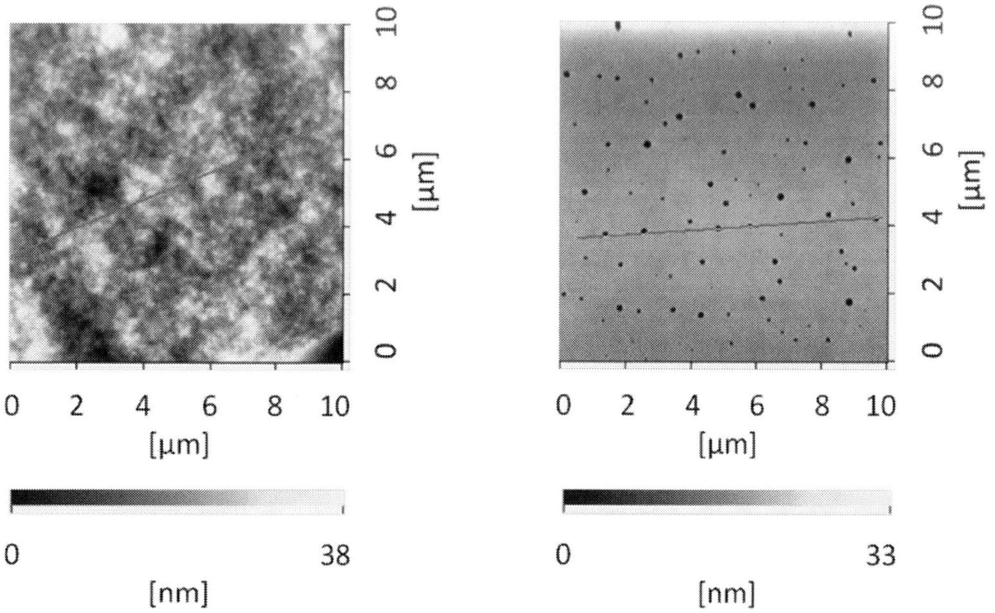

Figure 3. AFM images of (a) P3HT:PCBM and (b) PSBTBT:PCBM films.

(a)

(b)

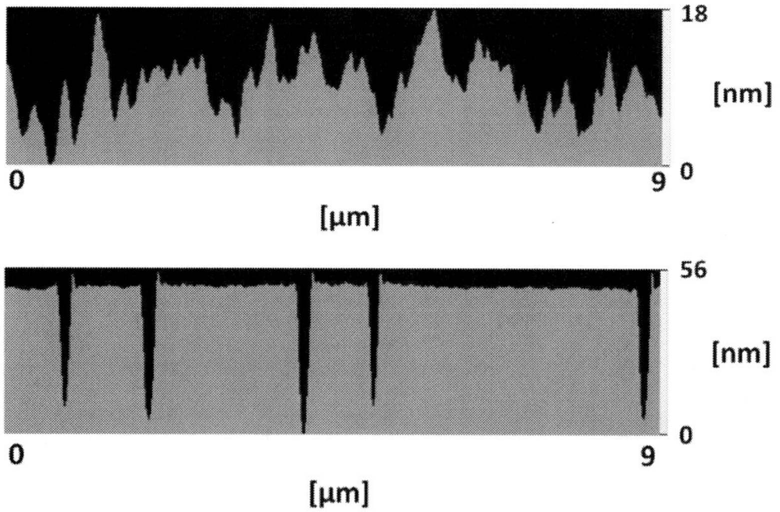

Figure 4. Cross-sectional AFM images of (a) P3HT:PCBM and (b) PSBTBT:PCBM thin films, corresponding lines as shown in Fig. 3 (a) and (b), respectively.

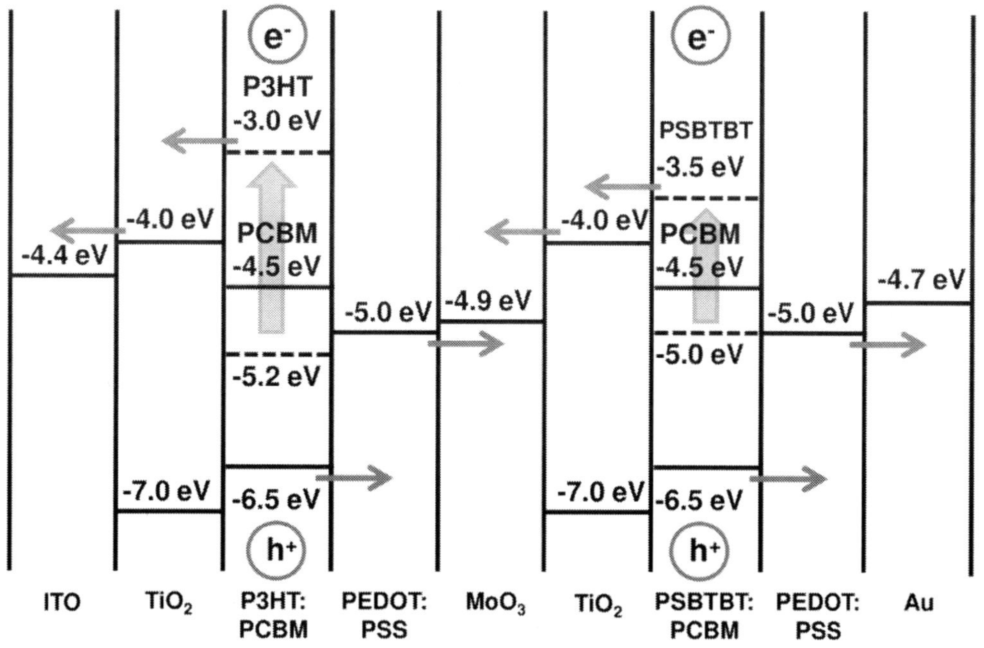

Figure 5. Schematic diagrams of energy levels of P3HT:PCBM + PSBTBT:PCBM tandem cell.

Schematic diagrams of energy levels of tandem cell at the highest occupied molecular orbital (HOMO) and the lowest unoccupied molecular orbital (LUMO) are shown in Fig. 5. Previously reported in references [3] for the energy levels of PEDOT:PSS, Au and ITO were used as shown in Fig. 5. Light irradiation promoted to generate charge separations. The excited electron in conducting band slightly diffused and charge-transferred from P3HT to PCBM on front active layer and Au electrode, and holes in valence band went through PEDOT:PSS layer, and arrived at the ITO substrate. V_{oc} of organic solar cells would be related with energy gap between HOMO of P3HT, PSBTBT and LUMO of PCBM. Control of the energy levels between HOMO and LUMO is important to improve the photovoltaic performance. In actual situation, recombination and loss of light-induced carrier near inner interface considerably generated energy loss, since Schottky energy barrier existed around the interface between P3HT, PSBTBT, PCBM and metal. The defect of crystal structure in microstructure would induce to generate recombination between carriers near the heterojunction interface under un-equilibrium of carrier-diffusion. Optimized microstructure near the interface will support to recover the carrier mobility, and will guide inter-diffusion of MoO_3 and TiO_2 between front and back layer in the tandem solar cells. The heat treatment will recover the crystal structure near the interface and then will improve the photovoltaic and optical properties.

4. Conclusion

Tandem solar cells combined with front and back cells were fabricated and characterized. The active layer based on thiophene block-copolymer PSBTBT:PCBM and P3HT:PCBM. PSBTBT was used for utilizing narrow band gap polymers. Optical properties showed a broad range of wavelength in the range of 300 to 800 nm. Surface morphology of active layers was observed by AFM. P3HT:PCBM thin layer had rough and irregular surface in continuous phase with phase separation. PSBTBT:PCBM thin layer had flat structure with several holes. Depth of these holes are in the range of 10 to 60 nm. The V_{OC} of tandem cell was obtained to be 0.69 V and that of single cell using P3HT:PCBM and

PSBTBT:PCBM were obtained to be 0.53 V and 0.10 V. The V_{OC} was increased by using the tandem structure.

Acknowledgement
This work was partly supported by a research fund from Takahashi Industrial and Economic Research Foundation.

References

[1] Chen H Y, Hou J, Hayden A E, Yang H, Houk K N and Yang Y 2010 *Adv. Mater.* **22** 371
[2] Hou J, Chen H Y, Zhang S, Li G and Yang Y 2008 *J. Am. Chem. Soc.* **130** 16144
[3] Sista S, Park M H, Hong Z, Wu Y, Hou J, Kwan W L, Li G and Yang Y 2010 *Adv. Mater.* **22** 380
[4] Chou C-H, Kwan W L, Hong Z, Chen L M and Yang Y 2010 *Adv. Mater.* **23** 1282
[5] Subbiah J, Amb C M, Reynolds J R and So F 2012 *Sol. Energy Mater. Sol. Cells* **97** 97
[6] Yang J, Zhu R, Hong Z, He Y, Kumar A, Li Y and Yang Y 2011 *Adv. Mater.* **23** 3465
[7] Kim H, shin M, Park J and Kim Y 2010 *ChemSusChem* **3** 476
[8] Marisol R-R, Kim K, Dewald J, Roma´n L S, Avadhanula A, Curran S and Carrol D L 2005 *Org. Lett.* **7** 5749
[9] Hau S K, O'Malley Kevin M, Cheng Y J, Yip H L, Ma H and Jen A K Y 2010 *IEEE. J. Sel. Top. Quant. Electron.* **16** 1665
[10] Moulé A J and Meerholz K M 2008 *Adv. Mater.* **20** 240
[11] Collins B A, Li Z, McNeill C R and Ade H 2011 *Macromolecules* **44** 9747
[12] Yuan Y, Huang J and Li G 2011 *Green* **1** 65
[13] Scharber M C, Koppe M, Gao J, Cordella F, Loi M A, Denk P, Morana M, Egelhaaf H J, Forberich K, Dennler G, Gaudiana R, Waller D, Zhu Z, Shi X and Brabec C J 2010 *Adv. Mater.* **22** 367
[14] Yang B, Cox J, Yuan Y, Guo F and Huang J 2011 *Appl. Phys. Lett.* **99** 133302
[15] Brunetti F G, Kumarb R and Wudl F 2010 *J. Mater. Chem.* **20** 2934

Fabrication and characterization of copper oxide-zinc oxide solar cells prepared by electrodeposition

Kazuya Fujimoto, Takeo Oku, Tsuyoshi Akiyama and Atsushi Suzuki
Department of Materials Science, The University of Shiga Prefecture, Hikone, Shiga 522-8533, Japan

E-mail: ze61kfujimoto@ec.ac.jp, oku@mat.usp.ac.jp

Abstract. Copper oxide-based heterojunction solar cells were fabricated by electrodeposition, and the electronic and optical properties were investigated. ZnO was used as an n-type semiconductor, and copper oxides were used as p-type semiconductors. A photovoltaic device based on an FTO/ZnO/Cu_2O heterojunction structure provided short-circuit current density of 2.7 $mAcm^{-2}$ and open circuit voltage of 0.28 V under an air mass 1.5 illumination. Microstructures of the solar cells were investigated by scanning electron microscope and X-ray diffraction, which indicated formation of copper oxides and ZnO. Energy levels of the solar cells were discussed.

1. Introduction

Copper oxide semiconductors are a promising alternative to silicon-based solar cells because Cu_2O possess high optical absorption and composed of low cost materials.Cu_2O is one of suitable materials for high efficiency solar cells because their band gap of ~1.5 and ~2.1 eV, respectively, are close to the ideal energy gap, and well matched with a solar spectrum. A maximum efficiency of ~3.8 % has been obtained for Cu_2O solar cells using high-temperature annealing and pulse laser deposition technique [1].

The purpose of the present work was to fabricate Cu_2O/ZnO and CuO/ZnO thin film solar cells by electrodeposition, and to investigate the effect of Cu_2O and ZnO layers on their electronic properties. ZnO has been used for a good electron acceptor and n-type semiconductor for Cu_2O/ZnO hetero junction solar cells [2-5]. The electrodeposition method has advantages of low cost and low temperature methods. Cu_2O and CuO thin films prepared by the electrodepotision method have also been reported previously [6-8]. Cu_2O or CuO/ZnO solar cells in the present study were investigated by measurements of optical absorption and photovoltaic activity.

2. Experimental

Cu_2O and CuO layers were prepared on pre-cleaned indium tin oxide (ITO) glass plate by electrodeposition using platinum counter electrode. Copper (II) sulfate ($CuSO_4$, 0.4 mol L^{-1}, Wako 97.5 %) and L-lactic acid (3 mol L^{-1}, Wako) were dissolved into distilled water. Electrolyte pH was adjusted to 12.5 by adding NaOH. The electrolyte temperature was kept at 65 °C during electrodeposition. Preparation of Cu_2O layers were carried out at current density at 1.5 mA cm^{-2} and quantity of electric charge of 4.0 C cm^{-2}. ZnO layers were also galvanostatically prepared at current density at 5 and 10 mA cm^{-2} from 8 mM aqueous solution of $Zn(NO_3)_2$ on Cu_2O layers. Electric charge was adjusted 1.5 C cm^{-2}. Gold (Au) metal contacts were vacuum deposited as top electrodes. Structure of heterojunction solar cells were shown in Figure 1 (a) and 1 (b), which were denoted as ITO or FTO/ZnO/CuO or Cu_2O/Au.

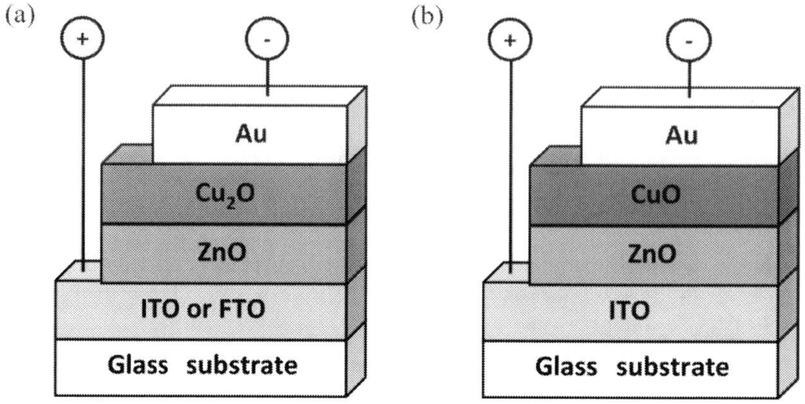

Figure 1. Structures of (a) ZnO/Cu_2O and (b) ZnO/CuO heterojunction solar cells.

Current density-voltage (J-V) characteristics (Hokuto Denko, HSV-110) of the solar cells were measured both in the dark and under illumination at 100 mW cm^{-2} by using an AM 1.5 solar simulator (San-ei Electric, XES-301S). The solar cells were illuminated through the side of the ITO substrates, and the illuminated area was 0.16 cm^2. Optical absorption of the solar cells was investigated by means of UV visible spectroscopy (Jasco, V-670). The incident photon to current conversion efficiency (IPCE) of solar cells of investigated (Peccell technologies, PEC-S20). The microstructures of the copper oxides were investigated by X-ray diffractometer (XRD, Philips, X'Pert-MPD System) with CuKα radiation operating at 40 kV and 40 mA, and scanning electron microscope (SEM) operating at 15 kV (Jeol, JEM-6380).

3. Results and discussion

J-V characteristics of ZnO/Cu_2O structure under illumination at 100 mW cm^{-2} by using an AM 1.5 solar simulator are shown in Figure 2. The photocurrent was observed under illumination, and the CuO/ZnO structure showed characteristic curves with short-circuits current and open-circuit voltage. A solar cell with a FTO/ZnO/Cu_2O structure provided power conversion efficiency (η) of 0.25 %, fill factor (FF) of 0.33, and short-circuit current density (J_{SC}) of 2.7 mA cm^{-2} and open-circuit voltage (V_{OC}) of 0.28 V. Measured parameters of the present cupper oxides-based solar cells are summarized in Table 1.

Table 1. Measured parameters of the present solar cells.

Sample	V_{OC} (V)	J_{SC} (mA cm^{-2})	FF	η (%)
ITO/ZnO(10 mA cm^{-2})/Cu_2O	0.063	0.45	0.26	7.3×10^{-3}
ITO/ZnO(5 mA cm^{-2})/Cu_2O	0.20	1.38	0.26	6.9×10^{-2}
FTO/ZnO(5 mA cm^{-2})/Cu_2O	0.28	2.7	0.33	0.25
ITO/ZnO(5 mA cm^{-2})/CuO	0.017	2.4×10^{-3}	0.25	1.0×10^{-5}

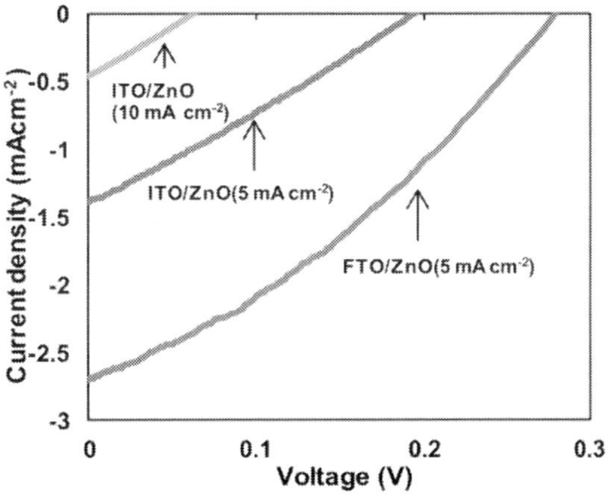

Figure 2. J-V characteristics of solar cells with ITO or FTO/ZnO(5 mA cm^{-2} or 10 mA cm^{-2})/Cu_2O(1.5 mA cm^{-2})/Au structures in the under AM 1.5 illumination.

Figure 3 shows optical absorption of the Cu_2O, ZnO and CuO thin films. Absorption peaks at ~360 nm and ~530 nm were due to CuO, and peaks at ~330 nm and ~470 nm were due to Cu_2O thin film, and a peak at ~350 nm corresponded to ZnO.

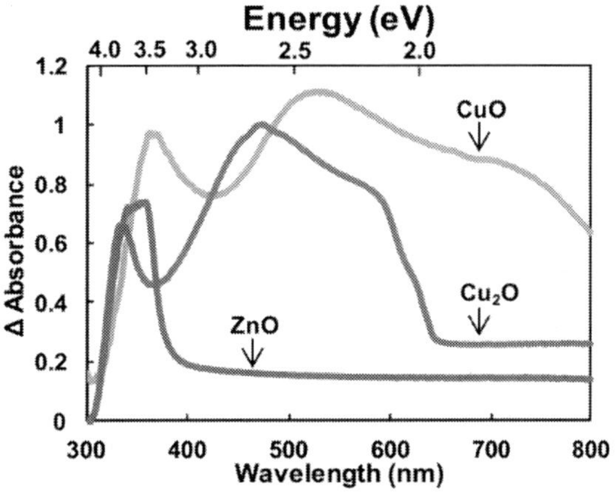

Figure 3. Absorption spectra of active layers

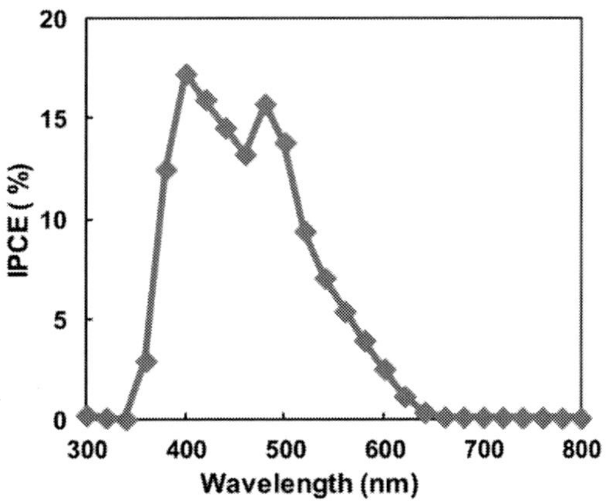

Figure 4. IPCE spectrum of solar cell with ZnO/Cu_2O structure.

An IPCE spectrum of a solar cell with the ZnO/Cu_2O structure is shown in Figure 4. The IPCE profile well corresponds to absorption spectra of ZnO and Cu_2O thin films, which indicated that excitons form effectively at active layers by illumination of light.

All the crystalline components in the Cu_2O, CuO and ZnO thin films were investigated by XRD, as shown in Figure 5 (a), (b) and (c). Diffraction peaks corresponding to Cu_2O and CuO and ZnO were observed for the Cu_2O and CuO and ZnO thin films, and they consisted of cuprite phase with cubic system (space group of Pn3m and lattice parameter of a = 0.427 nm) and cupric phase with monoclinic system (space group of C2/c and lattice parameter of a = 0.453 nm, b = 0.343 nm, c = 0.509 nm) and hexagonal system(space group of P63mc and lattice parameter of a = 0.326 nm c = 0.522 nm). As current density increased which ZnO (002) orientation raised. The fact observed influence of c axis growth by electrodeposition. The particle size was estimated using Scherrer's equation: D= 0.9 λ/B cos θ, where λ, B, and θ represent the wavelengths of the X-ray source, the full width at half maximum, and the Bragg angle, respectively. The crystallite sizes of Cu_2O and CuO and ZnO were determined to be 53.0 nm, 20.0 nm and 52.0 nm, respectively.

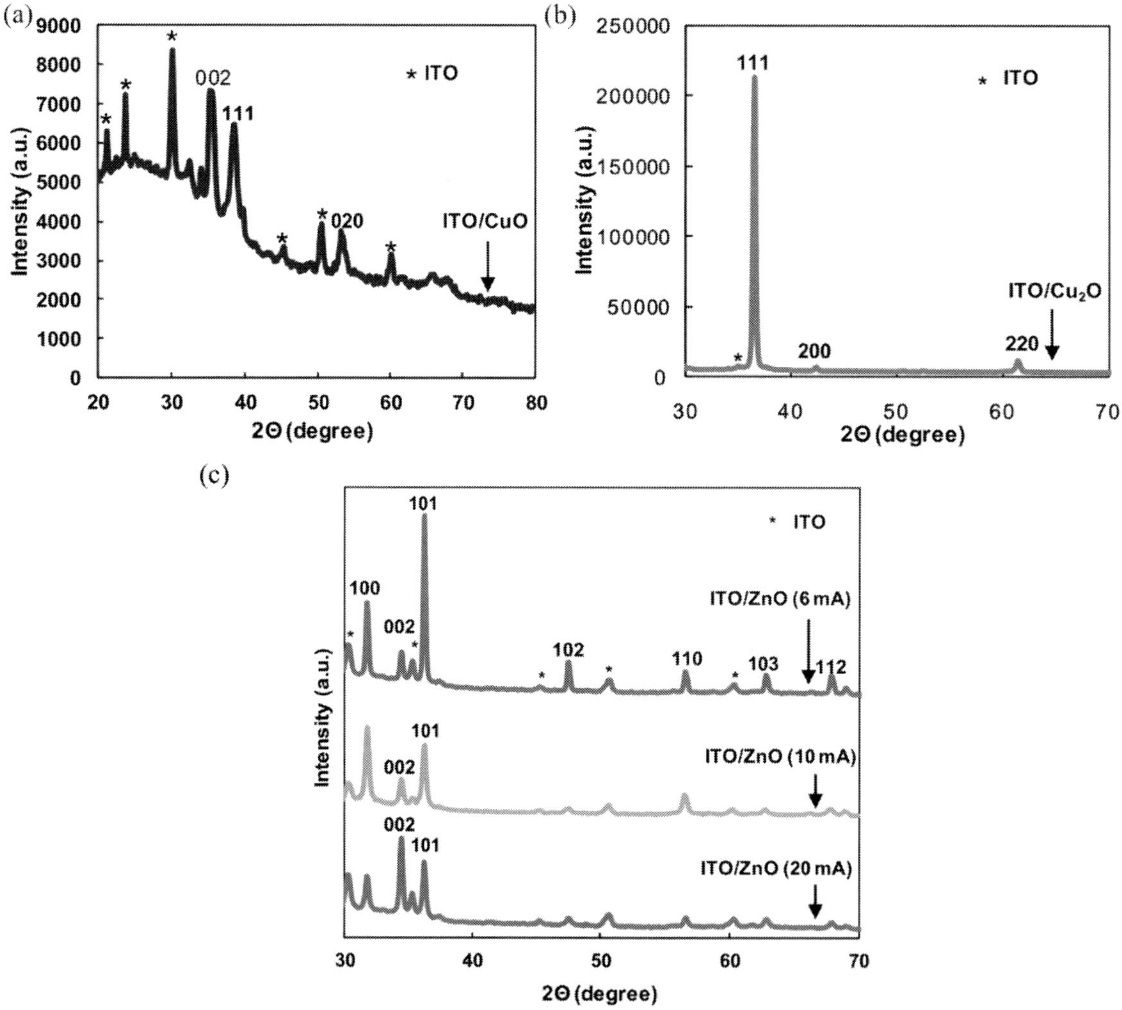

Figure 5. XRD pattern of (a) Cu_2O and (b) CuO thin films and (c) ZnO thin film prepared under different currents by electrodeposition.

Figures 6 (a) and (b) are SEM images of Cu_2O and ZnO thin films and (c) is cross-section image of FTO/ZnO/Cu_2O thin film, respectively. The SEM image indicated Cu_2O and growth of ZnO polycrystals with sizes of 2~5 μm and 1~2 μm, respectively. The cross-section indicated of Cu_2O and ZnO film thicknesses are ~2.5 μm and ~ 2.0 μm, respectively

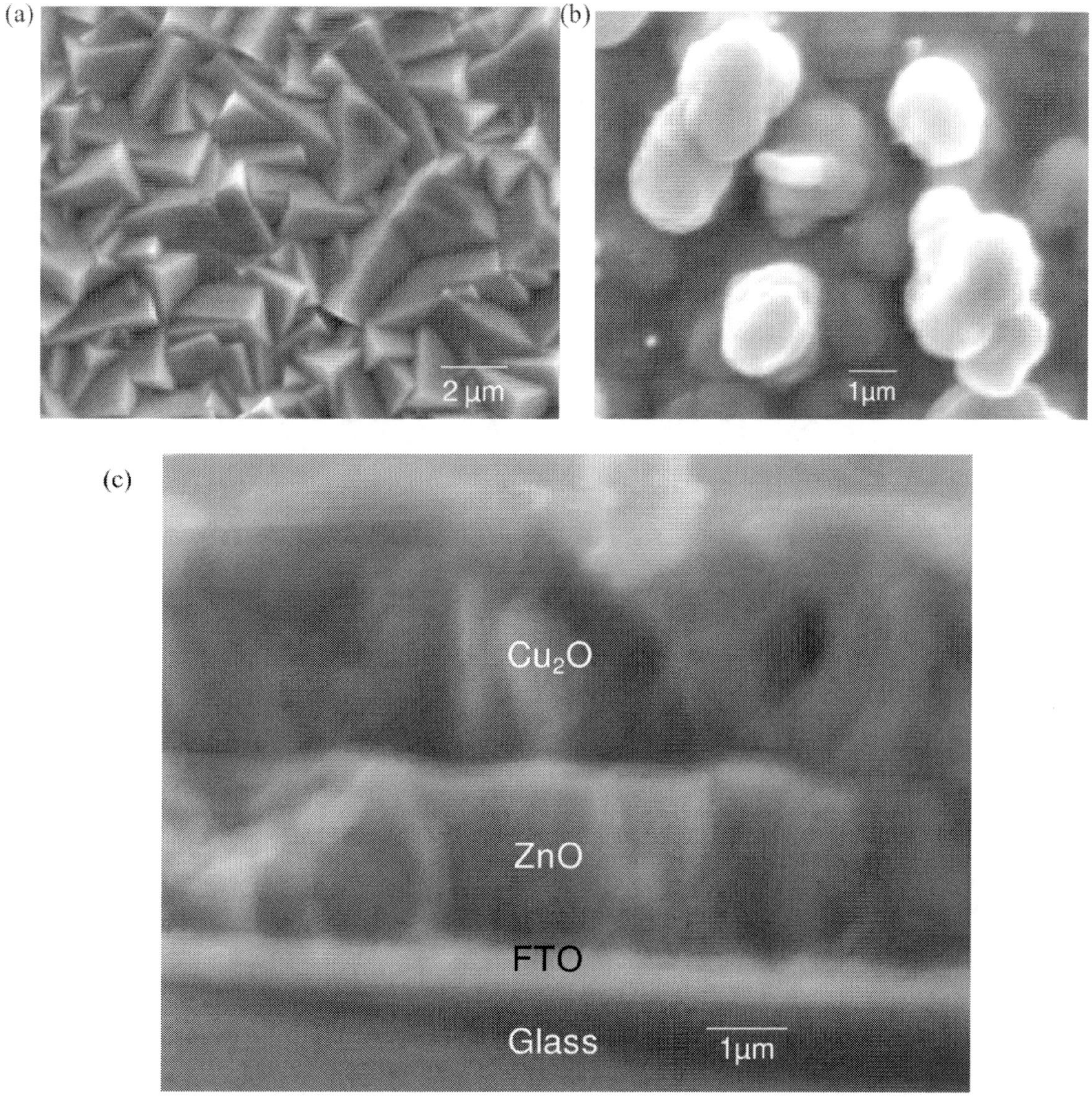

Figure 6. SEM image of (a) Cu_2O and (b) ZnO and (c) cross section of ZnO/Cu_2O thin films.

Energy level diagram of ZnO/Cu_2O and ZnO/CuO solar cells is summarized as shown in Fig. 7 (a) and 7 (b). Previously reported values were used for the energy levels [9-14]. It has been reported that V_{OC} is nearly proportional to the band gap of the semiconductors [15], and control of the energy levels is important to increase efficiency.

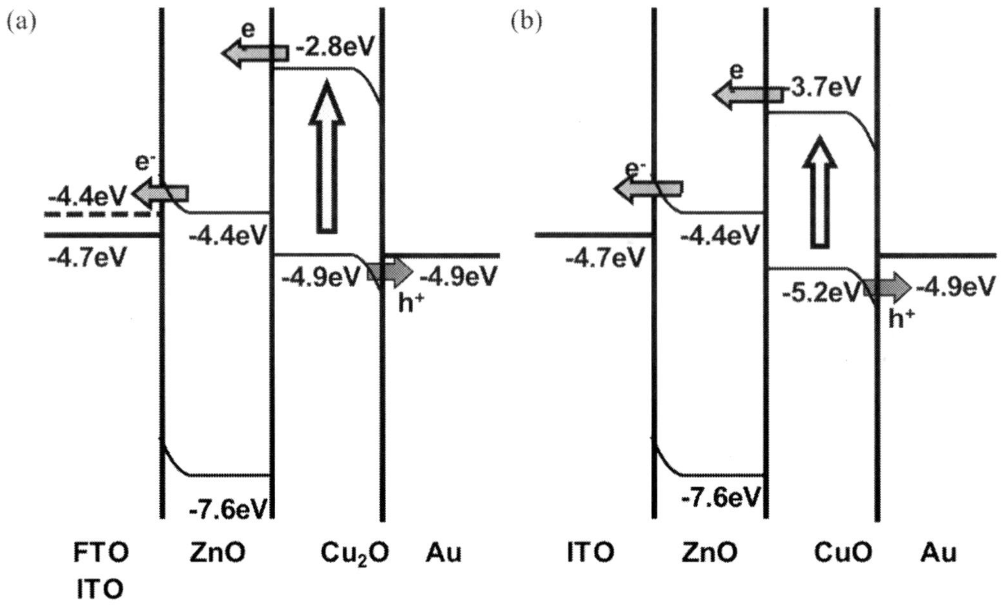

Figure 7. Energy level diagram of (a) ZnO/Cu_2O and (b) ZnO/CuO heterojunction solar cells.

Cu_2O and CuO based solar cells prepared by an electrodeposition method were investigated in the present work. The low conversion efficiency of the present solar cells would be due to large thickness of ZnO thin films. Optimization of thickness of active layers could increase the efficiency of the solar cells.

4. Conclusion

ZnO/Cu_2O/Au and ITO/ZnO/CuO/ heterojunction solar cells were produced and characterized. A device based on the ZnO/Cu_2O structure fabricated by an electrodeposition method which provided η of 0.25 %, FF of 0.33 , J_{sc} of 2.7 mAcm^{-2} and V_{oc} of 0.28 V. XRD indicated active layer deposited and orientation. As current density increased ZnO (002) orientation raised. The fact observed influence of c axis growth by electrodeposition. C axis growth would influence interfacial area lead to decrease of V_{oc}. Optimization of electorodeposition condition would improve the efficiencies of the solar cells.

Acknowledgements

The authors would like to thank Profs. B. Jeyadevan and H. Miyamura for support of XRD and SEM.

References

[1] Minami T, Nishi Y, Miyata T and Nomoto J 2011 *Appl. Phys. Exp.* **4** 062301
[2] Mittiga E, Salza E, Sarto F, Tucci M and Vasanthi R 2006 *Appl. Phys. Lett.* **88** 163502.
[3] Izaki M, Shinagawa T, Mizuno K, Ida Y, Inaba M and Tasaka A T 2007 *J. Phys. D* **40** 3326.
[4] Jeong S S, Mittiga A, Salza E, Masci A and Passerini S 2008 *Electrochim. Acta.* **53** 2226.
[5] Izaki M, Mizuno K, Shinagawa T, Inaba M and Tasaka A 2006 *J. Electrochem. Soc.* **153** C668.
[6] Teresa D G, Mark G S, Yanchun Z Rachel A V, Robert A V L and Jay A S 1996 *Chem. Mater.* **8** 2499.
[7] Nakaoka K and Ogura K 2002 *J. Electrochem. Soc.* **149** C579.

[8] Fujimoto K, Oku T and Akiyama T 2011 *J. Ceram. Soc. Jpn.* **119** 402.
[9] Oku T, Noma T, Suzuki A, Kikuchi K and Kikuchi S 2010 *J. Phys. Chem. Solids* **71** 551.
[10] Oku T, Motoyoshi R, Fujimoto K, Akiyama T, Balachandran J and Cuya J 2011 *J. Phys. Chem. Solids* **72** 1206.
[11] Domnina M I, Filatov S K, Zyuzyukia I I and Vergasova L P 1986 *Neorgan. Materialy* **22** 1743.
[12] Kidowaki H, Oku T and Akiyama T 2012 *J. Phys. Conf. Series* **352** 012022.
[13] Oku T, Takeda A, Nagata A, Fujimoto K, Akiyama T and Suzuki A 2012 *J. Phys. Conf. Ser.* 352 012024.
[14] Oku T, Takeda A, Nagata A, Kidowaki H, Kumada K, Fujimoto K., Suzuki A, Akiyama T, Yamasaki Y, and Osawa E 2013 *Mat. Tech.* **28** 21.
[15] Martin A G, Keith E, David L K, Yoshihiro H and Wilhelm W 2006 *Prog. Photovolt Res. Appl.* **14** 455.

Fabrication and characterization of silicon naphthalocyanine, gallium phthalocyanine and fullerene-based organic solar cells with inverted structures

Takeo Oku[1], Shigefumi Nose[1], Kazumi Yoshida[1], Atsushi Suzuki[1], Tsuyoshi Akiyama[1] and Yasuhiro Yamasaki[2]

[1] Department of Materials Science, The University of Shiga Prefecture, Hikone, Shiga 522-8533, Japan
[2] Orient Chemical Industries Co., Ltd., Department of New Business 8-1 Sanra-Higashi-Machi, Neyagawa, Osaka 572-8581, Japan

E-mail: oku@mat.usp.ac.jp

Abstract. Fullerene/phthalocyanine solar cells with inverted structures were fabricated, and the photovoltaic properties, optical absorption and microstructrues were investigated. Silicon naphthalocyanine, gallium phthalocyanine, and poly(3-hexylthiophene) were used for donor materials, and 6,6-phenyl C_{61}-butyric acid methyl ester was also used as an acceptor material. The solar cells with the inverted structure provided higher conversion efficiencies by addition of silicon naphthalocyanine, and had more stability compared to that with normal structure in the air. Nanostructures of the solar cells were investigated by transmission electron microscopy, which indicated the dispersion of nanocrystals in the active layers. Energy levels of the molecules were calculated, and a carrier transport mechanism was proposed.

1. Introduction

Organic thin film solar cells have an advantage for renewable energy resources because of their low cost, flexible, light weight and fabricate at low temperatures by spin-coating and printed method [1-3]. Recently, polymer /fullerene solar cells have been investigated, and the conversion efficiency of ~10 % was obtained [4]. Metal phthalocyanines (MPc) are a group of small molecules with Q-band absorption in the red to near-IR range. Since MPcs have chemical stability and photovoltaic properties, they are used for donor materials for organic thin film solar cells. The heterojunction solar cells using copper phthalocyanine and fullerene have been fabricated by evaporation method, and its conversion efficiency was ~3 % [5]. The characteristics such as electron conductivity and optical absorption are strongly dependent on the central metal atoms [6-9].

Organic thin film solar cells with normal structures have no stability in air. Aluminium (Al) metal has often been used as the back electrode of the organic solar cells with normal structures, due to its low work function. However, the Al is oxidized to insulating Al_2O_3 at the Al/organic interface, and the diffused Al into the active layer acts as a recombination site. An acidic poly(3,4-ethylenedioxylenethiophene): poly(4-styrene sulfonic acid) (PEDOT:PSS) would damage the device performance due to the corrosion to indium-tin-oxide (ITO). Both of which make lifetime of the cell very short, and one of the approach to solve the problems of stability is to use cells with an inverted structure. The cells with the inverted structure have a TiO_2 layer, which work as electron transport layer. There are some reports of inverted structure, and an improvement of stability has been reported [10-12].

The purpose of the present work is to fabricate and characterize MPc/fullerene solar cells with an inverted structure. Gallium phthalocyanine (GaPc), tetrakis (trimethylsilyl) [bis(trihexylsiloxy) silicon 2,3-naphthalocyanine] (SiNc) and poly(3-hexylthiophene) (P3HT) were used for donor materials, and 6,6-phenyl C_{61}-butyric acid methyl ester (PCBM) was also used as an acceptor material. Gold (Au) was used for the back electrodes instead of aluminium. TiO_2 thin films were fabricated by sol-gel method, and used as electron transfer layer. Photovoltaic mechanism, the light induced charge separation and charge transfer of the solar cells with the inverted structures were discussed on the basis of light-induced current density voltage (J-V) curves, and optical absorption. Nanostructures of the solar cells were investigated by transmission electron microscopy (TEM) and X-ray diffraction (XRD), and the energy levels of the molecules were calculated and discussed.

2. Experimental procedures

Solar of cells with an inverted structure were fabricated by using TiO_2 layer on transparent conducting oxide, as shown in Fig. 1. A heterojunction structure of PCBM/GaPc and a bulk heterojunction structure of PCBM:P3HT:SiNc were fabricated in the present work. The TiO_2 precursor solutions were prepared from titanium isopropoxide (TTIP), 2-methoxyethanol and acetylacetone. TTIP (0.46 ml) was add to 2-methoxyethanol (2.5 ml). After stirred for 1h, acetylacetone (0.61 ml) as the stabilizer was slowly added, and stirred for 12 h [12]. The TiO_2 precursor solution was spin-coated on fluorine dope tin oxide (FTO) substrate (Luminescence Technology, ~14 Ω/□).

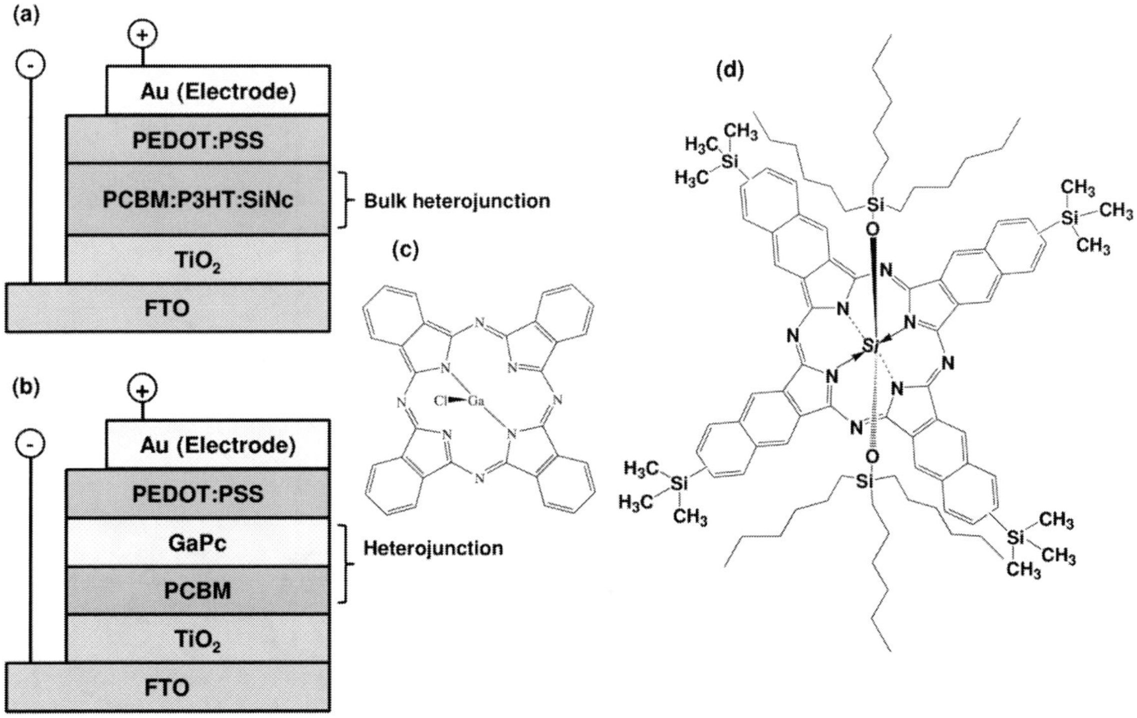

Figure 1. Structures of (a) PCBM/GaPc heterojunction and (b) PCBM:P3HT:SiNc solar cells. Molecular structures of (c) GaPc and (d) SiNc.

After annealing at 100 °C for 10 min in N_2 atmosphere, solution of PCBM in 1 ml chlorobenzene on a TiO_2 layer by spin-coat method. Then, gallium phthalocyanine layer were prepared on a PCBM layer by evaporation. Bulk heterojunction structures of PCBM:P3HT:SiNc were also prepared by spin-coating of solution of PCBM:P3HT:SiNc (8 mg, 10 mg, x mg) in 1 ml *o*-dichlorobenzene on a TiO_2 layer. Quantities of SiNc (x mg) were 2, 5, and 10 mg. A PEDOT:PSS (Sigma Aldrich) was spin-coated onto the active layers. Au metal contacts were evaporated as a top electrode and annealed at 140 °C for 10 min in N_2 atmosphere.

Current density-voltage (J-V) characteristics (Hokuto Denko Corp., HSV-110) of the solar cells were measured both in the dark and under illumination at 100 mW cm^{-2} by using an AM 1.5 solar simulator (San-ei Electric, XES-301S). The solar cells were illuminated through the side of the ITO substrates, and the illuminated area was 0.16 cm^2. Optical absorption of the solar cells was investigated by means of UV-visible spectroscopy (JASCO, V-670ST). The microstructures were investigated by X-ray diffractometer (XRD, PHILIPS X'Pert-MPD System) with CuKα radiation operating at 40 kV and 40 mA, and TEM observation was carried out by a 200 kV TEM (Hitachi H-8100) [13,14]. The molecular structures were optimized by CS Chem3D (Cambridge Soft) and molecular orbital calculations using Gaussian 09.

3. Results and discussion

J-V characteristics of PCBM:P3HT:SiNc solar cells under illumination at 100 mW cm^{-2} by using an AM 1.5 solar simulator are shown in Figure 2. The photocurrent was observed under illumination, and the PCBM:P3HT:SiNc structure showed characteristic curves with short-circuits current and open-circuit voltage. Measured parameters of the present solar cells are summarized as Table 1. A solar cell with a PCBM:P3HT:SiNc structure showed the highest power conversion efficiency (η) as the quantity of SiPc increased up to 10 mg. The PCBM:P3HT:SiNc(10 mg) solar cell provided η of 0.22 %, fill factor (FF) of 0.37, and short-circuit current density (J_{SC}) of 1.6 mA cm^{-2} and open-circuit voltage (V_{OC}) of 0.36 V.

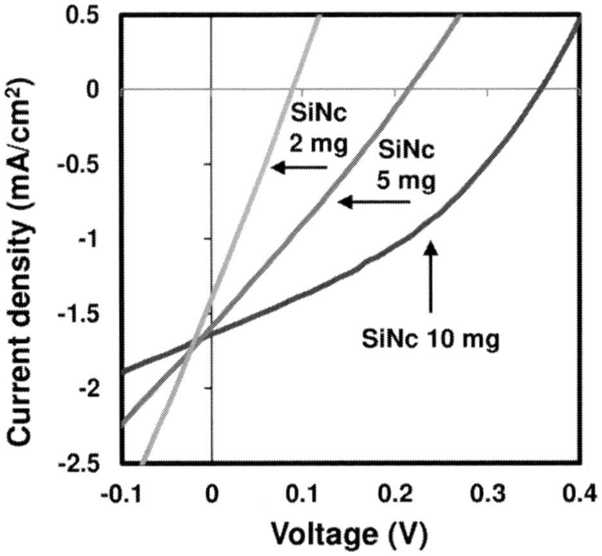

Figure 2. Measured J-V characteristic of PCBM:P3HT:SiNc solar cells under AM 1.5 illumination.

Table 1. Measured parameters of the present solar cells.

Sample	V_{OC} (V)	J_{SC} ($mAcm^{-2}$)	FF	η (%)
PCBM:P3HT:SiNc (2 mg)	0.085	1.4	0.27	0.032
PCBM:P3HT:SiNc (5 mg)	0.22	1.6	0.26	0.092
PCBM:P3HT:SiNc (10 mg)	0.36	1.6	0.37	0.22
PCBM/GaPc	0.56	0.44	0.24	0.059
PCBM/GaPc (After 2 month)	0.64	0.25	0.21	0.033

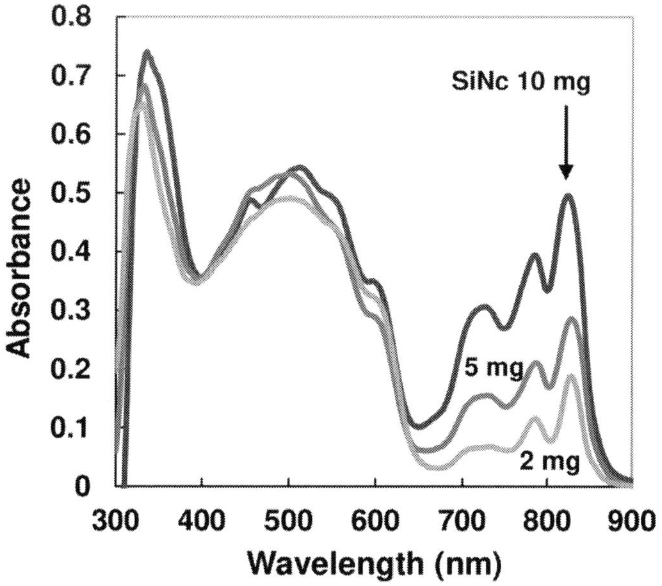

Figure 3. Absorption spectra of glass/TiO_2/SiNc:P3HT:PCBM thin films.

Although solar cells with a normal structure provided a conversion efficiency of 0 % after 24 h, a PCBM/GaPc solar cell with an inverted structure have more stability in air than that with a normal structure. Since PEDOT:PSS would prevented oxygen diffusion into active layers, active layers did not oxidized.

Figure 3 shows UV-visible absorption spectra of TiO_2/SiNc:P3HT:PCBM thin films. As the amount of SiPc increased, optical absorption increased in the range of 650 to 850 nm. Optical absorption at 330 nm corresponds to that of TiO_2 and PCBM. Absorption in the range of 400-600 nm corresponds to that of P3HT. Since the absorption was observed in the wide region, the sunlight would be efficiently absorbed.

An X-ray diffraction pattern of PCBM:P3HT:SiNc bulk heterojunction thin film is shown in Fig. 4. Diffraction peaks corresponding to SiNc and P3HT are observed as indicated by arrows, and slight shift of peaks were observed by mixing PCBM, P3HT and SiNc. No peak corresponding to PCBM was observed, which indicates the PCBM had an amorphous structure.

A TEM image and an electron diffraction pattern of PCBM:P3HT:SiNc thin film are shown in Fig. 5(a) and 5(b), respectively. The TEM image shows dispersion of P3HT and SiNc nanocrystals in the PCBM matrix, which would contribute the increase of the efficiencies. The electron diffraction pattern also indicates Debye-Scherer rings, which are due to the dispersion of nanocrystals

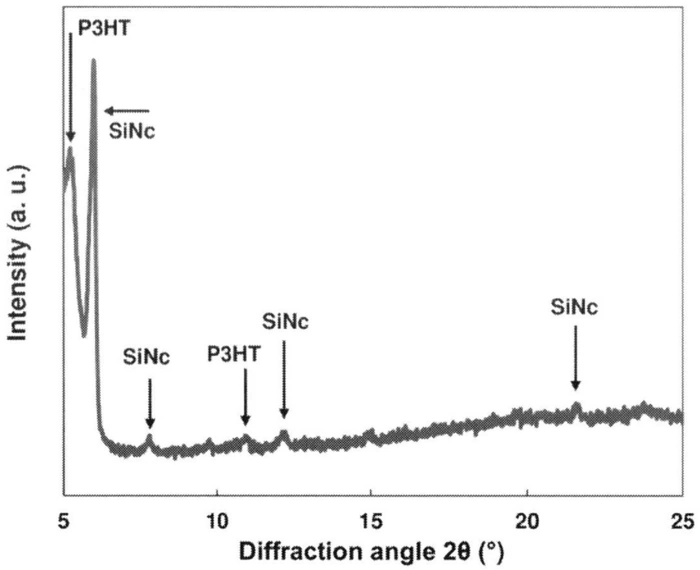

Figure 4. XRD pattern of PCBM:P3HT:SiNc thin film.

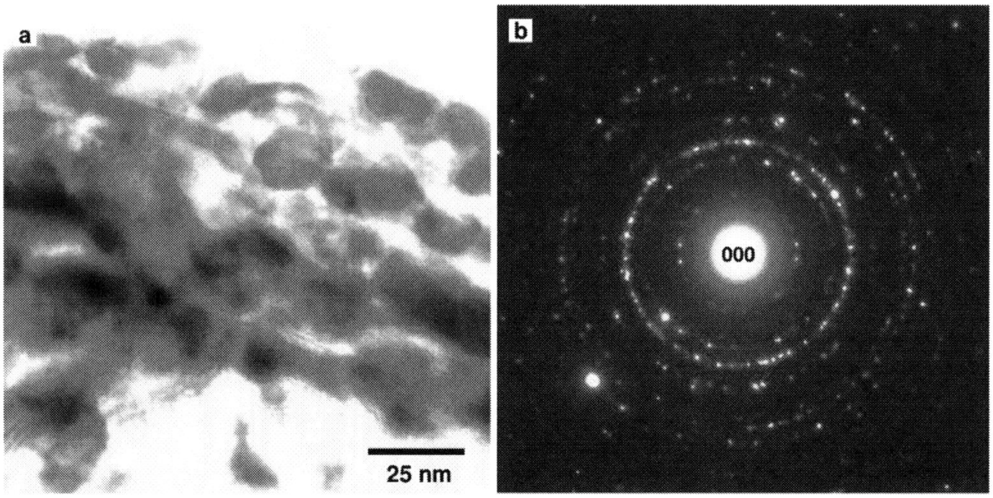

Figure 5. (a) TEM image and (b) electron diffraction pattern of PCBM:P3HT:SiNc thin film.

A TEM image, an electron diffraction pattern and a high-resolution image of TiO_2 thin films are shown in Figure 6(a), (b) and (c), respectively. The particle size of TiO_2 is in the range of 20-50 nm from the TEM image, and the electron diffraction pattern and high resolution image show formation of the TiO_2 anatase structure by annealing at 450 °C.

Energy level diagrams of the present PCBM/GaPc heterojunction and PCBM:P3HT:SiNc bulk heterojunction solar cells with the inverted structures were summarized as shown in Fig. 7. Previously reported values were used for the energy levels of the figures by adjusting to the present work [15-17]. Energy barrier would exist near the semiconductor/metal interface. In the cells with an inverted structure, electrons are transported to an FTO substrate, and holes are transported to an Al. When C_{60}

is used for the inverted structure, the energy barrier would be at the TiO_2/C_{60} interface. To reduce the energy barrier, PCBM with higher LUMO levels is suitable as used in the present work. Voc of organic solar cells is related with energy gap between HOMO of MPc and LUMO of PCBM, and control of the energy levels is important to improve the photovoltaic performance [15].

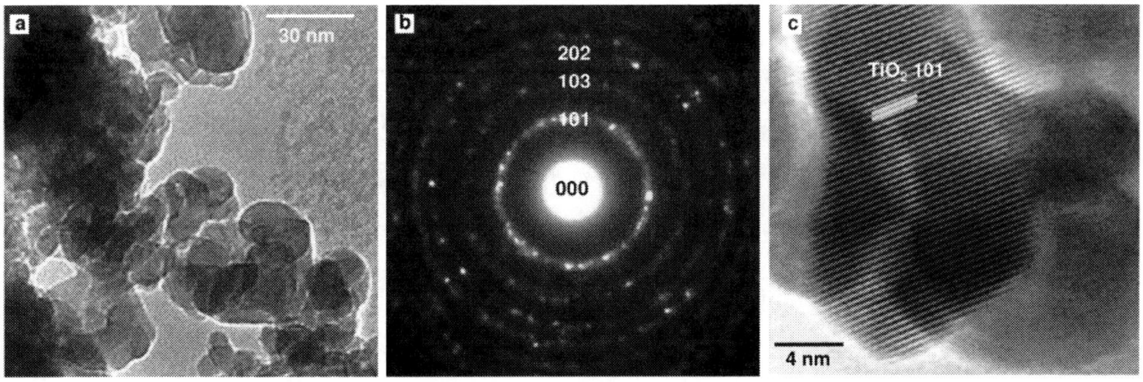

Figure 6. (a) TEM image, (b) electron diffraction pattern and (c) high-resolution image of TiO_2 thin film.

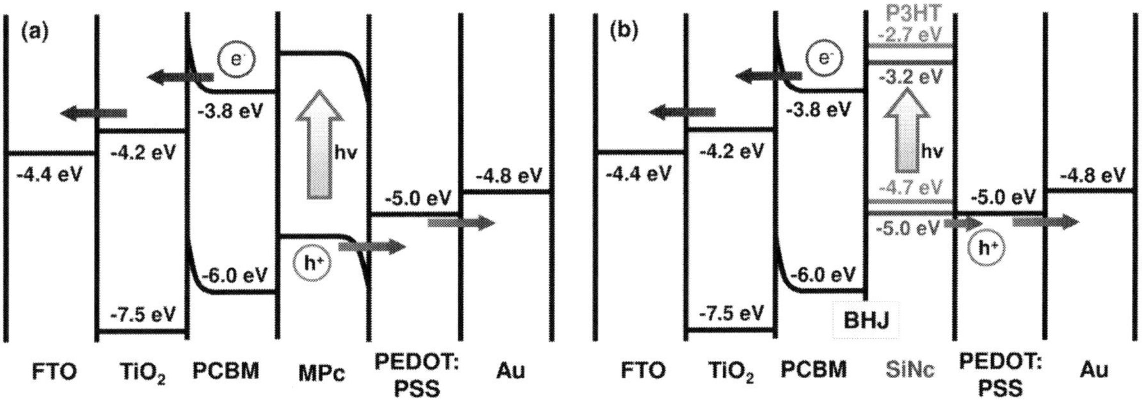

Figure 7. Energy level diagrams of (a) PCBM/GaPc and (b) PCBM/P3HT:SiNc solar cells.

4. Conclusion

Photovoltaic properties, optical absorption and microstructrues of PCBM/GaPc and PCBM:P3HT:SiNc solar cells with inverted structures were investigated. The PCBM:P3HT:SiNc solar cells with the inverted structure provided higher conversion efficiencies by addition of SiNc. TEM images, electron diffraction, and high-resolution images confirmed that the active layer had nanocrystalline structures, and the TiO_2 electron transport layer had an anatase structure. A device with inverted cell using GaPc/PCBM showed more stability in the air than that of a normal structure. Energy levels of the molecules were calculated, and a carrier transport mechanism was discussed.

Acknowledgement

The present work was partly supported by a Grant-in-Aid for Scientific Research from Japan Society for the Promotion of Science.

References

[1] Reyes-Reyes M, Kim K and Carroll D L, 2005 *Appl. Phys. Lett.* 87, 083506.
[2] Peumans P, Uchida S and Forrest S R, 2003 *Nature* **425** 158.
[3] Oku T, Takeda A, Nagata A, Noma T, Suzuki A, Kikuchi K, 2010 *Energies* **3** 671.
[4] Green M A, Emery K, Hishikawa Y, Warta W, Dunlop E D 2012 *Prog. Photovoltaics Res. Appl.* **20** 12.
[5] Peumans P and Forrest S R, 2001 *Appl. Phys. Lett.* **79** 126.
[6] Li L, Tang Q, Li H, Hu W, Yang X, Shuai Z, Liu Y and Zhu D, 2008 *Pure Appl. Chem.* **80** 2231.
[7] Bamsey N M, Yuen A P, Hor A M, Klenkler R, Preston J S, Loutf R O, 2011 *Sol. Energy Mater. Sol. Cells* **95** 1970.
[8] Bamsey N M, Yuen A P, Hor A M, Klenkler R, Preston J S, Loutf R O, 2011 *Sol. Energy Mater. Sol. Cells* **95** 2937.
[9] Kim D Y, So F and Gao Y, 2009 *Sol. Energy Mater. Sol. Cells* **93** 1688.
[10] Krebs F C, 2008 *Sol. Energy Mater. Sol. Cells* **92** 715.
[11] Kuwabara T, Nakayama T, Uozumi K, Yamaguchi T, Takahashi K, 2008 *Sol. Energy Mater. Sol. Cells* **92** 1476.
[12] Lin Y H, Yang P C, Huang J S, Huang G D, Wang I J, Wu W H, Lin M Y, Su W F, Lin C F, 2011 *Sol. Energy Mater. Sol. Cells* **95** 2511.
[13] Oku T, Noma T, Suzuki A, Kikuchi K and Kikuchi S, 2010 *J. Phys. Chem. Solids* **71** 551.
[14] Oku T, 2012 *Nanotechnol. Rev.* **1** 389.
[15] Oku T, Takeda A, Nagata A, Kidowaki H, Kumada K, Fujimoto K., Suzuki A, Akiyama T, Yamasaki Y, and Osawa E 2013 *Mat. Tech.* **28** 21.
[16] Oku T, Kumada K, Suzuki A, and Kikuchi K 2012 *Cent. Euro. J. Eng.* **2** 248.
[17] Honda S. Ohkita H, Benten H and Ito S, 2010 *Chem. Commun.* **46** 6596.

Effects of phosphorous doping to poly (methylphenyl silane) and fabrication of thin film solar cells

Junya Nakagawa[1], Takeo Oku[1], Atsushi Suzuki[1], Tsuyoshi Akiyama[1], Katsuhisa Tokumitsu[1], Masahiro Yamada[2] and Mika Nakamura[2]

[1] Department of Materials Science, the University of Shiga Prefecture, 2500 Hassaka, Hikone, Shiga 522-8533, Japan
[2] Osaka Gas Chemicals Co. Ltd., 3 Tiyozaki Nishiku, Osaka 550-0023, Japan

E-mail: oku@mat.usp.ac.jp

Abstract. Effects of phosphorus bromine (PBr_3) doping to polysilane were investigated for spin-coating thin films. Phosphorus doped poly (methylphenyl silane) (PMPS) provided n-type semiconductor behavior, which was confirmed by Hall effect measurements. Desorption of phenyl and methyl groups in doped PMPS thin films was observed after annealing at 300 °C from Raman scattering measurements and theoretical calculations. The band gap energy of PMPS was measured to be 3.2 eV. Decrease of photoluminescence intensity of PMPS was observed by phosphorus doping. Microstructures of the doped PMPS thin films were investigated by using X-ray diffraction, which indicated doped PMPS had an amorphous structure after annealing. A solar cell with PMPS(PBr_3):poly[3-hexylthiophene] bulk-heterojunction structure was fabricated, and provided a photovoltaic behavior. Formation mechanism and carrier transport mechanism of the doped PMPS thin films were proposed.

1. Introduction

Thin film solar cells using organic semiconductors have been developed as next generation systems. Organic thin film solar cells have advantages of low cost, flexible and light weight product, and photovoltaic and optical properties of the organic solar cells have been studied [1-4]. However, the power conversion efficiency of the organic cells is lower than inorganic cells. A primary cause of this is low mobility of charge carriers. Polysilanes a p-type semiconductor, and has been applied as electrical conductive materials and photovoltaic systems [5-7]. Polysilanes are known as σ-conjugate polymers, and hole mobility is 10^{-4} $cm^2V^{-1}s^{-1}$[5]. Although the polysilanes could be applied to p-type semiconductors on organic thin film solar cells for these characteristics, few studies on polysilane solar cells have been reported [6, 8-11]. Doped polysilanes as n-type semiconductor have been applied for organic solar cells [12].

The purpose of the present work is to fabricate photovoltaic solar cell by using mixture solution of polysilane doped with phosphorus, and to investigate the effect of annealing temperatures and doping of phosphorous (P) on their electronic properties and microstructures. It is expected that amorphous silicon doped with phosphorus would function as n-type semiconductor. Spin-coating is a low-cost method, and is essential for mass production of any solar cells. Light induced carrier separation with charge transfer was investigated by experimental measurements. Doping effect of phosphorus into polysilane was investigated by Hall effect measurement and Raman scattering spectra, and microstructure analysis was also investigated by X-ray diffraction (XRD).

2. Experimental

Poly (methylphenyl silane) (PMPS) was used in the present work. Figure 1 shows solar cell structures and molecular structures of PMPS, phosphorus bromine (PBr_3) and poly[3-hexylthiophene] (P3HT) (Sigma Aldrich). Indium tin oxide (ITO) grass prate (Geomatec, ~10Ω/□) was cleaned by an ultrasonic bath with acetone and methanol, and was dried by nitrogen gas. Then the sol-gel procedure for producing TiO_x is as follows: titanium tetra-isopropoxide (TTIP) was prepared as a precursor and mixed with 2-methoxyethanol and acetylacetone. Then the TiO_x precursor solution was spin-cast in air on the ITO substrate. The substrate was heated at 140 °C for 60 min in air. PBr_3 (0.02 mL, Wako) and PMPS (24 mg, Osaka Gas Chemicals Co., Ltd) were mixed in 1 mL o-dichlorobenzene, and then P3HT was mixed. The mixed solution was spin-cast (1000 rpm) on the resultant substrate, and then was heated at 300 °C for 10 min in an inert-atmosphere (N_2) glovebox. Highly conducting poly (3,4-ethylenedioxylenetthiophene): polystyrene sulfonic acid (PEDOT:PSS) was spin-cast (2000 rpm) and was heated at 140 °C for 60 min in a glove box. Gold contacts were evaporated as a top electrode. Finally, the devices were annealed at 140 °C for 30 min in N_2 atmosphere.

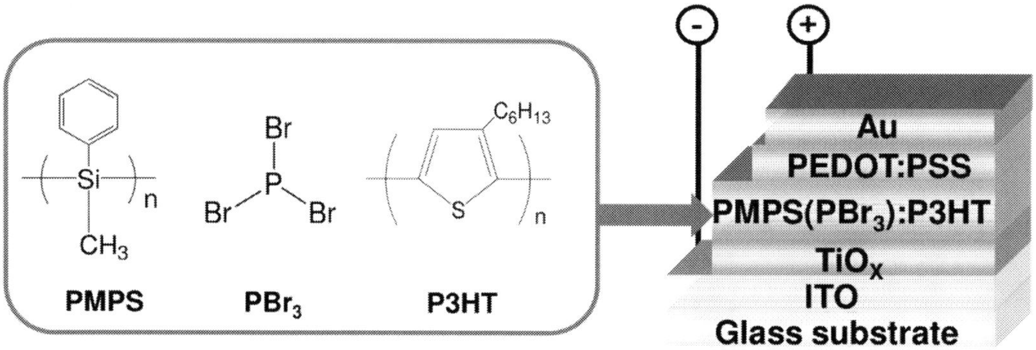

Figure 1. Structures of PMPS, PBr_3 and P3HT to fabricate solar cells PMPS (PBr_3):P3HT bulk-heterojunction solar cells.

Current density-voltage (J-V) characteristics (Hokuto Denko Corp., HSV-110) of the solar cells were measured both in the dark and under illumination at 100 $mWcm^{-2}$ by using an AM 1.5 solar simulator (San-ei Electric, XES-301S) in N_2 atmosphere. The solar cells were illuminated through the side of the ITO substrates, and the illuminated area is 0.16 cm^2. Optical absorption of the solar cells was investigated by means of UV-visible spectroscopy (JASCO, V-670ST).

The microstructures of the PMPS thin films were investigated by XRD (Philips X'Pert-MPD System) with CuKa radiation at 40 kV operating voltage and 40 mA operating current.

Surface resistivity, carrier concentration and mobility of doped thin film were measured by Hall effect measurement system (TOYO Corporation, Japan, Resti Test 8320).Raman scattering spectra were recorded with a laser Raman spectrometer (JASCO Co., Ltd., NRS-5100). Raman mode and optical image of the thin film after annealing was observed using excitation laser wavelength at 532 nm. The molecular structures of phosphorus bromine and polysilane monomers and phosphorus complex were optimized by CS Chem3D (Cambridge Soft). Molecular orbital calculations were carried out by MOPAC (Fujitsu Ltd.). The isolated molecular structures were optimized by *ab-initio* quantum calculation using density functional theory using B3LYP/6-31G (d) as basis function (Gaussian 03). Active modes in Raman scattering spectra were calculated by DFT/B3LYP/6-31G (d) using frequency mode.

3. Results and Discussion

Doping effects of phosphorus on the PMPS were investigated. Surface resistivity, carrier concentration and mobility of doped thin film with phosphorus are summarized in Table 1. Although non-doped PMPS films showed high electrical resistivity (~10^8 Ω cm), doped PMPS thin film showed n-type behaviour, low resistivity and high hole mobility. The experimental results at room temperature on doped film with phosphorus complex showed improvement of surface resistivity, carrier concentration and mobility of 0.15, 2.1×10^{19} cm^{-3} and 2.0 $cm^2V^{-1}s^{-1}$, respectively. After annealing at 300 °C, the electronic properties of the doped thin film were changed to be 0.26, 1.0×10^{19} cm^{-3} and 2.4 $cm^2V^{-1}s^{-1}$. However, after annealing at 500 °C, the electronic properties of the doped thin film were deteriorated. The doping of phosphorus at 300 °C provided a good performance for transport phenomena.

Table 1. Resistivity, carrier concentration and mobility of doped PMPS thin film

Temperature	Resistivity (Ω cm)	Carrier concentration (cm^{-3})	Mobility ($cm^2\,V^{-1}\,s^{-1}$)
R. T.	0.15	2.1×10^{19}	2.0
300 °C	0.26	1.0×10^{19}	2.4
500 °C	48	1.9×10^{17}	0.69

Figure 2 shows Raman scattering spectra of PMPS and the doped PMPS thin film. As comparison, the calculated Raman scattering spectra was also shown in Figure 2 for identifying each active mode of PMPS monomer. As shown doped PMPS, Raman active mode of the film was obtained as follows: 693, 854, 1645, 2394, 3178, 3185, 3192, 3199 and 3206 cm^{-1}. The middle peaks at and 693, 854, 1645 and 2394 cm^{-1} were identified as phenyl group vibration modes in PMPS. Strong peaks at 3178, 3185, 3192, 3199 and 3206 cm^{-1} were identified as methyl group vibration modes in PMPS [13-15]. Raman peaks observed in PMPS was conformed as vibration mode of carbon bond of phenyl and methyl group in PMPS monomer. After doping of PBr_3 with annealing at 300 °C, the peaks at 2394, 3178, 3185, 3192, 3199 and 3206 cm^{-1} disappeared. Peak around 1500 cm^{-1} appeared as a new mode of modified PMPS with phosphorus complex. In addition, the Si crystal peaks around 520 cm^{-1} was not observed in PMPS and doped PMPS thin film. Doping effect of phosphorus bromide on the PMPS was characterized.

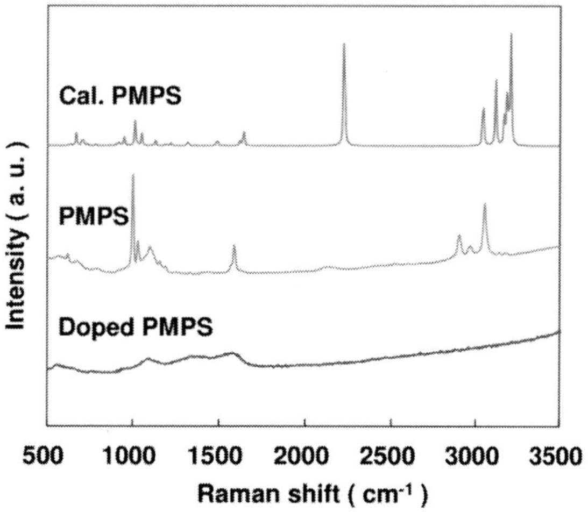

Figure 2. Raman scattering spectra of calculation PMPS, PMPS thin film and doped PMPS thin film.

Figure 3 shows XRD patterns of PMPS, doped PMPS thin films after annealing, and Glass substrate, respectively. There is no sharp diffraction peak in Figure 3. Peak observed in PMPS, as shown was polysilane specific crystalline peak. After doping of PBr_3 with annealing at 300 °C, the peak was disappeared. From this result, the structure of doped PMPS thin film which indicates formation of an amorphous structure from the crystalline structure. In addition, doping of phosphorus bromine into PMPS would enhance the decomposition of the PMPS to the amorphous structure.

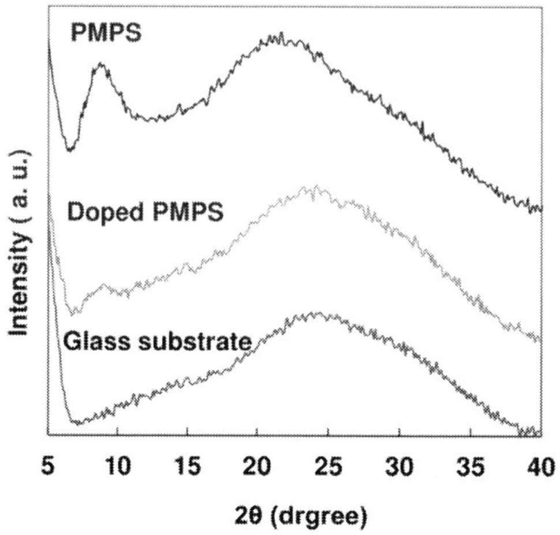

Figure 3. XRD patterns of PMPS, doped PMPS thin films after annealing at 300 °C and glass substrate.

Figure 4 shows UV-vis diffuse reflection absorption spectra of PMPS and doped PMPS. PMPS structure provided higher photo-absorption in the range of 300-400 nm. After doping of PBr_3 with annealing at 300 °C, the peak was decreased. From this result, the decrease of photo-absorption peak was affected of organic functional group of side chain of PMPS desorption.

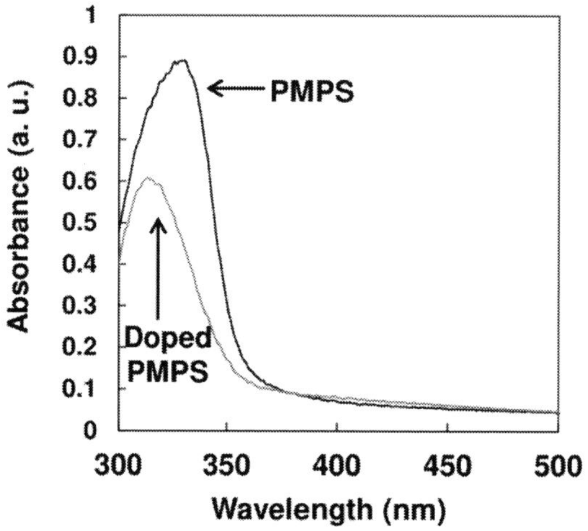

Figure 4. UV-vis diffuse reflection absorption spectra of PMPS and doped PMPS.

Figure 5 shows photoluminescence spectra of PMPS and doped PMPS. PMPS structure provided photoluminescence in the range of 300-450 nm. After doping of PBr_3 with annealing at 300 °C, the peak was decreased, significantly. Decrease of photoluminescence intensity of PMPS was observed by doping of phosphorus which was due to structural transition by doping of PBr_3 after annealing at 300 °C [16]. This characteristic was explained by the light-induced charge separation between PMPS units and phosphorus. Also, the band gap energy of doped PMPS was calculated from Figure 5 and the value was 3.2 eV.

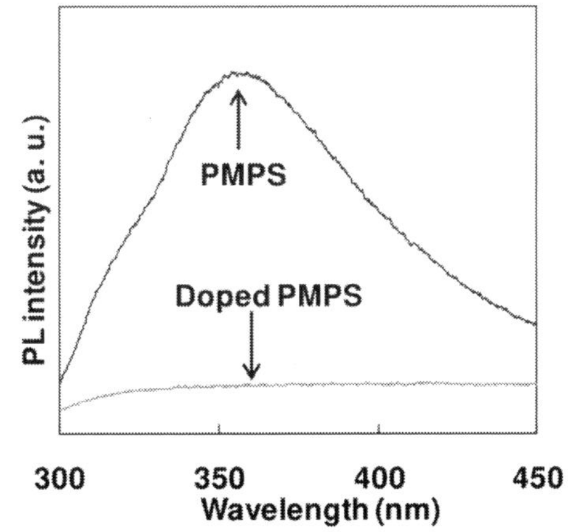

Figure 5. Photoluminescence spectra of PMPS and doped PMPS.

Figure 6 shows proposed a mechanism of PBr_3 doping to PMPS in the present work. Annealing from room temperature to 300 °C, several phenyl and methyl groups of side chain of PMPS desorbed and $-PBr_2$ bound of side chain of PMPS. In this case, methyl bromide and benzene bromide were generated, and an amorphous structure was formed from the main Si chain structure. Annealing at 300 °C, the structure closed in amorphous structure.

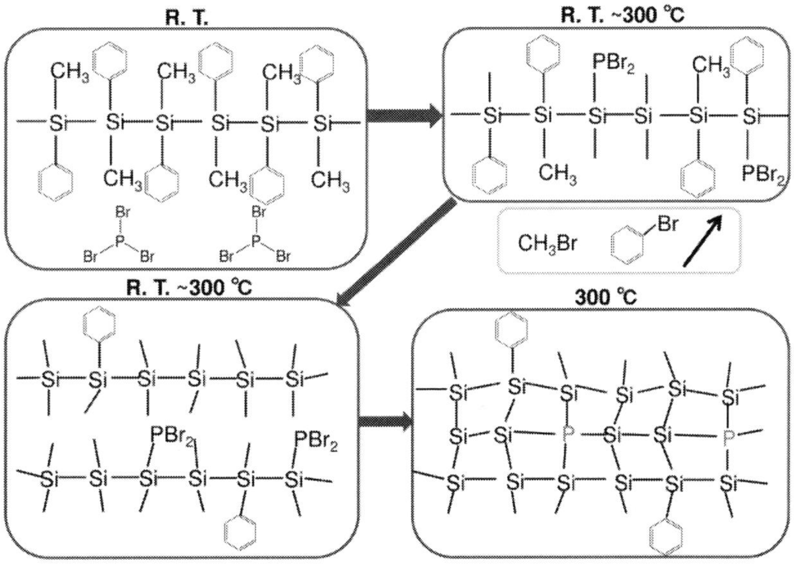

Figure 6. Microstructure change of PMPS (PBr_3) during annealing.

Measured parameters of the present solar cell using Doped PMPS and P3HT as photoactive layer are summarized in Table 2. Open-circuit voltage, short-circuit current density, fill factor and power conversion efficiency are denoted as V_{OC}, J_{SC}, FF and η, respectively. A solar cell with PMPS (PBr_3):P3HT structure was sure using doped PMPS. The present solar cells fabricated by spin cast have simple fabrication process and better cost performance. Further, the conversion efficiency improved using the present fabrication process.

Table 2. Measured parameters of the PMPS (PBr_3):P3HT bulk-heterojunction solar cells.

Sample	V_{OC} (V)	J_{SC} (μAcm^{-2})	FF	η (%)
PMPS(PBr_3):P3HT	0.064	3.8	0.36	8.8×10^{-5}

An energy level diagram of PMPS (PBr_3): P3HT bulk-heterojunction solar cell is summarized as shown in Figure 7. Previously reported values were used for the energy levels [17-18]. For the PMPS (PBr_3): P3HT bulk heterojunction solar cell, the combination of P3HT and doped PMPS would generate the carrier separation, which indicated in the increase of efficiency. The incident direction of light is from ITO side. Energy barrier would exist near the semiconductor/metal interface. Electronic charge is transferred by light irradiation from the ITO substrate side. Holes are transported to an Au electrode and electrons are transported to an ITO substrate.

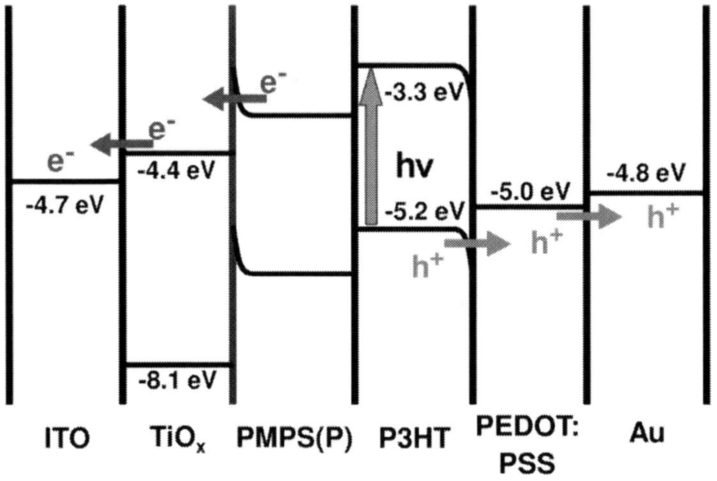

Figure 7. Energy level diagram of PMPS (PBr_3):P3HT bulk heterojunction solar cells.

4. Conclusion

Thin film of doping phosphorus to PMPS was fabricated by using spin-coating method and the effect was also investigated. Doped PMPS provided n-type semiconductor behavior and indicated a good performance for transport phenomena by a Hall effect measurement. From Raman scattering measurements and theoretical calculations, desorption of phenyl and methyl groups in doped PMPS thin film was observed after annealing at 300 °C. The band gap energy of doped PMPS was 3.2 eV and decrease of photoluminescence intensity of polysilane was observed by doping of phosphorus. This characteristic was explained by the light-induced charge separation between PMPS unit and phosphorus. Microstructures of the doped PMPS thin film were investigated by using X-ray diffraction, which indicated doped PMPS had an amorphous structure after annealing. A solar cell with PMPS (PBr_3):P3HT bulk-heterojunction structure was fabricated, and provided a photovoltaic behavior. Energy level of the molecules was suggested that the solar cell was provided a photovoltaic behavior. Investigation of percentage of doping concentration and optimization of the structure of doped PMPS would improve the cell performance using the present fabrication process.

References

[1] Bagienski W, Gupta M, 2011 *Sol. Energy Mater. Sol. Cells* **95** 933
[2] Tang C, 1986 *Appl. Phys. Lett.* **48** 183
[3] Mi S, Hyuk J, Jin J, 2010 *Sol. Energy Mater. Sol. Cells* **94** 152

[4] Zhao D, Ke L, Li Y, Tan S, 2011 *Sol. Energy Mater. Sol. Cells* **95** 921
[5] Lee J, Seoul C, Park J, York J, 2004 *Synth. Metals* **145** 11
[6] Silence S, Scott J, Hache F, Ginsbrug E, Jenkner P, Miller P, Twieg R, Moermer W, 2003 *J. Opt. Soc. Am. B***10** 2306
[7] Jung J, Ciesielski W, Ulanski J, 2006 *Mater. Sci. Poland* **24** 527
[8] Shimoda T, Matsuki Y, Furusawa M, Aoki T, Yudasaka I, Tanaka H, Iwasaki H, Wang D, Miyasaka M and Takeuchi Y 2006 *Nature* **440** 783
[9] Lee J, Seoul C, Park J and Youk J 2004 *Synth. Metals* **145** 11
[10] Nakagawa J, Oku T, Suzuki A, Akiyama T, Tokumitsu K, Yamada M and Nakamura M 2012 *J. Phys. Conf. Series* **352** 012019
[11] Kawashima A, Oku T, Suzuki A, Kikuchi K and Kikuchi S 2012 *Mater. Sci. Appl* **3** 557
[12] Masuda T, Sotani N, Hamada H, Matsuki Y and Simoda T 2012 *Appl. Phys. Lett.* **100** 253908
[13] Martin P H, Irmer G and Muller E 1998 *J. Eur. Ceram. Soc.* **18** 193
[14] Bong H B 2011 *J. Korea. Phys. Soc.* **59** 3192
[15] Bong H B 2011 *J. Korea. Phys. Soc.* **59** 3205
[16] Yoshino K, Hosoda K, Fujii A and Ishikawa M 1997 *J. Appl. Phys.* **36** 368
[17] Oku T, Motoyoshi R, Fujimoto K, Akiyama T, Balachandran J, Cuya J 2011 *J. Phys. Chem. Solids* **72** 1206
[18] Jin Y K, Sun H K, Hyun H L, Kwanghee L, Wanli M, Xiong G and Alan J H 2006 *Adv. Mater.* **18** 572

Cu_2O/ZnO Heterojunction Solar Cells Fabricated by Magnetron-Sputter Deposition Method Films Using Sintered Ceramics Targets

S Noda, H Shima and H Akinaga

Innovation Center for Advanced Nanodevices (ICAN), National Institute of Advanced Industrial Science and Technology (AIST), 1-1-1 Central 2, Umezono, Tsukuba, Ibaraki 305-8568, Japan

E-mail: s-noda@aist.go.jp

Abstract. Cu_2O/ZnO heterojucntion solar cells were successfully obtained by a magnetron-sputter deposition method. The Cu_2O thin film was deposited by the method using a sintered Cu_2O ceramics target. Crystalline phases of the films were controlled by adjusting an O_2 flow rate ratio ($O_2/(Ar+O_2)$) precisely during the sputtering process and Cu_2O single phase polycrystalline films were obtained at room temperature. The Cu_2O films qualities were improved by a rapid thermal annealing for 30 s in an Ar atmosphere of 1 atm. Hall mobility, carrier density, and resistivity of annealed films reached the values of 16.6 cm^2/V/s, $3.5x10^{15}$ cm^{-3}, and 107 Ωcm at 600 °C, respectively. The conversion efficiency of the Cu_2O/ZnO heterojucntion solar cell was 0.24 % with the open circuit voltage of 0.69 V.

1. Introduction

Cuprous oxide (Cu_2O) is a promising candidate of an all functional-oxide solar cell material as a p-type active layer because of its photoelectronic properties such as proper energy band gap of 2.1 eV, large optical absorption and high carrier mobility, further, environmentally friendly properties such as non-toxicity, abundance of resource and low material cost. As the solar cell, p-Cu_2O/n-ZnO heterojunction diode structures are most intensively investigated and several kinds of formation methods of the Cu_2O film have been proposed. Thermal oxidization of pure Cu metal sheet at a higher temperature above 1010 °C is one method which can produce the highest quality of Cu_2O film at present [1-4]. Typical Hall mobility (μ_H) is around 100 cm^2/V/s in this method and the highest conversion efficiency (η) of 3.8 % has been obtained using Al-doped ZnO/ZnO/Cu_2O structure recently [1]. Electrodeposition is a low-temperature high-speed method in liquid phase [5-8] and typical μ_H is 1 – 1.8 cm^2/V/s. The η value was 0.41 – 0.46 % using Cu_2O/ZnO structure. Recently the highest η of 1.28 % was obtained by optimizing the electrodeosition conditions [5]. This method can utilize ZnO nanopillars and η were improved to 0.88 % by enlarging the interface area [8]. Reactive sputter deposition using pure Cu metal target achieved μ_H of 30 cm^2/V/s at a substrate temperature of 500 °C [9]. The η value of 0.4 % was obtained by this method.

The conversion efficiency is expected to be 18% from the band gap energy of 2.1 eV [10]. However, the current highest efficiency is only 3.8 % even if sufficiently high quality Cu_2O films are used [1]. It is important to realize the precise photovoltaic mechanisms in this system and to estimate actual conversion limit. Although optical/vibrational properties and intrinsic mechanisms of electrical conduction in the Cu_2O films have been well studied [11], generation and recombination processes of photo-induced carriers in the bulk Cu_2O and at the junction interface have hardly understood yet. And furthermore, optimization of the cell structures is quite difficult because of the difficulty to control the electrical properties of the films. Such circumstances make it difficult to divide the problems existed in the film properties and device structures. Although η is not so improved, the reactive sputter deposition seemed to be most suitable for the applications of the electronic devices. We tried to use sintered ceramics copper oxide targets to lower the deposition temperature to expand application field.

2. Experimental

The Cu_2O films were formed on thermally oxidized Si (oxide), Pt coated oxidized Si (Pt), and synthetic quarts (glass) substrates by an rf magnetron sputtering at room temperature using Ar + O_2 mixture gas system. The oxide and the glass substrates were used to investigate basic characteristics of the films. The Pt substrates were used to fabricate solar cells. The sputtering targets used in this study were commercially available sintered ceramics of Cu_2O (2N) and CuO (3N). Typical gas pressure and rf power density were 0.5 Pa and 3.8 W/cm^2, respectively. The substrates were then annealed by a rapid thermal annealing (RTA) for 30 s in an Ar atmosphere at various temperatures up to 600°C.

Variations in Cu_2O film characteristics according to a change of an O_2 flow rate ratio *Ro* (O_2/(Ar+O_2)) during the sputtering process were investigated in detail. Crystal structure was evaluated by an X-ray diffraction (XRD) for the films deposited on the oxide and the Pt substrates. Optical properties were investigated from a spectral absorption coefficient α measured with an optical absorption spectrometer for the films deposited on the glass substrates. Electrical properties of the films were investigated from resistivity ρ, Hall mobility μ_H, and carrier density n_p. The ρ value was measured with four-point probe and the μ_H and n_p values were obtained by Hall-effect measurement at room temperature for the films on the oxide substrates. Spectral photoconductivity of the films was roughly estimated by measuring electric current between 2.8 mm-gap electrodes formed on the films under illumination. Grain morphology of the Cu_2O films was observed with an atomic force microscope (AFM) in intermittent contact mode for the films on the Pt substrate which are the same structure with the solar cells.

Figure 1 shows a structure of the solar cell fabricated in this study. The cell was basically a heterojunction p-i-n diode device. The p-type Cu_2O layer, an undoped ZnO, and an n$^+$-type Al doped ZnO (AZO) layer were sputter deposited on the Pt substrates in order. Effective cell area was defined by the area of the ZnO/AZO laminated pattern area as shown in Fig. 1. A number of the cell patterns are formed on the substrate using a shadow mask during the ZnO/AZO depositions. The cell area was varied from 0.0004 to 0.09 cm^2 (0.2 mm square to 3 mm square). The interface between the Cu_2O and the ZnO layers had to be formed successively in vacuo to obtain functioning p/n junction. Therefore, after the continuous deposition of 10 nm-thick ZnO layer on the Cu_2O, the substrates were taken out of the sputtering chamber and proceeded to the RTA treatment. Then the ZnO/AZO patterns were deposited by the magnetron sputtering using the shadow mask. The ZnO and the AZO layers were sputter deposited using sintered ceramics targets of ZnO (4N) and 2%-Al doped ZnO (4N), respectively. The ZnO and the AZO layer were deposited at room temperature and not annealed. The ρ values of the ZnO and the AZO films were 200 k Ωcm and 2.3 m Ωcm, respectively. Electrode pads of Au/Al for the probe contact were deposited by a vacuum evaporation using the shadow mask finally.

Basic properties of the solar cells were investigated from an I-V characteristic and a spectral response (external quantum efficiency: E.Q.E.). Photovoltaic performances of solar cells, that is, closed circuit current (I_{sc}), open circuit voltage (V_{oc}), conversion efficiency (η), fill facter (*FF*) and other values were evaluated with a solar simulator under air-mass (AM) 1.5 illumination.

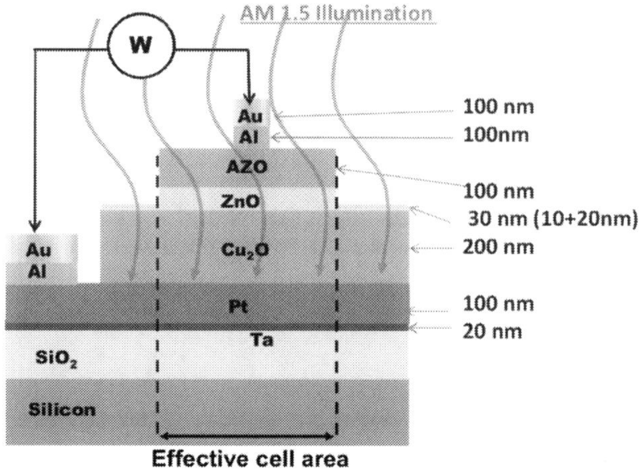

Fig. 1 Solar cell structure fabricated in this study.

3. Results and discussion

3.1. Characteristics of sputter deposited Cu_2O films

Figure 2 shows the XRD spectra of the sputter deposited films using (a) Cu_2O target and (b) CuO target on the oxide substrates at various *Ro* at the room temperature. Since all the films were thin polycrystalline structure and not oriented on the oxide substrates, the XRD measurement was made in a fixed grazing-angle (0.5°) – 2θ (out-of-plane) method. The variations in the film crystallinity were much different by the difference of the target material. In the case of the Cu_2O target (Fig. 2 (a)), the crystalline phase varied significantly by changing *Ro*. At the small *Ro* region below 2.3 %, the films showed Cu_2O crystalline phase only. The Cu_2O crystalline phase was transformed to CuO crystalline phase at large *Ro* region above 8 %. An intermediate phase (Cu_4O_3) was also observed at middle *Ro* region of 3.1 - 4 %. On the other hand, there was little difference in the spectra in the case of the CuO target and only CuO crystalline phase was observed regardless of *Ro* (Fig. 2 (b)). The polycrystalline structures were observed with a scanning electron microscope (SEM). All the films deposited at room temperature had a columnar structure of around 20 nm in diameter.

Figure 3 shows the spectral absorption coefficient of the films on the glass substrates under the same conditions as Fig. 2. In the case of the Cu_2O target-based films (Fig. 3 (a)), the spectrum profile varied significantly with changing *Ro* corresponding to the change of crystalline phase shown in Fig. 2 (a). Reference curves are reported data of single crystal Cu_2O [12] and polycrystalline CuO [13] prepared by a floating-zone (FZ) melting technique [14]. The spectrum profiles of the Cu_2O phase films obtained at small *Ro* region below 1.6 % were almost the same as that of the single crystal except for a small peak shown at low energy region around 1.2 eV. At the large *Ro* region above 2.3 %, the spectrum profile changed gradually to similar profile of the reported CuO with increasing *Ro*. However, the profile did not accord closely with the CuO curve. In the case of the CuO target-based films (Fig. 3 (b)), the spectrum profiles of all the films were very similar with that of the reference CuO curve regardless of *Ro*. It was found that the Cu_2O target was more useful to obtain Cu_2O crystalline phase film and had wider control margin than the CuO target from Figs. 2 and 3. It is considered that the difference in such behaviors is caused by the difference in oxidization reactivity between the two target materials.

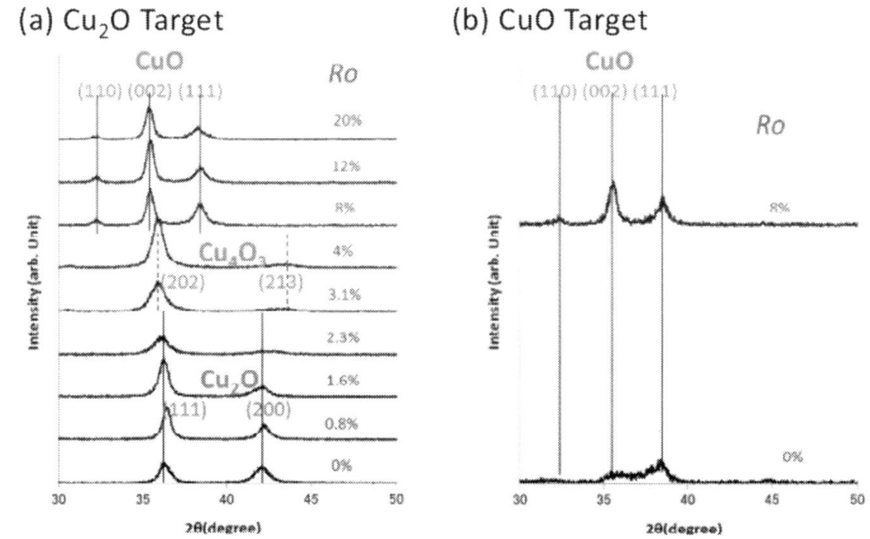

Fig. 2 Variations in crystalline phase of sputter deposited films with changing *Ro* using (a) Cu_2O target and (b) CuO target.

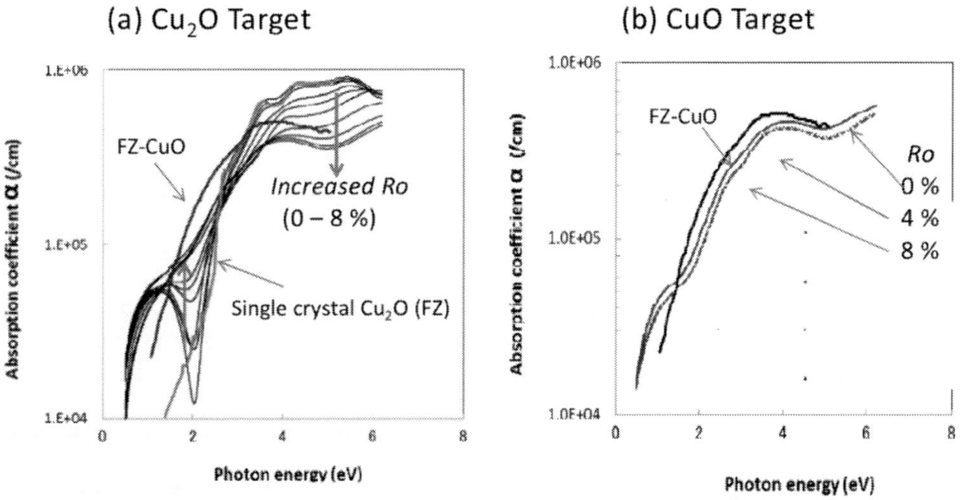

Fig. 3 Variations in optical absorption spectra of sputter deposited films with changing *Ro* using (a) Cu_2O target and (b) CuO target. Reference curves are cited from [12, 13].

The Cu_2O target-based films deposited at room temperature were expected to improve further by heat treatment. Figure 4 shows the XRD spectra of the films deposited at various *Ro* (i.e. of various crystalline phases) before (a) and after (b) the RTA at 600 °C. The spectrum peaks increased sharply and the three crystalline phases of Cu_2O, Cu_4O_3 and CuO were divided into two phases of Cu_2O and CuO by the RTA. It was found that each Cu_2O and CuO polycrystalline film improved itself keeping its initial crystalline phase and Cu_4O_3 polycrystalline film did not conserve its initial state. The Cu_2O films deposited on the Pt substrate was weekly oriented (111) surface and that was much improved by the RTA differently from the oxide substrates.

The improvement in the Cu_2O film crystallinity by the RTA was also confirmed from optical properties. Figure 5 shows Tauc plot of the absorption coefficient α before (a) and after (b) the RTA. The Cu_2O phase films deposited at small *Ro* region indicated a direct transition and the band gap energy of around 2.6 eV (Fig. 5 (a)). The slope of the absorption edge became steep and a fine structure appeared on the slope after the RTA (Fig. 5 (b)). These two edges were consistent to the E_{0C} ($\Gamma_7^+ \rightarrow \Gamma_8^-$) and E_{0D} ($\Gamma_8^+ \rightarrow \Gamma_8^-$) transition [12] (1s –Yellow band (in-direct; 2.03 – 2.17 eV) does not appear in this plot). It can be said that the Cu_2O phase polycrystalline films were much improved by the RTA.

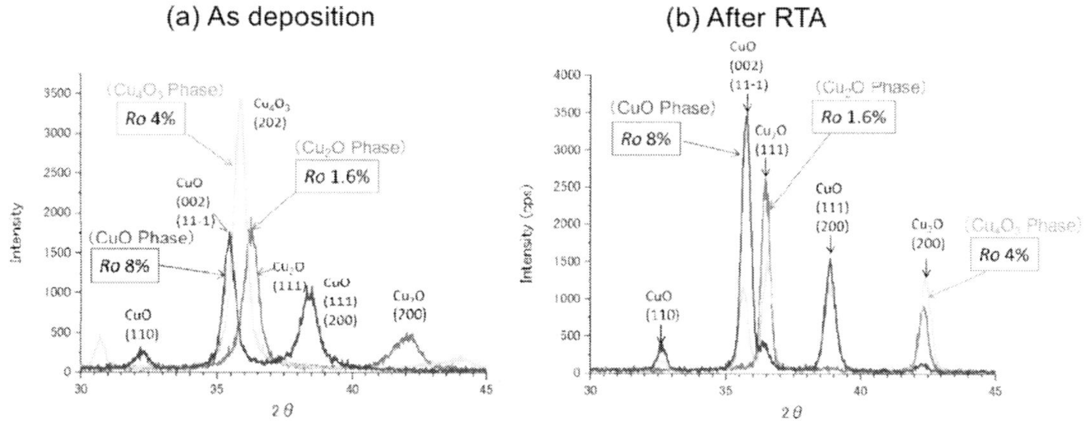

Fig. 4 XRD spectra of Cu_2O target-based films which have several kinds of crystalline phases (a) before and (b) after RTA for 30 s at 600 °C.

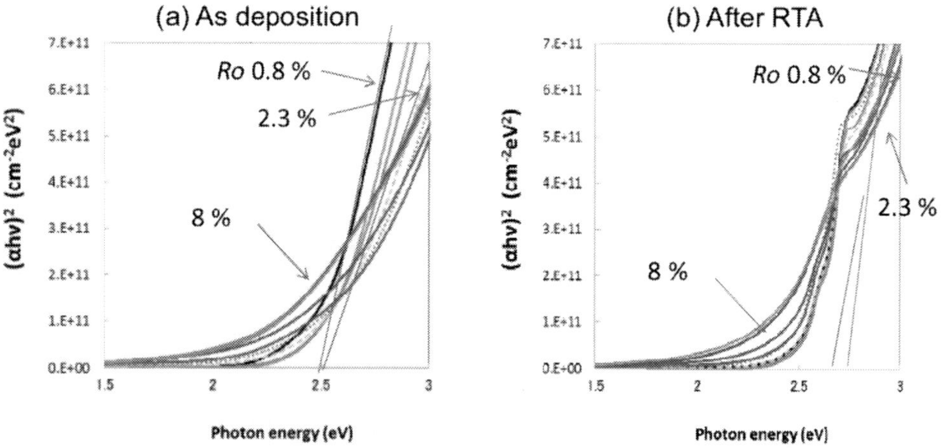

Fig. 5 Tauc plot of absorption coefficient *α* of Cu_2O target-based films (a) before and (b) after RTA for 30 s at 600 °C.

s at various RTA temperatures observed with the AFM. It was found that the polycrystalline grains began to grow around 400 °C and grew considerably large at 600 °C.

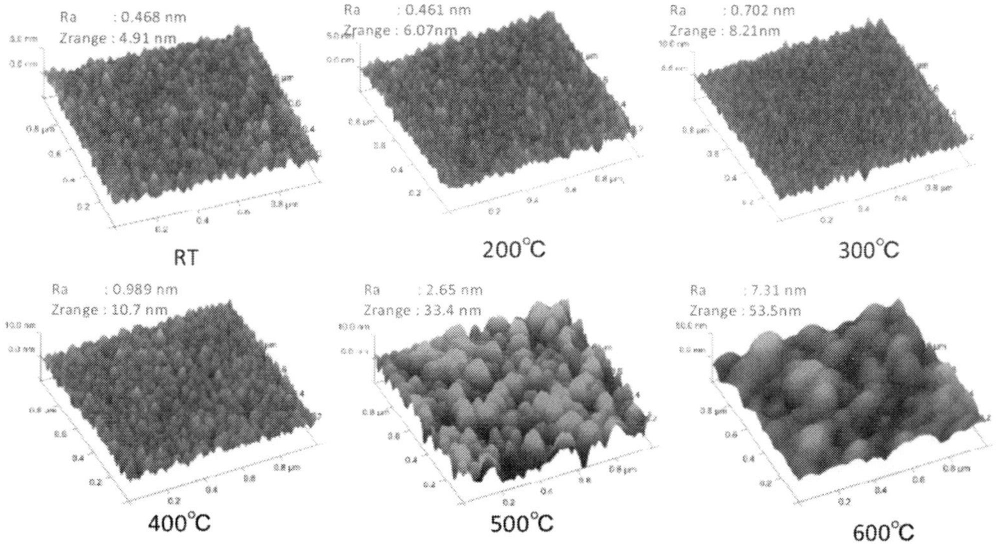

Fig. 6 Surface images of the annealed Cu_2O films on Pt substrates after RTA at various temperatures observed with the AFM.

Figure 7 shows the resistivity ρ of the films without the RTA deposited from the Cu_2O and the CuO targets as a function of *Ro*. Their dependences on *Ro* were significantly different. The ρ value of the Cu_2O target-based films varied complicatedly with increasing *Ro* corresponding to the changes in the crystalline phase as shown in Fig. 2. The ρ value of the Cu_2O phase films (*Ro*: 0 – 2.3%) decreased extremely with a very little increase in *Ro*. Then ρ increased with increasing *Ro* through the Cu_4O_3 phase film region (*Ro*: 2.3 – 8 %). Finally ρ decreased monotonously with increasing *Ro* in the CuO phase film region (*Ro*: 8% –). On the other hand, the CuO target-based films which had only CuO crystalline phase showed a simple dependence on *Ro*. The ρ value decreased monotonously with increasing *Ro*. The decreases of ρ with increasing *Ro*, which are observed in the Cu_2O and CuO phase films are considered to be due to an increase in a density of a Cu vacancy (V_{cu}) mainly. V_{cu} is considered to act as an acceptor in the p-type Cu_2O and CuO. Actually, it was confirmed that the carrier density of the CuO target-based film increased from 9.2 x 10^{16} cm^{-3} to 2.5 x 10^{18} cm^{-3} with increasing *Ro* from 0 to 10 %.

The ρ value of the Cu_2O target-based films varied significantly by RTA. Figure 8 shows the variations of ρ in each crystalline phase as a function of the RTA temperature. The dependences on the temperature were quite different. The ρ value of the highly resistive Cu_2O phase film (*Ro* 0.8 %) decreased largely while that of the low resistive Cu_2O phase film (*Ro* 1.6 %) did not varied so largely with increasing *Ro*. The ρ values of the Cu_4O_3 and CuO phase films increased with increasing *Ro*. Although the reason why ρ of the low resistive Cu_2O film did not changed was unclear, the feature is considered to be useful to control the film properties. Therefore the low resistive Cu_2O phase film (*Ro* 1.6 %) was evaluated for the solar cells.

Figure 9 shows the dependences of μ_H and n_p on the RTA temperature for the low resistive Cu_2O phase film (*Ro* 1.6 %). The μ_H value increased drastically above 400 °C and reached 16.6 $cm^2/V/s$ at 600 °C. At this time, n_p was changed from 3.3×10^{16} to 3.5×10^{15} cm^{-3}. At the RTA temperature region less than 400 °C, reliable μ_H and n_p could not obtained by Hall effect measurement. These dependences are well correspondent with the grain growth shown in Fig. 6.

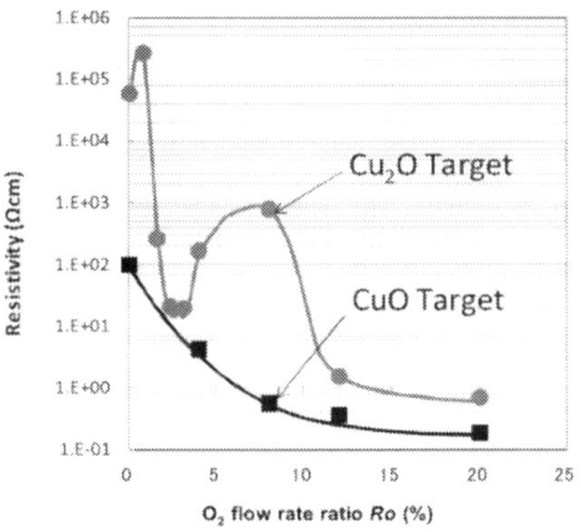

Fig. 7 Dependences of resistivity ρ of Cu_2O target-based film and CuO target-based film on O_2 flow rate ratio *Ro* before RTA.

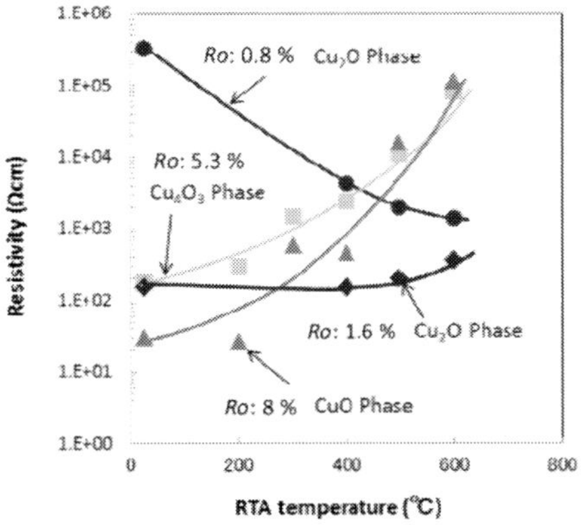

Fig. 8 Dependences of resistivity ρ of Cu_2O target-based films which have various crystalline phases on RTA temperature.

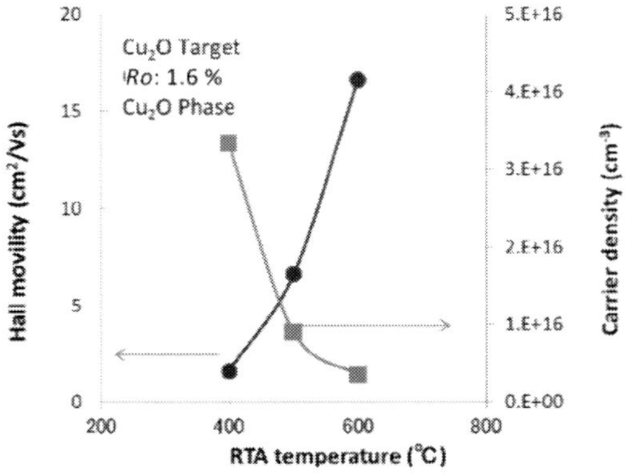

Fig. 9 Dependences of Hall mobility and carrier density on RTA temperature.

3.2. Characteristics of Solar cells

Figure 10 shows the I-V characteristics of the solar cells under a dark condition, which were made by the low resistive Cu_2O phase films (*Ro* 1.6 %) annealed at different temperatures. It was confirmed that a p/n junction was formed successfully even if no RTA was made on the Cu_2O films because good rectification effect was observed regardless of the RTA temperature. An ideality factor n was a little large and increased from 2.0 to 3.8 with increasing the RTA temperature. The relatively large n value is considered to be due to the high series resistance of the Cu_2O layer. Figure 11 shows the I-V curves of the same cells as Fig. 10 under the AM 1.5 illumination. It can be seen that an electric current and a voltage shift were generated by comparing with Fig. 10. The n values became nearly 1.0 in all cells due to generated photoconductivity.

Figure 12 shows the photovoltaic characteristics of the solar cells. The measurement was made for 0.016 cm^2 cells under the AM 1.5 illumination. The performance was improved with increasing the RTA temperature corresponding to the improvement of the film qualities described as above. The I_{sc}, V_{oc}, η and FF values increased to 1.2 mA/cm^2, 0.69 V, 0.24 %, and 0.28, respectively.

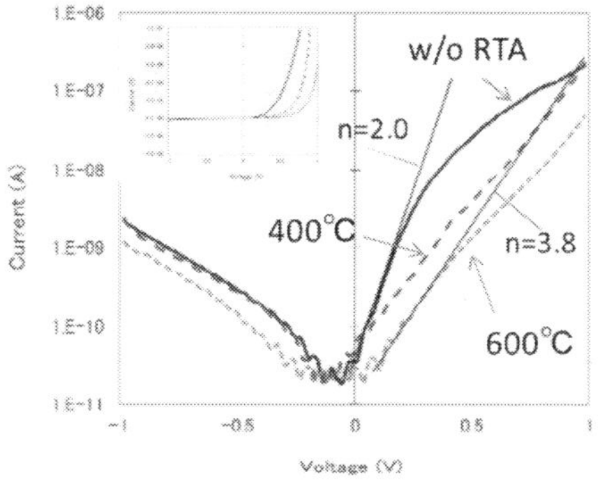

Fig. 10 I-V characteristics of solar cells under dark condition.

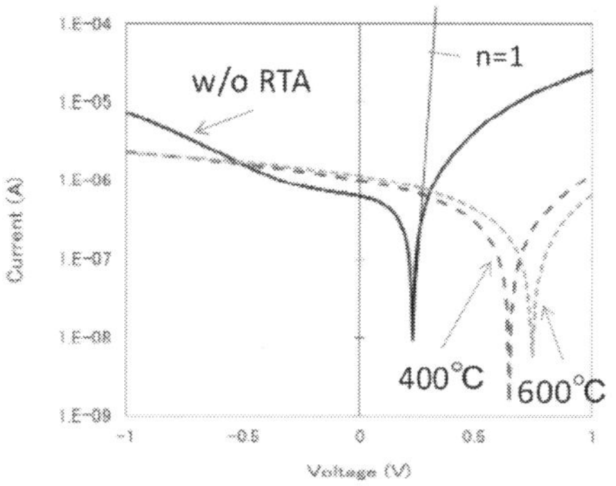

Fig. 11 I-V characteristics of solar cells under AM 1.5 standard illumination.

Fig. 12 Photovoltaic characteristics of solar cells which consist of Cu2O films annealed at various temperatures.

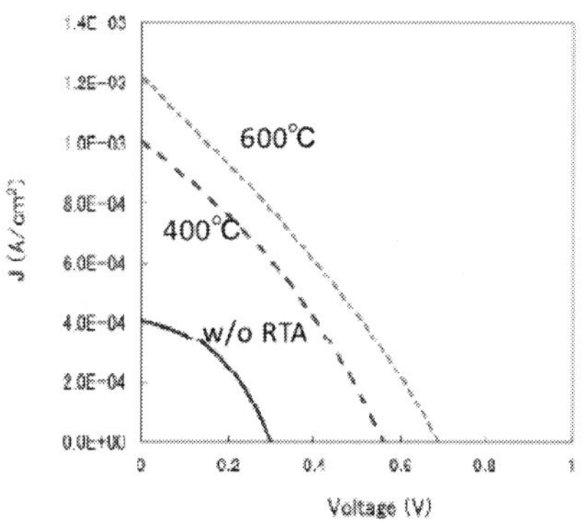

Figure 13 shows the spectral responses of the solar cells. The E.Q.E. improved remarkably by the RTA of 600 °C for the Cu_2O film however the spectral profile was not changed. The efficiency decreased steeply at around the wave length of 400 nm. Figure 14 shows the spectral photocurrent of the Cu_2O films after the RTA of 600 °C. It was observed that the carrier generation was sustained to long-wavelength side around 500 nm which was roughly consistent with the band gap energy of the Cu_2O film obtained in Fig. 5 (b). It was supposed that most of the photogenerated carriers in the wavelength range of 400 – 500 nm were recombined somewhere. Since the power density in the solar light spectrum increases greatly from 400 nm to visible wavelength region, improvement of the efficiency in this range is quite important. As for the short-wavelength side, the photocurrent was not detected in the Cu_2O film but the ZnO film showed the photocurrent instead as shown in the inserted figure in Fig. 14. It was understood that the outer circuit current of the Cu_2O/ZnO solar cell was the sum of the carriers generated in both Cu_2O and ZnO.

Although the efficiency depression in the 400 – 500 nm may be caused by a shortage of the lifetime of the carriers generated far from the interface, there is a possibility that a depletion layer does not exist in the Cu_2O-side, which generates driving field on the generated carriers. To verify this assumption, an effect on the carrier density of the ZnO layer in the p-i-n Solar cell structure was investigated. Precise carrier density control of the oxide semiconductor materials is difficult at present however the carrier density of the ZnO was roughly controlled by *Ro* and the RTA temperature. Figure 15 shows the results. The carrier density of Cu_2O (n_p) and ZnO (n_e) were estimated to be $3.5x10^{15}$ cm^{-3} and $6.6x10^{18}$ cm^{-3}, respectively. The efficiency in the 400 – 500 nm region was much improved by controlling a device parameter such as the carrier density. However V_{oc} decreased remarkably from 0.69 V to 0.14 V at the same time due to an increase in the junction leakage. As a result, η was reduced from 0.24 % to 0.047 %. There is a history that the ZnO layer was originally introduced to prevent the junction leakage and its resistivity must be large. When an ideal interface is realized, η will improve drastically. The controls of the device parameters [15, 16] and interface formation are as important as improving film qualities.

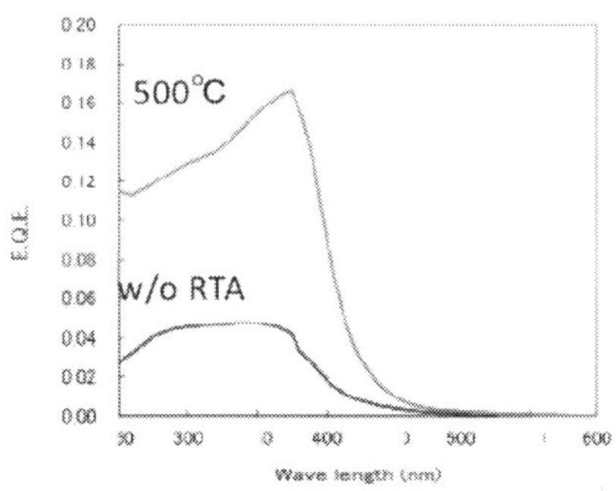

Fig. 13 External quantum efficiency of the solar cells with / without RTA

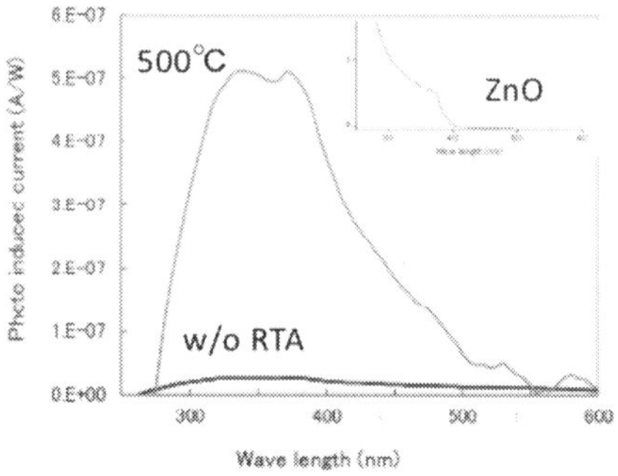

Fig. 14 Photocurrent of Cu_2O phase films. (Inserted figure is that of ZnO film.)

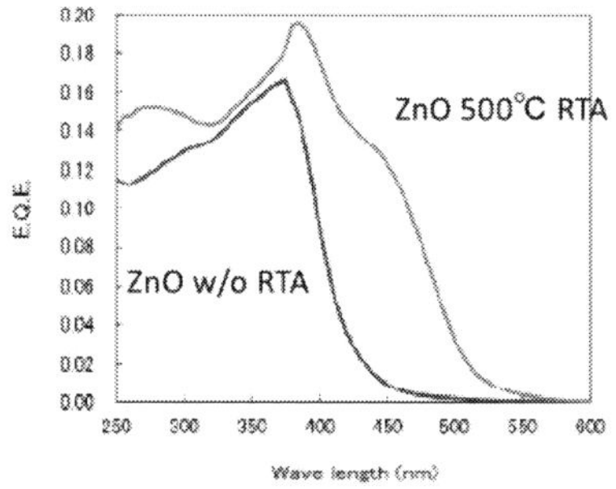

Fig. 15 Improvement in long-wavelength side efficiency by changing carrier density in ZnO layer.

Figure 16 shows cross sectional images of the solar cells (a) without RTA and (b) with 600 °C RTA observed by a transmission electron microscope (TEM). It was observed that the columnar polycrystalline grains were grown from about 20 nm to 200 nm in diameter and the crystallinity was improved. According to this improvement, various characteristics were improved as described above. However there is a trade-off between the film quality and the fabrication results of the solar cells. Actually the yield rate of the quality product decreased with increasing the RTA temperature. The decrease of the yield is considered to be caused by the electrical leakage related to the roughness at the interface. Further modifications such as device structures or fabrication processes are required.

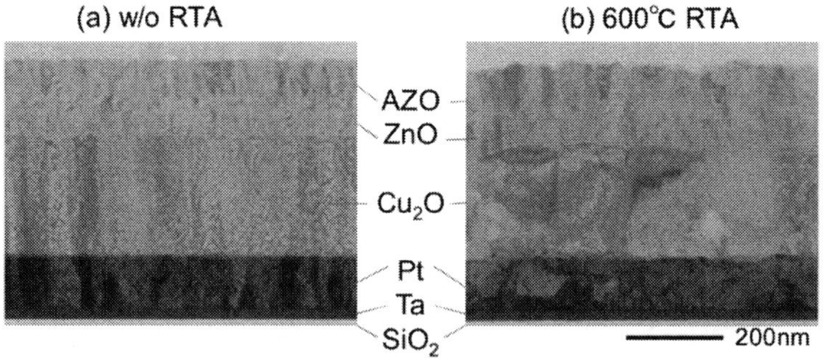

Fig. 16 TEM images of cross sectional view of solar cells (a) before (b) after RTA to 600 °C

4. Conclusion

The magnetron sputter deposition method of the Cu_2O film was investigated by utilizing the Cu_2O and CuO sintered ceramics target in detail for realizing the all functional-oxide solar cells. In the room temperature deposition, the Cu_2O target was more useful to control film properties and three kinds of single polycrystalline phase, Cu_2O, Cu_4O_3 and CuO were selectively formed. The Cu_2O target-based films are improved by the RTA for only 30 s. Especially the Cu_2O phase films were greatly improved in optical and electrical properties by the RTA at 600 °C. The absorption spectra were almost the same as those of the single crystal and the Hall mobility increased to 16.6 cm_2/V/s. By using the sputterd Cu_2O film, Cu_2O/ZnO/AZO solar cells were fabricated. The performance was also improved by the RTA. The conversion efficiency, short circuit current, open circuit voltage and fill factor were 0.24 %, 1.2 mA/cm^2, 0.69V and 0.28, respectively. The spectral response showed that the sensitivity at 400 – 500 nm region was much depressed. It was found that the depression were not due to the Cu_2O properties but were due to the problem of the optimization of device parameters. By the optimization of the device parameters and structure, the conversion efficiency will be expected to improve remarkably.

References

[1] Minami T, Nishi Y, Miyata T and Momoto J 2011 *Appl. Phys. Exp.* **4** 062301
[2] Minami T, Tanaka H, Shimakawa T, Miyata T and Sato H 2004 *Jpn. J. Appl. Phys.* **43** L917
[3] Mittiga A, Salza E, Sarto F, Tucci M and Vasanthi 2006 *Appl. Phys. Lett.* **88** 163502
[4] Tanaka H, Shimakawa T, Miyata T, Sato H and Minami T 2004 *Thin Solid Films* **469** 80
[5] Izaki M, Shinagawa T, Mizuno K, Ida Y, Inaba M and Tasaka A 2007 *J. Phys. D* **40** 3326
[6] Jeong S S, Mittiga A, Salza E Masci A and Passerini S 2008 *Electrochimica Acta* **53** 2226
[7] Hussain S, Cao C, Nabi G, Khan W S, Usman Z and Mahmood T 2011 *Electrochimica Acta* **56** 8342
[8] Cui J and Gibson U J 2010 *J. Phys Chem. C* **114** 6408
[9] Akimoto K, Ishizuka S, Yanagita M, Nawa Y, Paul G K and Sakurai T 2006 *Solar Eenergy* **80** 715
[10] Loferski J J 1956 *J. Appl. Phys.* **27** 777
[11] Shimada H and Masui T 1989 *J. Phys. Soc. Jpn.* **58** 1717
[12] Ito T, Kawashima T, Yamaguchi H, Masumi T and Adachi S 1998 *J. Phys. Soc. Jpn.* **67** 2125
[13] Ito T, Yamaguchi H, Masumi T and Adachi S 1998 *J. Phys. Soc. Jpn.* **67** 3304
[14] Ito T, Yamaguchi H, Okabe K and Masui T 1998 *J. Mater. Sci.* **33** 3555
[15] Liu Y, Turley H K, Tumbleston J R, Samulski E T and Lopez R 2011 *Appl. Phys. Lett.* **98** 162105
[16] Jiang T, Xie T, Zhang Y, Chen L, Peng L, Li H and Wang D 2010 *Phys. Chem. Chem. Phys.* **12** 15476

Acknowledgement

This work was conducted at Nano Processing Facility (NPF) supported by ICAN, AIST. A part of this work was performed at the Center for Integrated Nanotechnologies, a U.S. Department of Energy, Office of Basic Energy Sciences user facility at Los Alamos National Laboratory (Contract DE-AC52-06NA25396) and Sandia National Laboratories (Contract DE-AC04-94AL85000).

Experimental and Theoretical Investigation of Common Path Optical Coherence Tomography Using Conical Tip Lens Fiber

J Sugiyama[1] and K Taguchi[1]

[1]Department of Photonics, Ritsumeikan University, Kusatsu, Shiga 525-8577, Japan

E-mail: taguchi@se.ritsumei.ac.jp

Abstract. We proposed Common Path Optical Coherence Tomography (CPOCT) using gold coated conical tip lens fiber with a central wavelength of 1310 nm SLD to improve lateral resolution under the water. We already verified that conical tip lens fiber was good to improve the lateral resolution in CPOCT in air. However, CPOCT using conical tip lens fiber could not be functioning under the water solution because the value of back-reflected light field from the conical tip was too low. For this reason, we had to set reference intensity at a high level by coating the tip with a thin layer of gold. Then, we could have the right balance of reference intensity and sample intensity with gold coated conical tip lens fiber. Furthermore, the results of beam diameter, SNR, lateral resolution show that gold coated conical tip lens fiber is better than cleaved fiber. Then, we could observe two dimensional cross sectional scan of samples, for example, fresh leaves under the water by CPOCT using gold coated conical tip lens fiber.

1. Introduction

Optical Coherence Tomography (OCT) is an imaging system which can observe multiple sample's surface and inside with rapidly as real time microscopic optical imaging. OCT is also able to provide non-invasive imaging of living tissues and organisms with high-resolution. OCT is based on low-coherence optical interferometer principles. The way of imaging with OCT is using contrast of the light scattering properties of samples. OCT uses near-infrared region light source because that wavelength decreases scatter. In other words, OCT can image the deeper axial position of samples than using shorter wavelength. For this reason, OCT has already put to practical use in medical institution, so it is easy to expect to develop improvement medical instruments [1].

Original OCT technique is firstly developed as Time Domain OCT (TDOCT), however, Fourier Domain OCT (FDOCT) has been used extensively because of its advantages which are high sensitivity and high imaging speed. This advancement also made OCT be more high-resolution. Axial resolution depends on only the center wavelength and bandwidth of the light sources. Broad bandwidth light sources improve it in OCT because they can reduce the coherence length [2]. From this reason, we set up a goal that is to improve lateral resolution. That is needed to narrow down the beam. For that reason, using an optical lens is commonly way. However, there are problems that it makes OCT system be bigger and beam diameters spread very immediately.

High-resolution in OCT also has problems. It is that the difficulty of dispersion mismatch between sample and reference arm is caused by handling long optical fibers. That is to say, polarization fading, temperature, pressure, optical fiber bending and so on make problems. These problems can be resolved by controlling reference arm and sample arm reflectance independently. Therefore, we chose a common-path design OCT (CPOCT) whose sample and reference arm share the same optical path

[3]. Typical FDOCT takes reference beam from the reference mirror but CPOCT get it from the optical fiber tip. This CPOCT has many advantages [4]. Firstly, the dispersion and polarization mismatch of typical OCT's two optical paths (reference and sample arm) are automatically eliminated. Secondly, CPOCT can be a high-speed transport system and decrease the vibrations sensitivity. Finally, CPOCT is smaller than typical OCT because there is only one optical path to samples.

To improve the lateral resolution in CPOCT, we should narrow down the beam diameter from the optical fiber tip. On this account, we used a conical microlens that was chemically etched. We processed the optical fiber tip by a buffered hydrofluoric (BHF) etching solution, which was used in the selective-chemical etching technique. These processed optical fibers are called conical tip lens fiber. We already verified that conical tip lens fiber was good to improve the lateral resolution in CPOCT in air [5].

In this paper, we aimed to use CPOCT which was an invention with many potential uses in high refractive index media, like in vivo and blood vessel identification [6], [7]. Nonetheless, CPOCT using conical tip lens fiber could not be functioning in high refractive index media, such as water, because the value of back-reflected light field from the conical tip was too low. For this reason, we had to set reference intensity at a high level by coating the tip with a thin layer of gold [8], [9]. And so, we could have the right balance of reference intensity and sample intensity by this thin layer of gold coated conical tip lens fiber.

We also investigate the theory of conical tip lens fiber's beam propagation with the finite-difference time-domain (FDTD) method. The FDTD method is an efficient way to propose the full-wave analysis from fiber tip [10], [11]. Then, we demonstrated the conical tip lens fiber availability by CPOCT experiments and simulation of the FDTD method.

2. Experimental Setup

The proposed setup is shown in Figure 1. The light source, we used is, superluminescent diode (SLD) (DL-BX9-CS3089A, DENSELIGHT). The SLD generates light at center wavelength of 1310nm and full width at half maximum (FWHM) is 70nm. The light source is fed into the optical circulator (CIR-1310-50-APC, THORLABS) which allows light to travel in only one direction. And then, the light emitted from the optical circulator into the common-path interferometer through a single-mode optical fiber (P3-SMF28-FC-5, THORLABS). The optical fiber which goes on sample is fixed on a pole, and the pole is also on the Linear Stages (T-LS13M, Zaber Technologies Inc.) for scanning the sample in the x direction. The light from the sample arm is coupled into Fiber Optic Coupler (FC1310-70-50-APC, THORLABS), then is detected by the spectrometer (BTC 261E, B&W Tek Inc.) and Photodetector (PDA10CS, THORLABS). The spectrometer has 0.77nm wavelength resolution. We can get the OCT image by LabView on PC. Our experience is used the SLD which gives ~11μm axial resolution, however we focused on the lateral resolution by using conical tip lens fibers.

The SEM image in Figure 1 shows conical tip lens fibers and their tips. This conical tip lens fiber is processed by a buffered hydrofluoric (BHF) etching solution which is used in the selective-chemical etching technique with a volume ratio of NH_4F: HF, respectively as 2:1. This proceeded conical tip lens fiber has better lateral resolution [5]. We maintained a volume ratio of 2:1 and the solution temperature which is 40 degree Celsius constantly.

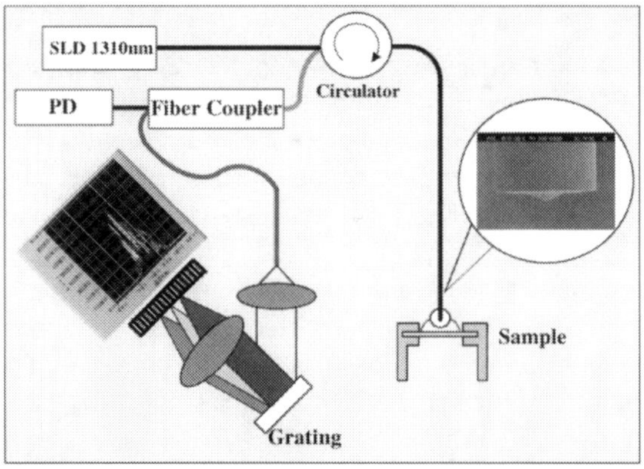

Figure. 1. Common-path Optical Coherence Tomography with conical tip lens fiber. The SEM picture is conical tip lens fiber's tip.

3. Theoretical Simulation

We also investigated the beam propagation with the FDTD method. The FDTD method has the direct solution to the Maxwell's time-dependent curl equations numerically, and, discretizes the space and time into rectangular cells and intervals, respectively. Then, the time variations of the field components at each cell.

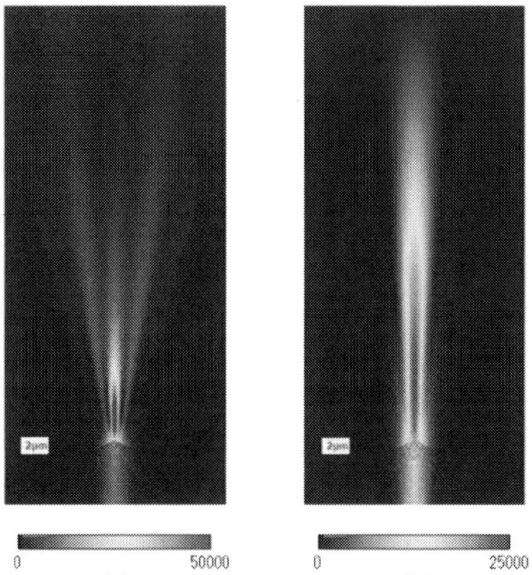

(a) (b)

Figure. 2. Calculated distribution of the electric field intensity from conical tip lens fibers: a in the air; b under the water.

To demonstrate FDTD computation, we considered that the simulation box consists of a 360,000 (400×900) grid in the x and y directions and the total number of the time steps is 3,500. We verified that conical tip lens fiber could be used instead of a conventional focusing lens to achieve both high lateral resolution and a long depth of focus simultaneously[12]. We could validate that conical tip lens fiber achieved a long depth of focus even under the water.

4. Results and discussion

We fabricated a conical tip lens fiber using a 2:1 BHF solutions yielding a cone angle of 129 degree. Before experiments of OCT imaging, we measured beam diameters from cleaved fiber and conical tip lens fiber under the water. Table 1 shows the beam diameters whose distances were beam waist, then, beam waist plus 10μm, 20μm and 30μm. According to these numerical values, conical tip lens fiber could narrow down the beam diameters than cleaved fiber. From these results, we used conical tip lens fiber for CPOCT in high refractive index media. Since the value of back-reflected light field from the conical tip was too low under the water, we coated the conical tip lens fiber with a thin layer of gold. Then, the value of back-reflected light field was about 155times as higher value, which was 0.46%, as non coated conical tip lens fiber.

Table. 1. The spot sizes of cleaved fiber and conical tip lens fiber under the water.

	Beam waist	+10μm	+20μm	+30μm
Cleaved fiber	**9.3μm**	**9.8μm**	**10.1μm**	**10.5μm**
Conical tip lens fiber	**7.2μm**	**7.2μm**	**7.6μm**	**8.6μm**

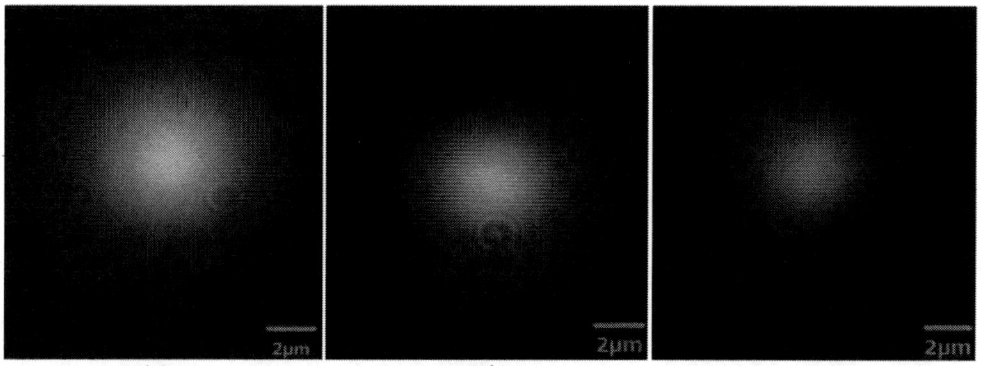

(a) (b) (c)

Figure. 3. CCD camera images of field distribution at the beam waist under the water of: (a) cleaved fiber; (b) conical tip lens fiber; (c) gold coated conical tip lens fiber, respectively. The central spot sizes are (a) 9.3μm, (b) 7.2μm and (c) 6.2μm with 1310nm wavelength light source, respectively.

We also measured the beam diameter of gold coated conical tip lens fiber as Figure. 3. They show the pictures of beam waist diameters. According to these pictures, the beam waist diameter of cleaved fiber is the largest (Figure. 3. (a)) and gold coated conical tip lens fiber is smallest (Figure. 3. (c)). That is to say, gold coated conical tip lens fiber could narrow down the beam diameter the most under the water. We did some experience with this gold coated conical tip lens fiber. Figure. 4 shows SNR whose axial positions are between the gold coated conical tip lens fiber tip and sample as mirror (150-500μm) under the water with constant power of 0.5mW at fiber tip. We could keep the SNR more than 30dB even at the position of 500μm. Therefore, gold coated conical tip lens fiber has right balance of reference intensity and sample intensity.

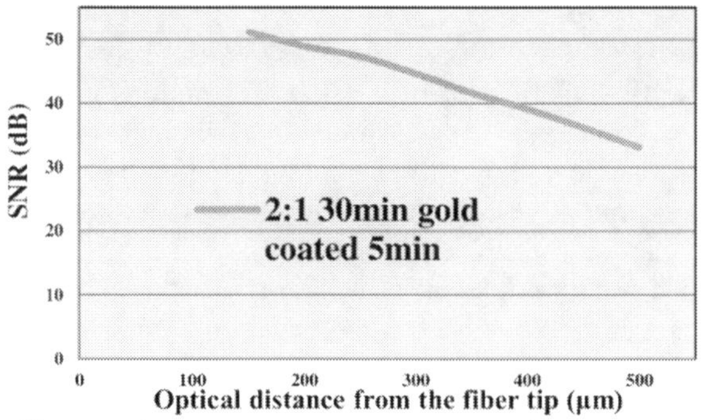

Figure. 4. SNR where the axial positions of 150-500μm under the water.

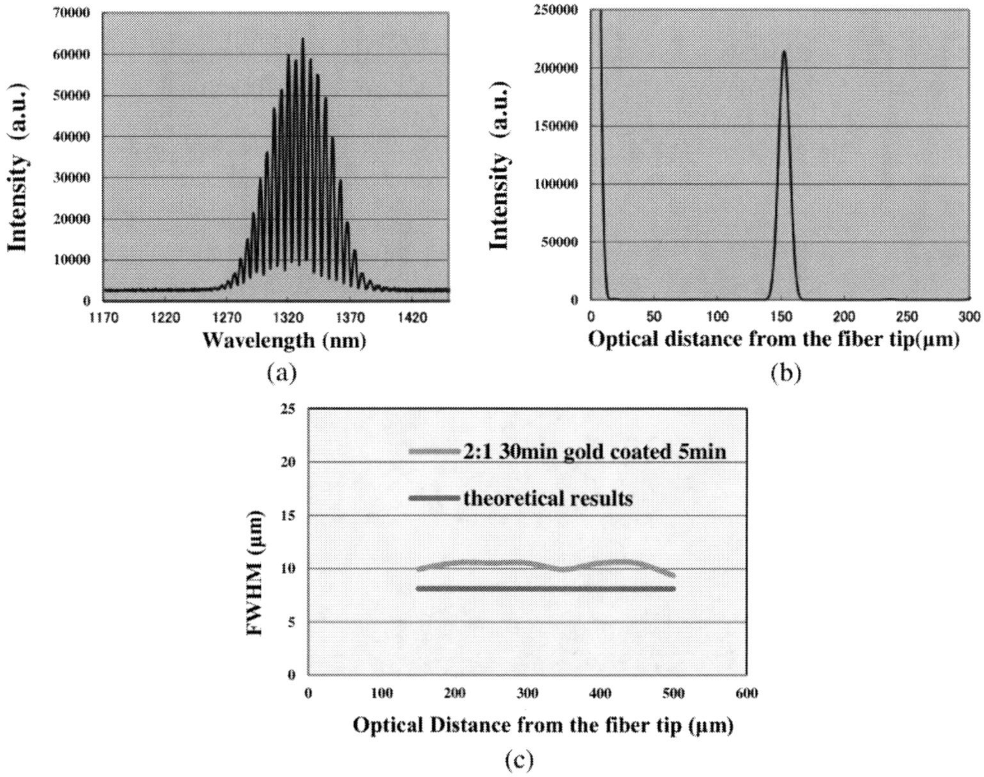

Figure. 5. Experimental results. (a) fringe pattern; (b) inverse Fourier transform of mirror; (c) Comparison of experimental and theoretical axial resolution results, respectively.

To compare experimental results and theory of axial resolution, we took data of a mirror. Figure. 5. (a) shows spectral fringes with mirror as the specimen. Figure. 5. (b) is an inverse Fourier transform of Figure. 5. (a), after performing k-space interpolation on Figure. 5. (a). Axial resolution measured with gold coated conical tip lens fiber under the water by Figure. 5. (b) was 9.95μm, and other axial position between optical fiber tip and a mirror is shown in Figure. 5. (c). We also calculated axial resolution by theoretical way and the equation is given by [13]

$$\delta z = \frac{2\ln 2}{\pi} \cdot \frac{{\lambda_0}^2}{n\Delta\lambda}, \tag{1}$$

where λ_0 is the center wavelength, $\Delta\lambda$ is the FWHM and n is the group refractive index. We used SLD whose light at center wavelength of 1310nm and FWHM is 70nm under the water (n=1.33). Then, as Figure. 5. (c) showing, the calculated theoretical axial resolution was 8.13μm. We considered that the difference between experiment and theory is due to optical aberrations in the optics. We note that these experiments were carried out in optical fibers. The resolution would be improved by a factor of n in the medium.

After measuring axial resolution, we compared with cleaved fiber and gold coated conical tip lens fiber by lateral resolution. We used 1951 USAF Resolution Target (NT38-257 Edmund optics) as a sample to examine lateral resolution. According to Figure. 6. (a), cleaved fiber could not scan 3rd element clearly, however, gold coated conical tip lens fiber could even at 5th element. That is to say, at the axial position of 30μm between the fiber tip and the sample, gold coated conical tip lens fiber has better lateral resolution than cleaved fiber. Figure. 6. (b) also shows that cleaved fiber could not any element while gold coated conical tip lens fiber could do 2nd element at the axial position of 150μm. These results mean that using gold coated conical tip lens fiber improved lateral resolution than cleaved fiber, even at the distant axial position.

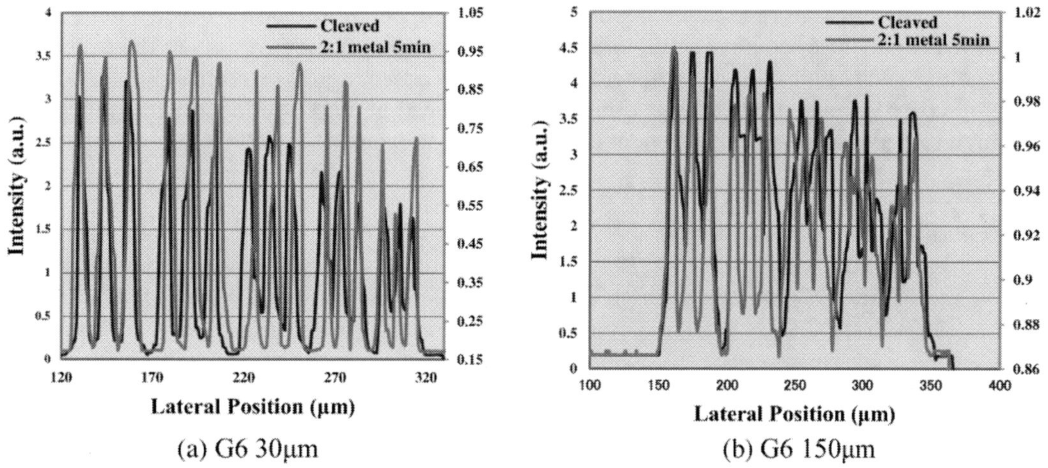

Figure. 6. Lateral resolution comparison with cleaved fiber and gold coated conical tip lens fiber: (a) the axial position which is from gold coated conical tip lens fiber to the sample was 30μm; (b) the axial position was 150μm, respectively.

We used this gold coated conical tip lens fiber to scan OCT image under the water. For example, we scanned a leaf which is under the water like Figure. 1's sample part. The result of scanning a leaf is shown in Figure 7. We could observe the leaf facing and its ducts clearly by CPOCT using gold coated conical tip lens fiber.

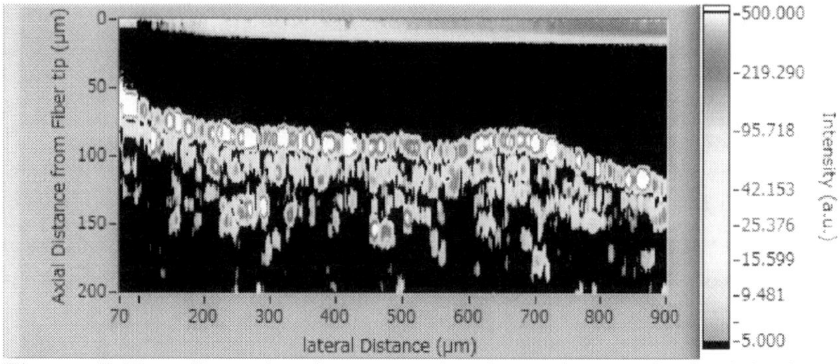

Figure. 7. Two dimensional cross sectional scan image of a fresh leaf under the water.

5. Conclusion

We proposed CPOCT using gold coated conical tip lens fiber with a wavelength of 1310 nm SLD to improve lateral resolution under the water. Using gold coated conical tip lens fiber, we had to set reference intensity at a high level, and we could have the right balance of reference and sample intensity. Furthermore, lateral resolution from our experimental results mean that gold coated conical tip lens fiber improved it clearly than cleaved fiber. Those results also show that our gold coated conical tip lens fiber has high lateral resolution and a long depth of focus. And then, we could observe two dimensional cross sectional scan of samples, for example, fresh leaves under the water by CPOCT using gold coated conical tip lens fiber. From our experimental and theoretical studies, it was found that gold coated conical tip lens fiber could be used instead of a conventional focusing lens to achieve both high lateral resolution and a long depth of focus simultaneously in CPOCT under the water.

References

[1] Vakoc J B, Fukumura D, Jain K R and Bouma E B 2012 Cancer imaging by optical coherence tomography: preclinical progress and clinical potential *NATURE REVIEWS CANCER* **12** 363-368

[2] P.D.L.Judson, K.M.Groom, D.T.D.Childs, M.Hopkinson, N.Krstajic, R.A.Hogg 2009 Maximising performance of optical coherence tomography systems using a multi-section chirped quantum dot superluminescent diode *Microelectronics Journal* **40** 588–591

[3] Fujiwara K and Matoba O 2011 High-speed cross-sectional imaging of valuable documents using common-path swept-source optical coherence tomography *APPLIED OPTICS* **50** 34

[4] Nikola Krstajić, Richard Hogg, Stephen J. Matcher 2011 Common path Fourier domain optical coherence tomography based on multiple reflections within the sample arm *Optics Communications* **284** 3168–3172

[5] K. M. Tan, M. Mazilu, T. H. Chow, W. M. Lee, K. Taguchi, B. K. Ng, W. Sibbett, C. S. Herrington, C. T. A. Brown, and K. Dholakia 2009 In-fiber common-path optical coherence tomography using a conical-tip fiber *OPTICS EXPRESS* **17** 2375-2384

[6] G. J. Tearney, B. E. Bouma, S. A. Boppart, B. Golubovic, E. A. Swanson, and J. G. Fujimoto 1996 Scanning single-mode fiber optic catheter-endoscope for optical coherence tomography *Opt. Lett* **21** 543–545

[7] G. J. Tearney, M. E. Brezinski, B. E. Bouma, S. A. Boppart, C. Pitris, J. F. Southern, and J. G. Fujimoto 1997 In vivo endoscopic optical biopsy with optical coherence tomography *Science* **276** 2037-2039

[8] J.-H. Han, I.K. Ilev, D.-H. Kim, C.G.Song, J.U.Kang 2010 Investigation of gold-coated bare fiber probe for in situ intra-vitreous coherence domain optical imaging and sensing *Applied Physics B Lasers and Optics* **99** 741–746

[9] Eduardo Fontana 2002 A Novel Gold-Coated Multimode Fiber Sensor *IEEE TRANSACTIONS ON MICROWAVE THEORY AND TECHNIQUES* **50** 82-87

[10] K.S. Yee 1966 Numerical solution of initial boundary value problems involving Maxwell's equations in isotropic media *IEEE TRATSACTIONS ON AXTEXXAS AND PROPAGATION* **AP-14** 302-307

[11] S. T. Chu and S. K. Chaudhuri 1989 A finite-difference time-domain method for the design and analysis of guided-wave optical structures *JOURNAL OF LIGHTWAVE TECHNOLOGY* **7** 2033-2038,

[12] J. Sugiyama, and K. Taguchi 2012 Common-path Optical Coherence Tomography (CPOCT) Using Axicon Fiber: A Comparison of Experiments and Theory proceeding of *38th International Micro & Nano Engineering Conference (MNE 2012)* spc0633(USB)

[13] S. H. Yun, G. J. Tearney, J. F. de Boer, N. Iftimia, and B. E. Bouma 2003 High-speed optical frequency-domain imaging *Opt Express* **11(22)** 2953–2963

Construction Tele-Robotics System with AR Presentation

K Ootsubo[1]*, T Kawamura[1] and H Yamada[1]
[1]Department of Human and Information Systems, Gifu University, 1-1 Yanagido, Gifu, 501-1193, Japan

*E-mail: otsubo@gifu-u.ac.jp

Abstract. Tele-Robotics system using bilateral control is an effective tool for task in disaster scenes, and also in extreme environments. The conventional systems are equipped with a few color video cameras captures view of the task field, and their video images are sent to the operator via some network. Usually, the images are captured only from some fixed angles. So the operator cannot obtain intuitively 3D-sense of the task field. In our previous study, we proposed a construction tele-robotics system based on VR presentation. The operator intuits the geometrical states of the robot presented by CG, but the information of the surrounding environment is not included like a video image. So we thought that the task efficiency could be improved by appending the CG image to the video image. In this study, we developed a new presentation system based on augmented reality (AR). In this system, the CG image, which represents 3D geometric information for the task, is overlaid on the video image. In this study, we confirmed the effectiveness of the system experimentally. Additionally, we verified its usefulness to reduction of the communication delay associated with a tele-robotics system.

1. Introduction

Recently, there is a need for system to perform task in areas where it is difficult for human such as disaster scenes and extreme environments; space, seabed, deep underground, nuclear plant and so on. Especially in Japan, a new system which is more effective compared with the conventional systems and can perform advanced task is strongly required after the Great East Japan Earthquake.

Tele-Robotics system using bilateral control is an effective tool for such task. In Japan, some conventional systems were used at real disaster scenes such as volcano eruption and earthquake. The development of the technique was started in order to avoid a secondary disaster in the recovery from the eruption of Mount Unzen in 1991. In this case, construction robot was tele-operated by feedforward control, and the operator performed tasks with a video image obtained from a stereovision camera. However, the image contains little spatial information of the task field, and the physical load was not negligible.

In the past, we investigated a tele-robotics system for a construction machine. The system consists of a servo-controlled construction robot, two joysticks for operating the robot from a remote place, and a 3-degree-of-freedom motion base. In order to improve the controllability of the system, we have examined the master and slave control method between joysticks and robot arms [1,2], the presentation method for the motion base [3,4], and the visual presentation of the task field to the operator [5]. Because the visual presentation is the information most essential to the operator, in this study we continued to focus on the presentation method for a remote operation field.

In our previous paper [6], we proposed a presentation method that used a mixture of stereographic video and the CG image of the robot, and clarified that the task efficiency was improved. In this system, the task object and the robot are presented by CG. These are simulated in real time in accordance with the operation, and the operator can obtain a view from a desired angle. So the operator intuits the 3D geometries of the object and the robot. However, most of the environmental elements were not represented. Since the visual information about the task field was poor compared with a video image, we think this drawback has a negative effect on the task efficiency. Additionally, the task viewing the multiple images putted a big load on the operator.

So we thought that the task efficiency and the load could be improved by appending the CG image to the normal color video image. In this study, we developed a new presentation system based on augmented reality (AR). In this system, the CG image, which represents 3D geometric information assists to perform a task, is overlaid on the video image. The operator can intuits spatial information about the task while obtaining the environmental information in the task field.

In this study, we confirmed the effectiveness of the system experimentally. Additionally, we verified it can deter decline of the task efficiency resulting from the communication delay associated with a tele-robotics system. In the experiments, the operators were asked to repeat a designed task, and the results demonstrated the effectiveness by using the task efficiency and their load.

2. Construction tele-robotics system

Figure1 shows an illustration diagram of our tele-operation system of construction robotics. This system is composed of joy stick (JS) for operation, camera for capturing the task field, construction robot, PC for controlling the robot (PC1) and PC for processing data from input devices (PC2). The control system is based on master-slave method, and displacement of the JS (master's side) corresponds to that of the piston (slave's side) which is fed back to master's system. An operator performs task looking the monitor connected to PC2.

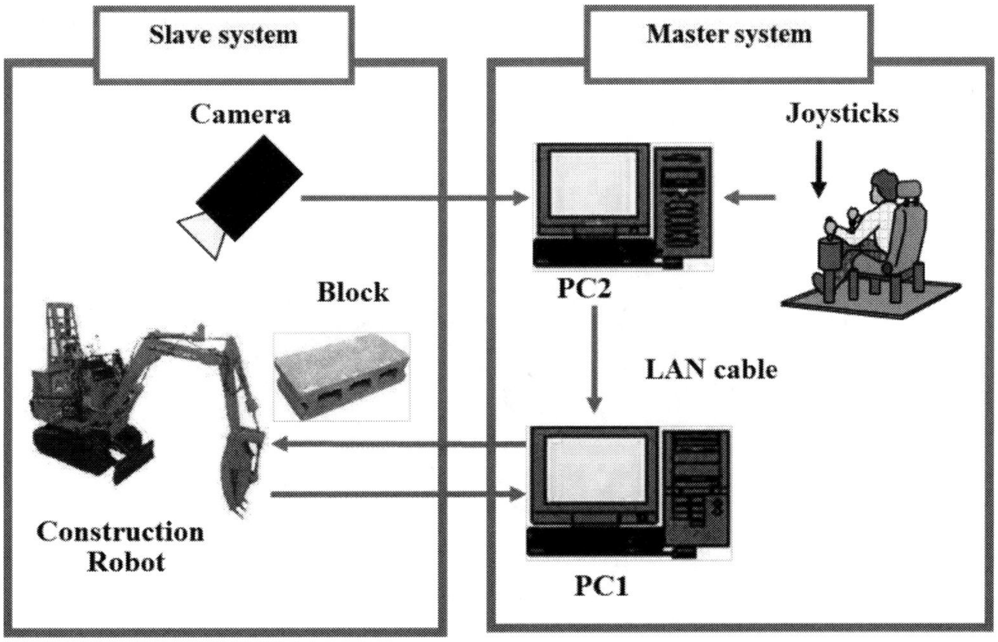

Figure 1. Construction Tele-Robotics System.

3. AR based presentation system

In this system, AR is achieved by overlaying the 3D-CG image, which supports the operation, on the video image.

The image contains 2-types of objects drawn as line segments, and overlaid on the video image by using the marker method. One represents the claw, and another represents a virtual gauge which indicates vertical distance to the other objects. In this method, the marker is a pattern image board with a square outline, and the system identifies it by image processing of the captured video image. Then the system calculates a rigid body transformation from the coordinate system of the marker to that of the video camera, and the CG objects are drawn onto the corresponding real objects by the transformation. In our system, the programming module of AR is based on the "ARToolKit" [7] which is a license-free library written in C for AR development.

In this study, we use two types of markers (Figure 2). One is a "floor marker" which is used to recognize the task field, and another is a "claw marker" which is used to recognize the claw's position and posture. As shown in figure 3, line segments are drawn from markers (a, b, c) to floor or claw's ends on the video image. This system allows the operator to intuit the opening degree of the claw and the distance between the robot and the other objects by looking at the image of the task field. In this study, communication delay is simulated by stopping the video image, and the CG image is updating at the time. Since the operator can continue the operation by relying on the image, we expected that improve the task efficiency as a result.

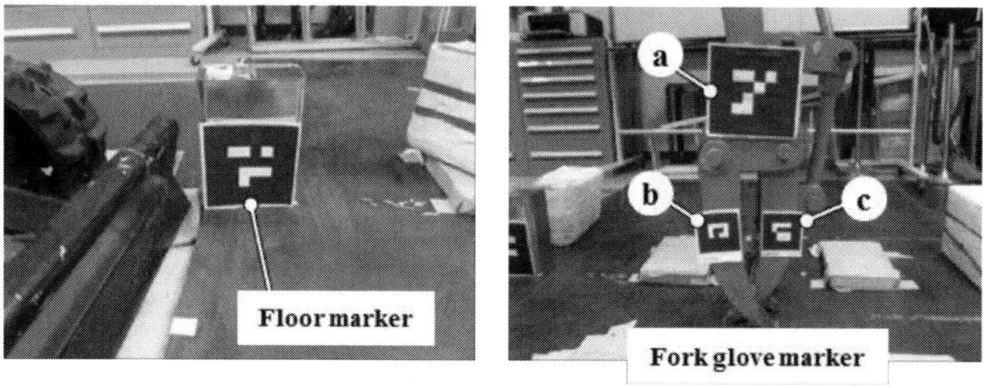

Figure 2. AR markers

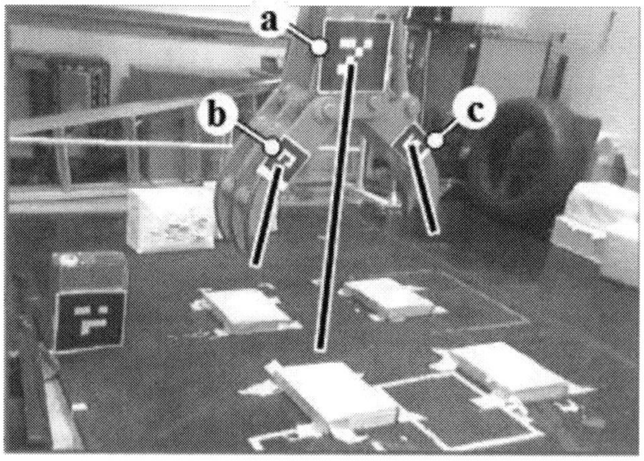

Figure 3. AR presentation

4. Experimentations and the results

In this study, we performed experiments described later in order to verify the usefulness of our system. In the experiments, the subjects were asked to iterate a work set composed of designed tasks.

We performed the quantitative evaluation by using the task efficiency and the number of failures. Here, the task efficiency is calculated as a ratio of total tasking time to the criterion time.

Also the qualitative evaluation is performed by using NASA-TLX and the questionnaire of free-description type as the indexes. The former is used to evaluate the task load, and the latter is used to evaluate the usability of our system. In the evaluation based on the former, the task load is defined by 6-items (table 1). In the first procedure, a subject decides the weights based on his/her individual importance by using pairwise comparison between the items. And after the experiments, decides the scores of the items according to degree of the loads. If a score is high, it means the load of the item is high for the operator. Finally, a numerical value is calculated by averaging the weighted scores. It indicates the comprehensive task load, and is called Weighted Work Load (WWL).

Table 1. 6-items of the task load in NASA-TLX.

Name	Abbr.	Description
Mental Demand	MD	How mentally demanding the task?
Physical Demand	PD	How physically demanding the task?
Temporal Demand	TD	How hurried or rushed was the pace of the task?
Performance	OP	How successful were you in accomplishing what you are asked to do?
Effort	EF	How hard did you have to work to accomplish your level of performance?
Frustration	FR	How insecure, discouraged, irritated, stressed, annoyed were you?

4.1. Experiment description

In the experiments, the operators were asked to grasp the center region of a block (figure 4), and carry it to a designated position (figure 5). A set of the tasks required transfers of blocks (1) from B1 to B2, (2) from B2 to B3, (3) from B3 to B4 and (4) from B4 to B1. The operators did 3-sets of tasks continuously, and we monitored the task time and the number of failures. Failure means the cases that the target block is not grasped at the center region, and the number is the total number in all the task time. The region is set near the center of gravity of the block. If the other region is grasped, the block falls easier during the transfer.

Figure 4. Blocks

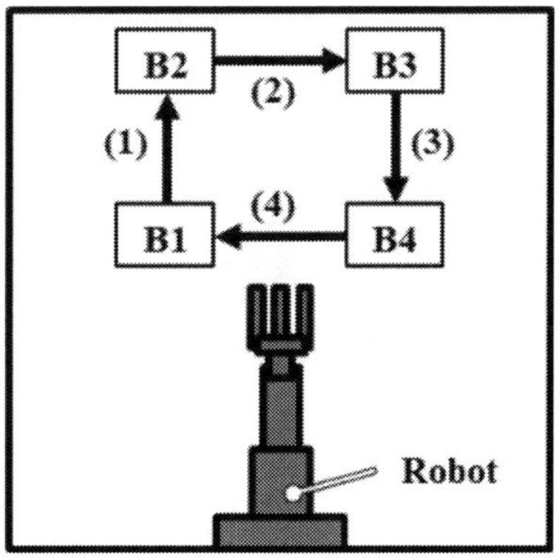

Figure 5. Transfer position in the task field

In the experiments, we tried to evaluate the usefulness of our presentation system (AR) by comparing with the conventional presentation system (Conv.) in different state of communication delay. The delay is given randomly as 1-3 seconds in each 3-10 seconds. The experiments were performed in the cases shown in table 2.

Table 2. Experimental cases

	Presentation type	Communication state
Case I	Conv.	Not delayed
Case II	AR	Not delayed
Case III	Conv.	Delayed
Case IV	AR	Delayed

4.2. Results
Here, the evaluation results based on quantitative index and qualitative index are shown in the following subsections.

4.2.1. Quantitative results
The task efficiencies given by 10-operators are shown in figure 6. Tasking times given by the operators in a condition are averaged, and the task efficiency is calculated as the ratio to that of the case I.

As shown in the figure, the task efficiency of the case III is lower than that of the case I, and also the same trend is seen between the case IV and the case II. Additionally, the difference between the latter cases is a little lower than that of the former cases. We think this is because that the operation was stopped more or less while the delay was occurred, and the time was shortened by using the CG image updated in real time.

The value of the case II is higher than that of the case I, and also the same trend is seen between the case IV and the case III. We think this is because that the recognition accuracy of the relative distance between the robot and the target object is improved by using AR. Additionally, the effect would be independent of the communication delay.

Next, the numbers of failure given by 5-operators are shown in figure 7.

As shown in the figure, the number of failure of the case II is less than half that of the case I. We think this is because that the recognition accuracy of the relative position of the claw and the center region is improved by using AR.

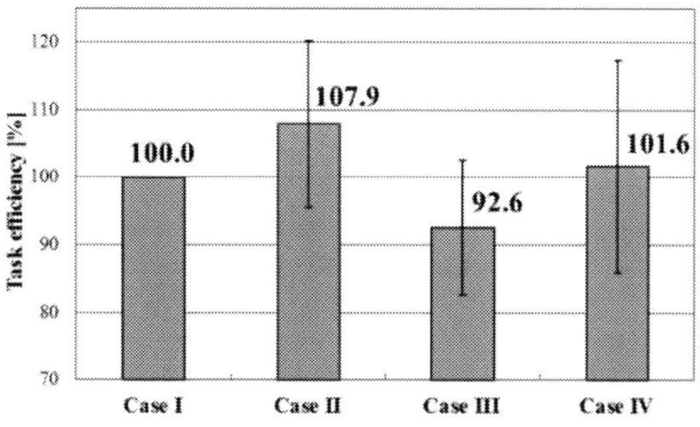

Figure 6. The task efficiency

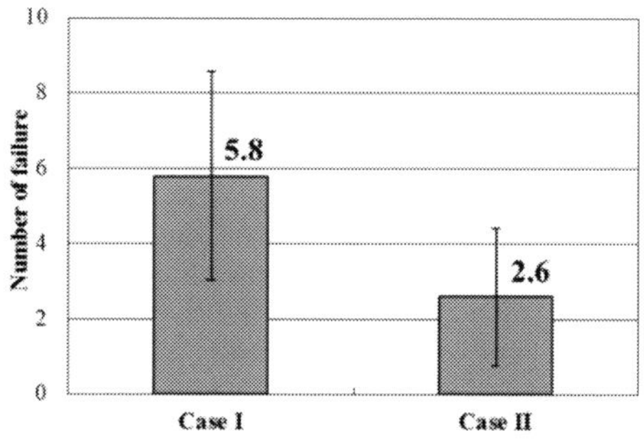

Figure 7. The number of failure

4.2.2. Qualitative results
The results based on the NASA-TLX are shown in figure 8. In the figure, (a) shows a result given by an operator experienced the operation using our system, and (b) shows that given by a non-experienced operator.

As shown in the figure, WWL of the case III and IV are higher than that of the case I and II. Especially, the differences of MD, EF and FR are larger than the other indexes, and these trends are seen in both of the results (a) and (b). We think this is because that randomly stopping of the video image increased time to recognize the present situation in the task field. As a result, it would have increased difficulty of the task.

WWL of the case II and IV are lower than that of the case I and III, and the trend is seen in both of the results (a) and (b). We think this is because that AR allowed the operator to intuit the relative geometry between the robot and the target object easily.

In the result (a), WWL of the case IV is lower than that of the case I, but the relation is reversed in the result (b). We think this means that the reduction effect of the task load can be improved with the experience.

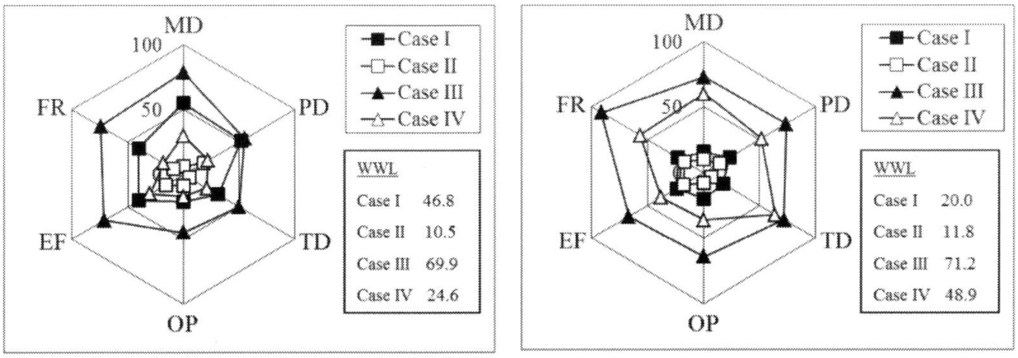

(a) Experienced operator (b) Non-experienced operator

Figure 8. NASA-TLX

At last, we describe the result of the questionnaire. The contents are shown in table 3, and the averaged result given by 10-operators is shown in figure 9. Each item was scored by 5-point scale.

When the item A is observed, you can see the score falls by communication delay but is improved by using AR. Also, the same trend is seen in the item B and C. From this result, we think that ease of understanding the relative position of the claw and the other objects is directly linked to ease of the grasping of the block. This result means our consideration about the quantitative evaluation is confirmed.

Additionally, you can see the scores of the item D and E have the same trend. This means that the stopping of the video image increases the mental load, and AR can be effective answer for the problem.

Table 3. Questions

ID	Item
A	Ease of the operation to grasp the block
B	Ease of understanding the relative position of the claw and the block
C	Ease of understanding the relative position of the claw, the floor and the transfer positions
D	Safety of the task
E	Comfort of the task

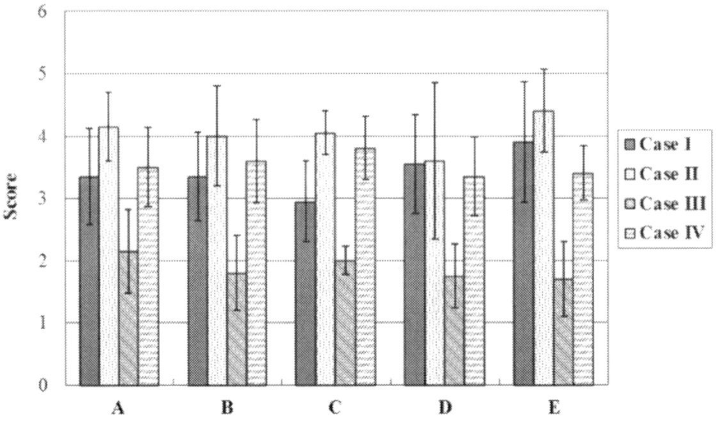

Figure 9. Result of the questionnaire

5. Conclusions

In our previous study, we proposed a VR based presentation system using multiple images. However, the simulated environment was limited, and the task load to look the multiple images was not negligible. So we thought that the task efficiency and the load could be improved by implementation of AR technique achieves overlay the CG image of the task on the video image.

In this study, we developed an AR based presentation system for tele-operation of construction robot. From the experimental results, we confirmed that the task efficiency and the load were improved by using the system. Also, the usefulness against the communication delay problem was confirmed.

In future works, we will try to further increase the task efficiency and the quality by implementing the following improvements to the system.

- Construction of accurate CG model and achievement of markerless AR by using a multiple-distance sensor with high resolution
- Development of new visual interface for supporting complex tasks

Acknowledgments

The authors would like to express their deep appreciation to Mr. Masashi HAMAGUCHI and Mr. Hidenori SATOH of the Graduate School at Gifu University for their extensive cooperation in performing the experiments.

6. References

[1] Yamada H, Muto T, Xia Y and Iwase T. 2000 *Tele-operated Construction Robot Using Virtual Reality (CG Presentation of Virtual Robot for Increasing Working Efficiency): Proc. of VSMM2000* pp 652-658.

[2] Yamada H, Muto T and Ohashi G. 1999 *Development of a Telerobotics System for Construction Robot Using Virtual Reality: Proc. of European Control Conference (ECC'99)* F1000-6.

[3] Zhao D, Yamada H, Muto T, Huang H and Kato H. 2001 *A New Method of Presenting Realistic Motions for a 3-Degree-of-Freedom Motion Base Applied in Virtual Reality System: Proc. of the 5th International Conference on Fluid Power Transmission and Control (ICFP' 2001)* pp254-258.

[4] Zhao D, Xia Y, Yamada H and Muto T. 2002 *Presentation of realistic motion to the operator in operating a tele-operated construction robot: Journal of Robotics and Mechatronics* 14-2 pp98-104.

[5] Yamada H, Muto T, Xia Y and Iwase T. 2000 Tele-operated Construction Robot Using Virtual Reality (CG Presentation of Virtual Robot for Increasing Working Efficiency). Proc. of VSMM2000: 652-658.

[6] Yamada H and Muto T. 2007 *Construction Tele-Robotic System with Virtual Reality(CG Presentation of Virtual Robot and Task Object using Stereo Vision System): International Journal of Control and Intelligent Systems* 35-3 pp195-201.

[7] Kato H, Billinghurst M. 1999 *Marker Tracking and HMD Calibration for a video-based Augmented Reality Conferencing System: Proc. of the 2nd International Workshop on Augmented Reality (IWAR 99)* pp85-94.

Journal of Physics: Conference Series **433** (2013) 012030 doi:10.1088/1742-6596/433/1/012030

Remote Sensing of Radiation Dose Rate by a Robot for Outdoor Usage

T. Kobayashi[1*], K.Doi[1], H. Kanematsu[2], Y. Utsumi[3] R. Hashimoto[4], and T. Takashina[5]

[1] Dep. of Electronics & Control Eng., Tsuyama National College of Technology, Okayama, Japan

[2] Dep. of Materials Science & Eng., Suzuka National College of Technology, Mie, Japan

[3] Laboratory of Advanced Science and Technology for Industry, University of Hyogo, Hyogo, Japan

[4] Collaborative Research Center, Hiroshima University, Hiroshima, Japan

[5] Int. Industry Academia Collaboration Division, Hiroshima University, Hiroshima, Japan

*E-mail: t-koba@tsuyama-ct.ac.jp; +81-(0)868-24-8215

Abstract. In the present paper, the design and prototyping of a telemetry system, in which GPS, camera, and scintillation counter were mounted on a crawler type traveling vehicle, were conducted for targeting outdoor usage such as school playground. As a result, the crawler type traveling vehicle can be operated smoothly in the school grounds of brick and asphalt. The results were as follows: (1) It was confirmed that the crawler type traveling vehicle can be operated smoothly in the school grounds of brick and asphalt (running speed: 17[m/min]). (2) It was confirmed that the location information captured by GPS is visible on the Google map, and that the incorporation of video information is also possible to play. (3)A radiation dose rate of 0.09[μSv / h] was obtained in the ground. The value is less than the 1/40 ([3.8μSv / h]) allowable radiation dose rate for children in Fukushima Prefecture.(4)As a further work, modifying to program traveling, the measurement of the distribution of the radiation dose rate in a school of Fukushima Prefecture, and class delivery on radiation measurement will be carried out.

1. Introduction

In hazardous environments robots used have been studied for nuclear industry and disaster relief, military usage for long time, especially robots have been actually used for many purposes in nuclear industries[1]-[3]. The most famous robot in the fields of military and disaster relief is "PackBot" manufactured by iRobot Co. more than 3000 units in Iraq and Afghanistan [4]. However, these robots used in extreme environments can't easily be used by anyone, further the price of a single PackBot is expensive at 120 thousand dollars [5]. The effects of the Tohoku-Pacific Ocean earthquake and tsunami that occurred on March 11, 2011, the cooling system of the Fukushima Daiichi Nuclear Power Plant, decontamination work has been carried out. However, radiation measurement has been still required in many places. A group of authors has been developed an inexpensive equipment by customizing the mass produced and commercially available robot for remote measurement of radiation dose rate. In the above system, an autonomous cleaning robot[6] and a micro computer were combined, and manual and programmed moving modes were additionally equipped to the robot which has an external signal connecter[7] in order to measure the distribution of radiation dose rate in low radiation level area, such an indoor space as school house. However, by using the microcomputer SH7125, they can send the measurement results of the rumba running control, calculation of movement position, and the radiation dose data acquisition. Moreover, they were tested to measurement test in three different modes, remote manual, remote programmed, and autonomous. In this paper, I was performed prototype and performance verification of remote control robot that is intended to measure the radiation dose rate distribution in the outdoors, such as school playgrounds.

2. Prototype

2.1Basic design

Fig.1 shows the system configuration. A crawler type dolly was selected as the driving system and the two crawlers were driven independently, since the purpose of this study is the measurement of radiation dose rate in the outdoor, with regard to the measuring system, a scintillation counter for measuring the radiation dose rate, a data logger for recording the signal and a driving recorder for capturing and monitoring the information of position visual image in real time, are equipped. It is equipped with a scintillation counter to measure an emission dose rate as system of measurement and carries the data logger together to record the measurements. In order to capture the position further information and images, to be monitored in real time, was equip with a drive recorder on a robot.

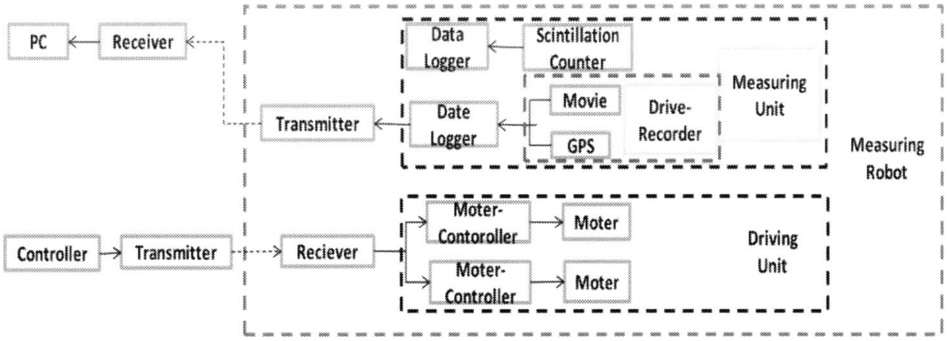

Fig.1 System diagram

2.2Equipment

2.2.1Scintillation counter

There are Geiger-Muller counter, scintillation counter, and go forth, as the equipment for the measurement of the radiation dose rate. A scintillation counter, which specifications are shown in Table1, was selected in the present study, as scintillation counter has one order of magnitude higher sensitivity than Geiger-Muller counter, further has much smaller the time constant.

Table1 Specification of survey meter

Measuring Ray	**γ – ray**
Measurable Range	**From background to 30.0 μSv/h**
Scintillater	**25.4φ×25.4mm NaI (T1)**
Accuracy	**Max. ±15%**
Output range	**DC +10mV (Full scale)**
Time constant	**3 seconds**

2.2.2Date logger

The measured signals in every second by the scintillation counter were recorded in a data logger. The data can be retrieved by USB memory and display the values in Excel at the present.
In future, the values will be transmitted to a computer in real time.

2.2.3Drive recorder

The movie taken by the CCD camera of the drive recorder can be displayed on a personal computer in real time. The position of the equipment captured by Global Positioning System

Journal of Physics: Conference Series **433** (2013) 012030 doi:10.1088/1742-6596/433/1/012030

(GPS) can be displayed on a google map, further it is possible to import the data into the data logger in the driving recorder.

3. Measurement test

3.1Analysis method of measurement results

3.1.1 Method for estimating the distribution of the radiation source from the measurement results

The usage of the present system is in a relatively wide play ground of school, therefore it will be simplified if the position or area of the radiation source can be estimated without measuring whole area of the ground. For example, assuming a planar radiation source produced from a puddle in which the radioactive substance was deposited and concentrated, the position and area of the radiation source are deduced as shown in Fig.2 and Fig.3. The measuring robot moves toward an assumed planar radiation source along the line2, line1, line0 in Fig.2. Then the distribution of the radiation dose rate will be measured as <2>, <1>, and <0> shown in Fig.3. The counter of the planar radiation source will be reduced on the line p･q where the maximum point of each distribution E,C,D is connected, and the center will be also on the line s-r which is a perpendicular bisector of the circle connected a contour constructed from the point C,B and A of each distribution <2>,<1> and<0> respectively. Next, the dimension of the planar radiation source *a* and *b* could be reduced by the spread of the distribution of the radiation dose using the equation mentioned next section.

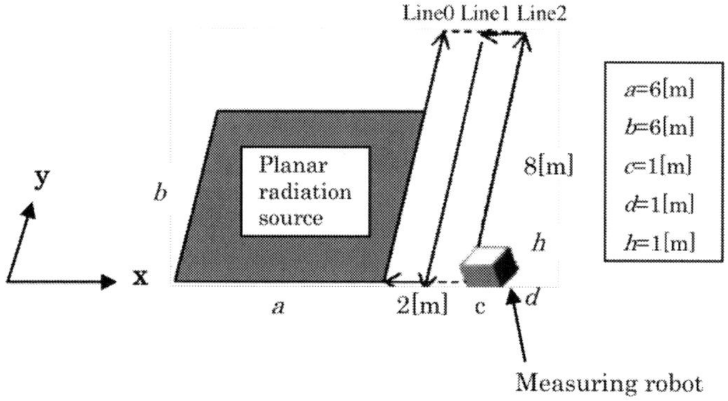

Fig.2 Measuring method

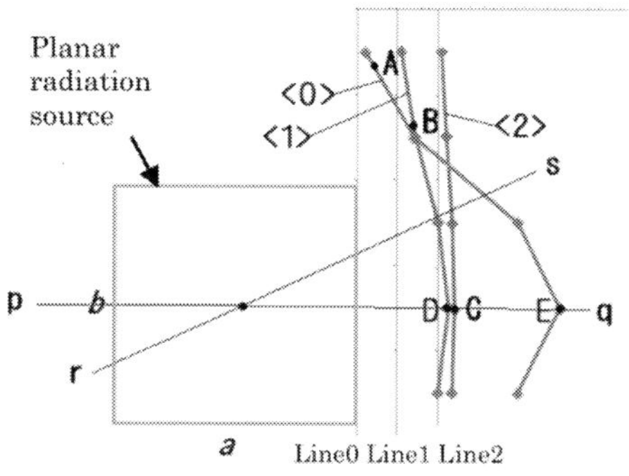

Fig.3 Method for measuring the position and area of the planar radiation source in which radioactive substance

3.1.2 Method calculating of the radiation dose rate distribution

A concept of the configuration factor used in the calculation of heat transfer was applied to the estimation of the distribution of the radiation source shown in section 3.1.1, since the radiation dose rate decreases in inverse proportion to the square of the distance. Definition equation of configuration factor is as follows,

$$F = \frac{1}{A_1}\int_{A_1}\int_{A_2}\frac{\cos\phi_1\cos\phi_2}{\pi l^2}dA_2dA_1 = \frac{h^2}{\pi}\int^a\left[\int^b\frac{1}{\{(x-u)^2+(y-v)^2+h^2\}^2}dy\right]dx$$

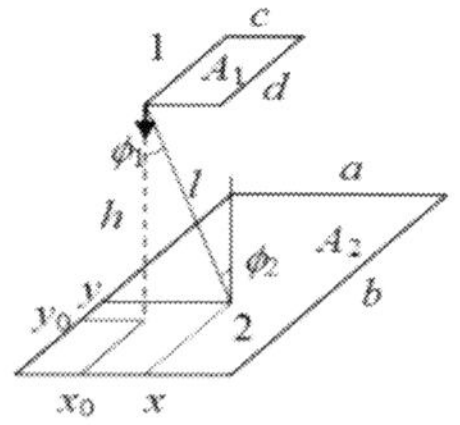

Fig.4 Geometric structure for the configuration factor

(a,b: Dimensions of the planar radiation source, c,d: Dimensions of the measuring instrument, h: Height of the measuring instrument)

The configuration factor was calculated under the condition in Fig.2, and the 3 dimensional distribution of the configuration factor obtained is shown in Fig.5.

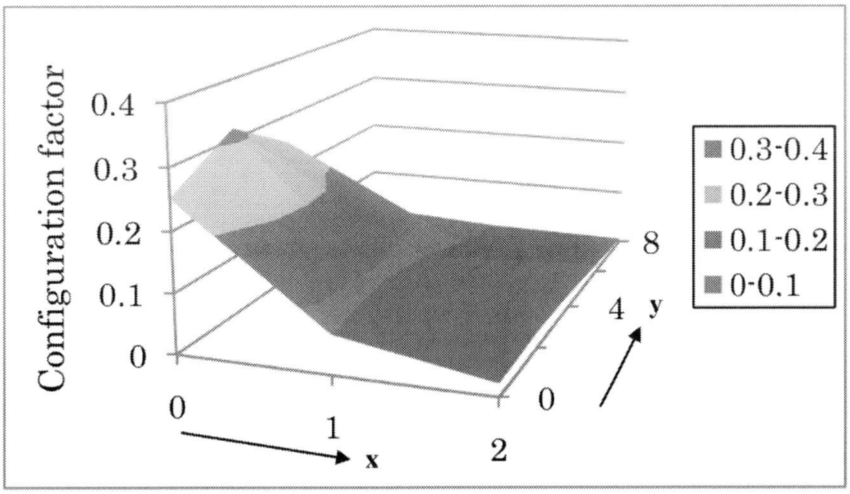

Fig.5 An example of the configuration factor calculated under the condition in Fig.2

3.2Measurements of radiation dose rate distribution

Fig.6 shows the prototyped remote control measuring robot. Movement speed is 17[m/min], Mileage is at about 200m, several times a test drive, I made the adjustment and confirmation of stable transmission and reception stability and stable running. Position information is captured in the drive recorder with GPS, it was confirmed that can be displayed on a google map after measurement. To chart the distribution of radiation dose rate that was loaded into the data logger, the result is shown in Fig.8. A, B, C of Fig.8 are corresponding to the position of the Fig.7. Range of scintillation counter when measured were carried out is 3 [μSv / h]. Radiation dose rate is 0.076 [μSv / h] ~ 0.104 [μSv / h], the average was 0.09 [μSv / h]. Further, in the measurement region of this distribution is considered particularly significant difference was not observed.

Fig.6 A photo of the measuring robot prototyped

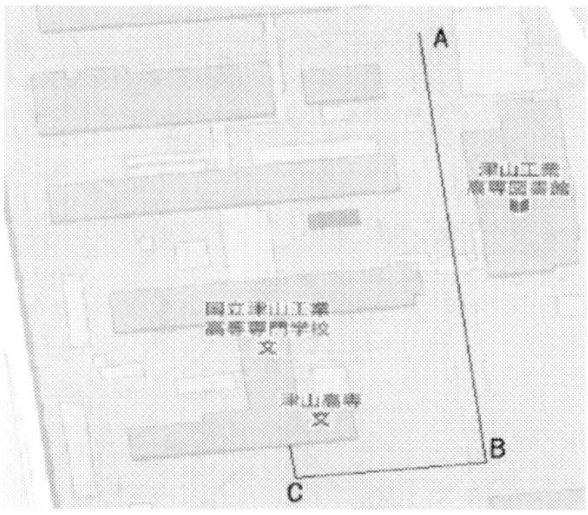

Fig.7 Track of the measuring robot in the campus of Tsuyama national college of technology

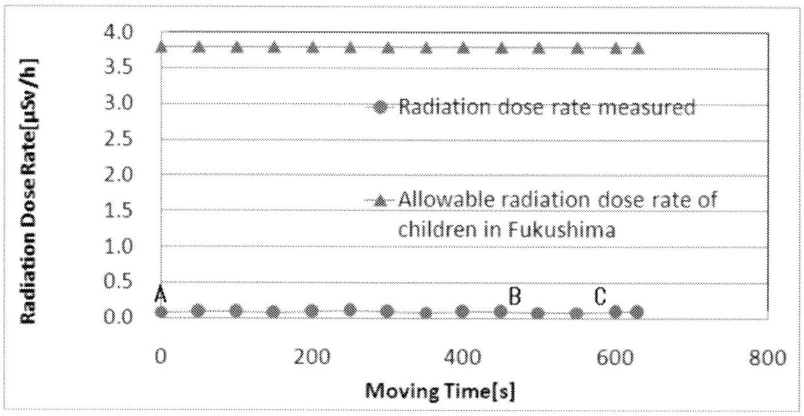

Fig.8 The distribution of radiation rate dose measured in the campus

4. Conclusion

(1) It was confirmed that the crawler type traveling vehicle can be operated smoothly in the school grounds of brick and asphalt. (Traveling speed: 17 [m / min])

(2) It was confirmed that the location information captured by GPS is visible on the Google map, and that the incorporation of video information is also possible to play.

(3)Fig.7 shows the results of measuring the radiation dose rate in the school grounds. The points A, B and C in Fig.8 correspond to the symbols in Fig.7. A radiation dose rate of 0.09[μSv / h] was obtained. The value is less than the 1/40 ([3.8μSv / h]) allowable radiation dose rate for children in Fukushima Prefecture.

(4)As a further work, modifying to program traveling, the measurement of the distribution of the radiation dose rate in a school of Fukushima Prefecture, and class delivery on radiation measurement will be carried out.

References

[1] Ohmichi T 1992 Operation Robot under Radiation Environment *Proc. Japan Welding Society*, **50**, p.19-25

[2] Oka K 2003 Robot for the Nuclear Energy Disaster: Development of a Radiation-Proof Robot <Special Issue> Disaster Response Robotics *Journal of the Japan Society of Mechanical Engineers* **106**, p.765-767.

[3] D'Annucci F and Lecour E 2010 Inspection of the Steam Generator Divider Plate. ICONE18-29457 *Proceedings of the 18th International Conference on Nuclear EngineeringICONE18, May 17-21, Xi'an, China*

[4] [http://www.webcitation.org/5xPANQOLV "iRobot Delivers 3,000th PackBot"]. Archived from the original on March 23, 2011. http://www.webcitation.org/5xPANQOLV.

[5] [http://ja.wikipedia.org/wiki/%E3%83%91%E3%83%83%E3%82%AF%E3%83%9C%E3%83%83%E3%83%88]

[6] http://eetimes.jp/ee/articles/1109/15/news035.html

[7] Morisato K and Kobayashi T 2011 *Proc, 41th the JSME students conference, Okayama, Feb.*

[8]Kobayashi T, Nakahara M, Morisato K, Takashina T and Kanematsu H 2012 Remote Sensing of Radiation Dose Rate by Customizing an Autonomous Robot *Journal of Physics: Conference Series* **352** 012033

Journal of Physics: Conference Series 433 (2013) 012031 doi:10.1088/1742-6596/433/1/012031

Biofilm Formation Derived from Ambient Air and the Characteristics of Apparatus

H. Kanematsu,[1] **H. Kougo, D. Kuroda, H. Itho, Y.Ogino, Y. Yamamoto**
Suzuka National College of Technology
Shiroko-cho, Suzuka, Mie 510-0294, Japan

E-mail: kanemats@mse.suzuka-ct.ac.jp

Abstract. Biofilm is a kind of thin film on solidified matters, being derived from bacteria. Generally, planktonic bacteria float in aqueous environments, soil or air, most of which can be regarded as oligotrophic environments. Since they have to survive by instinct, they seek for nutrients that would exist on materials surfaces as organic matters. Therefore, bacteria attach materials surfaces reversibly. The attachment and detachment repeat for a while and finally, they attach on them irreversibly and the number of bacteria on them increases. At a threshold number, bacteria produce polymeric matters at the same time by quorum sensing mechanism and the biofilm produces on material surfaces. The biofilm produced in that way generally contains water (more than 80%), EPS (Exopolymeric Substance) and bacteria themselves. And they might bring about many industrial problems, fouling, corrosion etc. Therefore, it is very important for us to control and prevent the biofilm formation properly. However, it is generally very hard to produce biofilm experimentally and constantly in ambient atmosphere on labo scale. The authors invented an apparatus where biofilm could form on specimen's surfaces from house germs in the ambient air. In this experiment, we investigated the basic characteristics of the apparatus, reproducibility, the change of biofilm with experimental time, the quality change of water for biofilm formation and their significance for biofilm research.

1. Introduction

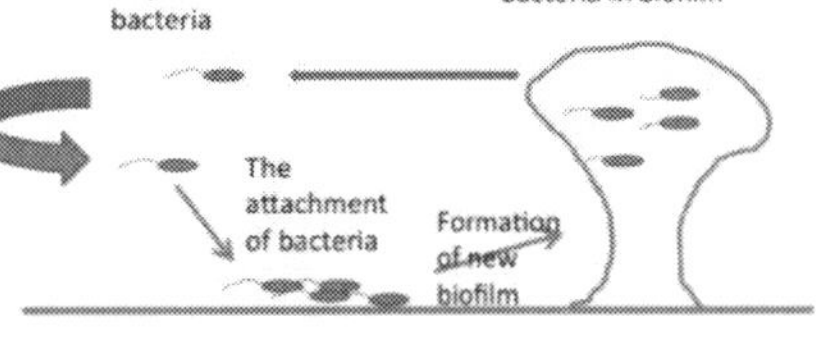

Fig. 1 Planktonic and biofilm bacteria

Biofilm[1-4] can be defined as the slimy matter formed on materials surface by bacterial activity. Strictly speaking, it is not any film, but just a thin matter on a solid-like matter. Actually, it is a mushroom-like or tower-like slimy matter composed of water, EPS (Exopolymeric substances), bacteria etc. Usually, many substances in water are incorporated into biofilm at the same time. Fig.1 shows the schematic illustration for biofilm formed on materials. It brings about many kinds of chronic diseases in our human bodies[5]. And it also leads to serious materials failures of structures from the mechanical or functional viewpoints, such as corrosion, cracks, fractures etc. of metallic or other

[1] Hideyuki Kanematsu, Suzuka National College of Technology, Shiroko-cho, Suzuka, Mie 510-0294, Japan

materials[6]. However, the biofilm has not only negative sides, but also many positive sides for our society[7]. Separation and recovery of metallic elements by biofilm reactors is one of such a promising possibility for sustainable circulating society in the near future[8]. At any rate, all of them would be related to biofilm. Therefore, it would be very important to investigate the biofilm behavior, structures etc. on laboratory scale and also to utilize it for various purposes practically. To fulfill the versatile purpose, an appropriate experimental apparatus on laboratory scale was needed.

Recently, the authors invented a certain type of biofilm reactor to investigate biofilm formation behaviour on laboratory scale from both of fundamental and practical-applicable viewpoints[9]. If it would work well for various purposes, lots of technical problems in practical industries could be solved and much more knowledge would be accumulated. However, various fundamental characteristics for the apparatus must be investigated and clarified to run it properly and also to apply the results to practical problems precisely and correctly. In this investigation, we produced and operated the biofilm formation reactor based on the cyclic water system from various viewpoints and discussed on the results.

2. Experimental

Fig.2 shows the schematic illustration for the biofilm reactor used for this investigation. It enables biofilm to be formed by planktonic germs in ambient air mainly. The apparatus is basically composed of a column, water tank, pump and pipes connecting those units. All of them constituted a cyclic water system. The column was made of transparent acryl cylinder. The clean water was accumulated at the bottom tank. Then it was pumped up through the pipe to the upper column. The water was flown into the column and out to the pipe at the other end. In the column, some specimens (sheets of austenitic stainless steels, carbon steels and glasses, 10x20mm, the thickness ranges from 0.5mm to 2mm.) were placed with the supporter. The flow direction in the column was parallel to the specimen's surface and specimens could occupy the upper and lower positions. The water flowing out of the column was introduced into the tank. However, a plate was placed just above the tank and the water was designed to run down on it before it was poured down into the tank. A fan was also set above the plate and the ambient air was designed to blow down on it from the fan. On the intermediate plate, air including planktonic germs was mixed into the water and the mixed water was circulated. The cycle was repeated during the test. The test term was 10 days at maximum.

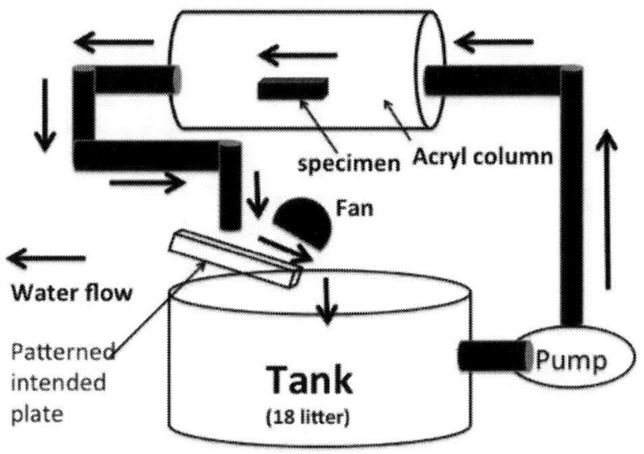

Fig.2 An apparatus for biofilm formation from germs in air.

The specimen after the exposure was taken out of the system properly and served for the following some observations and analyses. The specimen was observed by a low pressure SEM-EDX (Hitachi Miniscope, TM-1000) and also by an optical microscope (Keyence VW-9000). The former enabled us to analyse biofilms containing lots of water and organisms in biofilm. An the latter had a function to produce 3D images of samples where the observation stage moved up and down around a middle position slightly and automatically. The quality change of water in the system was checked by a multi pH meter periodically.

3. Results and Discussion

3.1 Biofilm formation and its reproducibility in the circulation system.

Firstly, we investigated if the biofilm formation could occur properly in the circulation system. Fig. 3 shows some results for the observation of an exposed specimen (carbon steel)[10]. Fig.3-1 shows the conventional optical-microscopic photo for carbon steel exposed in a week. Bubble-like matters

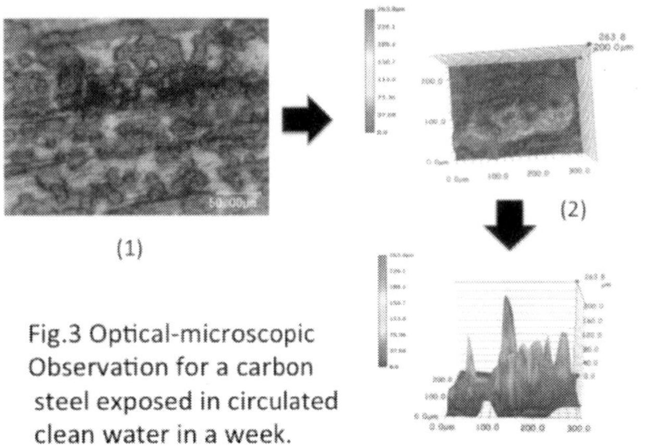
(1) (2) (3)

Fig.3 Optical-microscopic Observation for a carbon steel exposed in circulated clean water in a week.

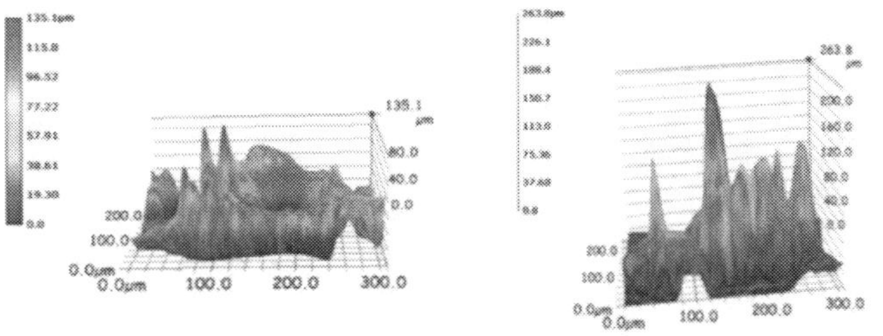
(1) In static water for one three days. (corrosion > biofilm) (2) In circulate water system for one week.

Fig.4 The carbon steels specimens in clean water for one week, being observed by the optical-microscopic 3D analysis.

were observed on the surface. By using the function for 3D analyses of the microscope, the color was assigned to the image according to the height of surface. The result was shown in Fig.3-2. The red color was given to the higher part, while blue to the lower one. The observation angle was changed virtually on the software and the result was shown in Fig.3-3. The final one shows the typical tower-like shape of biofilm.

Conventionally in our laboratory, the biofilm has been observed only by a low-pressure scanning electron microscope. Even though it could confirm the biofilm clearly, the difference between biofilm and corrosion was sometimes hard to be fixed. On the other hand, a 3D analysis by the optical microscope could clarify the difference to some extent, shown in Fig.4. Fig.4-(1), the corrosion occurred more remarkably, being confirmed by the observation of naked eyes. The typical red rust was scattered on the surface in the latter observation. In such a case, concave parts were observed more, even though many tower-like biofilms were observed. On the other hand, tower-like biofilms were much more remarkable in the case when the corrosion was controlled well (Fig.4-(2)). In this case, the concave parts were relatively few. Therefore, it suggests that the ratio and extent of tower-like films to the concavo parts could fix which would be dominant, corrosion or biofilm.

Fig.5 shows the specimen alignment in the column to investigate the effect of flow on biofilm formation. In our previous paper, we confirmed that the biofilm was hard to form without the flow parallel to the specimen surface and shear force as a result. In this investigation, we surveyed how the biofilm formation differed depending on the place from the viewpoint of flow strength. The part (a) in the figure 5 has stronger flow than at the part (b). Fig.6 corresponded to the results that the surfaces were observed by the 3D analysis of optical microscope. Fig.6-(a) corresponds to the result at the part (a) in Fig.5, while Fig.6-(b) to Fig.5-(b). When the specimen was placed at part (a), the biofilm was more remarkable, being compared with that at part (B). On the other hand, the extent of convex-concavo pattern at part (B) was much more remarkable, suggesting that the corrosion occurred dominantly.

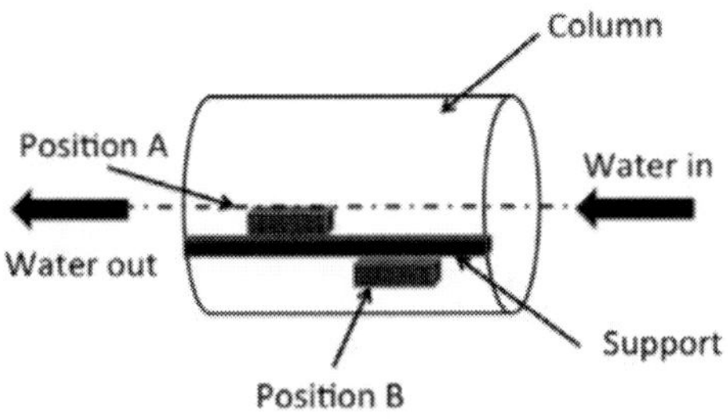

Fig.5 Specimen alignment

Fig. 7 shows the result to investigate the

Journal of Physics: Conference Series **433** (2013) 012031 doi:10.1088/1742-6596/433/1/012031

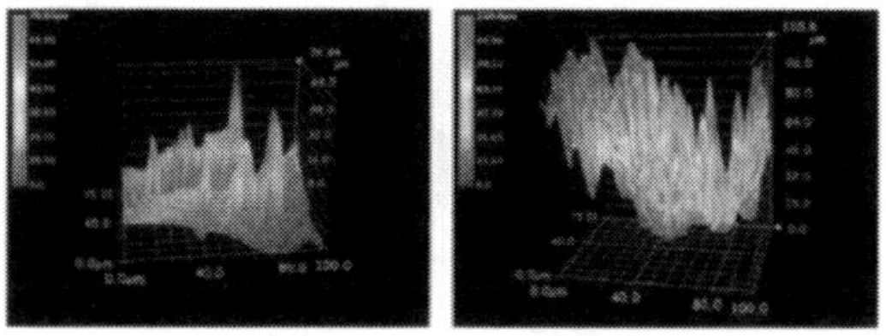

(a) Position A (b) Position B

Fig.6 Optical microscopic 3D analyses for Austenitic Stainless Steel after 7 days exposure.

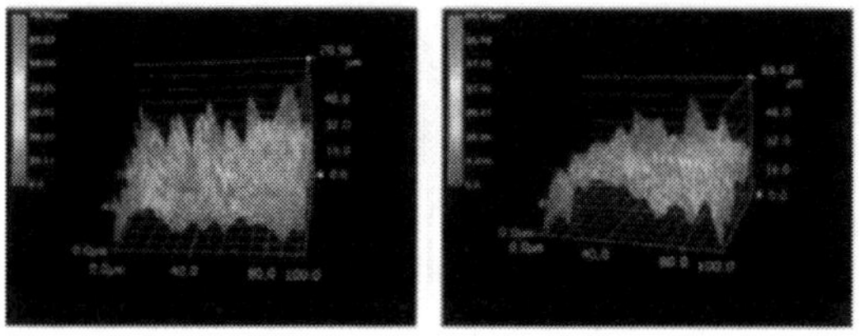

(1) Apparatus 1 (2) Apparatus 2

Fig.7 Biofilms formed on austenitic stainless steels exposed in 6 days in water circulation system.

reproducibility, when the different apparatuses having the same systems and structures were used to produce biofilms. Being compared each other, both specimens showed a similar extent of biofilm formation. As shown in Fig.1 schematically, the attachment of bacteria to material surfaces, the original cause of biofilm formation, would be the interfacial phenomenon with a certain probability between materials and organisms. Therefore, the biofilm shape and pattern might change for each opportunity. However, Fig.7 shows almost the similar amounts and thickness of biofilms were produced every time, when the experimental condition was the same.

3.2 Change of water quality and element condensation in biofilm

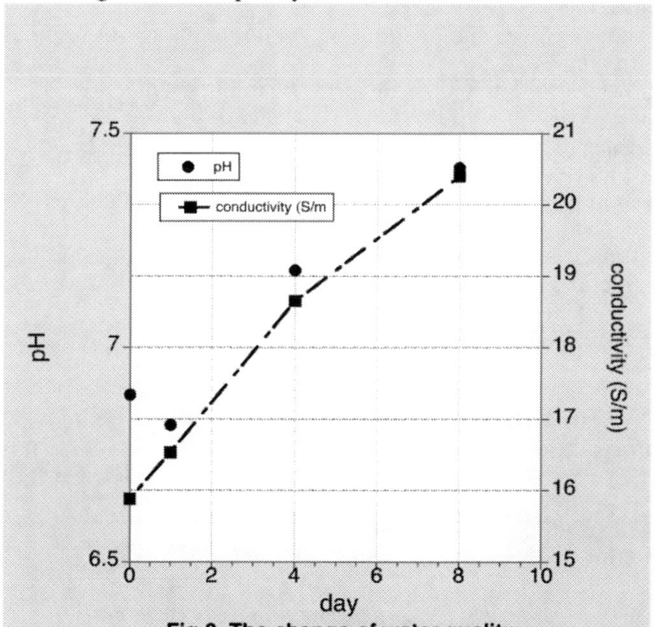

Fig.8 The change of water quality.

Fig.8 shows the change of pH and electric conductivity during the exposure test. As shown in Fig.8, the pH increased with time gradually, while the temperature was kept almost constant. The conductivity also increased with time, even though the plots varied widely to some extent. The increase of conductivity and pH suggest that various complex ions were produced with some metallic ion.

Fig.9 shows the results for biofilm formation on a glass specimen. The substrate glass was a commercial fluorine doped tin oxide. In the SEM photos for the specimen's surface, the dark part corresponds to the biofilm area. When the electron beam was irradiated in the region, various metallic elements were identified. Regarding indium, the surface concentration was kept constant and low. For the doped fluorine (that was not detected by the EDX analysis.) covered the under layer including indium etc. almost completely. Therefore, it suggests that the other elements reflected the concentration in biofilm on the FTO specimen. Silicon and calcium increased with the biofilm formation to form silicate, calcium carbonate, their complex ions, compounds etc. The condensation of the two elements in biofilms has been pointed out so far[11]. On the other hand, some metallic ions, iron and zinc were also concentrated in biofilm with time. At this point, it is not clear why these metallic elements were condensed in biofilm. We presume that those metallic ions would react with calcium compound or silicate to form complicated complex compounds. However, they might react with EPS (Exopolymeric substances), one of the main components of biofilm to form bioorganic compounds. The problem would be an important mission to be carried out from the viewpoint of biofilm utilization.

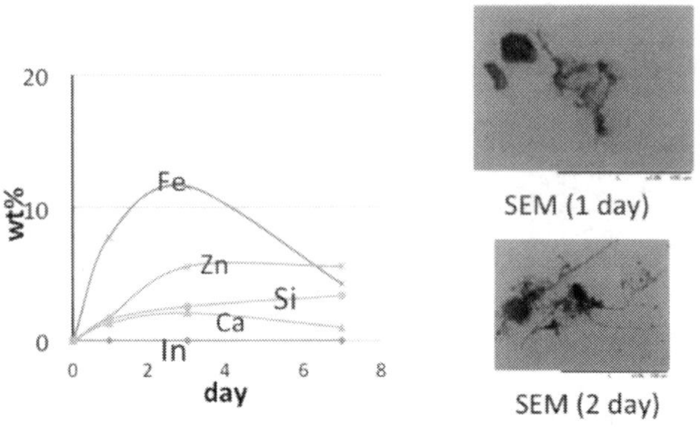

Fig.9 Concentration change of various elements
In biofilm formed on FTO glass.

4. Conclusions

A unique reactor based on the circulation water system was made and the characteristics were investigated as a preliminary research to investigate biofilm formation and to utilize it for practical application. The following results were obtained.
#1: Biofilm was produced on the specimen surface exposed to the clean water in the system.
#2: Typical tower-like biofilms were observed by the optical-microscopic 3D analyses.
#3:
#4: The biofilm was reproduced under the same experimental condition, even though the different apparatus with the same structure and system was used.
#5: The biofilm formation needed a certain flow strength and shear forth. Without them, corrosion dominated the biofilm on steel specimens.
#6: The pH and conductivity of circulated water increased with the exposure time. It might be attributed to the complex compounds.
#7: Silicon and calcium were concentrated in biofilm also for this investigation, just like the previous studies showed.
#8: Iron and zinc were concentrated in biofilms, even though the reason is not clear at this point. It should be solved in the near future for the practical application of biofilm.

REFERENCES

1. H.M. Lappin-Scott and J. William Costerton, *Microbial biofilms*1995, Cambridge, United Kingdom: Cambridge University Press.
2. Zbigniew Lewandowski and Haluk Beyenal, *Fundamentals of Biofilm Research*2007: CRC Press. 452.
3. J. William Costerton, *The Biofilm Primer*2007: Springer. 199.
4. Hans-Curt Flemming, Sriyutha.P Murthy, R. Venkatesan and K.E. Cooksey, *Marine and Industrial Biofouling*2009: Springer. 333.
5. Hajime Ikigai and Hideyuki Kanematsu, *Mechanism for antibacterial effects by metallic elements and their inhibition of bacterial infections*. Journal of Japanese Society for Biomaterials, 2011. **29**(4): p. 232-239.
6. Hideyuki Kanematsu, Hajime Ikegai and Daisuke Kuroda, *Biofilm and Metallic Materials*. Rust Prevention and Control, Japan, 2011. **55**(10): p. 369-377.
7. Hideyuki Kanematsu, Hajime Ikigai and Daisuke Kuroda, *Microbiofouling on Metallic Surfaces and Various Engineering Problems as a Result*. Journal of High Temperature Society of Japan, 2011. **37**(1): p. 17-24.
8. Hideyuki Kanematsu, Youko Miura, Hideo Itoh, Hajime Ikigai and Miho Tanaka, *Condenstation Behavior of Silicon into Biofilm Formed by Germns in Ambient Atmosphere in a Laboratory and Its Observation*. CAMP-ISIJ, 2012. **25**: p. 810-811.

9. Hideyuki Kanematsu, Daisuke Kuroda, Shun Koya and Hideo Itoh, *Development of Production Process on Labo Scale for Biofilm Formation by Immersion into Closed Circulation Water System*. Hyoumen Gijutsu (Journal of Surface Finishing Society of Japan), 2012. **63**(7): p. 459-461.
10. Hideyuki Kanematsu, Daisuke Kuroda, Hideo Itoh and Hajime Ikigai, *Biofilm Formation of a Closed Loop System And Its Visualization*. CAMP-ISIJ, 2012. **25**: p. 753-754.
11. Daisuke Kuroda, Nagisa Kamakura, Hideo Ito, Hajime Ikegai and Hideyuki Kanematsu, *Adhesion of Microorganisms on the Surfaces of Various Metallic Materials Immersed in a Cooling Water Tank of the Package Type Cooling Tower*. Tetsu-to-Hagané, 2012. **98**(4): p. 109-116.

Integrated approach of regulate global temperature rises and climate changes for sustainable planet.

N.D.G. Hettiarachchi

Natural Resources Management Center, Department of Agriculture, Sarasavi mawatha, Peradeniya, Sri Lanka.

E-mail: dhammikahetti@gmail.com

Abstract. This study emphasis the requirement of integrated approach for regulate the global temperature rises to protect the planet form climate changes and other system changes. It is clear that effect of global temperature increases, have already affected many physical and biological and human systems, This unbalances of global systems is more rapid after 1950' s increases of using environmental polluting energy sources with releasing the pollutants to global systems together with extra solar energy accumulation. How ever as a solution it is require to reduce pollutants and balance the energy level to optimum within the caring capacity. To protect the planet as a sustainable system with controlling temperature rises and other unbalances it is important to keep and protect the interrelationship of each and every system with protecting mechanisms. Therefore it is require integrated approach to establish protective mechanisms for each and every sectors of badly effecting with considering regional, national. international and global level.

1. Introduction

Global warming increases atmospheric, ground and sea temperatures and alters precipitation quantity and patterns, resulting in the increase of global average sea level, risks of coastal erosion and an increase of the severity of weather-related natural disasters. Changing water level, temperatures and flow will in turn affect food supply, health, industries, and transport and ecosystem integrity. Climate change will lead to significant economic and social impacts with some regions and sectors likely to bear greater adverse affects. Certain sections of society (the elderly, disabled, low-income households) are also expected to suffer more. Addressing climate change require two types of responses. Firstly, and importantly, we must reduce our greenhouse gas emissions (GHG) taking mitigation actions and secondly we must take adaptation strategies to deal with the unavoidable impacts. The EU's recently agreed climate change legislation puts in place the concrete measures to reduce emissions to 20% below till 1990 levels by 2020. However, even if the world succeeds in limiting and then reducing GHG emissions, our planet will take time to recover from the greenhouse gases already in the atmosphere. Thus we will be faced with the impact of climate change for at least the next 50 years. We need therefore to take measures to adaptation. [1]

Physical and biological systems on all continents and in most oceans are already being affected by recent climate changes, particularly regional temperature increases. Climatic effects on human systems, although more difficult to determine due to adaptation and non-climatic drivers, are emerging. Global-scale assessment of observed changes show that it is likely that anthropogenic warming over the last three decades has had a visible influence on many physical and biological systems [2]

There is exciting evidence, as presented in the IPCC reports , that the climate change will seriously affect the sustainable development challenges that we face not only in environmental issues but also in other areas. The climate change is not going to be country-selective but the fact remains that the some countries may suffer the most. Climate change will undermine the ability development goals and achieve sustainable development with all these negative impacts of climate change. The direct effects of the climate change will be through changes in temperature, rainfall, length of growing season and timing of extreme and critical threshold events relative to crop production. In the dry land of the tropics and subtropics, where crops are near their maximum temperature tolerance level, the yield will decline. In the regions where there is possibility of decreased rainfall, agriculture productivity will be seriously reduced..[3]

2.Review of Literature

Global temperature rising means the increase the average global temperature. It include air temperature, ocean temperature , earth surface temperature rising. Inter governmental panel of the climate change (IPCC) has identified it is as a result of the increasing green house gas concentration mainly because of the human activities like deforestation, fossil fuel burning, industries etc. It is more prominent since the middle of the 20th century. The IPCC further showed that it was contributed volcanic activities and solar energy as natural contributed factors. It is indicated that global surface temperature may rise 1.1 to 6.4 °C during 21st century. Most of the study identified global warming may continue beyond 2100 even if emission stop because of the large heat capacity of the ocean and long life time of carbon dioxide in the atmosphere. [4]

Increasing global temperature is effecting to melt the glaciers and sea level rise and effecting to change the pattern and intensify of rainfall and happen to extreme weather events. The most common indication of global warming is the increasing trend in global average temperature near the earth surface. This temperature rose by 0.74°C ±0.18°C over the period 1906-2005. The rate of warming over the last 50 years of that period was almost double that for the period as a whole (0.13°C ±0.03°C)per decade, versus 0.07°C ± 0.02°C per decade. Temperatures in the lower troposphere have increased between 0.12 and 0.22 °C (0.22 and 0.4 °F) per decade since 1979, according to satellite temperature measurements. Temperature changes vary over the globe. Since 1979, land temperatures have increased about twice as fast as ocean temperatures (0.25 °C per decade against 0.13 °C per decade). Ocean temperatures increase more slowly than land temperatures because of the larger effective heat capacity of the oceans and because the ocean loses more heat by evaporation. The Northern Hemisphere warms faster than the Southern Hemisphere because it has more land and because it has extensive areas of seasonal snow and sea-ice cover subject to the ice-albedo feedback. Although more greenhouse gases are emitted in the Northern than Southern Hemisphere this does not contribute to the difference in warming because the major greenhouse gases continue to long enough to mix between hemispheres. [5]

The greenhouse effect is the process by which absorption and emission of infrared radiation by gases in the atmosphere. The question is instead how the strength of the greenhouse effect changes when human activity increases the concentrations of greenhouse gases in the atmosphere. The major greenhouse gases are water vapor (not including clouds), which causes about 36–70 percent of the greenhouse effect; carbon dioxide (CO_2), which causes 9–26 percent; methane (CH_4), which causes 4–9 percent and ozone (O_3), which causes 3–7 percent. Human activity since the Industrial Revolution has increased the amount of greenhouse gases in the atmosphere, leading to increased irradiative forcing from CO_2, methane, troposphere ozone, CFCs and nitrous oxide. The concentrations of CO_2 and methane have increased by 36% and 148% respectively since the mid-1700s. These levels are considerably higher than at any time during the last 650,000 years, the period for which reliable data has been extracted from ice cores. Fossil fuel burning has produced about three-quarters of the increase in CO_2 from human activity over the past 20 years. Most of the rest is due to land-use changes, particularly deforestation.CO_2 concentrations are continuing to rise due to burning of fossil fuels and land-use change. The future rate of rise will depend on uncertain economic, sociological, technological, and natural factors. Accordingly, the IPCC Special Report on Emissions Scenarios gives a wide range of future CO_2 scenarios, ranging from 541 to 970 ppm by the year 2100.fossil fuel reserves are sufficient to reach these levels and continue emissions past 2100 if coal, tar sands or methane are extensively exploited.[4,14]

In addition to the raised temperature, effect to rise seawater level, which will have critical implications for countries whose lands are at elevations similar to sea level. For example, nearly 1,800 km of land on the north coast of Egypt and the Nile delta will be under seawater if there is a rise of 50 cm in the sea level, affecting 3.8 million people. The change in precipitation is already occurring. Wet areas are likely to become more wetter and dry areas will become more drier, increasing water scarcity in these dry areas..[3]

Climate change is going to have a drastic impact on dry land ecosystems and its almost 2.5 billion inhabitants. All the climate models used by the IPCC suggest that the dry areas will become dryer and more water stressed. The dry land areas (40% of world land surface)are home to over 2 billion people, accounting for 35% of the world's population. Some 55% of dry land inhabitants live in rural areas. More than 90% of dry land inhabitants are in the developing world and 70% in rural areas. Approximately half of the poorest people in the world live in the dry areas [3]

The Stern Review on the Economics of Climate Change states that climate change is the greatest and widest-ranging market failure ever seen, presenting a unique challenge for economics. The Review provides prescriptions including environmental taxes to minimize the economic and social disruptions. It's main conclusion is that the benefits of strong, early action on climate change far more important than the costs of not acting. The Review points to the potential impacts of climate change on water resources, food production, health, and the environment. According to the Review, without action, the overall costs of climate change will be equivalent to losing at least 5% of global gross domestic product (GDP) each year, now and forever. Including a wider range of risks and impacts could increase this to

20% of GDP or more. The Review proposes that one percent of global GDP per annum is required to be invested in order to avoid the worst effects of climate change. [6]

3.Review of Regulatory Mechanisms

Climate Change Mitigation

Mitigation include action taken to reducing the climate change. It need to reduce the radiative forces that effected on global warming with using regulatory mechanisms. It include reduce the concentration of green house gases that effected the energy level increases and improve the sinks that can absorb and capture the adequate energy to protect the global balance.[7] That need to reduce fossil fuel consumption as a energy source for industrial activities and electricity generation and need to move towards environmentally friendly energy sources such as wind power, solar energy , wave energy etc. Forest and marine plants can be used as good carbon sinks for reducing atmospheric carbon dioxide level. Further It is good path to improve emission friendly good and services market with low carbon technologies.[8]

Eventually mitigation appear effective only for preventing further warming, not at reversing existing warming. The Stern Review identifies several ways of mitigating climate change. These include reducing demand for emission-intensive goods and services, increasing use and development of low-carbon technologies, and reducing fossil fuel emissions.[6]

Stabilizing the green house gas concentration.

It is important to stabilizing the green house gas concentration of the atmosphere to optimum level that eco system can function well, agricultural production and climate change can adopt. The United Nations Framework Conservation on Climate Change (UNFCC) has emphasized this requirement. The most prominent green house gas (GHG) emitted by human activities is carbon dioxide (CO_2.). Stabilizing emissions of CO_2 at current levels would not lead to a stabilization in the atmospheric concentration of CO_2. In fact, stabilizing emissions at current levels would result in the atmospheric concentration of CO_2 continuing to rise over the 21st century and beyond .The reason for this is that human activities are adding CO_2 to the atmosphere faster than natural processes can remove it. Stabilizing the atmospheric concentration of the other greenhouse gases humans emit also depends on how fast their emissions are added to the atmosphere, and how fast the GHGs are removed. [9,10,11]

Reforestation and protecting existing forests

Trees capture and store atmospheric carbon. Stern reports indicate that deforestation has resulted to increasing atmospheric carbon dioxide level remarkably. Stern reviews 20% of total green house gas emission were from deforestation in 2007. Reforestation will help to reduce the atmospheric carbon level. It is showed that natural forest can capture sixty percent more carbon than forest plantation. Hence protection of existing forest is more important as a regulatory mechanism for control the global warming. [6,12]

Carbon capture and storage

Carbon capture and storage (CCS) is a system to mitigate climate change by capturing carbon dioxide (CO_2) from point sources such as power plants and subsequently storing it away safely instead of releasing it into the atmosphere. The Intergovernmental Panel on Climate Change says CCS could contribute between 10% and 55% of the cumulative worldwide carbon-mitigation effort over the next 90 years. The CCS is the most important single new technology for CO_2 savings in power generation and industry. Though it requires up to 40% more energy to run a CCS coal power plant than a regular coal plant, CCS could potentially capture about 90% of all the carbon emitted by the plant. [13]

Energy conservation technologies

Reducing energy use is seen as a key solution to the problem of reducing greenhouse gas emissions. Need to apply energy conservation technologies in sectors such as transport, housing and urban planning, building construction are important. [14,15]

Eliminating methane releasing

Methane is a significantly more powerful greenhouse gas than carbon dioxide. Methane can release to atmosphere form coal mines, oil wells, waste treatment plants, paddy fields etc, However, reducing the amount of waste methane produced in the first place and has an even greater beneficial impact.

Geological storage

This procedure important to capture the CO_2 without releasing to the atmosphere. Capture and geological storage of CO_2 provide a way to avoid emitting CO_2 into the atmosphere, by capturing CO_2 from the sources and transporting it by pipeline and injecting it into suitable deep rock formations. [16]

Adoption

Adoption include develop the capacity, strength and establishing mechanisms to face the global warming and climate change effects on global physical and biological systems. It is clear that need of adoption because of mitigation mechanisms need proper time to reduce the global temperature level and green house gases to tolerable optimum level to physical and biological systems.

Geo engineering

Geo engineering include large scale intervention for control of global warming, It is proposed solar radiation management techniques such as space sun shades, spaced based solar power generation techniques by using orbiting satellites to reduce the entering solar energy to the atmosphere. Hydrological geo engineering techniques such as diverting rivers to keep warm water away form the sea ice and control the melting the ice bergs adjusting thermohaline circulation. [15,16]

Flexible mechanisms

This include emission trading system, joint implementation of projects, clean development mechanism etc.

4.Conclusion and Recommendation

Global warming has been increased to global threatening level in past 5 decades because of industrial revolution with high rate of fossil fuel burning. At present it is clearly indicate the bad consequences of the past action. But however it cannot go back and have sudden changes in to the global systems. It require to steps to take precautions to regulate the further increasing of global warming as well as reduce the excess levels of green house gases in atmosphere . As a regulatory mechanisms we can use mitigation, geo engineering and flexible techniques appropriately. Presently it is clear global warming and energy level of the atmosphere is increasing. Energy balance of the sub systems are changing. Therefore we need to apply regulatory technologies appropriately balance the energy levels of the atmosphere, hydrosphere , biosphere and lithosphere. Effects of climate changes are harsh than we think and effects are more rapid. A requirement of action is clear as local, regional, national and international level. But bad effects will not recover soon. According to our remedial action it will take time to come back to optimum level. Hence adoption techniques are needed to tolerate and develop the strength to face the bad effects. It is now needed to integrated approach to including all the mechanisms of mitigation, geo engineering and flexible mechanisms with adoption strategies to protect the sustainability of the planet. Further it is our responsibility to regulate protecting mechanisms appropriate level.

References

[1] Adapting to climate change: Towards a European framework for action Commission of the European communities Brussels 1.4.2009
[2] Rosenzweig Cetal Assessment of observed changes and responses in natural and managed systems Pp-80-117
[3] BeltagyA.,Madkour M., Impact of climate change on arid land agriculture El-Beltagy and Madkour Agriculture and Food Security 2012- 1:3
[4] IPCC Special report 2012 Managing the Risks of Extreme Events and Disasters to Advance Climate Change Adaptation Cambridge university press New York USA.
[5] http://en.wikipedia.org/wiki/Global_warming
[6] Stern N 2006 Stern Review on The Economics of Climate Change (pre-publication edition). HM Treasury
[7] Energy balance model :Application and Background 2002 The shodor education foundation Inc Global climate research explorer National science foundation
[8] James E K.2003 To Control Climate Change Alternative Energy Technologies Must Be Developed University of Illinois at Urbana-Champaign
[9] Observed climate change impacts Linking climate change and water resources: impacts and responses Section 3

[10] Using climate system models 2007 National Center for Atmospheric Research Boulder Colorado.
[11] William A P 2006 Economic verses climate change Washington DC 20036
[12] Onyekuru N A, Marchant R 2011 Nigeria's response to the impact of climate change Journal of agricultural and environmental ethics
[13] IPCC 2005 Underground geological storage Report on Carbon dioxide Capture and Storage Cambridge University Press Cambridge New York USA.
[14] www. wikipedia.org Mass balance
[15] www.unep.org climate change adaptation
[16] http://en.wikipedia.org/wiki/Climate_change_mitigation

Proposal for an inhibitor of Alzheimer's disease blocking aggregation of amyloid-β peptides: *ab initio* molecular simulations

A Okamoto[1], K Nomura[1], A Yano[1], S Higai[2], T Kondo[2], S Kamba[2] and N Kurita[1]

[1] Department of Computer Science and Engineering, Toyohashi University of Technology, Tempaku-cho, Toyohashi, Aichi, 441-8580, Japan
[2] Murata Manufacturing Co., Ltd, 1-10-1 Higashikohtari, Nagaokakyo, Kyoto, 617-8555, Japan

E-mail: kurita@cs.tut.ac.jp

Abstract. Aggregation of amyloid-β (Aβ) peptides is believed to play a key role in the mechanism of molecular pathogenesis of Alzheimer's disease (AD). To inhibit the aggregation and prevent AD, numerous compounds have been synthesized. A previous experimental study elucidated that a triazine derivative AA3E2 has anti-amyloidogenic ability, while a triazine derivative AA3D2 having a different substituent has no inhibitory effect. However, the reason for this remarkable difference in the ability cannot be explained by the chemical structures of these derivatives. In the present study, we present stable structures of the solvated complexes with Aβ and AA3E2/AA3D2 obtained by classical molecular mechanics method. The specific interactions between Aβ and AA3E2/AA3D2 in the complexes are investigated by *ab initio* fragment molecular orbital calculations. Based on the results obtained, we attempt to propose new potent inhibitors for the Aβ aggregation.

1. Introduction

Recently, dementia resulting from Alzheimer's disease (AD) is a growing problem worldwide. The cause and progression of AD is considered to be associated with senile plaques in a brain [1], which are composed of the aggregation of amyloid β (Aβ) peptides. The Aβ peptides are produced by proteolytic cleavage of the amyloid precursor protein (APP) by β- and γ-secretases [2]. Because the γ-secretase can cleave APP at its several alternative sites, the produced Aβ peptides have a variety of lengths. The most abundant Aβ peptides existing in the senile plaques are Aβ(1-40) and Aβ(1-42)[2, 3]. Each of them has 40 or 42 amino acid residues, respectively. It was found in the experiments [4-6] that Aβ(1-42) aggregates more rapidly and comprises the major component of the senile plaque in a diseased brain, in comparison with the shorter Aβ(1-40).

The aggregation of Aβ peptides induces a cascade of events leading to a death of neuronal cells in a diseased brain [7]. Because the Aβ peptides have several hydrophobic amino acid residues, the peptides form strong aggregates in water due to the hydrophobic interactions between these residues, leading to a fibril formation [8]. Therefore, it is expected that compounds having a strong binding affinity to Aβ can inhibit the Aβ aggregation and be a potent inhibitor for the amyloidogenesis and pathogenesis of AD.

Numerous compounds have been synthesized [9-12] for producing potent inhibitors for the Aβ aggregation. It was found that a triazine derivative AA3E2 has anti-amyloidogenesis ability [9, 13-15], while a triazine derivative AA3D2 having a different substituent has no inhibitory effect [9, 15]. Because AA3E2 is classified as a β-sheet breaker [16], it can disrupt the hydrophobic interactions between Aβ peptides [17-19]. As a result, AA3E2 is expected to have a remarkable effect to prevent the Aβ aggregation [15]. On the other hand, the reason why AA3D2 has no inhibitory effect is not elucidated yet. And the reason for this remarkable difference in the ability between AA3E2 and AA3D2 cannot be explained by the chemical structures (figure 1) of these derivatives.

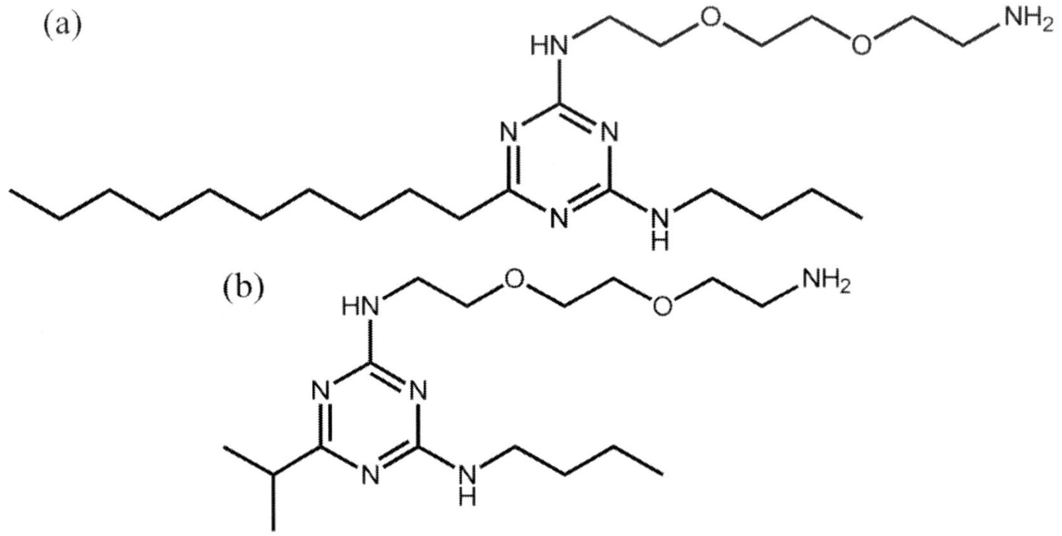

Figure 1. Chemical structures of (a) AA3E2 and (b) AA3D2.

In the present study, to elucidate the difference in specific interactions between Aβ and AA3E2/AA3D2, we obtained stable structures for the solvated complexes of Aβ with these triazine derivatives, by molecular simulations based on protein-ligand docking and classical molecular mechanics (MM) methods. For the most stable structure determined by *ab initio* fragment molecular orbital (FMO) [20-26] calculations, the specific interactions between each amino acid residue of Aβ and AA3E2/AA3D2 were investigated at an electronic level. From the results computed, we attempted to elucidate which amino acid residues of Aβ are important for the binding between Aβ and the triazine derivative, and which parts of the triazine derivative contribute to the binding. In addition, we proposed some new compounds and investigated their binding affinity to Aβ, in order to propose new potent inhibitors for the Aβ aggregation.

2. Details of molecular simulations

2.1. Optimization of solvated structures of Aβ + triazine complexes

In the present study, to elucidate how the specific interactions between Aβ and triazine are changed depending on the Aβ structure, we considered two types of Aβ structures, each of which is composed of α-helix or β-sheet conformation, and their initial three-dimensional structures were obtained from Protein Data Bank (PDB). The initial structure of the α-helix Aβ (α-Aβ) was constructed based on the NMR structure (PDB ID: 1Z0Q [27]). On the other hand, as for the β-sheet Aβ (β-Aβ) structure, there is a NMR structure for only the short Aβ(17-42) peptide (PDB ID: 2BEG [28]). We thus constructed a model structure for the part of the 1-16 residues, by using homology modeling program MODELLER 9.2 [29]. The PDB structures (PDB ID: 2BEG and 1BJB [30]) were employed as template structures, and 10 candidate structures of full-sized β-Aβ(1-42) were created. These structures were compared with the NMR structure (2BEG) to obtain the root mean square deviation (RMSD) between them. Among the 10 structures, we employed the structure having the smallest RMSD value as an initial structure for β-Aβ(1-42) monomer.

Aβ monomer has three His amino acid residues, and their protonation structures cannot be determined, because His has three types of the protonated structures depending on the pK_a value around it. We here evaluated the pK_a values of each His residue contained in Aβ by using PROPKA Web Interface 3.0 [31] and determined the His protonation based on the pK_a values. His residues with pK_a value larger than 6.0 have the Hip protonation, while those with pK_a value smaller than 6.0 have the Hid or Hie protonation. All His residues in Aβ were found to have the Hip protonation.

We added solvating water molecules with a 8 Å layer around the Aβ monomer structure and optimized the solvated structure by the classical MM and molecular dynamics (MD) simulation program AMBER9 [32], in which the AMBER99 force field [33] was used in combinations with the TIP3P water model [34]. The threshold value of the energy-gradient for the convergence in the AMBER9 optimization was set as 0.001 kcal/mol/Å. In addition, we performed a 1ns MD simulation at 300K for the solvated structure of β-Aβ monomer, to search for its stable structures widely. The snapshots obtained from the MD trajectory were classified into 10 clusters according to the RMSD value between each of the structures. We then optimized their structures by the MM method and determined the most stable structure of the solvated β-Aβ monomer by the *ab initio* FMO calculations.

As for the triazine derivatives AA3E2 and AA3D2, we optimized their structures *in vacuo* by using the *ab initio* MO calculations. The MP2/6-31G(d,p) [35] method implemented in the *ab initio* MO program package Gaussian03 (G03) [36] was used. As shown in figure 1, since the derivatives have a long polar group in the molecule, their dipole moments and charge distributions are likely to depend on the direction of the polar group. To elucidate the dependence of the dipole moment on the torsion angle of the polar group, we changed the angle and investigated dipole moment and charge distribution by using the MP2/6-31G(d,p) method in G03 to determine the most stable structure.

To obtained candidate structures for the complex of Aβ with the triazine derivative, we docked the derivative to a variety of sites around Aβ using the automated protein-ligand docking program Autodock 4.2 [37]. In the docking procedure for α-Aβ, the grid box was set as the 41.25 × 30.0 × 18.75 Å^3 centered on the gravity center of α-Aβ, and the spacing between the nearest neighboring grid points was set to 0.375 Å, which is the default value of Autodock 4.2. On the other hand, for the β-Aβ, the grid box and the space of grid points were set as the 50.0 × 27.8 × 32.6 Å^3 and 0.397 Å, respectively. The structure of Aβ was fixed, and all dihedral angles of the derivatives were freely rotated in the docking procedure to search for a variety of stable configurations for the derivative docked to Aβ. The RESP charges [38] obtained by the MP2/6-31G(d,p) calculation were assigned to each atom of the derivative. We here created 1,000 candidate structures of the complex by using the genetic algorithm of Autodock 4.2. They were classified into some clusters according to the RMSD value (10.0 Å) between each of the structures created. We created 2,000 candidate structures for the α-Aβ + AA3E2 complex, since the number of clusters for this complex obtained by AutoDock was few.

The representative structures in each cluster were fully optimized by the AMBER9-MM method. To consider the solvation effect on the complex properly, we added water molecules with a 8 Å layer around the complex and optimized the solvated structure by using AMBER9 [32], in which the AMBER99 [33] and TIP3P [34] force fields were assigned for the complex and water molecules, respectively. The threshold value of the energy-gradient for the convergence in the AMBER9 optimization was set as 0.001 kcal/mol/Å. Finally, total energies of these optimized structures were calculated accurately by the *ab initio* MP2/6-31G method in FMO [20-26], and the most stable structure of the solvated Aβ + triazine complex was determined.

2.2. *Ab initio FMO calculations for solvated Aβ + triazine complexes*

The electronic properties for the most stable structure of the solvated Aβ + triazine complex were investigated by the *ab initio* FMO calculation [20-26], to elucidate which amino acid residues in Aβ are important for the specific binding between Aβ and triazine derivative. In the FMO method, the target molecule is divided into units called "fragment", and the electronic properties of the target molecule are estimated from the electronic properties of the monomer and dimers of the fragments. The specific interactions between the fragments can be investigated from the interaction energies obtained by the FMO calculation.

In the previous docking study [15], the candidate structures for the complex of Aβ with the triazine derivative were created by using a new generation of Autodock, and a scoring function was used for determining the most preferable structure among the candidate structures. On the other hand, in the present study, we performed the *ab initio* FMO calculations for all the candidate structures to determine the most stable structure among them more accurately. In addition, from the results of the FMO calculations, it can be elucidated what amino acid residues of Aβ are important for the specific interactions between Aβ and the triazine derivative.

In the present FMO calculations, to investigate the relative stability among the solvated structures of the Aβ + triazine complexes, the number of solvating water molecules was set as 723 for all the complexes. Each amino acid residue of Aβ, triazine derivative and each water molecule were assigned as a fragment, because this fragmentation enables us to evaluate the interaction energies between the amino acid residue of Aβ, triazine derivative and solvating water molecules. The *ab initio* MP2/6-31G method was employed to investigate accurately the π-π stacking, NH-π and CH-π interactions between the amino acid residues of Aβ and the triazine derivative. We used the FMO calculation program ABINIT-MP Ver.4.3 [39]. In addition, by considering water molecules explicitly, we attempted to elucidate the influence of solvating water molecules on the specific interactions. In order to investigate the biding energy between Aβ and the derivative, the solvated Aβ + triazine structure was divided into the following four structural domains: Aβ + triazine complex containing solvating water molecules (Complex), Aβ containing solvating water molecules (Aβ + water), the triazine derivative containing solvating water molecules (Ligand + water) and solvating water molecules (Water). From total energies (T.E.) obtained by the FMO calculations, the binding energy (B.E.) between Aβ and the triazine derivative mediated by solvating water molecules was estimated as

$$\text{B.E.} = -\text{T.E. (Complex)} + \text{T.E. (A}\beta + \text{water)} + \text{T.E. (Ligand + water)} - \text{T.E. (Water)}.$$

3. Results and discussion

3.1. Optimized structures of solvated Aβ + triazine complexes

Based on the NMR structure of α-Aβ (PDB ID: 1Z0Q [27]), we constructed its solvated structure and optimized it by the AMBER9-MM method. To check the validity of the optimized structure (figure 2(a)), it was compared with the NMR structure. The RMSD between them is 1.0 Å, indicating that the optimized structure of the solvated α-Aβ is comparable to its NMR structure.

As for the β-Aβ structure, there are NMR structures for only the short Aβ(17-42) peptide (PDB ID: 2BEG [28]). We thus constructed 10 candidate structures for the full-length β-Aβ(1-42) peptide by using homology modeling MODELLER 9.2 [29], and chose a structure having the smallest RMSD value between the NMR structure [28]. Classical 1ns MD simulations were performed at 300K to obtain 10 clusters, and the representative structures of each cluster were selected to be optimized by AMBER9-MM. Their total energies evaluated by the *ab initio* FMO calculations are listed in table 1, indicating that the cluster 1 is at least 88 kcal/mol stable compared with the other structures. Therefore, we employed the cluster 1 as a solvated β-Aβ(1-42) structure, which is shown in figure 2(b). It is noted that two hydrogen bonds are formed between the side chains of Asp23 and Lys28 residues of β-Aβ(1-42). They are likely to contribute significantly to the stability of this structure.

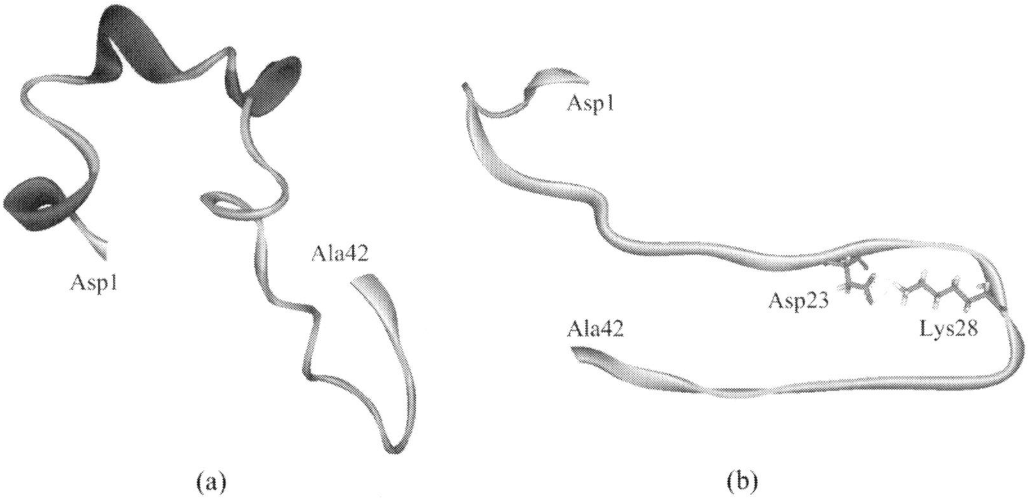

Figure 2. Structures of Aβ monomer optimized by AMBER99 in explicit solvating water molecules; (a) α-Aβ and (b) β-Aβ.

Table 1. Total energies (T.E.) (kcal/mol) for the β-Aβ(1-42) optimized in water by classical MM method based on AMBER99 and TIP3P force fields. The energies are evaluated by the *ab initio* MP2/6-31G method of the AMBINIT-MP 4.3 program [39].

Cluster	T.E.	ΔT.E.
1	-52902684.1	0.0
2	-52902085.8	598.3
3	-52902160.9	523.2
4	-52902298.5	385.5
5	-52902595.6	88.4
6	-52902144.5	539.6
7	-52902557.9	126.1
8	-52902256.5	427.5
9	-52902282.7	401.3
10	-52902365.7	318.3

Table 2. Total energies (kcal/mol) and dipole moments (Debye) for the triazine derivatives optimized *in vacuo* by the *ab initio* MP2/6-31G(d,p) method.

Structure	Total energy	Dipole moment
AA3E2 (Figure 3(a))	-865262.2	2.1
AA3E2 (Figure 3(b))	-865262.0	2.3
AA3E2 (Figure 3(c))	-865260.6	3.1
AA3D2 (Figure 3(d))	-693154.2	2.3
AA3D2 (Figure 3(e))	-693154.2	2.5
AA3D2 (Figure 3(f))	-693153.6	2.9

Structures of the triazine derivatives optimized by MP2/6-31G(d,p) are shown in figures 3(a) and 3(d). To elucidate the dependence of the dipole moment on the torsion angle of the polar group X in figure 3(a), we changed the angle as shown in figure 3(b), 3(c), 3(e) and 3(f) and optimized the structures. Table 2 lists their total energies and dipole moments, indicating that the dipole moment is significantly dependent on the angle, although the total energy does not depend so much. It is thus necessary to rotate all dihedral angles of the derivatives freely in searching for a variety of stable configurations of the derivative docked to Aβ.

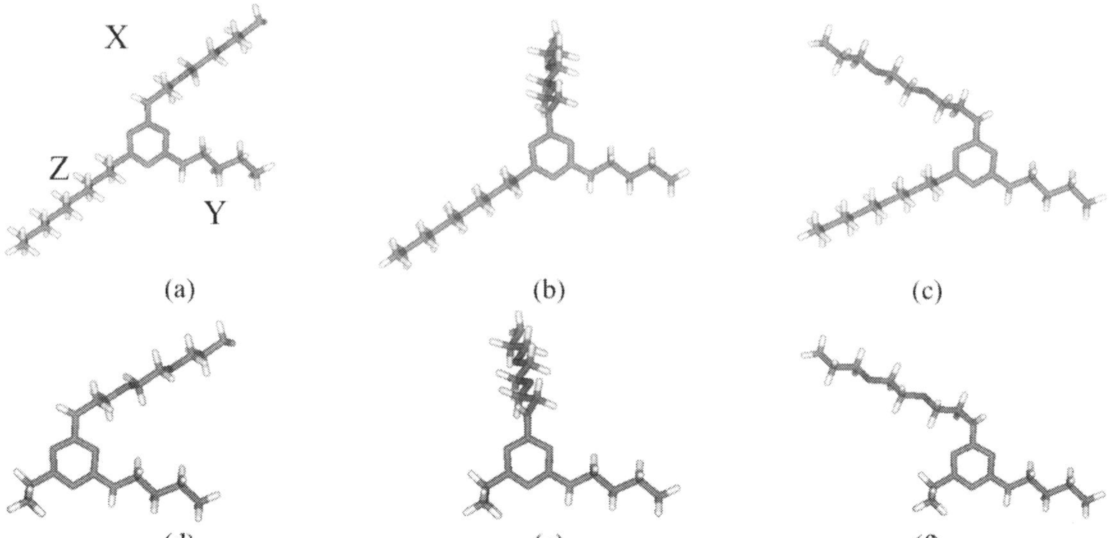

Figure 3. Structures of triazine derivatives optimized by *ab initio* MP2/6-31G(d,p) method; (a)-(c) AA3E2 and (d)-(f) AA3D2.

We first docked AA3E3 and AA3D2 triazine derivatives to α-Aβ to create the candidate structures of the complex. They were grouped into 20 (AA3E2) and 17 (AA3D2) clusters according to the RMSD between them, as listed in table 3. The representative structures of each cluster were optimized by AMBER9-MM, and their total energies were accurately evaluated by the *ab initio* FMO calculations to determine the most stable structure. Table 3(a) indicates that the cluster 1 and 5 are the most stable structures of the complex of α-Aβ with AA3E2 and AA3D2, respectively. Their structures are shown in figures 4(a) and 4(b). Both the derivatives bind to the region composed of the 19-42 residues of α-Aβ. In particular, the Phe19 and Phe20 of α-Aβ contribute to the binding of the derivatives.

In contrast, the triazine derivatives can dock to a variety of sites of β-Aβ, in the present docking process. As listed in table 3(b), there are 29 (AA3E2) and 43 (AA3D2) clusters of the candidate structures for the β-Aβ + triazine complexes. The most stable structures determined by the FMO calculations are shown in figures 4(c) and 4(d). The binding site of the triazine derivative on the β-Aβ differs significantly between AA3E2 and AA3D2. AA3E2 binds to the region near to the residues 17-21 and 35-41, while AA3D2 binds to the region of the residues 17-21 and 40-41. In the previous study [40], the hydrophobic region of the residues 16-20 (KLVFF) in Aβ were elucidated to be essential for the aggregation of Aβ. Our present molecular simulations indicate that both AA3E2 and AA3D2 can interact with these hydrophobic residues and can prevent the aggregation of Aβ peptides.

Table 3. Total energies (T.E.) (kcal/mol) for the Aβ + triazine complexes optimized in water by classical MM method based on AMBER99 and TIP3P force fields. The energies were evaluated by the *ab initio* ABINIT-MP 4.3 method.

Cluster	(a) α-Aβ + AA3E2/AA3D2 complex				(b) β-Aβ + AA3E2/AA3D2 complex			
	AA3E2		AA3D2		AA3E2		AA3D2	
	T.E.	△T.E.	T.E.	△T.E.	T.E.	△T.E.	T.E.	△T.E.
1	-45309479.3	0.0	-45137454.5	131.1	-45307862.7	208.6	-45135990.1	258.1
2	-45309252.7	226.6	-45137512.7	72.9	-45307726.4	344.9	-45135494.7	753.4
3	-45309201.9	277.4	-45137527.2	58.4	-45307654.0	417.3	-45135854.0	394.1
4	-45309272.7	206.6	-45137499.3	86.3	-45307893.2	178.1	-45135775.0	473.1
5	-45309331.5	147.9	-45137585.6	0.0	-45307476.5	594.8	-45135903.0	345.2
6	-45309309.2	170.1	-45137452.2	133.4	-45307557.9	513.4	-45135803.8	444.3
7	-45309397.7	81.6	-45137418.6	167.1	-45307433.6	637.7	-45135614.2	634.0
8	-45309340.9	138.4	-45137380.9	204.7	-45307976.3	95.0	-45135658.0	590.2
9	-45309278.4	200.9	-45137393.9	191.7	-45307553.2	518.1	-45135600.4	647.7
10	-45309204.1	275.2	-45137460.9	124.7	-45307570.6	500.7	-45136129.7	118.5
11	-45309333.4	145.9	-45137491.3	94.3	-45307653.4	417.9	-45135833.2	414.9
12	-45309343.9	135.4	-45137522.1	63.5	-45307396.7	674.6	-45136198.5	49.7
13	-45309270.6	208.8	-45137547.8	37.8	-45307861.4	209.9	-45135564.9	683.2
14	-45309261.5	217.9	-45137517.3	68.3	-45307663.5	407.8	-45135797.1	451.0
15	-45309261.4	217.9	-45137527.4	58.2	-45307715.4	355.9	-45135769.6	478.6
16	-45309211.1	268.2	-45137513.6	72.1	-45307729.2	342.1	-45136011.6	236.5
17	-45309391.8	87.5	-45137501.0	84.6	-45307823.0	248.3	-45135689.9	558.3
18	-45309206.9	272.4	-	-	-45307489.9	581.4	-45135786.9	461.2
19	-45309339.3	140.1	-	-	-45307986.1	85.2	-45136082.3	165.9
20	-45309277.2	202.1	-	-	-45307867.0	204.3	-45136053.6	194.6
21	-	-	-	-	-45307717.0	354.3	-45135942.6	305.5
22	-	-	-	-	-45307601.2	470.1	-45135417.7	830.4
23	-	-	-	-	-45307724.8	346.5	-45135650.8	597.3
24	-	-	-	-	-45307517.7	553.6	-45135722.4	525.7
25	-	-		-	-45307888.8	182.5	-45135515.5	732.6
26	-	-	-	-	-45307753.6	317.7	-45135777.3	470.8
27	-	-	-	-	-45308071.3	0.0	-45135749.1	499.0
28	-	-	-	-	-45307279.7	791.6	-45135725.7	522.4
29	-	-	-	-	-45307426.8	644.5	-45135530.8	717.4
30	-	-	-	-	-	-	-45136248.1	0.0
31	-	-	-	-	-	-	-45136044.2	204.0
32	-	-	-	-	-	-	-45136097.1	151.0
33	-	-	-	-	-	-	-45135666.6	581.5
34	-	-	-	-	-	-	-45135683.7	564.4
35	-	-	-	-	-	-	-45135874.9	373.3
36	-	-	-	-	-	-	-45136111.8	136.3
37	-	-	-	-	-	-	-45135887.7	360.4
38	-	-	-	-	-	-	-45136161.8	86.3
39	-	-	-	-	-	-	-45135835.9	412.2
40	-	-	-	-	-	-	-45135815.9	432.2
41	-	-	-	-	-	-	-45135620.6	627.5
42	-	-	-	-	-	-	-45135440.3	807.9
43	-	-	-	-	-	-	-45135579.7	668.5

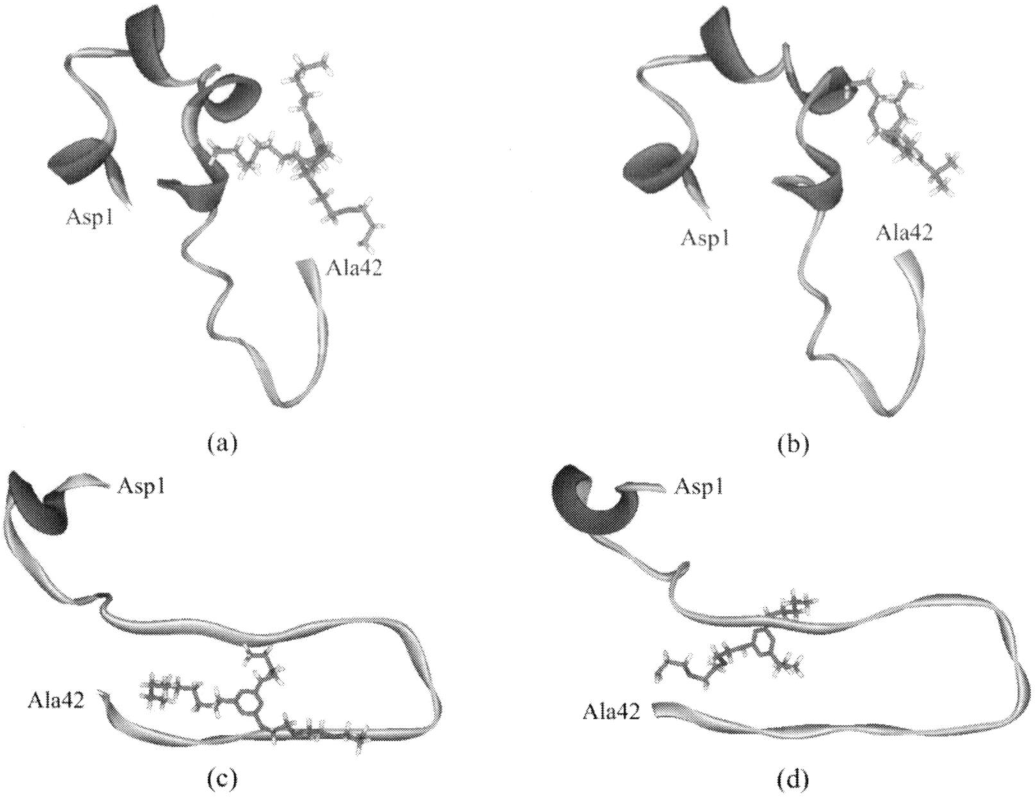

Figure 4. Most stable structures of the Aβ + triazine complexes optimized by AMBER99 in water; (a) α-Aβ + AA3E2, (b) α-Aβ + AA3D2, (c) β-Aβ + AA3E2 and (d) β-Aβ + AA3D2 complexes.

3.2. Specific interactions between Aβ and triazines in the solvated complexes

To elucidate the binding affinity between Aβ and the triazine derivatives, we evaluated the binding energy between Aβ and AA3E2/AA3D2 for the most stable structures (figure 4) determined by the FMO calculations. Table 4 lists the total energies and the binding energies estimated from these total energies. The AA3X2 derivative is our newly proposed derivative, whose structure and binding affinity to Aβ will be described in Section 3.3. Table 4 indicates that the binding energies between α-Aβ/β-Aβ and AA3E2 are larger than those for the Aβ + AA3D2 complexes, being consistent qualitatively with the experimental results [9, 15] showing the larger effect of AA3E2 than AA3D2. In addition, it was found that the binding energy between β-Aβ and the triazine derivative is larger than that for the α-Aβ + triazine complex. Therefore, β-Aβ is expected to bind the triazine derivatives more strongly than α-Aβ.

To elucidate the specific interactions between Aβ and the triazine derivatives, we furthermore investigated the interaction energies (I.E.) between each amino acid residue of Aβ and AA3E2/AA3D2 for the most stable structures of the complexes, by using the *ab initio* FMO method. As shown in figures 5(a) and 5(b), Phe20 of α-Aβ interacts most strongly with both the triazine derivatives AA3E2 and AA3D2, indicating that Phe20 is important for the binding between α-Aβ and the derivatives. Because the compounds having large binding affinity to the region of the 16-20 residues of Aβ are believed to inhibit the aggregation of Aβ peptides, AA3E2 and AA3D2 are expected to have the ability of inhibiting the aggregation of α-Aβ peptides. Therefore, our present result is consistent with the experiment results [9, 13-15] showing that AA3E2 prevents the Aβ aggregation.

Table 4. Total energies for solvated Aβ + triazine complexes (Complex), solvated Aβ (Aβ + water), solvated triazine derivative (Ligand + water) and solvating water molecules (Water), and estimated binding energies between Aβ and triazine derivative evaluated by the *ab initio* FMO calculations.

Complex	Total energy (kcal/mol)				Binding energy
	Complex	Aβ + water	Ligand + water	Water	(kcal/mol)
α-Aβ + AA3E2	-45309479.3	-44445619.8	-35407403.8	-34543575.9	31.7
α-Aβ + AA3D2	-45137585.6	-44445523.9	-35235485.6	-34543437.2	13.4
α-Aβ + AA3X2	-45431979.1	-44445425.8	-35529840.4	-34543320.2	33.1
β-Aβ + AA3E2	-45308071.3	-44444212.3	-35405061.6	-34541240.0	37.4
β-Aβ + AA3D2	-45136248.1	-44444220.9	-35233431.0	-34541438.6	34.9
β-Aβ + AA3X2	-45430838.0	-44444250.5	-35528131.1	-34541586.0	42.5

The interacting structures between the residues of α-Aβ and AA3E2 were investigated in detail to elucidate which part of AA3E2 and which residues of α-Aβ contribute to the binding between α-Aβ and AA3E2. As shown in figure 6(a), AA3E2 interacts with Phe19 and Phe20, in addition, the hydrocarbon chain of AA3E2 contacts with Val24 of α-Aβ. This contact does not occur in the Aβ + AA3D2 complex (figure 6(b)), because AA3D2 does not have the hydrocarbon chain. As indicated in figure 5(a), AA3E2 has also interacts strongly with Ala42 of α-Aβ. The terminal NH_2 group of AA3E2 forms hydrogen bonds with the oxygen atom of Ala42 backbone via a water molecule, as shown in figure 6(a). These specific hydrogen bonds contribute to the larger binding energy between α-Aβ and AA3E2 listed in table 4.

As for the β-Aβ + triazine complexes, the specific interactions between β-Aβ and the triazine are remarkably different for AA3E2 and AA3D2, as indicated in figures 5(c) and 5(d). AA3E2 interacts strongly with Val36 and Leu34, while AA3D2 interacts with Phe19, Val18, Vla40 and Ile41. All of these residues are hydrophobic. In the β-Aβ + AA3E2 complex shown in figure 6(c), AA3E2 forms hydrogen bonds with Leu34 and Val36 of β-Aβ. In the β-Aβ + AA3D2 complex, AA3D2 forms hydrogen bonds with Phe19 and Ile41 of β-Aβ, as shown in figure 6(d). The interactions between AA3D2 and Val18/Val40 are hydrophobic interactions. The β-Aβ has more extended structure in comparison with the α-Aβ as shown in figures 2(a) and 2(b). As a result, the side chains of each residue in β-Aβ are distributed more sparsely than those in α-Aβ. Accordingly, the triazine derivatives can contact more closely with β-Aβ, resulting in the larger binding energy compared with α-Aβ, as listed in table 4.

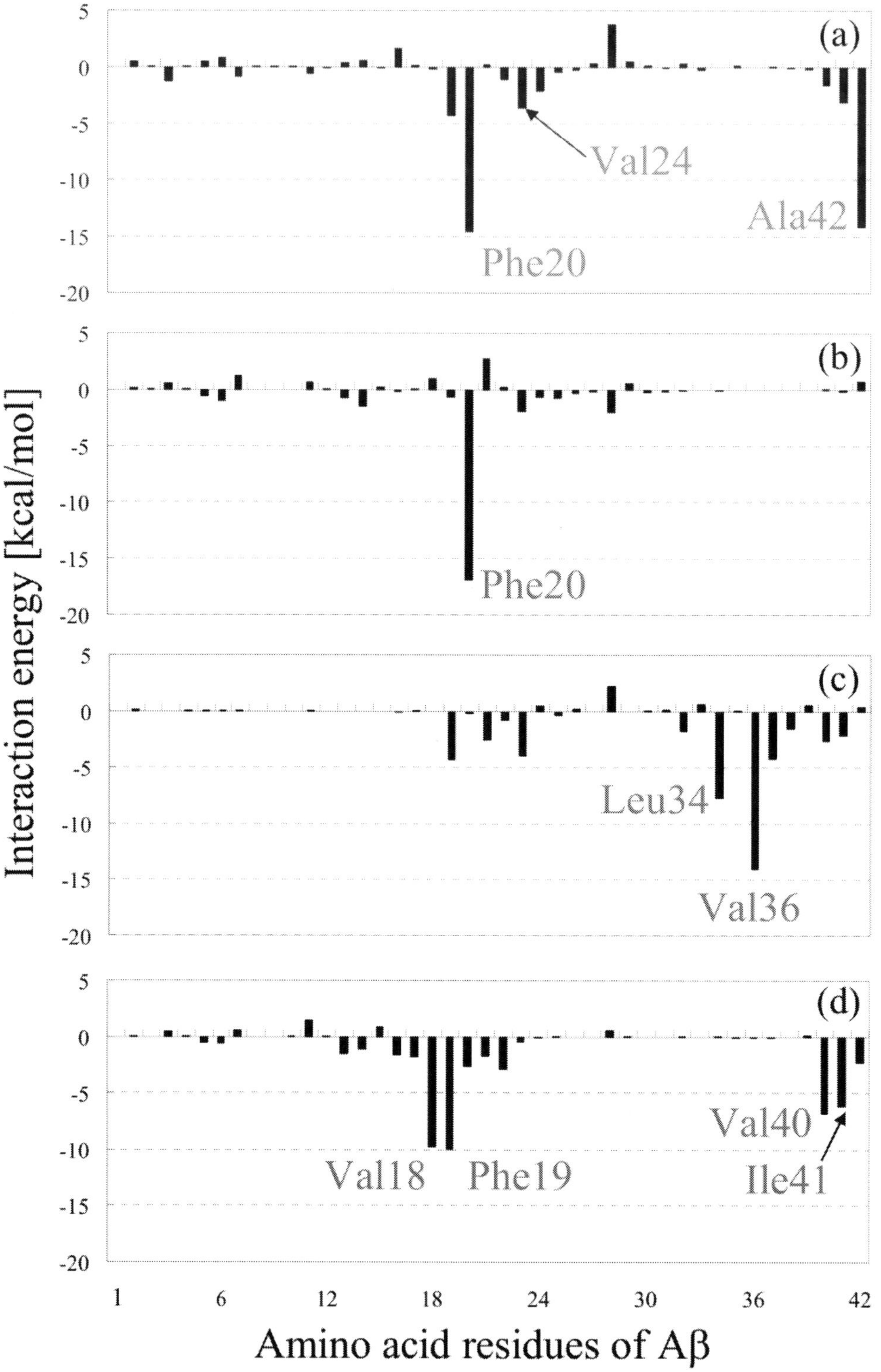

Figure 5. Interaction energies between triazine derivative and each amino acid residues of Aβ; (a) α-Aβ + AA3E2, (b) α-Aβ + AA3D2, (c) β-Aβ + AA3E2 and (d) β-Aβ + AA3D2.

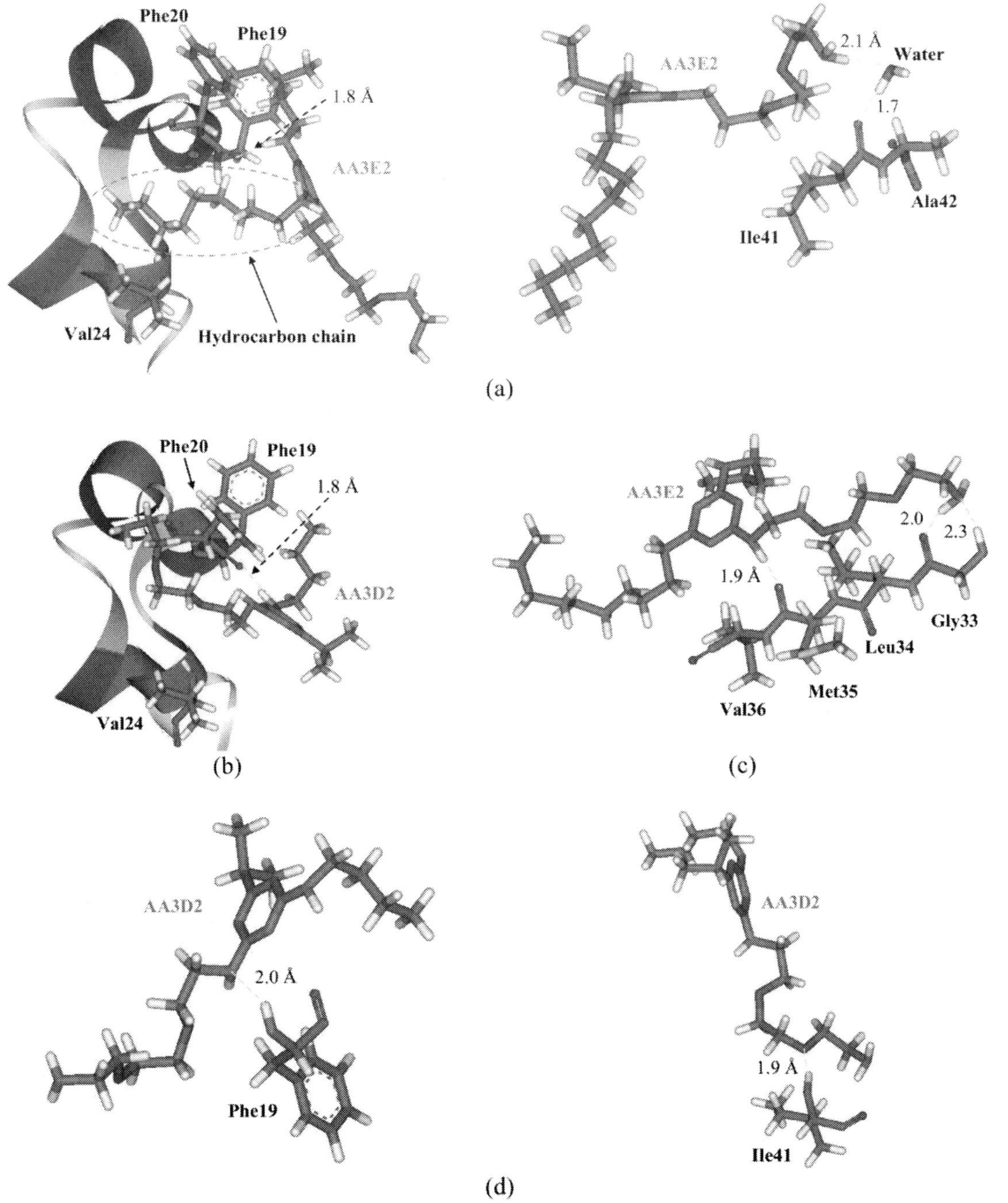

Figure 6. Interacting structures in Aβ + triazine complexes; (a) α-Aβ + AA3E2, (b) α-Aβ + AA3D2, (c) β-Aβ + AA3E2 and (d) β-Aβ + AA3D2.

3.3. Proposal for a potent inhibitor for Aβ aggregation

It is elucidated from the present simulations that AA3E2, which has a remarkable effect to prevent the Aβ aggregation [9, 13-15], interacts strongly with the hydrophobic amino acid residues of Aβ(1-42) and that the hydrocarbon chain of AA3E2 contributes to the interactions. In order to strengthen these interactions, we designed a novel potent compound AA3X2 shown in figure 7, by adding a hydrocarbon chain to AA3E2. The binding energy and the specific interactions between AA3X2 and Aβ were investigated by the *ab initio* FMO calculations.

We constructed the structures of Aβ + AA3X2 complexes based on the Aβ + AA3E2 complexes and optimized the solvated Aβ + AA3X2 complexes by using the AMBER9-MM method. The total energies and the binding energies were evaluated for the optimized structures by the *ab initio* FMO method. As listed in table 4, AA3X2 has larger binding energy than AA3E2. Therefore, AA3X2 is expected to have larger binding affinity to α-Aβ and β-Aβ and to inhibit the Aβ aggregations.

To elucidate the reason for this large binding affinity, we investigated the interaction energies and the structure around AA3X2 in the Aβ + AA3X2 complexes. As shown in figures 8(a) and 8(b), AA3X2 interacts more strongly with Phe20 of α-Aβ and Leu34 of β-Aβ, in comparison with AA3E2. Figure 9 shows the structures around the AA3X2 in the α-Aβ + AA3X2 and the β-Aβ + AA3X2 complexes. These structures indicate the hydrogen bonds between AA3X2 and Phe20 of α-Aβ (figure 9(a)) and between AA3X2 and Lue34 of β-Aβ (figure 9(b)). The comparison between the interaction energies for Aβ + AA3E2 (figures 5(a) and 5(b)) and those for Aβ + AA3X2 (figures 8(a) and 8(b)) elucidates that these interactions are larger for AA3X2 than AA3E2. Therefore, our proposed AA3X2 is expected to be a potent inhibitor for the Aβ aggregation.

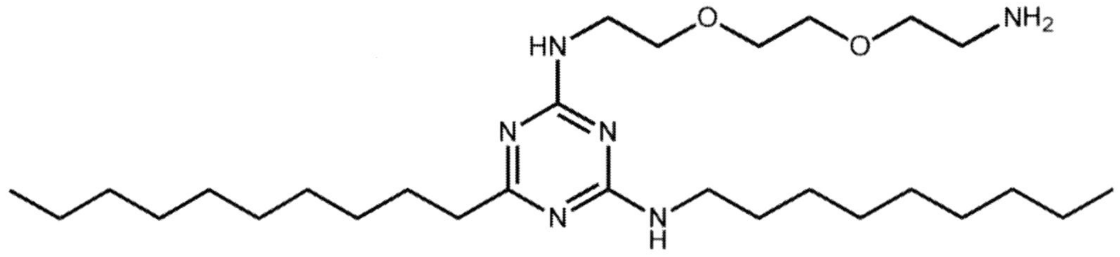

Figure 7. Chemical structure of our proposed triazine derivative AA3X2.

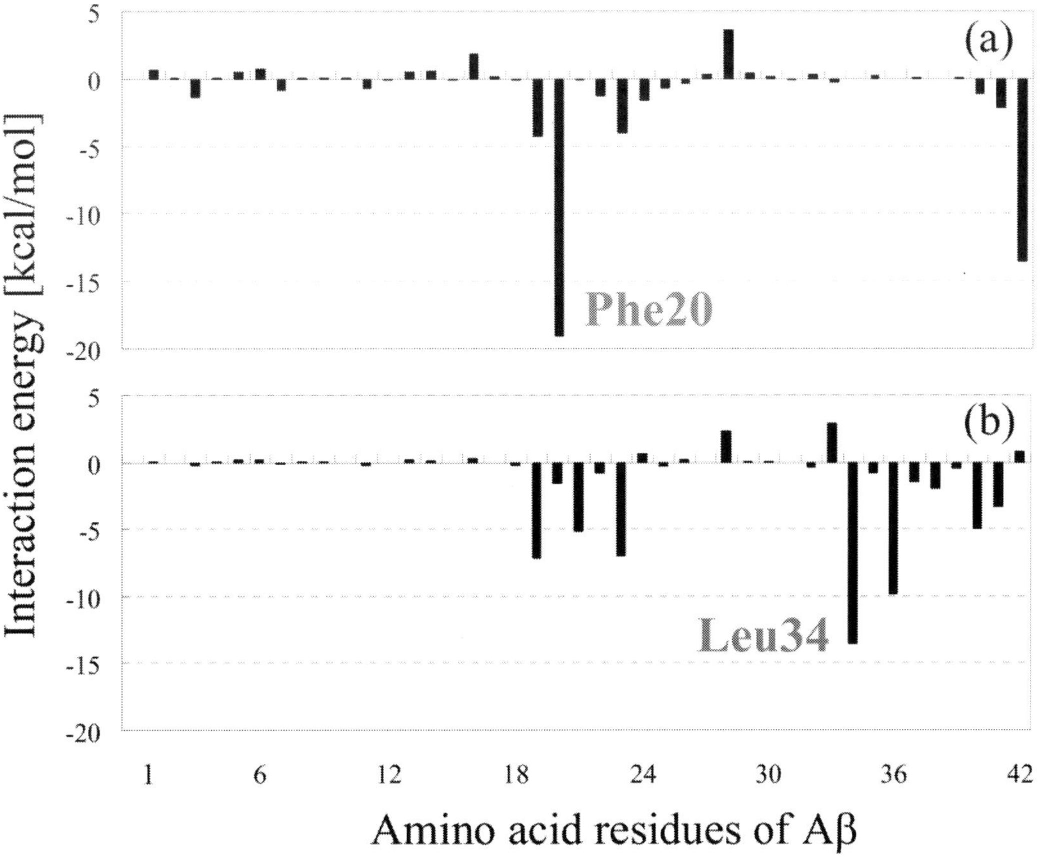

Figure 8. Interaction energies between AA3X2 and each amino acid residues of Aβ; (a) α-Aβ + AA3X2 and (b) β-Aβ + AA3X2.

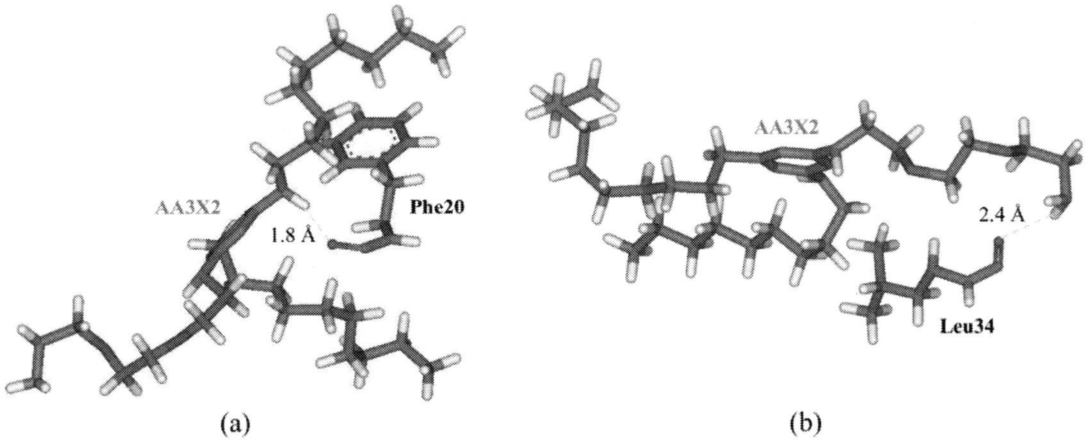

Figure 9. Interacting structures in Aβ + AA3X2; (a) α-Aβ + AA3X2 and (b) β-Aβ + AA3X2.

4. Conclusions

We here investigated the specific interactions between Aβ(1-42) and triazine derivatives for the two types of Aβ structures as well as the two kinds of triazine derivatives (AA3E2 and AA3D2) shown in figure 1, by *ab initio* molecular simulations, in which protein-ligand docking, classical molecular mechanics and *ab initio* fragment molecular orbital methods were used. The results elucidate the following points.

(1) Among the 42 amino acid residues of α-Aβ, the hydrophobic residue Phe20 is the most important for the biding between α-Aβ and AA3E2/AA3D2.
(2) Leu34 and Val36 of β-Aβ interact strongly with AA3E2.
(3) Val18 and Phe19 of β-Aβ interact strongly with AA3D2.
(4) Hydrophobic residues of Aβ contribute mainly to its binding to AA3E2/AA3D2.
(5) The hydrocarbon chain of AA3E2 enhances the interaction between AA3E2 and Val24 of α-Aβ.

Based on the above results, we furthermore proposed a novel compound AA3X2 (figure 7) and investigated its binding properties with Aβs. The binding energy evaluated indicates that AA3X2 binds more strongly to Aβ than AA3E2. Therefore, AA3X2 is expected to be a more effective inhibitor for the Aβ aggregation.

Acknowledgements
The present work was supported by JSPS Grant-in-Aid for Challenging Exploratory Reserch (No. 22650061), the grants from the Murata Science Foundation, the Iketani Science and Technology Foundation, the Tatematsu Foundation, the CASIO Science Promotion Foundation.

References

[1] Laird F M, Cai H, Savonenko A V, Farah M H, He K, Melnikova T, Wen H, Chiang H–C, Xu G, Koliatsos V E, Borchelt D R, Price D L, Lee H–K and Wong P C 2005 *J. Neurosci.* **25** 11693
[2] Selkoe D J 2001 *Physiol. Rev.* **81** 741
[3] Rak M, Bigio M R D, Mai S, Westaway D and Gough K 2007 *Biopolymers* **87** 207
[4] Gravina S A, Ho L, Eckman C B, Long K E, Otvos Jr. L, Younkin L H, Suzuki N and Younkin S G 1995 *J. Biol. Chem.* **270** 7013
[5] Roher A E, Lowenson J D, Clarke S, Woods A S, Cotter R J, Gowing E and Ball M J 1993 *Proc. Natl. Acad. Sci. U.S.A.* **90** 10836
[6] Jarrett J T, Berger E P and Lansbury Jr. P T 1993 *Biochemistry* **32** 4693
[7] Findeis M A 2007 *Pharmacol. Ther.* **116** 266
[8] Pimplikar S W 2008 *Int. J. Biochem. Cell Biol.* **41** 1261
[9] Kim W, Kim Y, Min J, Kim D J, Chang Y–T and Hecht M H 2006 *ACS Chem. Biol.* **1** 461
[10] Mason J M, Kokkoni N, Stott K and Doig A J 2003 *Curr. Opin. Struc. Biol.* **13** 526
[11] Miura Y, Yasuda K, Yamamoto K, Koike M, Nishida Y and Kobayashi K 2007 *Biomacromol.* **8** 2129
[12] Rizzo S, Rivière C, Piazzi L, Bisi A, Gobbi S, Bartolini M, Andrisano V, Morroni F, Tarozzi A, Monti J–P and Rampa A 2008 *J. Med. Chem.* **51** 2883
[13] Yazdanparast R and Shaykhalishahi H 2009 *Toxicol. In Vitro* **23** 1277
[14] Shaykhalishahi H, Yazdanparast R, Ha H–H and Chang Y–T 2009 *Eur. J. Pharmacol.* **622** 1
[15] Shaykhalishahi H, Taghizadeh M, Yazdanparast R and Chang Y–T 2010 *Chem.-Biol. Interact.* **186** 16
[16] Soto C, Kindyc M S, Baumannb M and Frangioneb B 1996 *Biochem. Biophys. Res. Commun.* **226** 672
[17] Hardy J 1997 *Trends Neurosci.* **20** 154

[18] Yazdanparast R, Esmaeili M A and Khodagholi F 2007 *Int. J. Biol. Macromol.* **40** 126
[19] Yazdanparast R and Khodarahmi R 2005 *Biotechnol. Appl. Biochem.* **41** 157
[20] Fedorov D and Kitaura K 2007 *J. Phys. Chem. A* **111** 6904
[21] Kitaura K, Ikeo E, Asada T, Nakano T and Uebayasi M 1999 *Chem. Phys. Lett.* **313** 701
[22] Kitaura K, Sawai T, Asada T, Nakano T and Uebayasi M 1999 *Chem. Phys. Lett.* **312** 319
[23] Nakano T, Kaminuma T, Sato T, Akiyama Y, Uebayasi M and Kitaura K 2000 *Chem. Phys. Lett.* **318** 614
[24] Nakano T, Kaminuma T, Sato T, Fukuzawa K, Akiyama Y, Uebayasi M and Kitaura K 2002 *Chem. Phys. Lett.* **351** 475
[25] Kitaura K, Sugiki S, Nakano T, Komeiji Y and Uebayasi M 2001 *Chem. Phys. Lett.* **336** 163
[26] Ito M, Fukuzawa K, Ishikawa T, Mochizuki Y, Nakano T and Tanaka S 2008 *J. Phys. Chem. B* **112** 12081
[27] Tomaselli S, Esposito V, Vangone P, van Nuland N A, Bonvin A M, Guerrini R, Tancredi T, Temussi P A and Picone D 2006 *ChemBioChem* **7** 257
[28] Luhrs T, Ritter C, Adrian M, Riek-Loher D, Bohrmann B, Dobeli H, Schubert D and Riek R 2005 *Proc. Natl. Acad. Sci. USA* **102** 17342
[29] Eswar N, Marti-Renom M A, Webb B, Madhusudhan M S, Eramian D, Shen M, Pieper U and Sali A 2006 *Curr. Protoc. Bioinformatics* **15** 5.6.1
[30] Poulsen S–A, Watson A A and Craik D J 2000 *J. Struct. Biol.* **130** 142
[31] Olsson M H M, Søndergard C R, Rostkowski M and Jensen J H 2011 *J. Chem. Theory Comput.* **7** 525
[32] Case D, Darden T, Cheatham III T, Simmerling C, Wang J, Duke R, Luo R, Merz K, Pearlman D and Crowley M 2006 *AMBER9* University of California, San Francisco
[33] Cornell W, Cieplak P, Bayly C, Gould I, Merz K, Ferguson D, Spellmeyer D, Fox T, Caldwell J and Kollman P 1995 *J. Am. Chem. Soc.* **117** 5179
[34] Jorgensen W, Chandrasekhar J, Madura J, Impey R and Klein M 1983 *J. Chem. Phys.* **79** 926
[35] Lee C, Yang W and Parr R 1988 *Phys. Rev. B* **37** 785
[36] Frisch M J, Trucks GW, Schlegel H B, Scuseria G E, Robb M A, Cheeseman J R, Montgomery Jr. J A, Vreven T, Kudin K N, Burant J C, Millam J M, Iyengar S S, Tomasi J, Barone V, Mennucci B, Cossi M, Scalmani G, Rega N, Petersson G A, Nakatsuji H, Hada M, Ehara M, Toyota K, Fukuda R, Hasegawa J, Ishida M, Nakajima T, Honda Y, Kitao O, Nakai H, Klene M, Li X, Knox J E, Hratchian H P, Cross J B, Bakken V, Adamo C, Jaramillo J, Gomperts R, Stratmann R E, Yazyev O, Austin A J, Cammi R, Pomelli C, Ochterski J W, Ayala P Y, Morokuma K, Voth G A, Salvador P, Dannenberg J J, Zakrzewski V G, Dapprich S, Daniels A D, Strain M C, Farkas O, Malick D K, Rabuck A D, Raghavachari K, Foresman J B, Ortiz J V, Cui Q, Baboul A G, Clifford S, Cioslowski J, Stefanov B B, Liu G, Liashenko A, Piskorz P, Komaromi I, Martin R L, Fox D J, Keith T, Al-Laham M A, Peng C Y, Nanayakkara A, Challacombe M, Gill P M W, Johnson B, Chen W, Wong M W, Gonzalez C and Pople J A 2004 *Gaussian 03, Revision E. 01* Gaussian, Inc., Wallingford, CT
[37] Morris G M, Huey R, Lindstrom W, Sanner M F, Belew R K, Goodsell D S and Olson A J 2009 *J. Comput. Chem.* **30** 2785
[38] Besler B, Merz K and Kollman P 1990 *J. Comput. Chem.* **11** 431
[39] Yoshioka A, Fukuzawa K, Mochizuki Y, Yamashita K, Nakano T, Okiyama Y, Nobusawa E, Nakajima K and Tanaka S 2011 *J. Mol. Graphics Modell.* **30** 110
[40] Tjernberg L O, Näslund J, Lindqvist F, Johansson J, Karlstrçm A R, Thyberg J, Terenius L and Nordstedt C 1996 *J. Biol. Chem.* **271** 8545

A proposal of potent inhibitor for cancer metastasis blocking the pocket of urokinase receptor: *ab initio* molecular simulations

Tomoyo Kasumi[1], Kohta Araki[1], Tatsuroh Mizushima[1], Hiroshi Kobayashi[2] and Noriyuki Kurita[1*]

[1]Department of Computer Science of Engineering, Toyohashi University of Technology, Tempaku-cho, Toyohashi, Aichi, 441-8580, Japan
[2]Department of Obstetrics and Gynecology, Nara Medical University, 840 Shijio-cho, Kashihara, Nara, 634-8521,Japan

[*]E-mail : kurita@cs.tut.ac.jp

Abstract. Recent biochemical experiments have elucidated that a variety of proteases play important roles in cancer invasion and metastasis. In particular, binding of urokinase-type plasminogen activator (uPA) to uPA receptor (uPAR) existing on the surface of a cancer cell is considered to be a trigger for cancer invasions. Therefore, the blocking of the binding is expected to inhibit cancer invasion. In previous experiments, several peptides of amino acids were proposed as a potent inhibitor for blocking the binding. In the present study, we obtained stable structures of the solvated complexes with uPAR and the peptides and investigated the specific interactions between uPAR and the peptides by *ab initio* molecular simulations. Base on these results, we clarified which peptide can bind more strongly to uPAR and proposed a novel potent peptide which can inhibit the uPAR-uPA binding efficiently.

1. Introduction

Recently, the number of deaths due to cancer has been increasing continuously. The main cause for the death of cancer patients is cancer metastasis. Cancer can spread rapidly to organs in the body by metastasis and robs the patient of basis functions necessary for supporting the life. Therefore, the inhibition of cancer metastasis is expected to improve the patient's prognosis dramatically. However, there is no effective treatment for inhibiting the metastasis, and the development of effective inhibitors for the metastasis is a pressing research issue to solve [1]. In general, it is difficult for cancer patients to take strong medicine with side effect and to visit a hospital every day for taking injection drugs. It is thus necessary to develop medicines of tablet that inhibit cancer metastasis with relatively few side effects.

During the processes of invasion and metastasis, cancer cells produce various proteases to dissolve cell tissues such as extracellular matrix (ECM) and basement membrane and facilitate invasion into blood vessels. The cancer cell passes through the blood vessel to reach the endothelial cells of other organs and invades these cells to develop a new metastases focus [2]. By repeating this process, cancer cells grow proliferously. Therefore, it is expected that the inhibition of cancer invasion can suppress cancer metastasis and lead to a good prognosis of the cancer patients.

The recent experiment [3] elucidated several proteases playing essential roles in the cancer invention. In particular, urokinase-type plasminogen activator (uPA) [4, 5] has been found to play a key role in cancer invasion and metastasis. The specific binding of the amino-terminal fragment (ATF) of uPA to the uPA receptor (uPAR) [6] existing on the surface of cancer cell is considered a trigger for promoting cancer invasions [7]. Consequently the blocking of binding between uPA and uPAR is expected to inhibit cancer invasions and metastasis. Based on this idea, Kobayashi *et al.* synthesized a chimeric protein ATF-HI-8 [8], which is composed of bikunin existing in amniotic fluid and the ATF of uPA. The effectiveness of this chimeric protein on inhibiting cancer invasion has been confirmed by the biological studies using laboratory animals [1]. In order to propose which parts of ATF-HI-8 are important for the specific interactions between ATF-HI-8 and uPAR, we investigated the stable structure and its electronic properties of ATF-HI-8 by semiempirical molecular simulations [9, 10].

As for the specific interactions between uPAR and uPA, the previous structure analyses [11, 12] for the complex of uPAR with uPA elucidated that the amino acid residues 1-132 of uPA bind specifically to uPAR. In particular, the Ω-loop (residues of 19-31) in the ATF of uPA was found to be responsible for the high-affinity interactions between uPAR and uPA [13]. On the other hand, we investigated the specific interactions between uPAR and ATF at electronic level by *ab initio* molecular orbital calculations and elucidated that the positively charged Lys46, Lys61, and Lys98 residues as well as the Ω-loop of ATF contribute to the binding between uPAR and ATF [14].

Many types of peptide drugs inhibiting the uPAR-uPA binding were synthesized based on the Ω-loop structure of ATF. In the ATF around the Ω-loop, two disulfide bonds (Cys11-Cys19 and Cys13-Cys31) are formed, indicating that the distance between Cys19 and Cys31 is affected by the positions of Cys11 and Cys13. In fact, the distance between Cys19 and Cys31 is 6.1 Å, which is significantly longer than those (5.2 Å) of the Cys11-Cys19 and Cys13-Cys31 disulfide bonds. Considering this flexibility of Cys19 and Cys31 positions, two types of the 19-31 peptides with and without the disulfide bond between Cys19 and Cys31 were synthesized. Their effects on the inhibition were investigated experimentally to find that both the peptides have an equal inhibition rate for uPAR. Based on the result, 8 kinds of ring-shaped peptides having a disulfide bond were proposed as a peptide drug inhibiting the uPAR-uPA binding [15]. However, their effects on the inhibition and their specific interactions with uPAR are not investigated yet. In order to develop potent peptide inhibitors, it is necessary to clarify the specific binding properties between uPAR and the peptides.

In the present study, to elucidate the specific interactions between uPAR and the 8 types of peptides proposed in the previous experiment [15], we obtained stable structures for the solvated complexes with uPAR and these peptides by classical molecular mechanics (MM) simulations. In addition, the specific interactions between uPAR and the peptides were investigated at an electronic level by *ab initio* fragment molecular orbital (FMO) calculations. From these results, we clarified which peptide can bind more strongly to uPAR and proposed a new peptide with large binding affinity to uPAR. These results will become useful for the development of novel potent inhibitors.

2. Details of molecular simulations

2.1. Construction and optimization of solvated structure for uPAR-peptide complexes

In our current molecular simulations, we employed the X-ray crystal structure of the ternary complex of uPAR-ATF-SMB, in which SMB is somatomedin B domain of vitronectin. This structure is registered in the Protein Data Bank (PDB ID: 3BT1 [16]), and we used it because it has only missing two amino acids 83 and 84 in uPAR. This structure is of uPAR (residues 1-275) in complex with the ATF (residues 8-132) of uPA, SMB (residues 2-41) and other co-factors. We extracted only the uPAR structure from this PDB structure and predicted the structures of these missing residues in uPAR, by using the homology modeling program SWISS-MODEL [17]. The N- and C-terminals of uPAR were terminated by NH_3^+ and COO^-, respectively.

Since the X-ray crystal structure of uPAR has no information about hydrogen atoms, hydrogen atoms should be added to the original PDB structure in an appropriate manner. In particular, it is not unique to determine the positions of hydrogen atoms for His amino acid, because His has some different protonated structures depending on the p*K*a value around the His. We here evaluated the p*K*a values of His residues contained in uPAR by using the PROPKA Web Interface3.0 [18-20] and determined the protonation based on the p*K*a values. His amino acid has three types of protonated structures: Hid has a hydrogen atom at the δ-site of its imidazole ring, Hie has a hydrogen atom at the ε-site of the imidazole ring, and positively charged Hip has hydrogen atoms at both the δ- and ε-sites. His residues with p*K*a value larger than 6 have the Hip protonation, while those with p*K*a value smaller than 6 have the Hid or Hie protonation. uPAR has 12 His residues and their p*K*a values are listed in Table 1, indicating that 7 His residues have p*K*a values smaller than 6 in uPAR. These His residues have Hid or Hie protonation. Because it is not practical to consider all the combinations of Hid/Hie for the 7 His residues included in uPAR, we here considered that the 7 His residues have Hie protonation, while the other 5 His residues have Hip protonation. As for the other ionizable amino acid residues contained in uPAR, the ionized state was used.

Table 1. p*K*a values and protonation states of each His residue in uPAR.

Amino acid	p*K*a	Protonation
His47	4.18	Hie
His110	5.23	Hie
His128	4.40	Hie
His143	6.18	Hip
His160	3.90	Hie
His166	1.14	Hie
His203	6.38	Hip
His229	5.05	Hie
His249	6.64	Hip
His251	5.26	Hip
His260	6.40	Hip
His273	6.32	Hip

We added solvating water molecules in an 8 Å shell around uPAR and considered them explicitly in our classical MM simulations. This solvated uPAR structure was optimized using the classical MM program AMBER9 [21], in which the Parm99SB [22] and TIP3P [23] force fields were used for uPAR and the water molecules, respectively. The threshold value of the energy-gradient for the convergence of the AMBER9 optimization was set as 0.001 kcal/mol/Å.

The structures of the 8 kinds of the peptides, each of which have amino acid sequence listed in Table 2, were optimized in solvating water molecules by the AMBER9 program [21]. The N- and C-terminals were terminated by an acetyl group (-CH_3CO) and an amine group (-$NHCH_3$), respectively. We considered that the His residues of the peptides have Hie protonation, in order to unify the total charge of all peptides.

Table 2. Amino acid sequences of the peptides investigated in the present study. The common sequence is underlined.

Peptide	Amino acid sequence (from N-terminal to C-terminal)	Number of residues
P1	Cys, Asn, Ser, Phe, Tyr, Lys, Cys	7
P2	Trp, Cys, Asn, Ser, Phe Tyr, Lys, Cys	8
P3	Trp, His, Cys, Asn, Ser, Phe, Tyr, Lys, Cys	9
P4	Trp, Cys, Asn, Ser, Phe, Tyr, Lys, Asn, Cys	9
P5	Trp, Cys, Ile, Asn, Ser, Phe, Tyr, Lys, Cys	9
P6	Trp, Cys, His, Ile, Asn, Ser, Phe, Tyr, Lys, Asn, Cys	11
P7	Trp, Cys, His, Ile, Asn, Ser, Phe, Tyr, Lys, Asn, Ser, Cys	12
P8	Trp, Cys, Ile, Asn, Ser, Phe, Tyr, Lys, Asn, Ser, Cys	11

To obtained candidate structures for the complex of uPAR-peptide, we docked the peptide to a variety of sites around uPAR using the automated protein-ligand docking program Autodock 4.2 [24]. In the docking procedure, the grid box was set as the 40.0 × 126.0 × 40.0 Å^3 centered on the gravity center of uPAR, and the spacing between the nearest neighboring grid points was set to 0.375 Å, which is the default value of Autodock 4.2. The structure of uPAR was fixed, and all dihedral angles of the peptide were freely rotated in the docking procedure to search for a variety of stable configurations for the peptide docked to uPAR. We here created 200 candidate structures of the complex by using the genetic algorithm of Autodock 4.2. They were classified into some clusters according to the RMSD value (5.0 Å) between each of the structures created.

The representative structures in each cluster were fully optimized by the AMBER9-MM method. To consider the solvation effect on the complex properly, we added water molecules with a 8 Å layer around the complex and optimized the solvated structure by using AMBER9 [21], in which the Parm99SB [22] and TIP3P [23] force fields were assigned for the complex and water molecules, respectively. The threshold value of the energy-gradient for the convergence in the AMBER9 optimization was set as 0.001 kcal/mol/Å.

2.2. *Ab initio FMO calculations for the solvated structures of uPAR-peptide complexes*

The electronic properties for the most stable structure of the solvated uPAR-peptide complexes were investigated by the *ab initio* FMO calculation [25], to elucidate which amino acid residues in uPAR and peptide are important for the specific binding between uPAR and peptide. In the FMO method, the target molecule is divided into units called "fragment", and the electronic properties of the target molecule are estimated from the electronic properties of the monomers and dimers of the fragments. The specific interactions between the fragments were elucidated from the interaction energies obtained by the FMO calculation.

In the present study, we used the FMO calculation program ABINIT-MP Ver.4.3 [26]. In the FMO calculation, *ab initio* MP2[27,28]/6-31G method was employed to investigate accurately the π-π stacking, NH-π and CH-π interactions between the amino acid residues of uPAR and peptides. Each amino acid residue of uPAR and peptide, each water molecule were assigned as a fragment, because this fragmentation enables us to evaluate the interaction energies between the amino acid residue of uPAR and peptide. From the comparison of the interaction energies, it is possible to propose which amino acid residues of uPAR and peptide have large contribution to the specific interactions between uPAR and peptide. In addition, we considered the water molecules existing within a 5 Å from the peptide explicitly, in order to elucidate the influence of solvating water molecules on the specific interactions. The number of solvating water molecules was set as 100 for all the complexes to investigate the relative stability among the solvated structures.

In order to investigate the biding energy between uPAR and peptide, the solvated uPAR-peptide complex was divided into the following four structural domains: uPAR-peptide containing solvating water molecules (uPAR-peptide+water), uPAR containing solvating water molecules (uPAR+water), the peptide containing solvating water molecules (peptide+water) and solvating water molecules (water). From total energies (T.E.) obtained by FMO calculations, the binding energy (B.E.) between uPAR and the peptide mediated by solvating water molecules was estimated as B.E. = – T.E. (uPAR-peptide+water) + T.E. (uPAR+water) + T.E. (Peptide+water) – T.E. (water).

3. Results and discussion

3.1. *Stable structures of the solvated complexes of uPAR-peptide*

To check the validity of the solvated structure of uPAR optimized by AMBER9, we compared it with the uPAR structure in the X-ray structure of the uPAR-uPA-SMB complex registered in PDB (PDB ID: 3BT1 [16]). The RMSD for all Cα atoms of uPAR is 1.3 Å, indicating that the uPAR structure is changed significantly by the optimization in water. We furthermore investigated the displacement of Cα atom for each amino acid residue of uPAR. As shown in Figure 1, among the 275 residues of uPAR, 10 residues have displacement larger than 2.5 Å. Among them, 6 residues exist on the surface of uPAR to be affected by the solvation. In contrast, the other 4 residues exist near the 83rd and 84th residues, whose structures are missing in the X-ray structure. Therefore, the effect of solvation is likely to cause the change in structure of these residues, resulting in the RMSD of 1.3 Å. In the previous biochemical experiment [13], the residues 24-70, 122-171 of uPAR were found to contribute to the specific interactions between uPAR and uPA. Figure 1 indicates that the Cα positions of these residues are not affected significantly by the optimization in water. It is thus elucidated that the structure of uPA binding site of uPAR optimized in water by the AMBER9-MM method is reasonable.

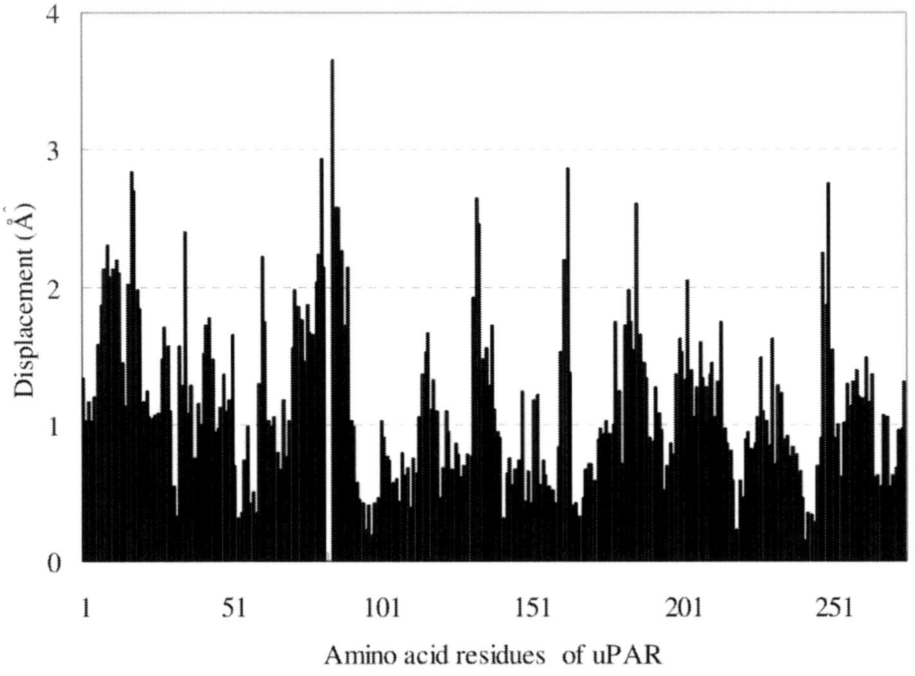

Figure 1. Displacement of each amino acid residue of uPAR between the optimized structure in water and the X-ray structure [17].

The peptides were docked to uPAR by the docking program AutoDock4.2 [24] to create the 200 candidate structures for each uPAR-peptide complex. For the peptides P1, P6 and P7, all of the 200 structures were grouped into a single cluster, while for the other peptides, the created structures were grouped into some clusters. The representative structures in each cluster were fully optimized by the AMBER9-MM method, and their total energies calculated by the *ab initio* MP2/6-31G method in FMO [26] are listed in Table 3.

Table 3. Total energies (T.E.) (kcal/mol) for the optimized structures of uPAR-peptide complexes evaluated by *ab initio* FMO method, and their ranking in stability. P1 and C1 mean the peptide 1 and the cluster 1, respectively.

Structure	T.E.	Δ T.E.	Ranking
uPAR-P1	-79706094.5	0.0	1
uPAR-P2-C1	-80087576.3	0.0	1
uPAR-P2-C2	-80086969.4	606.9	2
uPAR-P3-C1	-80382434.2	0.0	1
uPAR-P3-C2	-80382276.3	157.9	2
uPAR-P4-C1	-80347053.4	547.0	3
uPAR-P4-C2	-80347600.4	0.0	1
uPAR-P4-C3	-80347451.0	149.4	2
uPAR-P5-C1	-80315784.2	0.0	1
uPAR-P5-C2	-80315454.9	329.3	2
uPAR-P5-C3	-80315405.7	378.5	3
uPAR-P6	-80870742.6	0.0	1
uPAR-P7	-81071961.0	0.0	1
uPAR-P8-C1	-80777063.6	0.0	1
uPAR-P8-C2	-80776590.0	473.6	2

To clarify the reason for the difference in total energy listed in Table 3, we analyzed the optimized structures of the uPAR-peptide complexes in detail as shown in Figure 2, where the boundary between the inside and the outside of the ligand-binding pocket of uPAR is defined by drawing a line between the Cα atoms of Glu34 and Glu132, because both the residues exist near the entrance of the pocket. Based on Figure 2, we attempted to elucidate how the peptide enters into the pocket of uPAR.

There are three candidate structures for the uPAR-P4 complex as listed in Table 3. The most stable structure is at least 150 kcal/mol more stable than the other ones. In this structure, P4 goes deeply into the pocket of uPAR and binds strongly to the residues (Thr127 and Asp140) of uPAR, which exist at the bottom of the pocket. As a result, almost all parts of P4 enter into the pocket as shown in Figure 2d. In the other two clusters of the uPAR-P4 complex, P4 exists around the entrance of the pocket and is not stabilized enough. For the uPAR-P5 complex, the situation is the same as the uPAR-P4 complex, and almost all parts of P5 enter into the pocket as shown in Figure 2e. In contrast, in the complexes with uPAR and the other peptides, the peptides are stabilized at the intermediate position of the pocket as shown in Figure 2. In the uPAR-uPA complex, the Ω-loop of uPA composed of the residues 19-31 enters completely into the pocket and binds strongly to uPAR [16]. Consequently, the present molecular simulations elucidate that only P4 and P5 can bind to uPAR in the similar way as the uPAR-uPA complex.

To elucidate the reason why only P4 and P5 can enter deeply into the pocket of uPAR, we compared the structures of all peptides. They are stabilized by the disulfide bond between Cys residues and some hydrogen bonds between residues. In the P4 and P5, the amino group of the Lys side chain and the oxygen atom of the backbone between Lys and Tyr are hydrogen bonded each other. As a result, P4 and P5 are stabilized into a compact form to enter easily into the pocket. In contrast, in the other peptides, the Lys side chain extends outward. And it seems that this side chain inhibits the peptide to enter into the pocket. Therefore, it is clarified that the conformation of the Lys side chain in the peptide has large influence on the binding position of the peptides in the uPAR pocket.

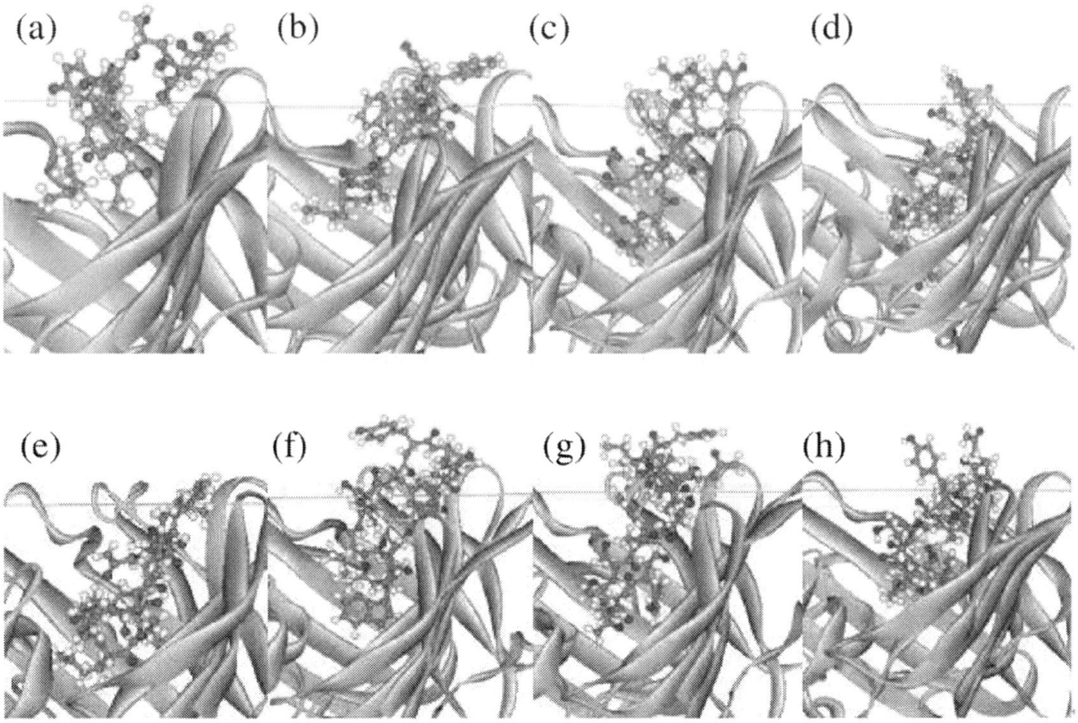

Figure 2. Close-up views around the uPA-binding pocket of uPAR and peptide for the optimized structure of the solvated uPAR-peptide complexes; (a) uPAR-P1, (b) uPAR-P2, (c) uPAR-P3, (d) uPAR-P4, (e) uPAR-P5, (f) uPAR-P6, (g) uPAR-P7 and (h) uPAR-P8. The horizontal line, which connects the Cα atoms of the Glu34 and Glu132 residues of uPAR, indicates a boundary between the inside and the outside of the uPA-binding pocket of uPAR.

3.2. Electronic properties of the solvated complexes of uPAR-peptide

We first evaluated the binding energy between uPAR and the peptide by *ab initio* FMO calculations and predicted the binding affinity between them. As listed in Table 4, P4 has the largest binding energy among the 8 peptides investigated here. To elucidate which amino acid residues of P4 are important for this binding energy, the interaction energy between each amino acid residue of P4 and uPAR was investigated by FMO. As shown in Figure 3a, Lys has the strongest attractive interaction with uPAR. We furthermore investigated the interaction energy between Lys and each residue of uPAR. Table 5 lists the residues of uPAR having strong attractive interaction with Lys in the decreasing order of the magnitude of interaction. The positively charged Lys residue of P4 has attractive interactions with negatively charged Asp and Glu residues of uPAR by the electrostatic interactions. In fact, as shown in Figure 4a, the amino group of the Lys side chain of P4 and the COO group of Asp140 of uPAR are hydrogen bonded, in addition, a hydrogen bond is formed between the amino group and the OH group of the Thr127 side chain. As a result, non-charged Thr127 has strong attractive interaction with the Lys of P4. It is thus elucidated that Asp140 and Thr127 as well as the negatively charged Asp and Glu residues of uPAR contribute to the strong binding between uPAR and P4.

In our previous molecular simulations for the uPAR-uPA complex [14], Lys23 included in the Ω-loop of uPA interacts strongly with the Asp140 and Thr127 residues of uPAR. These residues are located on the bottom of the ligand-binding pocket of uPAR. In addition, the interacting structure between Lys of P4 and Asp140/Thr127 of uPAR shown in Figure 4a is similar to that in the uPAR-uPA complex. Accordingly, P4 is expected to be a potent inhibitor for the uPAR-uPA binding.

Table 4. Total energies (T.E.) for each component of the optimized structures of uPAR-peptide complexes, and estimated binding energies (B.E.) between uPAR and peptide. The B.E. was evaluated as B.E. = – T.E.(uPAR-peptide+water) + T.E.(uPAR+water) + T.E.(peptide+water) – T.E.(water).

Structure	T.E.(kcal/mol)				B.E.
	uPAR-peptide+water	uPAR+water	peptide+water	Water	(kcal/mol)
uPAR-P1 (7)*	-79706094.5	-77374004	-7109012.4	-4777168.3	246.3
uPAR-P2 (8)	-80087576.3	-77374103.1	-7490053.3	-4777250.4	670.3
uPAR-P3 (9)	-80382434.2	-77374065.6	-7785173.4	-4777269.0	464.1
uPAR-P4 (9)	-80347600.4	-77374146.5	-7749953.9	-4777260.3	760.4
uPAR-P5 (9)	-80315784.2	-77374227.3	-7718130.2	-4777206.1	632.8
uPAR-P6 (11)	-80870742.6	-77373849.4	-8273380.8	-4777200.4	712.7
uPAR-P7 (12)	-81071961.0	-77373917.7	-8474900.5	-4777265.9	408.8
uPAR-P8 (11)	-80777063.6	-77374017.6	-8179758.8	-4777292.6	579.7

* The number of residues contained in each peptide is shown in parenthesis.

Table 5. Interaction energies (I.E.) (kcal/mol) between Lys of the peptide 4 and the amino acid residues of uPAR in the uPAR-P4 complex. The residues having strong attractive interaction are listed in the decreasing order of the magnitude of interaction.

Residues	I.E.
Asp140	-90.4
Asp254	-45.2
Glu68	-38.8
Asp141	-29.9
Thr127	-29.2
Glu42	-27.3
Glu183.	-27.1
Asp124	-26.1
Glu134	-25.8
Glu106	-24.1

As listed in Table 4, P6 has the second largest binding energy to uPAR. The difference in specific interactions between uPAR and peptide was investigated for P4 and P6. Figure 3b shows the interaction energies between each residues of P6 and uPAR. The attractive interaction between Lys and uPAR is about 100 kcal/mol weaker compared with the uPAR-P4 complex shown in Figure 3a. To elucidate the reason for this difference, the position of Lys in both the uPAR-Peptide complexes were compared in Figure 5. In the uPAR-P4 complex, Lys exists in the pocket of uPAR, while the Lys of P6 exists out of the pocket, resulting in the weaker interaction between uPAR and Lys. We considered that this difference in the Lys position is caused from the effect of the charged residues around Lys139 existing at the ligand-binding pocket of uPAR. As shown in Figure 6, negatively charged residues Glu34, Glu36, Asp140, Asp141, Glu132, Glu134 and Glu135 exist around Lys139 to surround the pocket of uPAR. Positively charged parts are marked in blue, while negatively charged parts are marked in red, indicating that there is a large polarization around Lys139. Because the peptides have only one positively charged Lys residue, this residue feels sensitively the electrostatic potential of the charged residues existing around the pocket of uPAR, when the peptides approach to the pocket. Accordingly, the position of Lys and the conformation of its side chain in the peptide is seems to have large effect on the binding properties of peptide to uPAR. In fact, Lys of P6 cannot enter the pocket of uPAR as shown in Figure 5b. In this complex, Asn existing at the opposite side of Lys interacts with Asp140 and Lys139 of uPAR, as shown in Figure 4b, playing an important role in the binding between uPAR and P6. Therefore, Asn as well as Lys in the peptides are expected to be important for the binding to uPAR.

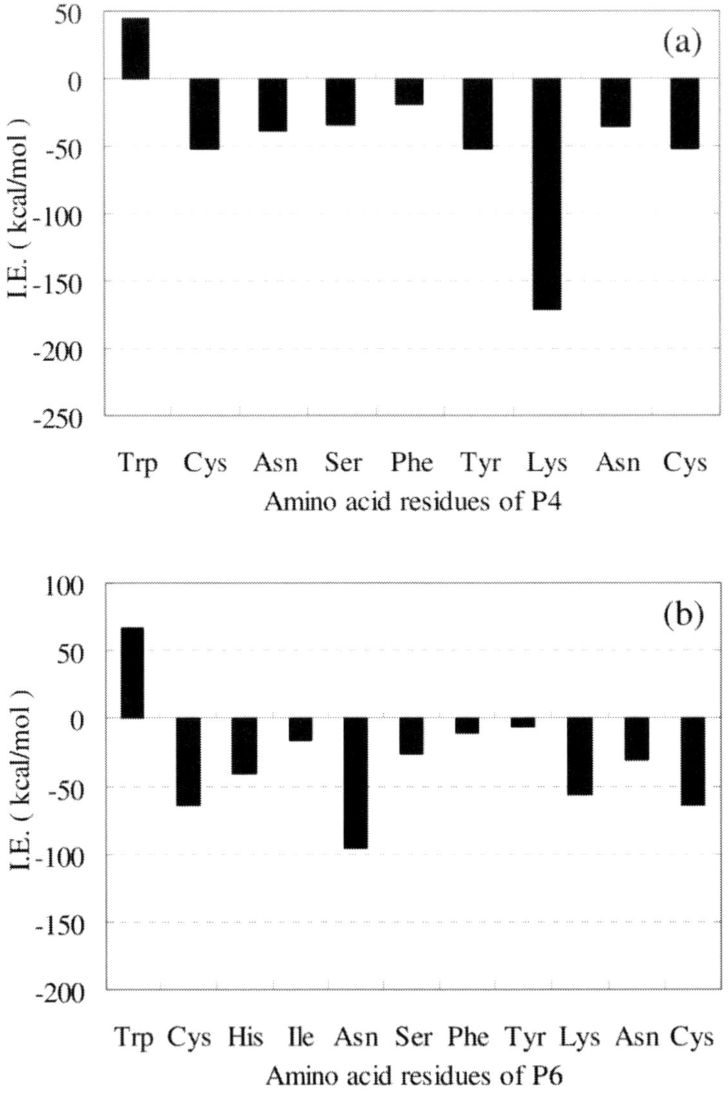

Figure 3. Interaction energies (I.E.) between each amino acid residue of peptide and uPAR for (a) uPAR-P4 and (b) uPAR-P6 complexes.

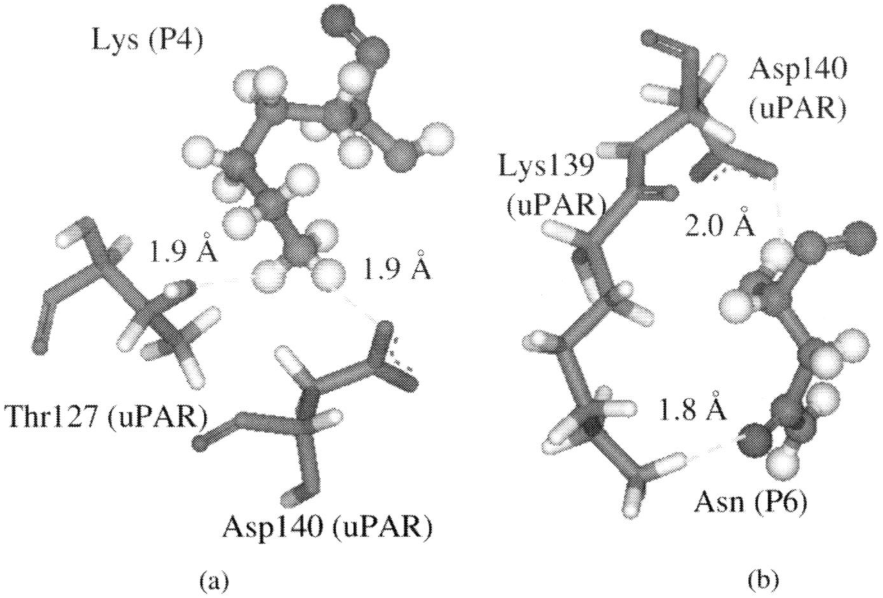

Figure 4. Hydrogen bonding structures and distances (Å) between amino acid residues of uPAR and (a) Lys of Peptide 4 for the uPAR-P4 complex, (b) Asn of Peptide 6 for the uPAR-P6 complex.

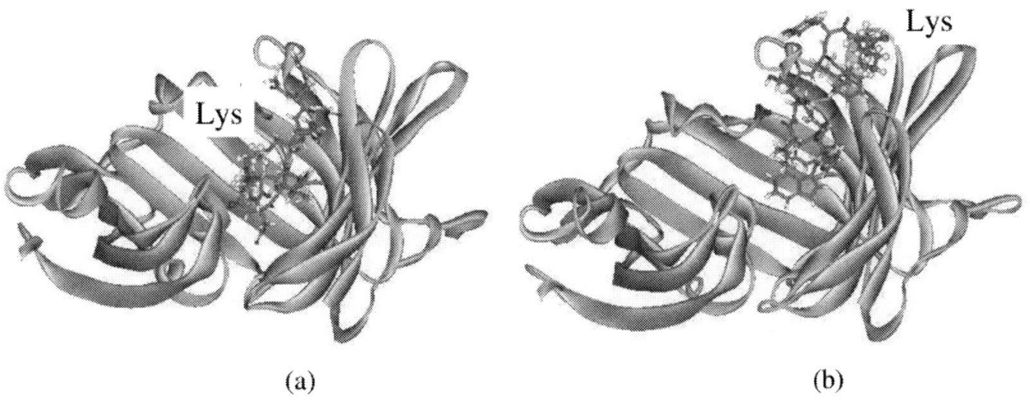

Figure 5. Position of Lys residue of peptide in the uPAR-peptide complex: (a) uPAR-P4 and (b) uPAR-P6 complexes.

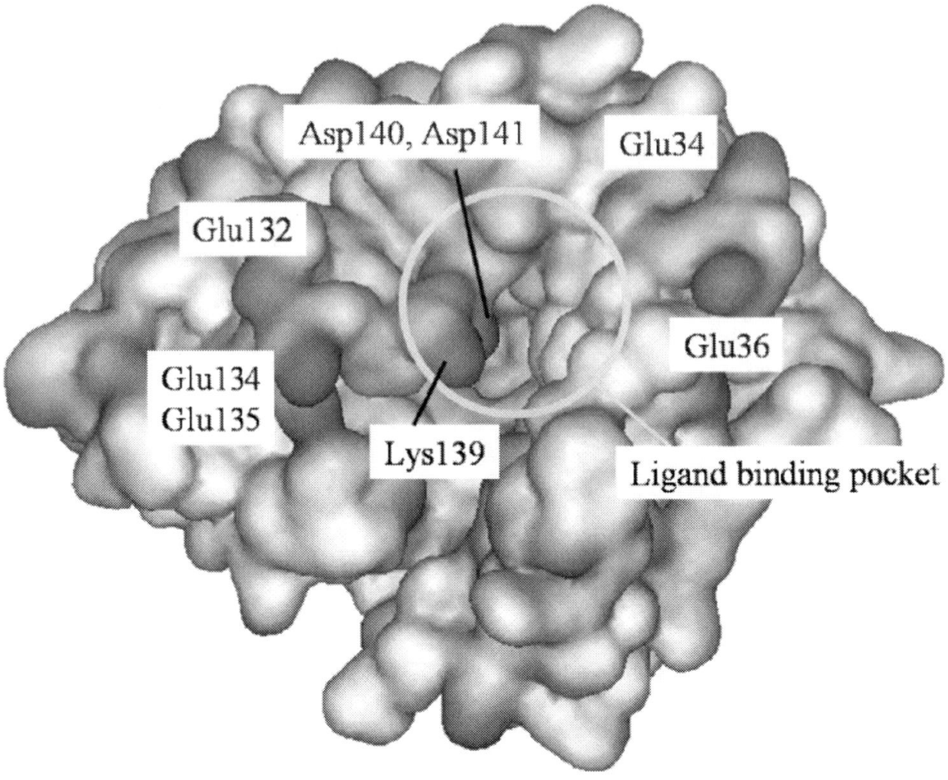

Figure 6. Surface electric charge distribution around the ligand binding pocket of uPAR. Positively charged parts are marked in blue, while negatively charged parts are marked in red.

3.3. Proposal of potent inhibitor for uPAR-uPA binding

The 8 kinds of peptides proposed in the previous experiment [15] have a common sequence of Asn-Ser-Phe-Tyr-Lys, as listed in Table 2. This sequence comes from the central 5 residues of the 7 residues (N-terminal-Ile-Asn-Ser-Phe-Tyr-Lys-Asn-C-terminal) constructing the Ω-loop of uPA. The Ω-loop is considered to be flexible and contribute mainly to the uPAR-uPA binding.

Our present calculations elucidate that P4 and P6 bind strongly to uPAR as listed in Table 4. These peptides have the common sequence of Lys-Asn-Cys on the C-terminal. This Asn residue corresponds to the terminal Asn of the 7 residues of the flexible Ω-loop. To investigate the effect of this residue on the binding between uPAR and peptide, we compared the results for P2 and P4, because P2 has the same sequence as P4 except for the missing of Asn, as shown in Table 2. The evaluated binding energy for P2 is 90 kcal/mol smaller than that for P4, as listed in Table 4. Therefore, it can be concluded that the Asn residue on the C-terminal is important for strong binding of peptide to uPAR.

On the other hand, from the comparison between the results for P2 and P5 listed in Table4, it was elucidated that the Ile residue on the N-terminal of the 7 residues weakens the binding between uPAR and peptide. The binding energy of P5 is 37.5 kcal/mol smaller than that of P2, indicating that the addition of Ile to P2 causes the weakening of the uPAR-peptide binding. From the above comparison of the binding energies listed in Table 4, we concluded that the peptides having the sequence such as (N-terminal)-Xn-Lys-Asn-Cys-(C-terminal) may bind strongly to uPAR and inhibit the uPAR-uPA binding.

Based on these results obtained by the present calculations, we attempted to propose some potent inhibitors for blocking the binding between uPAR and uPA and investigated their binding properties to uPAR. Because P4 has the largest binding energy, we mutated some residues of P4 by the other type of residue to improve the binding affinity to uPAR.

We first constructed a peptide by mutating Asn of the N-terminal of P4 by the negatively charged Asp. As shown in Figure 3b and Figure 4b, this Asn residue plays important roles in the binding between P6 and uPAR, although Asn is an uncharged residue. In order to enhance the interaction between this residue and uPAR, we mutated it by the negatively charged Asp residue, which has similar side chain as Asn. This peptide is defined as P4-ND. Figure 7a shows the optimized structure of the uPAR-P4-ND complex, indicating that P4-ND enters completely into the pocket of uPAR and binds to Thr127 and Asp140 in the similar way as P4. In addition, the mutated Asp residue of P4-ND is hydrogen bonded with Lys139 of uPAR. As a result, the interaction energies between each residue of P4-ND and uPAR are slightly enhanced by the mutation, as shown in Figure 8a. Therefore, it is expected that our proposed P4-ND peptide can bind strongly into the ligand-binding pocket of uPAR.

We furthermore constructed a new peptide by mutating Ser of P4 by the other reside, since the interaction energies between each residue of P4 and uPAR shown in Figure 3a indicates that Ser interacts weakly with uPAR. This Ser residue was mutated by Thr, which has a similar side chain as Ser, and the specific interactions between uPAR and the mutated peptide were investigated. This mutated peptide is defined as P4-ST. As shown in Figure 7b, P4-ST can enter into the pocket of uPAR completely. This result comes from the compact structure of P4-ST, in which the amino group of Lys side chain is hydrogen bonded to the two oxygen atoms of the backbones between Lys and Tyr and between Trp and Cys. In addition, the side chain of Asn is hydrogen bonded to the nitrogen atoms of the backbones between Trp and Thr. These hydrogen bonds make P4-ST compact, and P4-ST enters easily into the pocket of uPAR. However, P4-ST dose not have enough side chains for interacting with uPAR, so that the interaction energies between each residue of P4-ST and uPAR are significantly smaller than those for the uPAR-P4 complex, as shown in Figure 8b. Therefore, P4-ST is not expected to be a potent inhibitor for the uPAR-uPA binding.

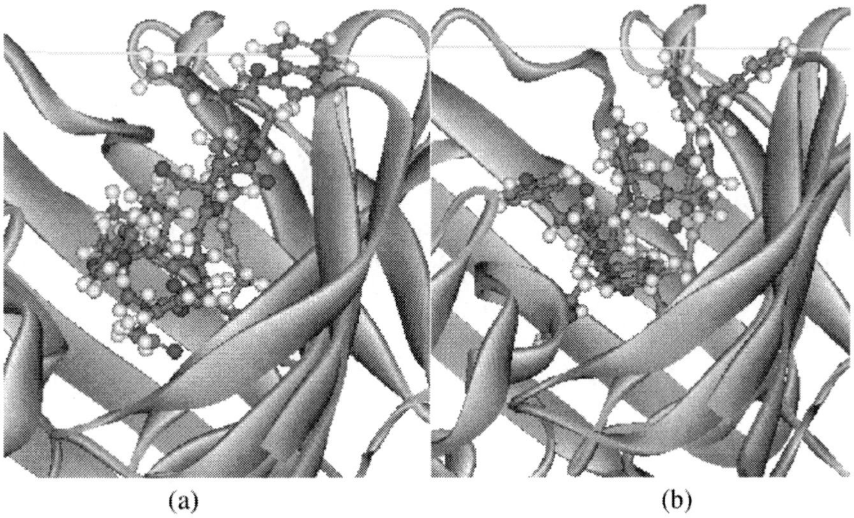

(a) (b)

Figure 7. Stable structures of the solvated complexes with uPAR and our proposed peptides optimized by AMBER9; (a) uPAR-P4-ND and (b) uPAR-P4-ST.

Journal of Physics: Conference Series **433** (2013) 012034 doi:10.1088/1742-6596/433/1/012034

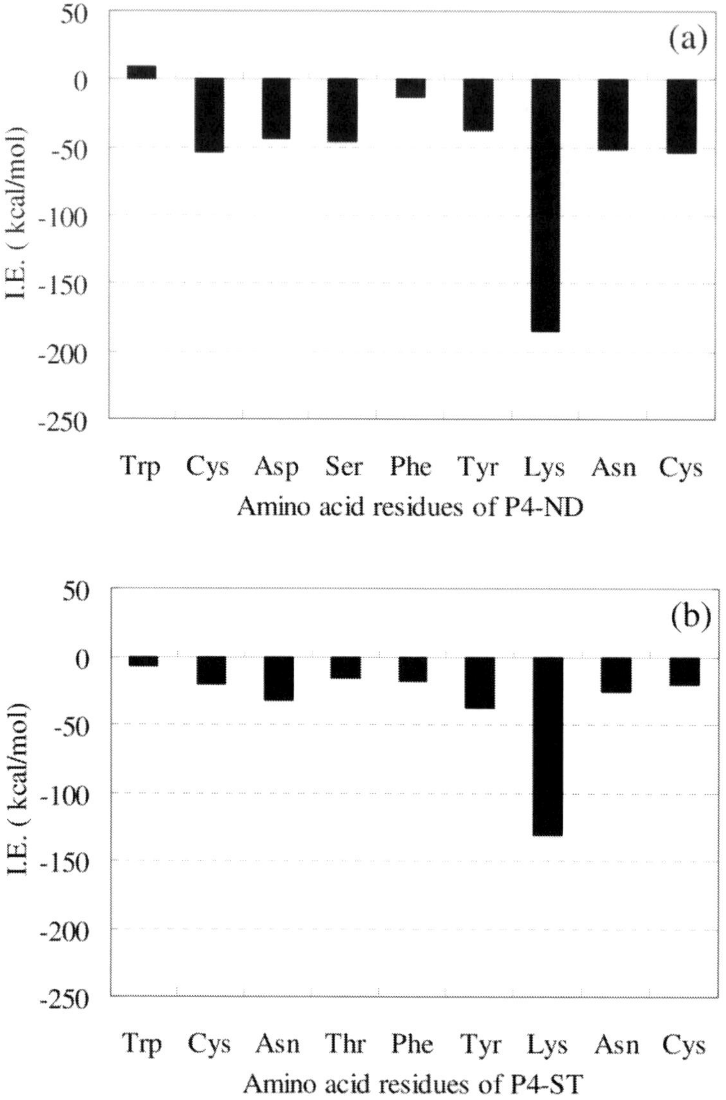

Figure 8. Interaction energies (I.E.) between each amino acid residue of peptide and uPAR for (a) uPAR-P4-ND and (b) uPAR-P4-ST complexes.

4. Conclusions

In the present study, we have investigated at atomic and electronic levels the binding properties between uPAR and the 8 kinds of peptides proposed in the previous experiment [15] and elucidated the following points.

(1) The peptide with the Trp-Cys-Asn-Ser-Phe-Tyr-Lys-Asn-Cys sequence has the largest binding energy with uPAR and can be a potent inhibitor for the uPAR-uPA binding.
(2) Lys and Asn residues included in the above peptide are important for its binding to uPAR.
(3) The peptides having the sequence of N-terminal-Xn-Lys-Asn-Cys-C-terminal can bind strongly to uPAR.
(4) In the uPAR, Thr127, Lys139 and Asp140 contribute mainly to the uPAR-peptide binding.

Based on these results, we furthermore proposed new peptides and investigated their binding properties to uPAR. The results indicate that the peptide with the Trp-Cys-Asp-Ser-Phe-Tyr-Lys-Asn-Cys sequence binds strongly to uPAR and can be a potent inhibitor for the uPAR-uPA binding.

Acknowledgments

This work was supported in part by the grants from Iketani Science and Technology Foundation, Tatematsu Foundation, CASIO Science Promotion Foundation and Toukai Foundation for Technology.

References

[1] Kobayashi H, Gotoh J, Hirashima Y, Fujie M, Sugino D and Terao T 1995 *The Journal Biological chemistry* **270** 8361-8366
[2] Ruoslahti E 1996 *Scientific American* **9** 48-55
[3] Baker E A and Leaper D J 2003 *European Journal of Cancer* **39** 981-988
[4] Skelly M M, Troy A, Duffy M J, Mulcahy H E, Duggan C, Connell T G, O'Donoghue D P and Sheahan K 1997 *Clinical Cancer Research* **3** 1837-1840
[5] Herszènyi L, Plebani M, Carraro P, Paoli M De, Roveroni G, Cardin R, Tulassay Z, Naccarato R and Farinati F 1999 *Cancer* **86** 1135-1142
[6] Stephens R W, Nielsen H J, Christensen I J, Thorlacius-Ussing O, Sorensen S,Dano K and Brunner N 1999 *Journal of the National Cancer Institute* **91** 869-874
[7] Nagase K, Kobayashi H, Nagata S, Yamamoto H, Natsume T, Dedachi K, Tsukamoto T and Kurita N 2007 *Journal Computer Aided Chemistry* **8** 75-84
[8] Kobayashi H, Terao T, Sugino D and Okushima M 1997 The United States Patent and Trademark Office WO97/25422
[9] Sugiura H, Tsumura N, Kobayashi H and Kurita N 2004 *Journal Computer Aided Chemistry* **5** 62-69
[10] Tsumura N, Sugiura H, Kobayashi H and Kurita N 2004 *Journal Computer Aided Chemistry* **5** 70-76
[11] Appella E, Robinson E A, Ullrich S J, Stoppelli M P, Corti A, Cassani G and Blasi F 1987 *The Journal of Biological Chemistry* **262** 4437-4440
[12] Bdeir K, Kuo A, Sachais B S, Rux A H, Bdeir Y, Mazar A, Higazi A A and Cines D B 2003 *Blood* **102** 3600-3608
[13] Barinka C, Parry G, Callahan J, Shaw D E, Kuo A, Bdeir K, Cines D B, Mazar A and Lubkowski J 2006 *Journal of Molecular Biology* **363** 482-495
[14] Tsuji S, Kasumi T, Nagase K, Yoshikawa E, Kobayashi H and Kurita N 2011 *Journal of Molecular Graphic and Modelling* **29** 975-984
[15] GmbH R D, Burgle M, Graeff H, Kessler H, Magdolen V, Konig B, Koppitz M, Riemer C, Schmitt M, and Weidle U 1998 WO98/46731

[16] Huai Q, Zhou A, Lin L, Mazar A P, Callahan J, Shaw D E, Furie B C and Huang M 2008 *Nature Structure & molecular Biology* **15** 422-423
[17] Schwede T, Kopp J, Gues N and Peitsch M C 2003 *Nucleic Acids Res* **31** 3381-3385
[18] Li H, Robertson A D and Jensen J H 2005 *Proteins* **61** 704-721
[19] Delphine C B, Rogers D M and J. H. Jensen J H 2008 *Proteins* **73** 765-783
[20] Olsson M H M, Søndergard C R, Rostkowski M and Jensen J H 2011 *Journal of Chemical Therory and Computation* **7** 525-537
[21] Case D A, Cheatham III T E, Darden T, Gohlke H, Luo R, Merz Jr. K M, Onufriev A, Simmerling C, Wang B and Woods R J 2005 *Journal of Computatinal Chemistry* **26** 1668-1688
[22] Wang J, Cieplak P and Kollman P A 2000 *Journal of Computatinal Chemistry* **21** 1049-1074
[23] Jorgensen W L, Chandrasekhar J, Madura J and Klein M L 1983 *Journal of Chemical Physics* **79** 926-935
[24] Morris G M, Huey R, Lindstrom W, Sanner M F, Belew R K, Goodsell D S and Olson A I 2009 *Journal of Computer Chmistry* **16** 2785-2791
[25] Kitaura K, Sawai T, Asada T, Nakano T and Uebayashi M 1999 *Chemical Physics Letter* **312** 319-324
[26] Ito M, Fukuzawa K, Ishikawa T, Mochizuki Y, Nakano T, and Tanaka S 2008 *The Journal of Physical Chemistry B* **112** 12081-12094
[27] Mochizuki Y, Nakano T, Koikegami S, Tanimori S, Abe Y, Nagashima U and Kitaura K 2004 *Theoretical Chemistry Accounts* **112** 442-452
[28] Mochizuki Y, Koikegami S, Nakano T, Amari S and Kitaura K 2004 *Chemical Physics Letter* **396** 473-479

Effect of BrU on the transition between wobble Gua-Thy and tautomeric Gua-Thy base-pairs: ab initio molecular orbital calculations

Kazuya Nomura[1], Ryota Hoshino[1], Yasuhiro Hoshiba[1], Victor I. Danilov[2], and Noriyuki Kurita [1,*]
[1] Department of Computer Science and Engineering, Toyohashi University of Technology, Tenpaku-cho, Toyohashi, Aichi 441-8580, Japan
[2] Institute of Molecular Biology and Genetics, National Academy of Sciences of Ukraine, 150, Zabolotnogo Str., Kyiv-143, Ukraine, 03680

*E-mail: kurita@cs.tut.ac.jp

Abstract. We investigated transition states (TS) between wobble Guanine-Thymine (wG-T) and tautomeric G-T base-pair as well as Br-containing base-pairs by MP2 and density functional theory (DFT) calculations. The obtained TS between wG-T and G*-T (asterisk is an enol-form of base) is different from TS got by the previous DFT calculation. The activation energy (17.9 kcal/mol) evaluated by our calculation is significantly smaller than that (39.21 kcal/mol) obtained by the previous calculation, indicating that our TS is more preferable. In contrast, the obtained TS and activation energy between wG-T and G-T* are similar to those obtained by the previous DFT calculation. We furthermore found that the activation energy between wG-BrU and tautomeric G-BrU is smaller than that between wG-T and tautomeric G-T. This result elucidates that the replacement of CH_3 group of T by Br increases the probability of the transition reaction producing the enol-form G* and T* bases. Because G* prefers to bind to T rather than to C, and T* to G not A, our calculated results reveal that the spontaneous mutation from C to T or from A to G base is accelerated by the introduction of wG-BrU base-pair.

1. Introduction

It was revealed from the studies of the genetic codes that the formation of the canonical Watson-Crick (W-C) base-pairs (i.e. G-C and A-T) for DNA is strictly observed only for the first two base-pairs in the codon-anticodon interactions [1]. At the third codon position, many types of base-pairs including both the canonical W-C and non-canonical W-C base-pairs are involved. Based on this fact, Crick [1] proposed that the base-pair at the 3' end of the codon can deviate or wobble within arbitrary limits that are consistent with the additional pairing interactions suggested by the code. Since then, specific base-pairings in the codon-anticodon interaction have been investigated for understanding of the basis on which particular base-pairs are selected at the wobble position [2].

High resolution nuclear magnetic resonance (NMR) and ethidium bromide binding studies [3] demonstrated that polymers of deoxynucleotide G and T form ordered double-helical structures at low temperature. In these structures, G and T bases are hydrogen bonded together in a wobble base-pair form, while alternative hydrogen bonding schemes involving the tautomeric form of either G or T are eliminated. In addition, the X-ray diffraction analysis [4] at 2.5 Å resolution for the deoxyoligomer d(CGCGAATTTGCG) duplex showed that the G-T base-pairs in the duplex adopt a wobble structure, in which T and G bases project into the major and minor groove, respectively. The wobble G-T (wG-T) base-pair can also be accommodated in the octameric A-form DNA fragment [5] as well as the Z-form DNA with the sequence of d(CGCGTG) [6]. *Ab initio* molecular orbital (MO) calculations [7] demonstrated that the double proton transfer (DPT) in the wG-T base-pair is a potential pathway for the generation of the rare tautomers of guanine. Physical-chemical mechanism of transformation of the wobble DNA base-pairs into the tautomeric base-pairs was suggested by Brovarets and Hovorun [8]. This transition mechanism, involving the tautomeric form of either G or T was investigated by means of *ab initio* MO and DFT methods. The results showed that the structure of the transition state (TS) between wG-T and the tautomeric G-T* base-pairs is planar. On the other hand, the TS between wG-T and enol-form G (G*)-T base-pairs is non-planar, and this TS has larger activation energy than that between wG-T and G-T*, indicating that the transition between wG-T and G*-T is unfavorable in DNA duplex structure energetically and structurally. In addition, Brovarets & Hovorun [9] performed the analogical MO and DFT calculations for the wobble G-BrU (wG-BrU) and G-BrU base-pairs to elucidate the effect of BrU on the transition mechanism. The effect of the other halogenated derivatives of U on the DNA replication errors was also investigated by *ab initio* MO calculations [10].

In order to search for more stable and planar TS structures for these transitions, we investigated in detail the transition mechanism between wG-T and the G-T base-pairs involving the enol-form of either G or T by using the *ab initio* MP2 method. This method takes into account correlation of electrons, which is important for determining stable structures and evaluating activation energies of reactions more accurately. In addition, to reveal the effect of the BrU substitution on the transition mechanism, we performed the same MO calculations for the wobble and the tautomeric G-BrU base-pairs and evaluated the activation energies for the transitions between these base-pairs. From the comparison of the results obtained for both the G-T and G-BrU base-pairs, we attempted to demonstrate that introduction of Br atom to the 5th position of T increases the probability of the tautomerization reaction, finally leading to spontaneous mutations of DNA/RNA bases.

2. Details of ab initio MO calculations

We first optimized the structures of the wG-T, G*-T and G-T* base-pairs by using the MP2/6-31G(d,p) method of the *ab initio* MO program Gaussian09 (G09) [11]. The structures of the TSs between these base-pairs were searched for by the STQN method [12] of G09, and the obtained TS structures were confirmed to have a single imaginary frequency by the vibrational analysis of G09. Afterwards, the reaction path of proton transfer between bases was followed by performing an intrinsic reaction coordinate (IRC) calculation [13], in order to make sure that the obtained TS structure really is TS structure connecting the reactant and the product structures. The potential energy surface for the transitions between the G-T base-pairs was evaluated by including the zero-point vibrational energy (ZPE), and the free energies at 298.15 K and 1 atmosphere of pressure were also evaluated by MP2/6-31G(d,p). The same procedure was adapted for the analysis of the transition mechanism between the G-BrU base-pairs.

In addition, to confirm the validity of the above procedure based on MP2/6-31G(d,p), we preformed the same calculations employing the meta-hybrid DFT method based on generalized gradient approximation. The M06 functional [14] and the 6-311++G(d,p) basis-set were employed, because the M06 functional has good performance for calculating activation energies and thermochemical kinetics.

3. Results and discussion

3.1. Optimized structures for wG-T, G-T, G*-T and their TS structures*

We first optimized the structures of wG-T, G-T* and G*-T in gas phase by the MP2/6-31G(d,p) and M06/6-311++G(d,p) methods. The structures optimized, the hydrogen bond lengths between bases and total energies are shown in Figure 1. The structures optimized by the MP2 and M06 methods are planar and almost the same to each other. In addition, the distance between the nitrogen atoms connecting to the sugar parts of DNA duplex is 8.83 (wG-T), 8.92 (G-T*) and 8.98 Å (G*-T), respectively, in the MP2/6-31G(d,p) calculations. These distances are near to those for the Watson-Crick base-pairs of the B-form DNA duplex: 9.04 (G-C) and 9.08 Å (A-T). Therefore, the base-pairs shown in Figures 1a, 1b and 1c are expected to fit into the B-form DNA duplex. In addition, these base-pairs are planar to be stacked to each other in the DNA duplex.

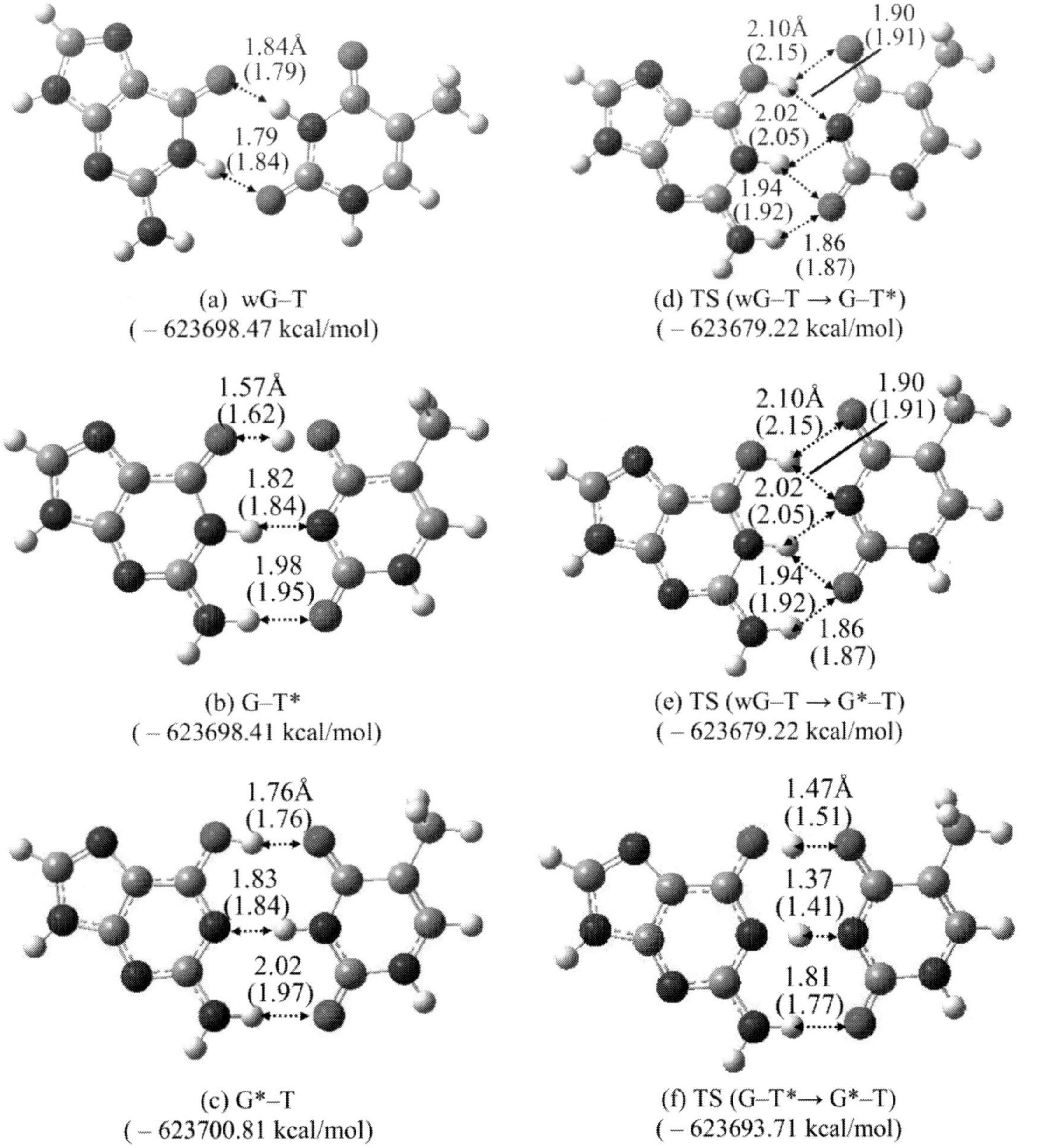

Figure 1. Structures and total energies of (a) wG-T, (b) G-T*, (c) G*-T, (d) TS between wG-T and G-T*, (e) TS between wG-T and G*-T, and (f) TS between G-T* and G*-T optimized by MP2/6-31G(d,p). The distances of hydrogen bonds are shown. The structures optimized by M06/6-311++G(d,p) are almost the same, and the distances of hydrogen bonds are shown in parentheses.

To confirm these structure are energy minimum structure, we performed vibrational analysis using MP2/6-31G(d,p). These structures have no imaginary frequency and six nearly zero frequencies related to the rotation and the translation of the whole structure. In Table 1, the total energy including zero point vibrational energy (ZPE) and free energy are listed. Among the three structures, G*-T is the most stable. Although the energy values obtained by the MP2 and the M06 methods are remarkably different to each other, the relative energies are not so different. The size of ZPE is about 150 kcal/mol, and the entropic effect included in free energy is about 30 kcal/mol. Therefore, their effect should be considered in the comparison of relative stability among the base-pairs.

In the previous DFT study [8], the structures of wG-T, G-T* and G*-T base-pairs were optimized by the B3LYP/6-311++G(d,p) method. The optimized structures and hydrogen-bond lengths between bases obtained by our MP2/6-31G(d,p) and M06/6-311++G(d,p) calculations are shown in Figures 1a, 1b and 1c. These bond lengths are identical to those obtained by the DFT method in the previous work [8] within the error of 0.02 Å. Therefore, the comparison of our present and the previous DFT results shows a good enough consent between them. Hereafter, we will describe mainly the results of MP2/6-31G(d,p) calculations.

To obtain TS structure between wG-T and G-T* base-pairs, we performed the QST2 calculation using the initial structures shown in Figures 1a and 1b and the MP2/6-31G(d,p) or the M06/6-311++G(d,p) method. Figure 1d shows the obtained TS structure, which has only one imaginary (-170 cm-1) frequency, as listed in Table 2. In addition, the IRC analysis starting with the TS structure obtained the reactant and the product structures similar to wG-T and G-T* shown in Figures 1a and 1b. Consequently, the TS structure shown in Figure 1d is confirmed to be a real TS structure connecting between wG-T and G-T* base-pairs. This TS structure is identical to that obtained by the previous B3LYP/6-311++G(d,p) calculation [8] within an error of 0.06 Å in hydrogen bonds.

Table 1. Total energy (TE) (kcal/mol), total energy with zero-point vibrational energy (ZPE), and total free energy (FE) at 298.15 K and 1 atmosphere of pressure for the optimized structures of wG-T, G-T*, G*-T base-pairs, and the transition states between them evaluated by (a) MP2/6-31G(d,p) and (b) M06/6-311++G(d,p) methods.

(a) MP2/6-31G(d,p)

	wG-T	TS(wG-T→G-T*)	G-T*	TS(G-T*→G*-T)	G*-T	TS(wG-T→G*-T)
TE	-623698.5	-623679.2	-623698.4	-623693.7	-623700.8	-623679.2
(Relative energy)	0.0	19.3	0.1	4.8	-2.3	19.3
TE + ZPE	-623550.0	-623532.0	-623550.0	-623548.7	-623552.1	-623532.0
(Relative energy)	0.0	18.0	0.0	1.3	-2.1	18.0
FE	-623580.2	-623560.7	-623578.8	-623576.9	-623581.2	-623560.7
(Relative energy)	0.0	19.4	1.3	3.3	-1.0	19.4

(b) M06/6-311++G(d,p)

	wG-T	TS(wG-T→G-T*)	G-T*	TS(G-T*→G*-T)	G*-T	TS(wG-T→G*-T)
TE	-625243.1	-625226.4	-625243.6	-625237.7	-625244.7	-625226.4
(Relative energy)	0.0	16.6	-0.6	5.3	-1.6	16.6
TE + ZPE	-625097.5	-625081.4	-625097.7	-625095.2	-625098.7	-625081.4
(Relative energy)	0.0	16.1	-0.2	2.3	-1.2	16.1
FE	-625127.8	-625109.9	-625126.7	-625125.2	-625128.9	-625109.8
(Relative energy)	0.0	17.9	1.0	2.6	-1.1	17.9

Table 2. Vibrational frequencies (cm-1) of the lowest 10 modes for the TS structures between wG-T, G-T* and G*-T base-pairs evaluated by MP2/6-31G(d,p).

Mode	wG-T →G-T*	wG-T →G*-T	G-T* →G*-T
1	-170.8	-170.8	-900.0
2	21.2	21.2	32.4
3	28.6	28.5	37.7
4	76.0	76.0	81.8
5	100.8	100.8	110.5
6	107.0	107.0	115.4
7	133.8	133.8	135.1
8	146.8	146.8	139.4
9	171.0	171.0	155.0
10	173.2	173.2	184.2

We furthermore investigated the change in structure during the transition according to the path obtained by the IRC analysis. In the wG-T base-pair, G and T bases are displaced each other and have only two hydrogen bonds. To obtain the G-T* structure having three hydrogen bonds, some alterations of hydrogen bonds are necessary. As a first step of the transition, the proton bound to the nitrogen atom of T in wG-T is transferred to the G side, making a weak hydrogen bond between the oxygen atoms of G and T bases. At the same time, both G and T rotate to the anti-clock direction on the plane shown in Figure 1a. As a result, two weak hydrogen bonds between the nitrogen atoms of G and T and between the NH2 group of G and the oxygen atom of T are made as shown in Figure 1d. In this structure, three protons exist on the G side, so that G is a proton donor, while T is a proton acceptor. Moreover, the two oxygen atoms of G and T come closer to each other and finally the proton is transferred from G to T, leading to the G-T* structure shown in Figure 1b. Therefore, it is elucidated that the proton between the oxygen atom of G and the nitrogen atom of T in wG-T (Figure 1a) moves frequently between G and T bases during the transition from wG-T to G-T*. On the other hand, the other proton between G and T in wG-T remains at the G base to bind strongly to the nitrogen atom of G. This fact is likely to be related with the result that there is no direct path between wG-T and G*-T, as will be shown below.

As for the transition structure between wG-T and G*-T, the previous B3LYP calculation [8] predicted the non-planar TS structure. To elucidate whether more stable and planar TS structure exists or not, we performed the QST2 calculation using the initial structures of wG-T and G*-T shown in Figures 1a and 1c. The TS structure obtained (Figure 1e) is planar and almost identical to the TS structure between wG-T and G-T*. This result is greatly different from the TS structure obtained by the previous DFT study [8]. We consider that this result may come from the close similarity of G-T* and G*-T structures (Figures 1b and 1c) in the comparison with the wG-T structure (Figure 1a). In fact, the positions of two protons contributing to the hydrogen bonds between the G and T bases are slightly different (about 0.1 Å), while the positions of the other atoms are the same to each other. In the QST2 calculation, only the structures of the reactant and the product are used for proposing the TS structure. Accordingly, the QST2 results for the transition states between wG-T and G-T* and between wG-T and G*-T are almost the same each other.

In order to search for the other TS structures between wG-T and G*-T, we furthermore performed QST3 calculations using the structures of wG-T and G*-T as well as the candidate TS structure. In fact, we constructed the six types of the candidate structures for TS based on the structure (Figure 1d) of the TS between wG-T and G*-T obtained by the QST2 calculation. To obtain the structure similar to G*-T, three protons contributing to the hydrogen bonds were shifted. By using these initial structures, we performed the six types of QST3 calculations, three of them obtained the converged structure. However, the obtained TS structures are almost the same to the TS structure between wG-T and G-T* shown in Figure 1d. We also confirmed that this TS structure has only one imaginary frequency as listed in Table 2. Consequently, it can be concluded that the TS between wG-T and G*-T is almost the same as that between wG-T and G-T*. This result is completely different from that obtained by the previous B3LYP/6-311++G(d,p) calculation [8], which predicted that the activation energy from wG-T to G*-T to be 39.21 kcal/mol is much larger than that (17.51 kcal/mol) for the wG-T to G-T* transition. In contrast, our QST2 and QST3 calculations based on MP2/6-31G(d,p) and M06/6-311++G(d,p) revealed that the activation energies for the transitions of wG-T → G-T* and wG-T → G*-T are the same to be 19.4 (MP2) and 17.9 kcal/mol (M06). Therefore, in the view point of the activation energy, our obtained TS structure for the transition between wG-T and G*-T is more preferable than that obtained by the previous DFT study [8]. In addition, our predicted TS structures are planar to be fitted into the DNA duplex, while the TS structure obtained by the previous study is non-planar. It is thus concluded that our predicted TS structure is more realistic in the view points both of energy and structure.

To confirm that the TS structure (Figure 1e) for the wG-T → G*-T transition can connect the wG-T and G*-T structures or not, we performed the IRC analysis. The analysis obtained the structures of wG-T and G-T*, indicating that there is no direct transition between wG-T and G*-T base-pairs. As mentioned above, the proton bonded to the nitrogen atom of G in wG-T (Figure 1a) remains to strongly bind to the same atom during the transition from wG-T to the TS of wG-T → G-T*, while the other proton contributing to the hydrogen bonds between G and T bases moves easily between the bases. This means that G in wG-T can not release the proton bonded to the nitrogen to tautomerize into G*. Consequently, wG-T can not be transferred directly into the tautomeric G*-T.

Finally, to elucidate the transition mechanism between G-T* and G*-T base-pairs, we performed QST3 calculation based on MP2/6-31G(d,p) and M06/6-311++G(d,p) methods. The obtained TS structure and its vibrational property are shown in Figure 1f and Table 2, respectively. This structure has one imaginary (-900 cm-1) frequency to be confirmed a TS structure. In addition, by the IRC analysis starting with the TS structure, we confirmed that the TS structure connects the structures of G-T* and G*-T. It should be noted that during the transition, there are antiparallel migrations of two protons between the oxygen atoms of each base and between the nitrogen atoms of each base.

As mentioned above, it is not an easy task to search for correct TS structures. However, finding of the correct TS structure between a reactant and its product is critically important for predicting the kinetic constant of the chemical and biochemical reactions.

3.2. Optimized structures for G-BrU, G-BrU, G*-BrU and their TS structures*

By the same procedure as for the wG-T base-pair, we optimized the structures of wG-BrU, G-BrU* and G*-BrU base-pairs in gas phase. The structures and the hydrogen bond distances optimized by the MP2/6-31G(d,p) or M06/6-311++G(d,p) method are shown in Figures 2a, 2b and 2c. The effect of substitution of CH_3 group of T by Br is only at most 0.02 Å displacement of protons contributing to the hydrogen bonds. The total energies and free energies for wG-BrU, G-BrU* and G*-BrU are listed in Table 3. G*-BrU is the most stable among the base-pairs. In the previous DFT study [9], the structures of wG-BrU, G-BrU* and G*-BrU base-pairs were optimized by the B3LYP/6-311++G(d,p) method. The lengths of hydrogen bonds between G and BrU bases are identical to those obtained by our present MP2 and M06 calculations within the difference of 0.02 Å. Therefore, the structures of these G-BrU base-pairs obtained by both the MP2 and M06 methods are confirmed to be comparable to the previous B3LYP results [9].

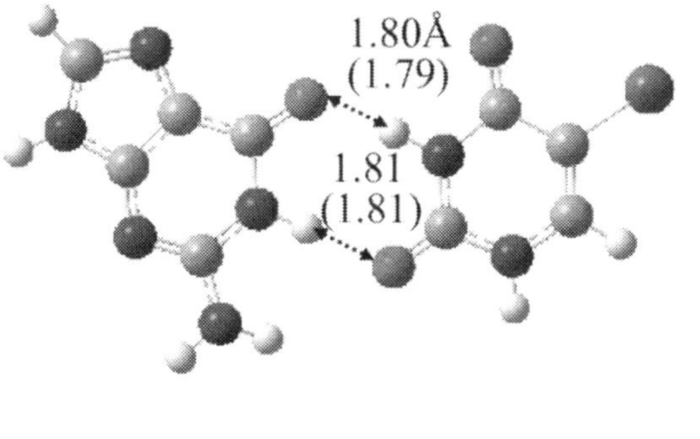

(a) wG–BrU
(– 2211433.79 kcal/mol)

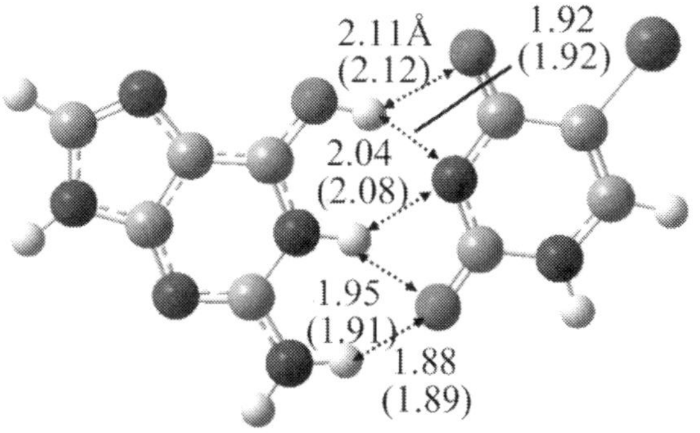

(d) TS (wG–BrU → G–BrU*)
(– 2211417.11 kcal/mol)

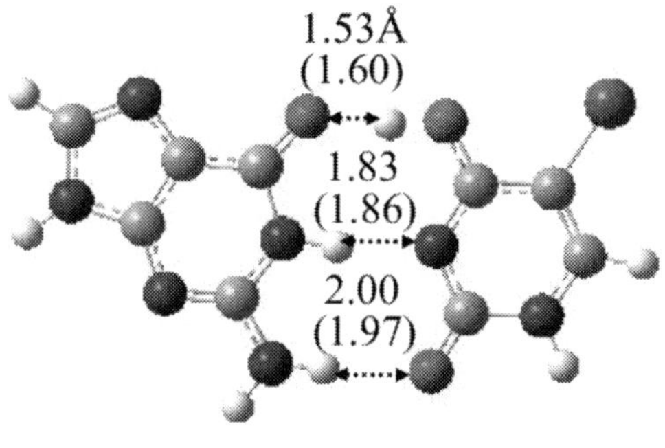

(b) G–BrU*
(– 2211433.18 kcal/mol)

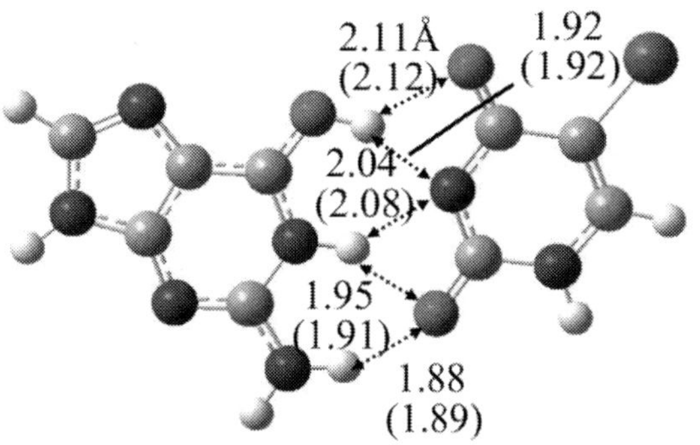

(e) TS (wG–BrU → G*–BrU)
(– 2211417.11 kcal/mol)

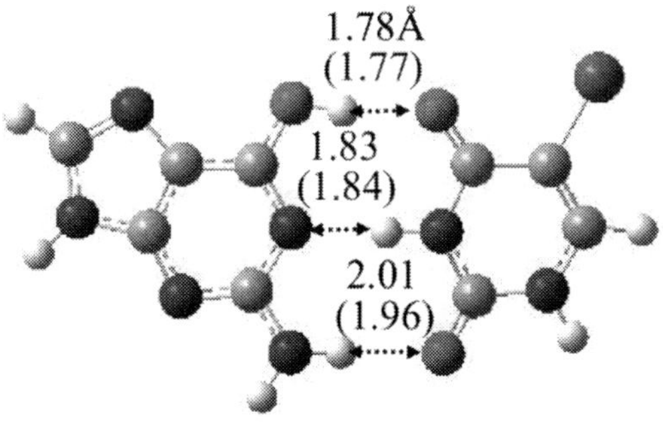

(c) G*–BrU
(– 2211435.62 kcal/mol)

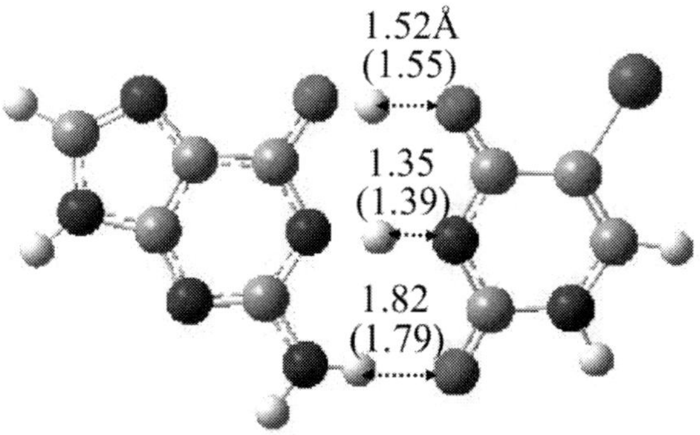

(f) TS(G–BrU*→G*–BrU)
(– 2211429.70 kcal/mol)

Figure 2. Structures and total energies of (a)wG-BrU, (b)G-BrU*, (c)G*-BrU, (d)TS between wG-BrU and G-BrU*, (e)TS between wG-BrU and G*-BrU, and (f)TS between G-BrU* and G*-BrU optimized by MP2/6-31G(d,p). The distances of hydrogen bonds are shown. The structures optimized by M06/6-311++G(d,p) are almost the same, and the distances of hydrogen bonds are shown in parentheses.

Table 3. Total energy (TE) (kcal/mol), total energy with zero-point vibrational energy (ZPE), and total free energy (FE) at 298.15 K and 1 atmosphere of pressure for the optimized structures of wG-BrU, G-BrU*, G*-BrU base-pairs, and the transition states between them evaluated by (a) MP2/6-31G(d,p) and (b) M06/6-311++G(d,p) methods.

(a) MP2/6-31G(d,p)

	wG-BrU	TS (wG-BrU→G-BrU*)	G-BrU*	TS (G-BrU*→G*-BrU)	G*-BrU	TS (wG-BrU→G*-BrU)
TE	-2211433.8	-2211417.1	-2211433.2	-2211429.7	-2211435.6	-2211417.1
(Relative energy)	0.0	16.7	0.6	4.1	-1.8	16.7
TE + ZPE	-2211309.8	-2211294.2	-2211309.4	-2211308.9	-2211311.5	-2211294.2
(Relative energy)	0.0	15.6	0.4	0.9	-1.7	15.6
FE	-2211340.7	-2211323.7	-2211339.0	-2211337.9	-2211341.2	-2211323.7
(Relative energy)	0.0	17.0	1.7	2.9	-0.5	17.0

(b) M06/6-311++G(d,p)

	wG-BrU	TS (wG-BrU→G-BrU*)	G-BrU*	TS (G-BrU*→G*-BrU)	G*-BrU	TS (wG-BrU→G*-BrU)
TE	-2215424.7	-2215410.2	-2215425.1	-2215420.2	-2215425.8	-2215410.2
(Relative energy)	0.0	14.5	-0.4	4.5	-1.1	14.5
TE + ZPE	-2215302.7	-2215288.8	-2215302.9	-2215301.0	-2215303.7	-2215288.8
(Relative energy)	0.0	13.9	-0.2	1.7	-1.0	13.9
FE	-2215333.7	-2215318.7	-2215332.9	-2215330.3	-2215332.9	-2215318.6
(Relative energy)	0.0	15.1	0.9	3.4	0.9	15.1

The TS structure between wG-BrU and G-BrU* obtained by the QST2 calculations based on MP2 and M06 methods is shown in Figure 2d, which has only one imaginary frequency (-154 cm-1) as listed in Table 4. In addition, the IRC calculation starting with the TS structure obtained the structures of wG-BrU and G-BrU*. Therefore, the structure of Figure 2d is confirmed to be a TS structure between wG-BrU and G-BrU*. This TS structure is identical to that obtained by the previous B3LYP/6-311++G(d,p) calculation [9] within a difference of 0.06 Å in hydrogen bonds. The mechanism of transition between wG-BrU and G-BrU* is almost the same as that between wG-T and G-T*, which was described in detail in the section 3.1.

In the same way, we obtained the TS structure between wG-BrU and G*-BrU. As shown in Figure 2e and Table 4, this TS structure and its vibrational properties are almost the same as those between wG-BrU and G-BrU*. It should be noted that there is no direct path between wG-BrU and G*-BrU as for the transition between wG-T and G*-T. The TS structure between G-BrU* and G*-BrU was obtained by the QST3 calculation. The structure and its vibrational properties are shown in Figure 2f and Table 4, respectively. This structure has only one imaginary frequency (-999.5 cm-1) and was confirmed to be a TS structure between G-BrU* and G*-BrU by the IRC analysis. This TS structure is identical to the TS structure obtained by the previous B3LYP/6-311++G(d,p) calculation [9] within the difference of 0.03 Å in hydrogen bonds. Therefore, it was elucidated that the TS structures between wG-T and tautomeric G-T base-pairs are not affected by the introduction of Br atom into T base.

Table 4. Vibrational frequencies (cm-1) of the lowest 10 modes for the TS structures between wG-BrU, G-BrU* and G*-BrU base-pairs evaluated by MP2/6-31G(d,p).

Mode	wG-BrU →G-BrU*	wG-BrU →G*-BrU	G-BrU* →G*-BrU
1	-154.1	-154.0	-999.5
2	17.4	17.4	27.1
3	28.3	28.3	37.5
4	68.8	68.8	71.8
5	77.5	77.7	89.2
6	94.5	94.5	96.2
7	133.8	133.8	118.5
8	155.7	155.7	138.1
9	170.6	170.6	182.3
10	183.1	183.1	183.9

3.3. Free energy surfaces for G-T and G-BrU base-pairs

Schematic free energy diagrams of the transition in gas phase between wG-T, G-T* and G*-T base-pairs are shown in Figure 3a. The relative free energies (kcal/mol) evaluated by MP2/6-31G(d,p) are shown, and the M06/6-311++G(d,p) results are shown in parentheses. The activation energy for the wG-T→G-T* transition is evaluated to be 19.4 and 17.9 kcal/mol by the MP2 and M06 methods, respectively. These values are similar to that (17.51 kcal/mol) obtained by the previous B3LYP/6-311++G(d,p) calculation [8]. As for the activation energy for the G-T*→G*-T transition, the B3LYP calculation obtained 0.73 kcal/mol, while our results obtained 2.0 (MP2) and 1.6 kcal/mol (M06), respectively. Therefore, our MP2 and M06 calculations result in the larger activation energy between G-T* and G*-T base-pairs, in comparison with the previous B3LYP calculation. The evaluated activation energy by our calculations is larger than 800 K, so that the transition between G-T* and G*-T seems improbable at a room temperature.

It is noted that the activation energy for the wG-T→G*-T transition is the same as that for the wG-T→G-T* transition in our MP2 and M06 calculations. This fact is a new finding different from the previous B3LYP result [8] and seems reasonable in considering almost the same structures of G-T* and G*-T in the comparison with wG-T. The energy barrier of the double proton transfer (DPT) between the Watson-Crick G-C base pair has been predicted at various MO calculations to be in the range between 15 and 19 kcal/mol [15]. Accordingly, our calculated results indicate that the reactions between wG-T and tautomeric G-T base-pairs are energetically accessible as the DPT between G-C base-pair.

In order to reveal the effect of BrU on the isomerization reaction between wG-T, G-T* and G*-T base-pairs, we investigated the schematic free energy diagrams of the transition in gas phase between wG-BrU, G-BrU* and G*-BrU base-pairs. As indicated in Figure 3b, the activation energy for the wG-BrU→G-BrU* transition is evaluated to be 17.0 and 15.1 kcal/mol by MP2/6-31G(d,p) and M06/6-311++G(d,p), respectively. These values are similar to that (15.35 kcal/mol) obtained by the previous B3LYP/6-311++G(d,p) calculation [9]. Therefore, it was elucidated that the activation energy for the transition between the wG-T and G-T*/G*-T base-pairs is reduced about 2.4-2.8 kcal/mol by the introduction of BrU. The activation energy for the G-BrU*→G*-BrU transition is evaluated to be 1.2 (MP2) and 2.5 kcal/mol (M06), respectively. From the comparison of the results for the G-T and G-BrU base-pairs, we can conclude that the introduction of BrU base into base-pair brings forward the transition reaction from wG-T to the tautomeric G-T* and G*-T. Since G* is rather stable and pairs with T instead of C, T is introduced into the site complementary to G at the time of DNA replication. As a result, a mutation from G-C to A-T base-pair is generated at the subsequent duplex formation. The present theoretical study reveals a possible origin for the spontaneous mutation induced by the introduction of BrU into wobble G-T base-pair.

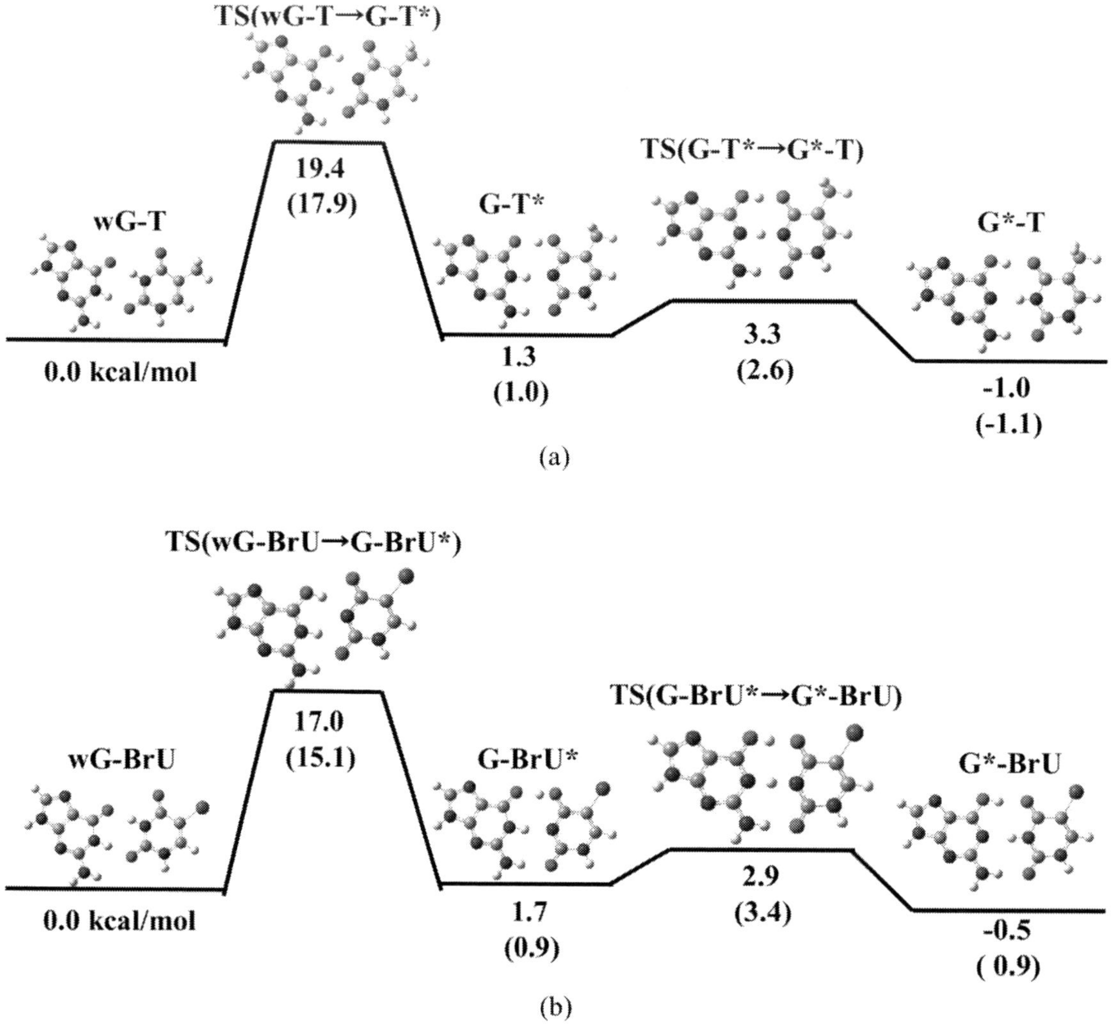

Figure 3. Schematic free energy diagrams of the transition in gas phase between (a) wG-T, G-T* and G*-T base-pairs, and (b) wG-BrU, G-BrU* and G*-BrU base-pairs. The relative free energies (kcal/mol) are evaluated by MP2/6-31G(d,p) at 298.15 K and 1 atmosphere of pressure. The results obtained by M06/6-311++G(d,p) are shown in parentheses.

4. Conclusions

It is experimentally known that 5-bromouracil (BrU) induces spontaneous mutations in DNA sequence. We here investigated the effect of BrU on the transition reactions between wobble G-T (wG-T) and tautomeric G-T base-pairs by *ab initio* molecular orbital and DFT calculations. The evaluated activation energy between wG-BrU and its tautomeric G-BrU base-pairs is smaller than that between wG-T and its tautomeric G-T base-pairs, indicating that the introduction of BrU into base-pair accelerates the transition reaction producing the tautomeric G* and T* bases. Because the tautomerized G* prefers to bind to T rather than to C, the BrU introduction is expected to enhance the spontaneous mutation from C to T base.

Acknowledgments

This work was supported in part by grants from the JSPS Grant-in-Aid for Challenging Exploratory Research (No. 22650061), the Murata Science Foundation, the Iketani Science and Technology Foundation, the Tatematsu Foundation and the CASIO Science Promotion Foundation.

References

[1] Crick F H C 1967 *Proceedings of the Royal Society B* **167** 331-347
[2] Topal M D and Fresco J R 1976 *Nature* **263** 285-289
[3] Early T A, Olmsted III J, Keams D R and Lezius A G 1978 *Nucleic Acids Research* **5** 1955-1970
[4] Hunter W N, Brown T, Kneale G, Annad N N, Rabinovich D and Kennard O 1987 *Journal of Biological Chemistry* **262** 9962-9970
[5] Brown T, Kennard O, Kneale G and Rabinovich D 1985 *Nature* **315** 604-606
[6] Ho P S, Frederick C A, Quigley G J, van der Marel G A, van Boom J H, Wang A H -J and Rich A 1985 *EMBO Journal* **4** 3617-3623
[7] Padermshoke A, Katsumoto Y, Masaki R and Aida M 2008 *Chemical Physics Letters* **457** 232-236
[8] Brovarets O O and Hovorun D M 2009 *Ukrainica Bioorganica Acta* **2** 12-18
[9] Brovarets O O and Hovorun D M 2009 *Ukrainica Bioorganica Acta* **2** 19-23
[10] Brovarets O O and Hovorun D M 2010 *Bulletin of University of Kyiv Series: Physics & Mathematics* **N2** 239-242.
[11] Frisch M J, et al. 2009 *Gaussian 09, Revision A.1* Gaussian Inc. Wallingford CT
[12] Peng C and Schlegel H B 1993 *Israel Journal of Chemistry* **33** 449-454
[13] Gonzalez C and Schlegel H B 1989 *Journal of Chemical Physics* **90** 2154-2161
[14] Zhao Y and Truhlar D G 2008 *Theoretical Chemical Accounts* **120** 215-241
[15] Gorb L, Podolyan Y, Dziekonski P, Sokalski W A and Leszczynski J 2004 *Journal of American Chemical Society* **126** 10119-10129

The Irago Conference 2012 IOP Publishing
Journal of Physics: Conference Series **433** (2013) 012036 doi:10.1088/1742-6596/433/1/012036

Nitrate removal properties of solid-phase denitrification processes using acid-blended poly(L-lactic acid) as the sole substrate

T Yamada[1*], H Matsuoka[1], J Sun[1], S Yoshikawa[2], H Tsuji[1], and A Hiraishi[1]

[1]Department of Environmental and Life Sciences, Toyohashi University of Technology, Toyohashi 441-8580, Japan

[2] Toyo Seikan Group Corporate R&D, 240-0062, Kanagawa, Japan

*Corresponding author. E-mail: tyamada@ens.tut.ac.jp

Abstract. The large amount of waste that is discharged along with the diffusion of poly(L-lactic acid) (PLLA) articles in use is persistent concern. Previously, we studied solid-phase denitrification (SPD) processes using PLLA to establish an effective re-use of PLLA waste. We found that PLLA with a weight-average molecular weight (M_w) of approximately 10,000 was suitable for SPD processes; however, the recycling of PLLA waste consumes a high energy. A new PLLA plastic including 5% poly(ethylene oxalate) (PEOxPLLA) as a blend material has attracted attention because recycling of PEOxPLLA consumes less electricity than that of PLLA. In this study, our main objectives were to evaluate whether PEOxPLLA can be used for SPD processes by changing its M_w and to investigate the bioavailability for denitrification of hydrolysates released from PEOxPLLA. The predicted hydrolysates, including oxalic acid, ethylene glycol, and lactate, are abiotically released, leading to different biological nitrate removal rates. Consequently, the nitrate removal rate of PEOxPLLA ranged from 0.9–4.1 mg-NO_3^--N·g-MLSS·h^{-1} by changing the M_w in the range of 8,500–238,000. In culture-dependent approaches, denitrifying bacteria using each substrate as an electron donor are found in activated sludge, suggesting that all hydrolysates functioned in the SPD processes using PEOxPLLA.

1. Introduction

Poly(L-lactic acid) (PLLA) has attracted much attention as a biodegradable plastic material that can be employed for general use, as an alternative to petrochemical plastics. The diffusion of PLLA plastics in use would bring about a discharge of large amount of PLLA waste, which calls for an effective way of using the waste as an earth-friendly resource circulation system. A promising method for this subject involves the use of PLLA instead of externally added soluble substrate (e.g., methanol) as reducing power for denitrification during nitrogen removal from wastewater.

Thus far, biological denitrification with biodegradable plastics, such as poly (3-hydroxybutyrate) (PHB) [1,2,3], poly [(R)-3-hydroxybutyrate-*co*-hydroxyvalerate] (PHBV) [4,5], and poly (ε-caprolactone) (PCL) [6,7], termed solid-phase denitrification (SPD) [8], has been intensively studied. However, much less effort has been directed toward the use of PLLA in SPD processes, because of its relatively low biodegradability. In fact, nitrate removal during SPD processes with commercially available PLLA has been shown to be much lower than that with PHBV or PCL [8]. For practical use of PLLA as the source of electron donor in the SPD processes, it is particularly important that PLLA be modified for enhancing hydrolysis and degradability. One of the effective and simple ways to achieve this is the adjustment of the weight-average molecular weight (M_w) to enhance physicochemical hydrolysis and degradability of PLLA. Our previous study has shown that good nitrate removal efficiency takes place with lactate abiotically released from the PLLA when its M_w is adjusted to 10,000 [9]. However, a persistent problem is the concern that the adjustment of M_w requires large energy consumption. A new PLLA plastic including 5% poly(ethylene oxalate) (PEOxPLLA) as a blend material has attracted much interest as an environmentally-friendly PLLA resin. Physicochemical hydrolysis and degradation of PEOxPLLA are faster than those of PLLA in the environment, because oxalate and ethylene glycol are first released, enhancing the hydrolysis PLLA region in PEOxPLLA materials. Considering the physicochemical characteristics, PEOxPLLA can be made with less energy, and serve as an electron donor for SPD processes. However, little information has so far been available on SPD processes with PEOxPLLA.

The main objectives of this study were to evaluate whether PEOxPLLA can serve as a solid substrate for denitrification by changing its M_w, and to investigate whether the oxalate, ethylene glycol, and lactate released from PEOxPLLA could be utilized as electron donors by denitrifying bacteria. Also, viable denitrifying bacteria that can utilize each hydrolysate were measured and compared with the total cell count and the total viable count in activated sludge.

2. Materials and methods

Preparation and characterization of PEOxPLLA For the production of poly(ethyleneoxalate) (PEOx), 354 g (3.0 mol) of dimethyl oxalate, 223.5 g (3.6 mol) of ethylene glycol, and 0.30 g of tetrabutyl titanate were introduced into a 1 L separable flask equipped with a mantle heater, a stirrer, a nitrogen inlet, and a condenser. The flask was heated under nitrogen stream until the inside temperature increased from 110 to 170°C, while methanol was distilled off. After the reaction proceeded for 9 h, 210 mL of methanol was distilled off at the end of the reaction. Thereafter, stirring was performed for 1 h at an inside temperature of 150°C and at a reduced pressure of 0.1 to 0.5 mmHg. After the reaction proceeded for 7 h at an inside temperature of 170 to 190°C, the PEOx products were taken out from the flask. Oval PEOxPLLA pellets, measuring 5 mm in length and 4 mm in width and consisting of poly(lactic acid)/poly(ethylene oxalate) (95/5% [v/v]), were melted and blended at 200°C using a twin-screw extruder (ULT Nano 05-20AG, Technovel Corporation, Osaka, Japan). The M_w of the PEOxPLLA pellets was approximately 238,000 (M_w/number-average molecular weight (M_n)=2.0). The pellets were autoclaved at 121 or 130°C under high pressure of 1.05 or 1.9 kgf cm^{-2} to decrease their M_w (Table 1).

Table 1 Autoclave conditions for production of PEOxPLLA and PLLA having different M_w applied for SPD processes

Biodegradable plastics	Temperature (℃)	Pressure (kgf cm $^{-2}$)	Time (min)	M_n
PLLA	120	1.05	0	193,300
			120	45,100
			240	14,100
			270	11,500
			300	9,900
			360	7,900
			480	5,300
			600	3,800
PEOxPLLA	121	1.05	0	238,000
			5	99,600
			15	43,000
			30	28,400
			120	15,500
	130	1.9	60	8,500
			90	5,000

The M_w of the PEOxPLLA pellets was determined by HLC-8120 gel permeation chromatography (GPC) (Tosoh, Tokyo, Japan) with a TSK gel Super HM-H column (Tosoh). For comparison, oval PLLA pellets (5 mm long and 4 mm wide) which had an initial M_w of approximately 238,000 (M_w/number-average molecular weight (M_n)=1.74), were also autoclaved at 121°C under high pressure of 1.05 kgf cm $^{-2}$. The M_w of the autoclaved PLLA pellets was determined by GPC (Tosoh, Tokyo, Japan) as described [10].

Measurement of hydrolysis release rate and nitrate removal rate The hydrolysate release rate was measured by monitoring the hydrolysates (oxalate, ethylene glycol, and lactate) from PEOxPLLA pellets having differentM_w . This measurement was performed using vials containing 50 ml RM2 medium [11] supplemented with 20 mM KNO_3 and PEOxPLLA pellets (1% [v/v]). The concentration of oxalate and lactate was determined by HPLC with an organic acid column (Waters, Milford, MA, USA), as described [4]. Ethylene glycol was measured with a GC353B gas chromatograph (GL science, Tokyo, Japan) equipped with a CP-porabond Q fused silica column (Crawford Scientific, Scotland, UK) and a flame ionization detector. To measure the nitrate removal rate with PEOxPLLA having different M_w, an activated sludge sample was collected from the main aerobic treatment tank of a sewage treatment plant in Toyohashi, Japan. The sludge sample was washed twice with 50 mM phosphate buffer (pH 7.0) and subjected to the nitrate removal assay. The washed sludge suspension at a final concentration of 2,000 mg L^{-1} was introduced into rubber-plugged test tubes (75 mL capacity) containing 50 mL RM2 medium supplemented with 20 mM KNO_3 and PEOxPLLA pellets (1% [v/v]). Nitrate removal tests were also conducted using the same medium with 20 mM oxalate, 30 mM ethylene glycol, and 20 mM lactate to determine the bioavailability of each hydrolysate. The nitrate removal rate was measured by monitoring the change in the concentration of nitrate by ion chromatography as described previously [4,9].

Numeration of total cell count, total viable count, and culturable denitrifying bacteria The total cell count in the activated sludge samples was measured by the SYBR Green-based total cell counting method [5]. Using the 1/10 TSA agar-medium (tryptone, 1.7g·L^{-1}; soytone, 0.3 g·L^{-1}; glucose, 0.25 g·L^{-1}; NaCl, 0.5 g·L^{-1}; K_2HPO_4, 0.25 g·L^{-1}; agarose, 15 g·L^{-1}, pH7.0), diluted sludge samples were plated and incubated under aerobic conditions at 25°C for 1 week to measure the total viable count in the activated sludge. Culturable denitrifying bacteria in the sludge were measured by a pour-plating method with agar medium containing mineral base RM2 [11], 20 mM KNO_3, and 1.8 % (w/v) agar as the basal medium, to which 20 mM oxalate, 30 mM ethylene glycol, or 20 mM lactate as the sole carbon/energy source was added. On the media, diluted sludge samples were plated and cultivated under anaerobic condition at 25°C for 3 weeks. In all cases, plates recovering colonies in a range of 20–200 were taken for counting of colony-forming units (CFU).

3. Results and discussion

PEOxPLLA pellets with different M_w were produced by autoclaving at different temperatures and times and compared with PLLA pellets treated under similar conditions. PEOxPLLA pellets with low M_w could be made with less energy, because the autoclaving time for varying the M_w of the PEOxPLLA pellets was much shorter than that for PLLA (Table 1). This can be explained by inferring that water easily approaches the pore where oxalate and ethylene glycol as the acid-blended materials in PEOxPLLA are released during the first step of autoclaving, thereby stimulating the hydrolysis of the PLLA region in PEOxPLLA.

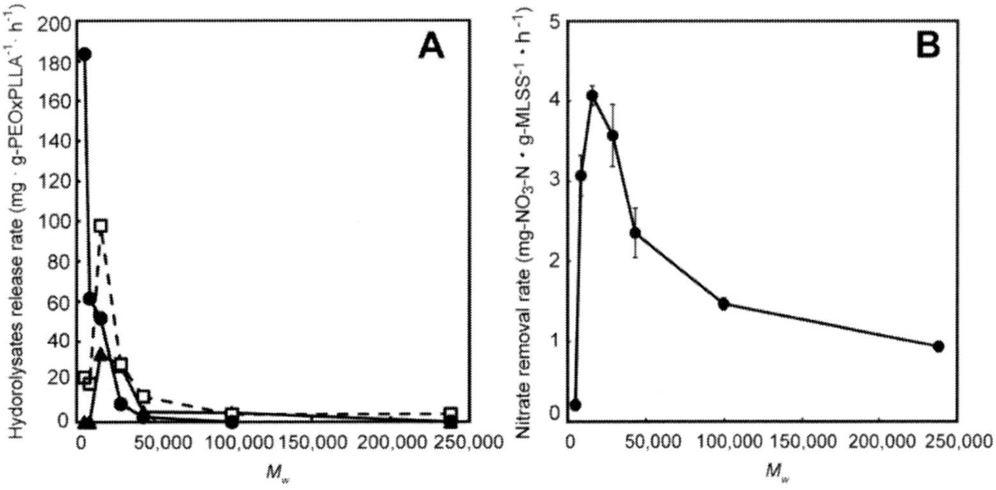

Fig. 1 Changes in abiotic hydrolysate-release rates for PEOxPLLA having different M_w (A) and biological nitrate removal rates using them as the substrate (B). Symbols for hydrolysates (A): oxalate (closed triangles), ethylene glycol (open squares), and lactate (closed circles).

By using PEOxPLLA with different M_w, we measured the abiotic hydrolysate release rate by detecting lactate, oxalate, and ethylene glycol. The lactate release from the PEOxPLLA increased with lower M_w (Fig. 1A) whereas oxalate and ethylene glycol release rates showed different tendencies in terms of M_w. The oxalate release was observed in a range of approximately 15,500 to 100,000 of M_w, but no hydrolysis release occurred in M_w less than approximately 8,500 or more than 100,000 (Fig. 1A). On the other hand, ethylene glycol was released from all M_w-type PEOxPLLA tested (Fig. 1A).

Using PEOxPLLA with different M_w noted above and activated sludge, the nitrate removal rate was measured to confirm the bioavailability of PEOxPLLA as electron donor for SPD processes. The highest nitrate removal rate (4.1 mg-NO_3^--N·g-MLSS·h^{-1}) was shown with the PEOxPLLA with a M_w of approximately 15,500 (Fig. 1B). This nitrate removal rate was similar to that in PLLA-added SPD processes achieved in our previous study [9], suggesting that PEOxPLLA may be utilized as an electron donor for denitrification. Interestingly, SPD using the PEOxPLLA of high M_w (>15,500) still exhibited nitrate removal rates (0.9 to 3.6 mg-NO_3^--N·g-MLSS·h^{-1}), contrasting with our previous finding that no nitrate removal took place with such M_w of PLLA [9]. The nitrate removal rate of the PEOxPLLA with high M_w might result from the use of hydrolysates other than lactate as electron donors (Fig. 1B). These findings suggested that the nitrate removal could be engineeringly controlled with the PEOxPLLA at a range wider than that with PLLA [9].

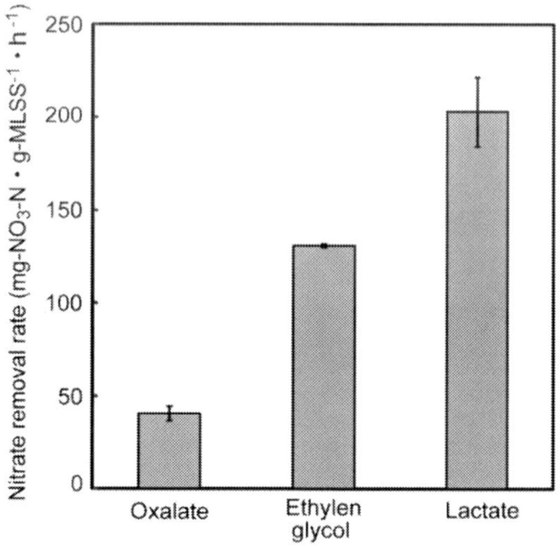

Fig. 2 Comparative nitrate removal rates of activated sludge using liquid substrates that PEOxPLLA is released.

The nitrate removal tests with PEOxPLLA showed that all hydrolysates released could be utilized as electron donors for SPD, although the difference in nitrate removal rates was noted among the substrates used (Fig. 2). To explain this difference, denitrifying bacteria that can utilize oxalate, ethylene glycol, and lactate as the hydrolysates from PEOxPLLA were measured in the seed activated sludge. In the sludge, the total cell count and total viable count were 9.9×10^8 cell·mL^{-1} 4.9×10^7 CFU·mL^{-1}. The plate counts of denitrifying bacteria with oxalate, ethylene glycol, and lactate as the sole substrate were 4.2×10^4, 4.7×10^5, and 2.4×10^6 CFU·mL^{-1}, respectively, accounting for 0.09, 0.9, and 4.9% of the total viable count. These results suggest that the difference of nitrate removal rates was affected by the absolute number of denitrifying bacteria in the seed sludge rather than by the difference in bioavailability of the substrates.

Acknowledgements

This study was supported in part by a Grant-in-Aid for Scientific research (No.21510084) from the Ministry of Education, Culture, Sports, Science, and Technology, Japan. This study was also supported by the Toukai Foundation for Technology and Kurita Water and Environmental Foundation (24083).

References

[1] Bernat K and Wojnowska-Baryla I 2008 *Environ. Technol.* **29** (1) 81–89

[2] Boley A, Muller W -R and Haider G 2000 *Aquacul. Eng.* **22** (1) 75–85

[3] Gibbs B M, Shephard L R, Third K A and Cord-Ruwisch R 2004 *Wat. Sci. Technol.* **50** (10) 181–188

[4] Karn S T, Horiba Y, Yamamoto M and Hiraishi A 2002 *Appl. Environ. Microbiol.* **68** (7) 3206–3214

[5] Karn S T, Horiba Y, Takahashi N and Hiraishi A 2007 *Microbes Environ.* **22** (1) 20–31

[6] Boley A, Mergaert J, Muller C, Lebrenz H, Cnockaert M C, Müller W -R, and Swings J 2003 *Acta Hydrochim. Hydrobiol.* **31** (3)195–203

[7] Horiba Y, Karn S T and Hiraishi A 2005 *Microbes Environ.* **20** (1) 25–33

[8] Hiraishi A and Karn S T 2003 *Appl. Microbiol. Biotechnol.* **61**(2) 103–109

[9] Takahashi M, Yamada T, Tanno M, Tsuji H and Hiraishi A 2011 *Microbes Environ.* **26**(3) 212–219

[10] Tsuji H and Tenzuka Y 2005 *Macromol. Biosci.* **5** 135–148

[11] Hiraishi A and Kitamura H 1984 *Bull. Jpn. Sci. Soc. Fish.* **50** 1929–1937

Relationships between 16S-23S rRNA gene internal transcribed spacer DNA and genomic DNA similarities in the taxonomy of phototrophic bacteria

K. Okamura[1,2*], T. Hisada[1], K. Takata[1], and A. Hiraishi[1,2]

[1] Department of Environmental and Life Sciences and

[2] Electronics-inspired Interdisciplinary Institute (EIIRIS), Toyohashi University of Technology, Toyohashi 441-8580, Japan

E-mail: okamura@recab.tut.ac.jp

Abstract. Rapid and accurate identification of microbial species is essential task in microbiology and biotechnology. In prokaryotic systematics, genomic DNA-DNA hybridization is the ultimate tool to determine genetic relationships among bacterial strains at the species level. However, a practical problem in this assay is that the experimental procedure is laborious and time-consuming. In recent years, information on the 16S-23S rRNA gene internal transcribed spacer (ITS) region has been used to classify bacterial strains at the species and intraspecies levels. It is unclear how much information on the ITS region can reflect the genome that contain it. In this study, therefore, we evaluate the quantitative relationship between ITS DNA and entire genomic DNA similarities. For this, we determined ITS sequences of several species of anoxygenic phototrophic bacteria belonging to the order *Rhizobiales*, and compared with DNA-DNA relatedness among these species. There was a high correlation between the two genetic markers. Based on the regression analysis of this relationship, 70% DNA-DNA relatedness corresponded to 92% ITS sequence similarity. This suggests the usefulness of the ITS sequence similarity as a criterion for determining the genospecies of the phototrophic bacteria. To avoid the effects of polymorphism bias of ITS on similarities, PCR products from all loci of ITS were used directly as genetic probes for comparison. The results of ITS DNA-DNA hybridization coincided well with those of genomic DNA-DNA relatedness. These collective data indicate that the whole ITS DNA-DNA similarity can be used as an alternative to genomic DNA-DNA similarity.

1. Introduction

There is general agreement that taxonomic information about microorganisms is incorporated in their whole genome [1, 2]. The current concept of the bacterial species – a 'genomically coherent' group of strains that share many common traits [3] – depends evidently on the relatedness of bacterial genomes measurable by available molecular techniques. Next-generation sequencing technology has made it possible to relatively rapidly get whole genome sequence information on a microorganism and is giving a new insight into our understanding of the biodiversity and ecology of microorganisms [4, 5]. Nevertheless, genomic DNA-DNA hybridization still remains the 'gold standard' to discriminate bacterial species [1, 2]. Compared to PCR-based techniques, quantitative DNA-DNA hybridization is a laborious and time-consuming technique, because it requires relatively large-scale of cultures to prepare sufficient quantity and high quality of genomic DNA. Also, the DNA-DNA reassociation assay has limitation in quantitativity and reproducibility. This situation calls for alternatives to DNA-DNA hybridization experiments for taxonomic purpose [6-9]

An alternative approach to quantification of genome relatedness is to compare selected DNA sequences that can reliably represent whole genomes of different organisms. An ad hoc committee for the re-evaluation of the species definition in bacteria pointed out that an analysis of at least five genes of diverse chromosomal loci and wide distribution could provide sufficient information to distinguish a bacterial species from related taxa [2]. It is an open question how much information any given gene sequences can reflect the genome that includes them. Sequence differences in a given gene between closely related organisms are presumably low, because continual acquisition of random mutations has resulted from natural selection. On the other hand, such non-encoding regions as spacers on genome are functionally much less important and, therefore, detectable mutations in these regions are mostly neutral. Sequence differences between related organisms are higher in given spacers than in given genes.

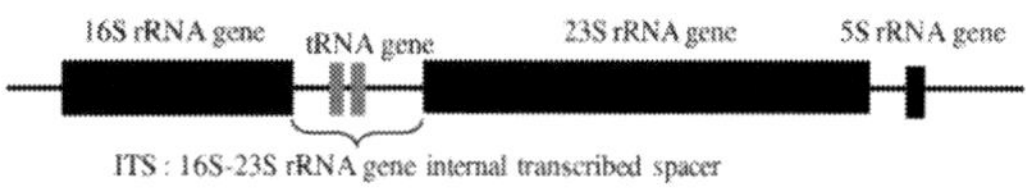

Fig. 1 Schematic structure of bacterial *rrn* operon.

Information on the 16S-23S rRNA gene internal transcribed spacer (ITS) region, in which hypervariable transcribed sequences with a number of deletion and insertion are found among species, has increasingly been used to discriminate bacterial strains at the species and intraspecies levels [10]. In general, multiple copies of the rRNA operon occur per genome, and the ITS region shows variation in length and sequences not only among species but also among different rRNA gene loci within a single organism. The ITS region also includes tRNA genes depending upon the copies and species. Since ITS DNA is easily PCR-amplified with a universal primer set, the experiment can start with single colonies or small-scale cultures of a wide variety of species. The ITS parameters used for taxonomic purpose are direct electrophoretic patterns of PCR products, their restriction fingerprints, and sequences of one or more copies. The ITS sequence analysis is an easy and reproducible tool for the discrimination of closely related strains and species. In *Bradyrhizobium* strains and related organisms, some positive relationships between ITS sequence and genomic DNA-DNA similarities were found, although there was a less correlation between the two parameters in closely related groups [11, 12].

The aim of the present study is to elucidate the relationship between ITS sequence and genomic DNA similarities and to determine whether ITS relatedness can serve as an alternative to genome relatedness in bacterial systematics. For this, we first compared the G+C content of all loci of ITS regions and entire genomic DNAs of prokaryotic organisms whose whole complete genome information has been available. Second, we determined ITS sequences of several species of the phototrophic bacteria belonging to the order *Rhizobiales*, and examined the relationships between pairwise ITS sequence similarities and DNA-DNA relatedness. Finally, we performed ITS DNA-DNA hybridization and genomic DNA-DNA hybridization and evaluated the relationships between the two genetic parameters.

2. Materials and methods

Bacterial strains and cultivation. Thirty-five strains belonging to species of the genera *Afifella*, *Blastochloris*, *Rhodobium*, *Rhodoplanes*, and *Rhodopseudomonas* (*Rps.*) were used in this study (Table 1). The strains with ATCC and DSM numbers were obtained from the American Type Culture Collection, Manassas, U.S.A. and from the Deutsche Sammlung von Milroorganismen und Zellkulturen GmbH, Braunschweig, Germany, respectively. Cultivation was performed anaerobically in screw-cap test tubes or bottles filled with medium [13-18] under incandescent illumination (2,000 lx). Incubation was at 30°C for mesophilic strains and at 40°C for thermotolerant strains [13].

Table 1. Bacterial strains used in this study.

Species	Source or designated as:	Reference
Blastochloris viridis	DSM 133^T, DSM 134	[14]
Blastochloris sulfoviridis	DSM 729^T	[14]
Blastochloris sp.	5 strains from New Zealand hot spring	[15]
Rhodobium orientis	MB312^T, MB303	[16]
Afiffela marina	DSM 2698^T and other 5 strains	[16]
Rhodoplanes elegans	AS130^T	[17]
Rhodoplanes roseus	DSM 5909^T	[17]
Rhodoplanes cryptolactis	DSM 9987^T	[13]
Rhodoplanes sp.	5 strains from New Zealand hot spring	[15]
Rhodopseudomonas palustris	DSM 123^T	[18]
Rhodopseudomonas rhenobasencis	Rb^T	[18]
Rhodopseudomonas faecalis	g-c^T	[18]
Rhodopseudomonas sp.	8 strains from New Zealand hot spring and Japanese pond	[15]

Amplification, cloning and sequencing of the ITS region. Crude cell lysates as a DNA source were prepared for PCR use as described [18]. The ITS region was amplified with two pair sets of PCR primers, SS1512f and LS23r or LS117r [18], which corresponded to positions 1512-1531 of *Escherichia coli* 16S rRNA and positions 21-37 or 117-137 of *E. coli* 23S rRNA, respectively. The ITS region of several *Rhodopseudomonas* strains were not amplified by using primers SS1512f and LS23r. For these strains, we used another reverse primer, LS23r-p (5'-AAGGCATCCGTCGAACA-3'), which was designed based on the conserved region of 23S rRNA gene of the strains. Amplified fragments were separated by agarose gel electrophoresis, purified with the Wizard Gel and PCR Clean-up System (Promega), and subcloned using a pT7Blue Perfectly Blunt ™ Cloning kit (Novagen) according to the manufacturer's instructions. At least 5 clones per strain were subjected to sequencing. Inserts were sequenced using plasmid primers T7 and M13 and a Dye Terminator Cycle Sequencing kit, and analyzed using a PRISM 3100 DNA sequencer (Applied Biosystems).

Sequence analysis. Sequence data were complied with the GENETYX-MAC ver.13 program

(Software Development). The multiple alignment of sequence and the calculation of the corrected evolutionary distance was performed using the CLUSTAL X program [19]. Searching for tRNA genes and predicting the secondary structure of tRNA were also performed using the GENETYX-MAC program.

Comparison between GC contents of ITS and genomic DNA. The guanine + cytosine (G+C) contents of the genomic DNA of 215 prokaryotic organisms whose whole genomes have been completely sequenced were calculated by using the 'GC Plot' program from Genome Information Broker of DDBJ database (http://gib.genes.nig.ac.jp/) [20]. The G+C contents of ITS regions within genomes were calculated using the GENETYX-MAC program. For this, all loci of ITS sequences within genomes were retrieved from Genome Information Broker or Sequence Retrieval System [http://srs.ddbj.nig.ac.jp] of DDBJ. Statistical analyses of all numerical data including calculation of the Pearson correlation coefficient *r* were performed using the Microsoft Excell.

Genomic DNA-DNA hybridization. Genomic DNA was extracted and purified by the method of Marmur [21]. DNA-DNA hybridization studies were performed by the dot-blot hybridization method with alkaline phosphatase labeling and chemiluminescence detection using an Amersham-Pharmacia AlkalPhos kit, as described [18]. Labelling and detection were performed according to the manufacturer's instructions. Hybridizations were performed at 50 °C. Hybridization signals were detected with an Amersham-Pharmacia ECL mini-camera, and their intensity was measured by the NIH image program available at the web site [http://rsb.info.nih.gov/nih-image/].

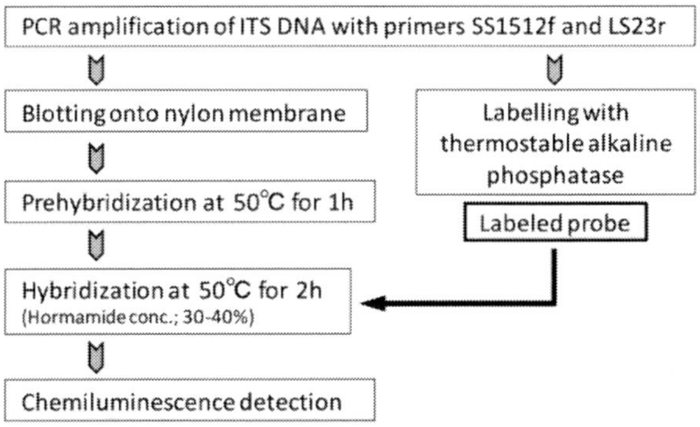

Fig. 2. Simplified procedure for ITS DNA-DNA hybridization.

ITS DNA-DNA hybridization. PCR products of ITS regions were obtained as noted above and detected by agarose gel electrophoresis. In the case that a single band was observed on the gel, the DNA fragment was purified by using the PCR clean up system (Takara). When the non-specific amplification was observed, the DNA was purified with the Wizard Gel and PCR Clean-up System (Promega). All DNA samples were prepared at a concentration of 100 ng/μl. To eliminate the effect of conserved DNA regions, *i.e.*, primer-annealing regions and tRNA genes, on hybridization signals, a PCR product of *Escherichia coli* ITS regions was used as competitor DNA. The competitor DNA was purified and digested with *Dra*I (Takara) and *Mbo*II (Takara) prior to use and blotted onto nylon membrane at a concentration of 50 times (w/w) of a total blotted DNA. The ITS DNA-DNA hybridization assay was performed according to the protocol for genomic DNA-DNA hybridization with slight modifications. The blotted volume of test DNA solutions (10 ng/μl) was 2 μl. Prehybridization was performed at 50°C for 1 h, in the presence of denatured competitor DNA and 30% deionized formamide in hybridization buffer. Then, hybridization was performed at 50°C for 2 h. The hybridized membranes were washed twice at 50°C with primary wash buffer according to the manufacturer's instructions. Hybridization signals were detected as described above for genomic

DNA-DNA hybridization.

3. Results

Relationship between the G+C content of ITS regions and whole genomes. We first studied the relationships between the G+C content of ITS regions and genomic DNAs of 215 prokaryotic organisms including mesophiles and moderately thermophiles of archaea and bacteria (Fig. 3). Although ITS DNA had a lower G+C content than that of genomic DNA in general, there was a strong correlation between the two variables (slope of the regression line =0.614, $r = 0.75$, $P < 0.01$). An exceptional case was that, in hyperthermophilic archaea and bacteria, ITS G+C ratios were higher than genomic G+C ratios (slope of the regression line =0.913, $r = 0.593$, $P < 0.01$). These results suggest that ITS regions can be used as genetic markers reflective of the entire genome.

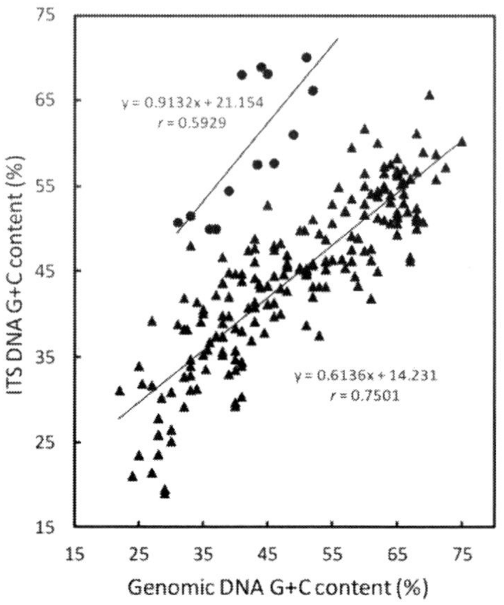

Fig. 3. Correlation between G+C contents of whole ITS DNAs and Genomic DNAs in 215 prokaryotic organisms. Symbols: closed triangles, mesophilic archaea and bacteria; closed circles, thermophilic archaea and bacteria.

ITS sequence analysis. ITS sequences were analyzed for 35 strains of the phototrophic bacteria belonging to the genera *Afifella*, *Blastochloris*, *Rhodobium*, *Rhodoplanes*, and *Rhodopseudomonas*. From most of the test strains, ITS regions could be amplified with a PCR primer set of SS1512f and LS23r. In the case of the New Zealand hot spring and Japanese pond isolates (see Table 1), however, ITS regions were amplified only using a primer set of SS1512f and LS117r. Since most of the test strains gave a single band of PCR amplicons upon agarose gel electrophoresis, direct sequencing of the PCR products without subcloning was possible. Also, the PCR products from these strains were subcloned, and at least five clones per strain were sequenced. As a result, all the clones from a single strain had identical sequences. Therefore, the test bacteria from which a single band of amplicon was produced may have one single type of rRNA operon, although the number of the operon copies was unknown. On the other hand, three different clones of ITS having different sequences were obtained from *Rps. rhenobacensis* strain Rb^T.

The ITS sequences determined varied in length among the test strains as follows: 787 to 839 nucleotides in the *Blastochloris* strains, 724 to 875 nucleotides in the *Rhodobium* strains, 937 to 1021 nucleotides in the *Rhodoplanes* strains, and 761 to 1085 nucleotides in the *Rhodopseudomonas* strains. All of the detected ITS regions contained the $tRNA^{Ile}$ and $tRNA^{Ala}$ genes. The primary and secondary structures of the $tRNA^{Ile}$ and a $tRNA^{Ala}$ predicted were the same in all test strains within a single genus.. Therefore, the ITS-tRNA structures may be used as a genus-specific genetic maker.

An alignment analysis of ITS sequences showed that there were great sequence variations in non-encoding regions with insertions and deletions among the test species. Therefore, it was much more difficult to make correct alignments of ITS sequences than of rRNA gene sequences.

ITS sequence similarities were deduced from these alignments with gaps.

Relationship between pairwise ITS sequence similarities and genomic DNA-DNA relatedness. The relationship between pairwise ITS sequences similarities and DNA-DNA relatedness in the phototrophic species is shown in Fig. 4. Pairwise similarities to the corresponding ITS sequence of *E. coli* and DNA-DNA hybridization levels against *E. coli* were used as the control to make a starting point of the regression line. There is a strong correlation between the two genetic markers (r = 0.8448, P < 0.01). Based on this relationship, 70% DNA-DNA relatedness was found to correspond to a 92% level of the ITS sequence similarity. Also, 60-80% ITS sequence similarities corresponded to 20-65% DNA-DNA similarities in the phototrophic bacteria, although these relationships were different to some extent from species to species. The aforementioned results suggest the usefulness of the ITS sequence similarity with gaps as a criterion for determining genospecies of the phototrophic bacteria.

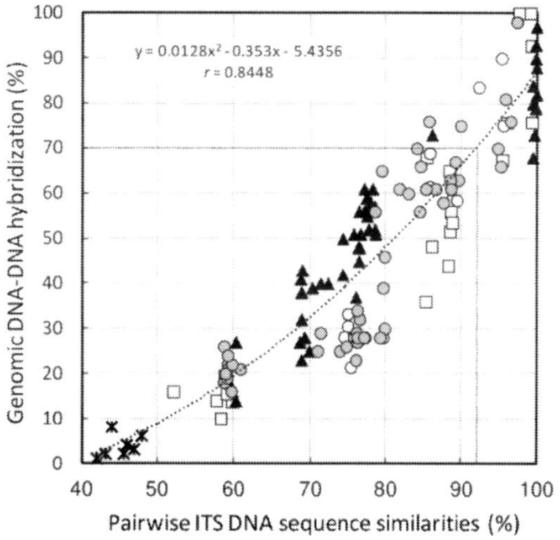

Fig. 4. Correlation between pairwise ITS DNA sequence similarity and genomic DNA-DNA hybridization values in the phototrophic bacteria. Alignment positions of ITS with gaps were taken into consideration to calculate the similarity. Symbols: *Afifella*, *Blastochloris* strains (●), *Rhodopseudomonas* strains (○), *Rhodoplanes* strains (▲), *Rhodobium* strains (□), and *E. coli* strain K-12 vs. phototrophic bacteria.

Relationship between ITS-ITS hybridization and DNA-DNA hybridization values. Finally, we applied the ITS DNA-DNA hybridization method using PCR-amplified DNA from all loci of ITS per genome as well as genomic DNA-DNA hybridization for genetically closely related strains of the phototrophic bacteria within each genus. Figure 5 shows the correlation between pairwise ITS DNA-DNA similarity and genomic DNA-DNA relatedness. A liner positive relationship between the two parameters was observed with a significant correlation coefficient (r=0.900, P < 0.01). Based on this relationship, 70% DNA-DNA relatedness was estimated to correspond to an 82% level of ITS-ITS similarity

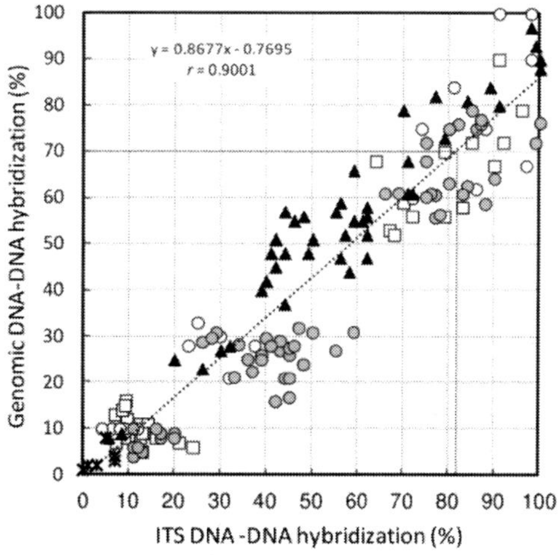

Fig. 5. Correlation between ITS DNA-DNA hybri -dization and genomic DNA-DNA hybridization values in the phototrophic bacteria. Each symbols shows the values of *Affifella* strains *Blastochloris* strains (○), *Rhodopseudomonas* strains (●), *Rhodoplanes* strains (▲), *Rhodobium* strains (□), and *E. coli* strain K-12 vs. phototrophic bacteria.

4. Discussion

The differences in genomic G+C content may have been caused by directional mutation pressure, and such mutation pressure seems to have been exerted uniformly on the entire genome during evolution [22]. However direction and magnitude of this pressure vary among the phylogenetic lines, as the G+C content of bacterial DNA actually ranges from 25 to 75% [23] with low intraspecific heterogeneity [24-27]. Thus, the G+C content of entire genomic DNA is one of the most important characteristics for the description of the bacterial genotype [28]. In this study, the G+C content analysis of the ITS and genomic DNA of 215 prokaryotic species showed that there is a high correlation between the two, this suggesting that the ITS DNA similarities reflects entire genome similarities for taxonomic purpose. Muto and Osawa [29] demonstrated that the G+C content of various parts of the bacterial genome has a positive linear correlation with that of the whole genomic DNA. The G+C ratios of both spacers and protein-encoding genes had a strong liner correlation with genomic G+C contents.

This study has also shown that ITS sequence similarities significantly correlate with DNA-DNA relatedness in species of the four phototrophic genera belonging to the *Rhizobia* (Fig. 2). We have shown that the 92.0% level of ITS sequence similarities corresponds to 70% of DNA-DNA relatedness. Willems *et al.* [11, 12] found that, in *Bradyrhizobium* strains and related organisms, a 95.5% level of ITS sequence similarities corresponded to 60% DNA-DNA relatedness. The G+C contents of entire genome and whole ITS regions in *Affifella*, *Bradyrhizobium*, *Blastochloris*, *Rhodobium*, *Rhodoplanes*, and *Rhodopseudomonas* species, all of which belong to the order *Rhizobiales*, are similar to each other (genomic DNA, 64-66 mol%; ITS, 53.2-56.6 mol%). The mutation pressure for ITS and genomic DNA of these organisms during evolution may be in the same direction, resulting in a high correlation between the two genetic markers.

A single bacterium has usually multilocus rRNA operons including respective different sequences of ITS regions. ITS-based phylogenetic relationships among the organisms may be affected by which ITS copies analyzed are used. Therefore, the organisms under investigation should not be classified by using only an ITS sequence of one locus. Also, it is actually difficult to align all sequences of different ITS copies from different species for phylogenetic analysis, because ITS sequences are hypervariable. An approach to overcome this problem is to estimate the sequence similarity of whole ITS regions directly by the hybridization technique using their PCR products. As reported herein, the results of ITS-ITS hybridization assays were well agreement with those of the conventional DNA-DNA hybridization; 70% of DNA-DNA relatedness as the threshold of species discrimination roughly corresponded to an 82% level of ITS-ITS relatedness. These results indicate that ITS-ITS relatedness can be used as an alternative to genomic DNA-DNA relatedness to discriminate bacterial strains at the species level. An advantage of the ITS-ITS hybridization method is that it can be applied to multi-copies of ITS regions having different sequences.

In conclusion, the ITS-ITS hybridization method using PCR products is more rapid and simpler than the conventional DNA-DNA hybridization technique. This method is applicable to the DNA microarray analysis not only for rapid identification of bacteria but also for characterization of microbial communities. Information about the ITS region has already been applied to microbial community analyses [30-33]. To more definitely evaluate ITS-ITS similarity as a genetic marker, however, further study is necessary with more strains of wide variety of species.

Acknowledgements

This work was supported by a Grant-in-Aid for Young Scientists ((B) 23710091), the Ministry of Education, Culture, Sports, Science and Technology, Japan.

References

[1] Stackebrandt E *et al.* 2002 *Int. J. Syst. Evol. Microbiol.* **52** 1043–1047
[2] Wayne L G *et al.* 1987 *Int. J. Syst. Bacteriol.* **37** 463–464
[3] Rosselló-Mora R and Amann R 2001 *FEMS Microbiol. Rev.* 25 39–67
[4] Trevors J T and Masson L 2010 *Antonie van Leeuwenhoek* **98** 249–262.
[5] Dini-Andreote F *et al.* 2012 *Microb. Ecol.* **64** 1–7
[6] Cho J C and Tiedje J M 2001 *Appl. Environ. Microbiol.* **67** 3677–3682
[7] Coenye T *et al.* 2005 *FEMS Microbiol. Rev.* **29** 147–167
[8] Gevers D *et al.* 2005 *Nat. Rev. Microbiol.* 3 733–739
[9] Goris J et al. 2007 *Int. J. Syst. Evol. Microbiol.* **57** 81–91
[10] Gürtler V and Stanisich V A 1996 Microbiology **142** 3–16
[11] Willems A *et al.* 2001 *Int. J. Syst. Evol. Microbiol.* **51** 623–632
[12] Willems A *et al.* 2003 *Syst. Appl. Microbiol.* **26** 203–210
[13] Okamura K *et al.* 2007 *J. Gen. Appl. Microbiol.* **53** 357–361
[14] Hiraishi A 1997 *Int. J. Syst. Bacteriol.* **47** 217–219
[15] Hisada T *et al.* 2007 *Microbes Environ.* **22** 405–411
[16] Hiraishi A *et al.* 1995 *Int. J. Syst. Bacteriol.* **45** 226–234
[17] Hiraishi A and Ueda Y 1994 *Int. J. Syst. Bacteriol.* **44** 665–673
[18] Okamura K *et al.* 2009 *J. Gen. Appl. Microbiol.* **55** 469–478
[19] Thompson J D *et al.* 1997 *Nucleic Acids Res.* **24** 4876–4882
[20] Fumoto M *et al.* 2002 *Nucleic Acids Res.* **30** 66-68.
[21] Mamur J 1961 *J. Mol. Biol.* **3** 208–218
[22] Sueoka N 1962 *Proc. Natl. Acad. Sci. USA* **48** 582–592
[23] Lee K Y *et al.* 1956 *Ann. Inst. Pasteur* **91** 212–224
[24] Sueoka N *et al.* 1959 *Nature* **183** 1429–1431
[25] Rolfe R and Messelson M 1959 *Proc. Natl. Acad. Sci. USA* **44** 1039–1043
[26] Sueoka N 1961 *Proc. Natl. Acad. Sci. USA* **47** 1141–1149
[27] Schildkraut C L *et al.* 1962 *J. Mol. Biol.* **4** 430–43
[28] Barbu E *et al.* 1956 *Ann. Inst. Pasteur.* **91** 212–224
[29] Muto A and Osawa S 1987 *Proc. Natl. Acad. Sci. USA* **84** 166–169
[30] Gonzalez N *et al.* 2003 *J. Microbiol. Methods.* **55** 91–97
[31] Cook K L *et al.* 2004 *J. Microbial. Methods.* **57** 79–93
[32] Nübel U *et al.* 2004 *FEMS Microbiol. Lett.* **15** 215–223
[33] Gunther S *et al.* 2006 *J. Microbiol. Methods.* **65** 226–236

Future prospect 2012-2025
- How will our business change for the next 10 years -

Sakae Tanaka [1]
[1]President, Aquabit Corporation,
3F, Shuukou Kensetu Bldg., 1-12-14, Nakano-Ku, Tokyo, 165-0026, Japan

*E-mail: sakaet@aquabit.co.jp

Abstract. The purpose of this lecture is to discuss about the "Future". How our business will change in the next 10 years? I believe the key is 3 mega-trends "Sustainability", "Cloud Computing" and "Life Innovation". With the development of social environment, the required business will change, too. The future would be invisible if you shut yourself up in your single industry. It is important to see various business fields horizontally, and recognize various key changes stereoscopically such as demographics, economy, technology, sense of value and lifestyle, when you develop mid-and-long term strategy. "Cloud" is silent, but the revolution of personal computing. It will bring the drastic changes in every industry. It will make "voice" and "moving image" possible to use as the interface to access your computer. Cloud computing will also make the client device more diversified and spread the application range widely. 15 years ago, the term "IT" was equivalent to "personal computer". Recently, it rather means to use smartphone and tablet device. In the next several years, TV and car-navigation system will be connected to broadband and it will become a part of personal computing. The meaning of personal computing is changing essentially year by year. In near future, the universe of computing will expand to the energy, medical and health-care, and agriculture etc. It passed only 20 years since we use "Computer" in a full scale operation. Recently, computer has start understanding our few words and talking in babble like a baby. The history of computing has just started.

Recent Progress of Seismic Observation Networks in Japan

Y Okada[1]

National Research Institute for Earth Science and Disaster Prevention (NIED), 305-0006, Japan

*E-mail: okada@bosai.go.jp

Abstract. Before the occurrence of disastrous Kobe earthquake in 1995, the number of high sensitivity seismograph stations operated in Japan was nearly 550 and was concentrated in the Kanto and Tokai districts, central Japan. In the wake of the Kobe earthquake, Japanese government has newly established the Headquarters for Earthquake Research Promotion and started the reconstruction of seismic networks to evenly cover the whole Japan. The basic network is composed of three seismographs, i.e. high sensitivity seismograph (Hi-net), broadband seismograph (F-net), and strong motion seismograph (K-NET). A large majority of Hi-net stations are also equipped with a pair of strong motion sensors at the bottom of borehole and the ground surface (KiK-net). A plenty of high quality data obtained from these networks are circulated at once and is producing several new seismological findings as well as providing the basis for the Earthquake Early Warning system. In March 11, 2011, "Off the Pacific coast of Tohoku Earthquake" was generated with magnitude 9.0, which records the largest in the history of seismic observation in Japan. The greatest disaster on record was brought by huge tsunami with nearly 20 thousand killed or missing people. We are again noticed that seismic observation system is quite poor in the oceanic region compared to the richness of it in the inland region. In 2012, NIED has started the construction of ocean bottom seismic and tsunami observation network along the Japan Trench. It is planned to layout 154 stations with an average spacing of 30km, each of which is equipped with an accelerometer for seismic observation and a water pressure gauge for tsunami observation. We are expecting that more rapid and accurate warning of earthquake and tsunami becomes possible by this observing network.

[1] The following slides are from the presentation given by Dr Yoshimitsu Okada at the Irago Conference 2012. Each slide is accompanied with an explanatory notes.

Figure 1. Recent progress of seismic observation networks in Japan.

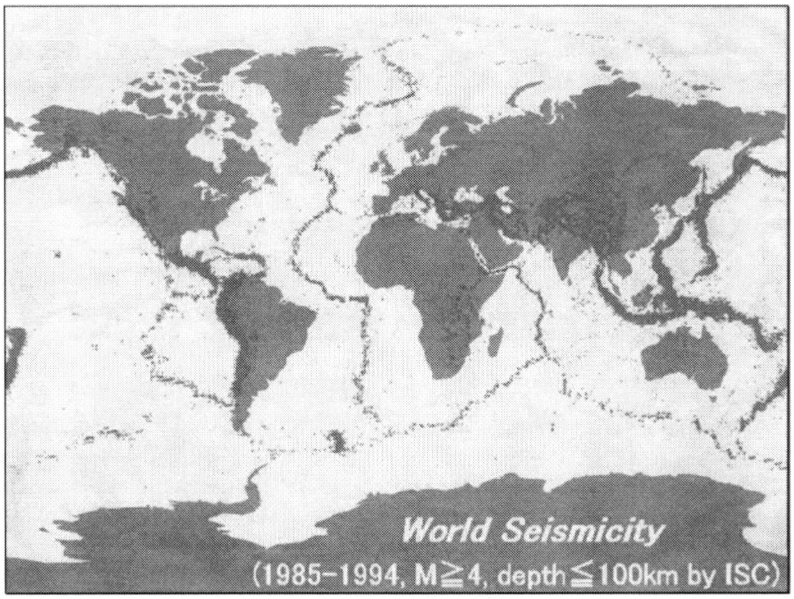

Figure 2. This figure shows the world distribution of earthquakes of magnitude 4 or greater and focal depth shallower than 100km, which were generated within recent 10 years (1985-1994). Seismic belts correspond to plate boundaries and we can hardly recognize the shape of Japanese islands.

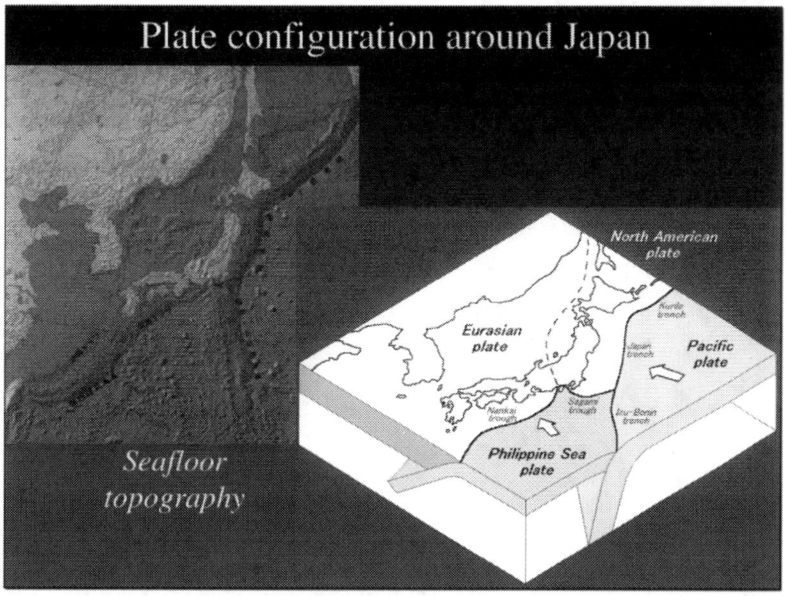

Figure 3. As can be seen in the seafloor topography, tectonic setting around Japanese islands is very complex. Beneath the Japan, the Pacific plate is subducting from east to west and another Philippine Sea plate is subducting from south-east to north-west. Japanese islands itself is considered to belong to the Eurasian plate in its southwestern part, while it belongs to the North American plate in its northeastern part.Such a complex tectonic situation is producing a large number of earthquakes and volcanoes in Japan.

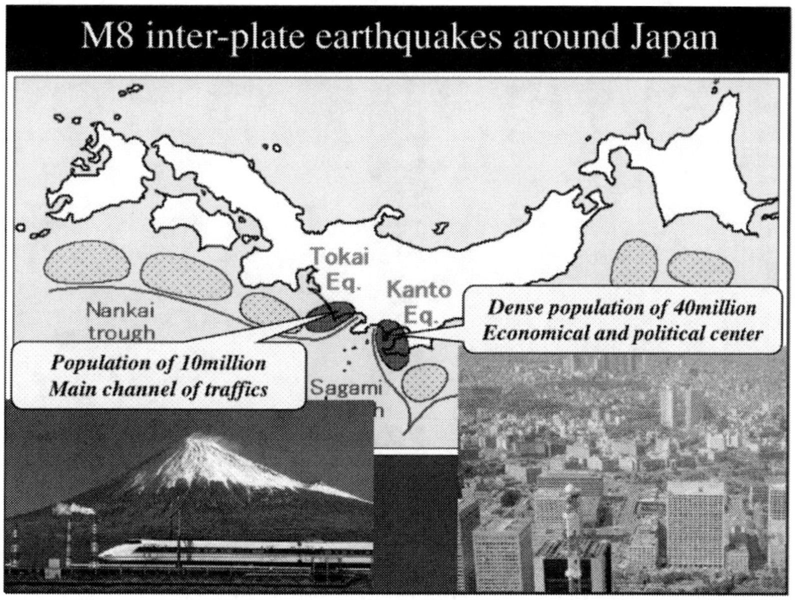

Figure 4. Along the Japanese islands, it is known the aligned regions of M8-class inter-plate earthquakes. Among them, the Tokai and Kanto earthquakes are rather special in the sense that their focal area is landing inside of the Japanese islands. The Kanto area has a dense population of 40 million and is bearing an economical and political center of Japan, while the Tokai area has an another population of 10 million and is bearing a main channel of traffics connecting eastern and western Japan.

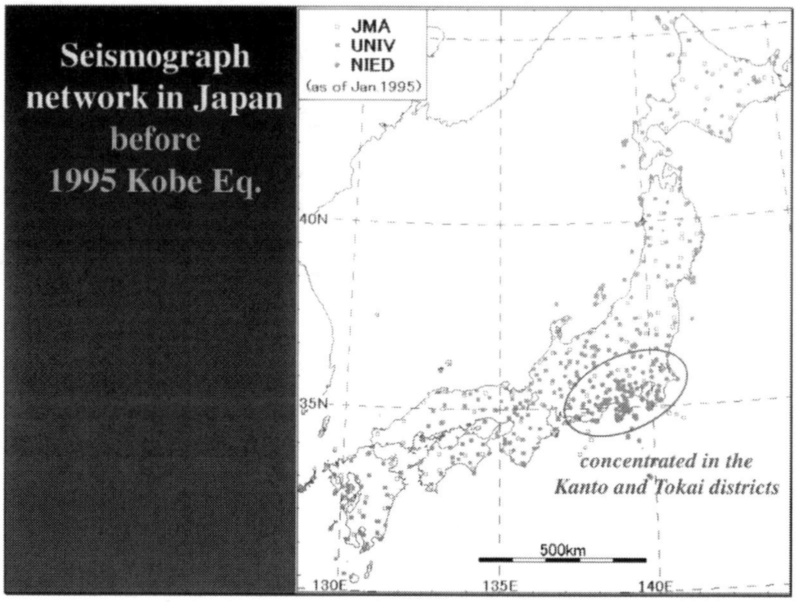

Figure 5. This figure shows the seismograph network in Japan before the 1995 Kobe earthquake was happened.By the circumstance as already stated in the previous slide, a lot of seismic stations were concentrated in the Kanto and Tokai districts.

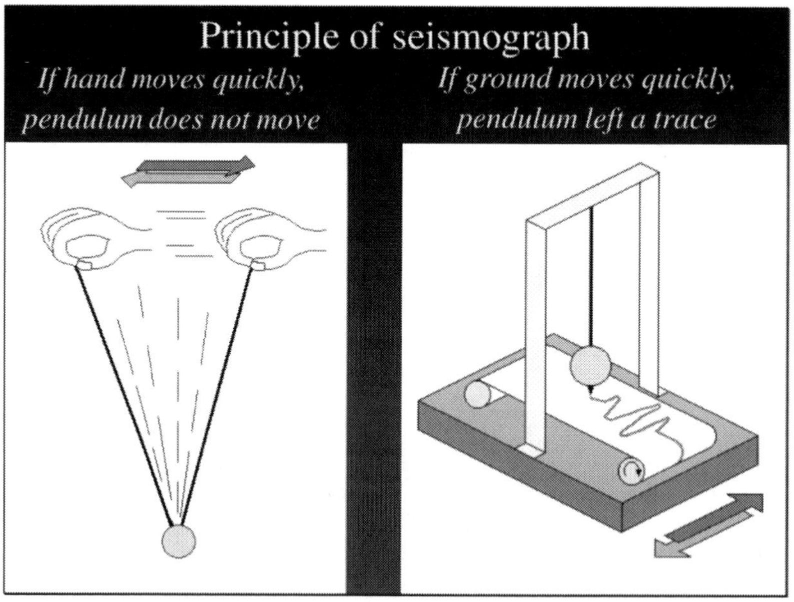

Figure 6. Here, let me explain the principle of seismograph.It is simply using the law of inertia. If we move hand very quickly, the hanged pendulum does not move and hold its original position. So, when we construct such a system with pendulum and rolling paper, we can get a trace of ground vibration if the ground moves quickly.

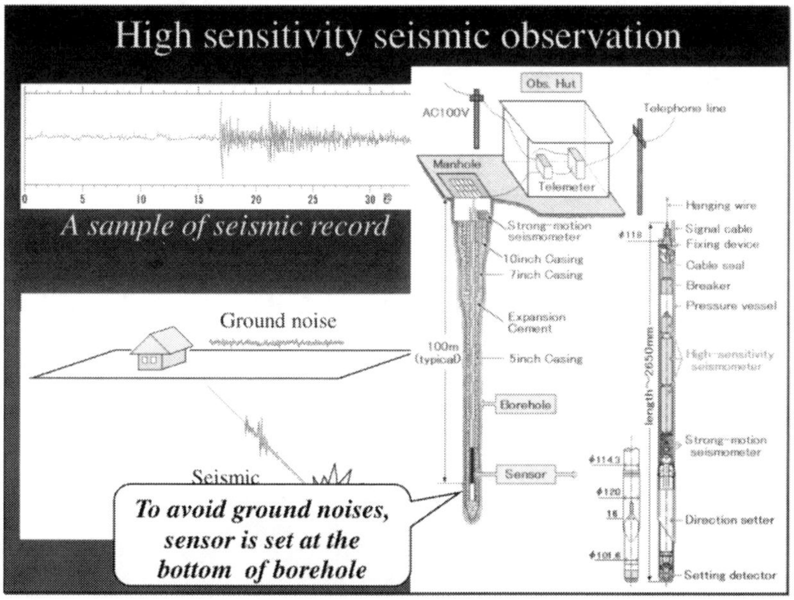

Figure 7. If we set a seismograph on the ground, we will soon get such a seismic record. This is composed of a seismic wave and ambient ground noise. When we want to detect very small earthquake signals, the ground noise becomes main obstacle for the observation. So, for a high sensitivity seismic observation, we usually set the sensor at the bottom of a borehole to avoid the ground noises. The depth of the borehole is usually around 100-200m.

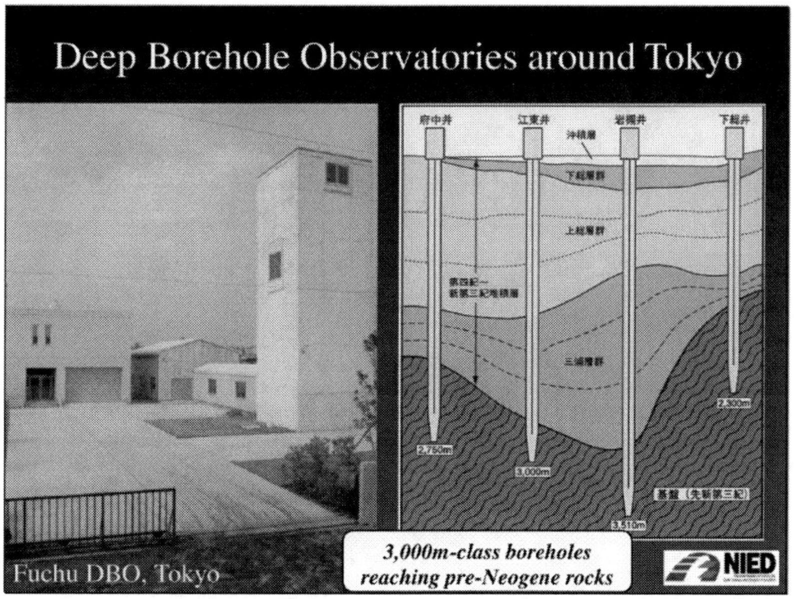

Figure 8. However in the metropolitan region such as Tokyo, the ground noise is so high that the borehole depth of 100m or so is not enough to achieve high sensitive seismic observation. In such a case, we need very deep borehole of around 3,000m depth penetrating a soft sedimentary layer and reach to the hard basement rock. The left picture shows the outlook of the Fuchu Deep Borehole Observatory in western Tokyo.

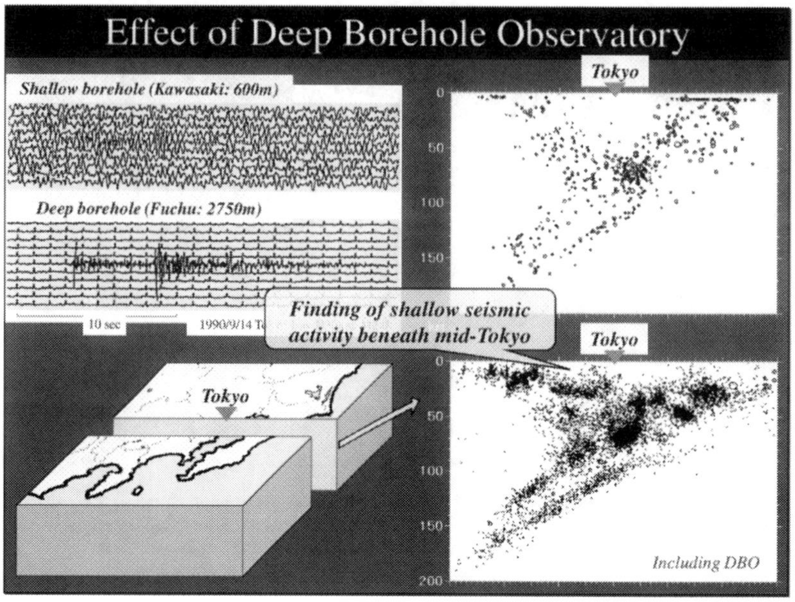

Figure 9. The effect of the Deep Borehole Observatory is enormous. This figure is comparing the seismic records of a micro-earthquake of M2 obtained at a shallow borehole station and a deep borehole station. We can clearly identify P- and S-phases using the deep borehole record. Using such a high quality data, we can now catch a precise image of hypocenter distribution under Tokyo Metropolis. The right figure compares the east-west cross section beneath Tokyo down to 200km. The top is an image obtained by a conventional seismic network and the bottom is the one including the data from Deep Borehole Observatories. Although the seismic zone corresponding to the subduction of the Pacific plate can be seen in the both images, the shallow seismic activity corresponding to the subduction of the Philippine Sea plate can be recognized only in the bottom image.

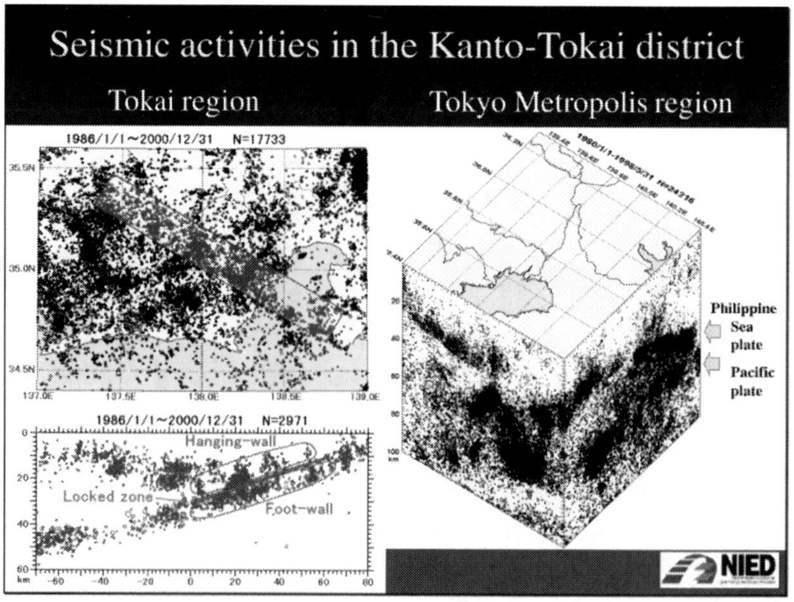

Figure 10. Using such a dense observation network, seismic activities around Kanto and Tokai

(continued) districts were investigated in details. The left figure shows the hypocenter distribution in the Tokai region and we can identify a locked zone in the cross sectional view along the subduction of the Philippine Sea plate. This locked zone is believed to generate hypothetical Tokai earthquake in future. On the other hand, the right figure shows a three-dimensional hypocenter distribution beneath the Tokyo Metropolis showing complex structure of the overlaid Philippine Sea plate and the Pacific plate.

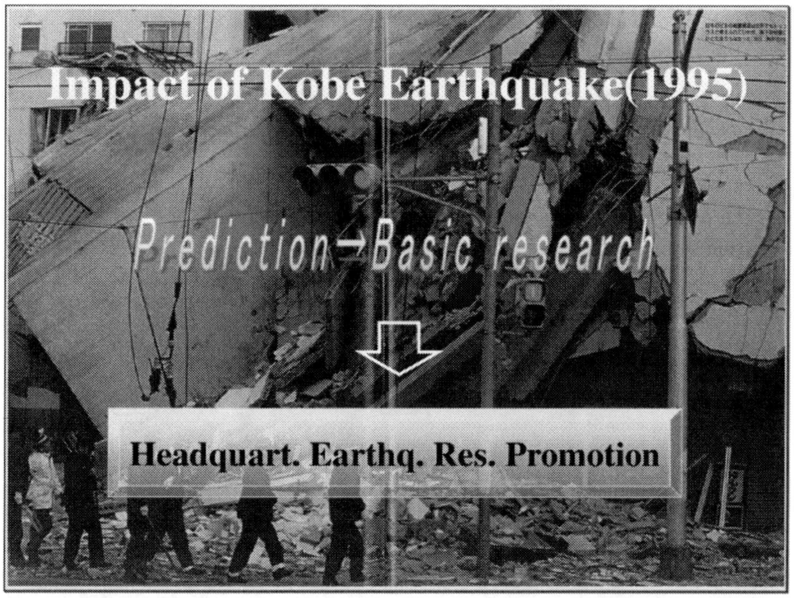

Figure 11. The Kobe earthquake of 1995 gave a great impact to the system of earthquake research in Japan. The trend of prediction-oriented study was changed to direct more basic researches, and the Headquarters for Earthquake Research Promotion was newly established in the Government.

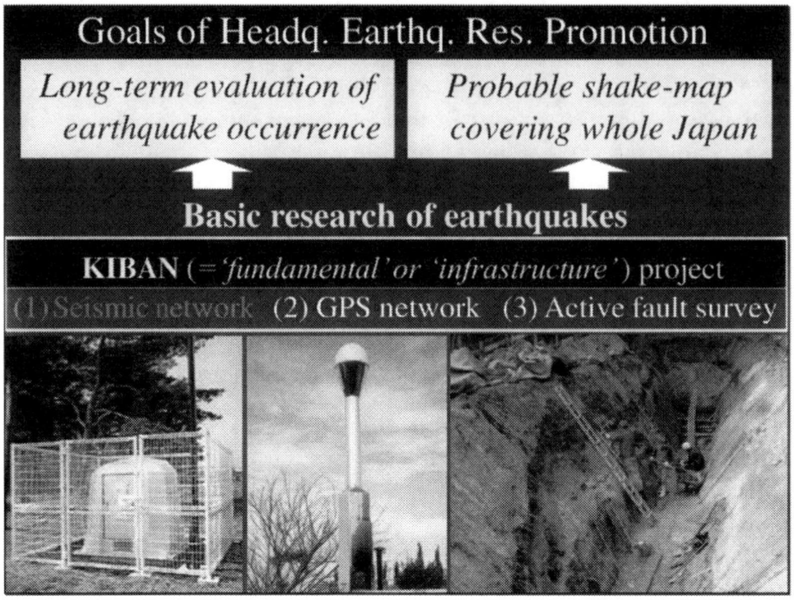

Figure 12. The goals of this Headquarters were to evaluate the long-term earthquake occurrence and to make up a probable shake-map covering the whole of Japan through basic research of earthquakes. To achieve this goal, the KIBAN project was started. KIBAN is a Japanese word to mean fundamental or infrastructure. It was composed of three components, construction of a nation-wide seismic network, construction of a nation-wide GPS network, and systematic active fault survey covering the whole of Japan. Hereafter, we refer only to the seismic network.

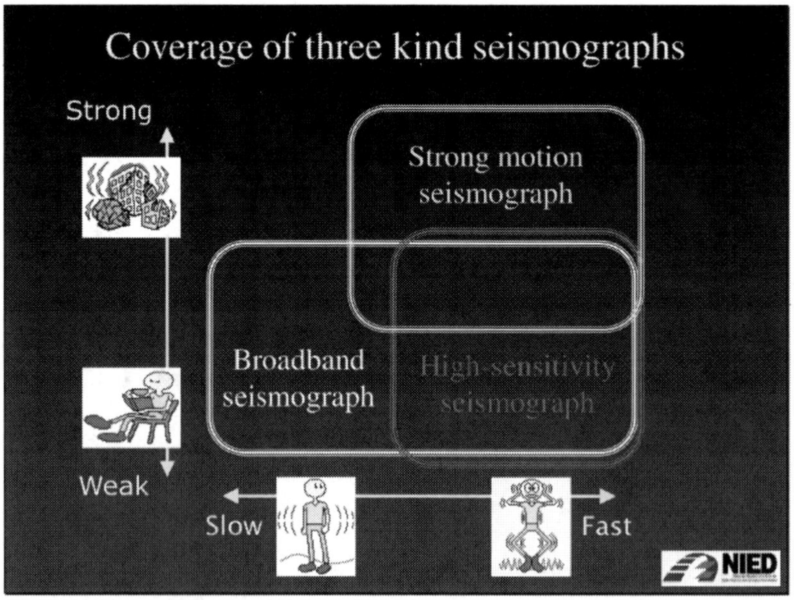

Figure 13. Since the ground motion has wide spectrum both in amplitude, and shaking period, we need three kind seismographs to cover all the ground motions, i.e. strong motion seismograph, broadband seismograph and high-sensitivity seismograph.

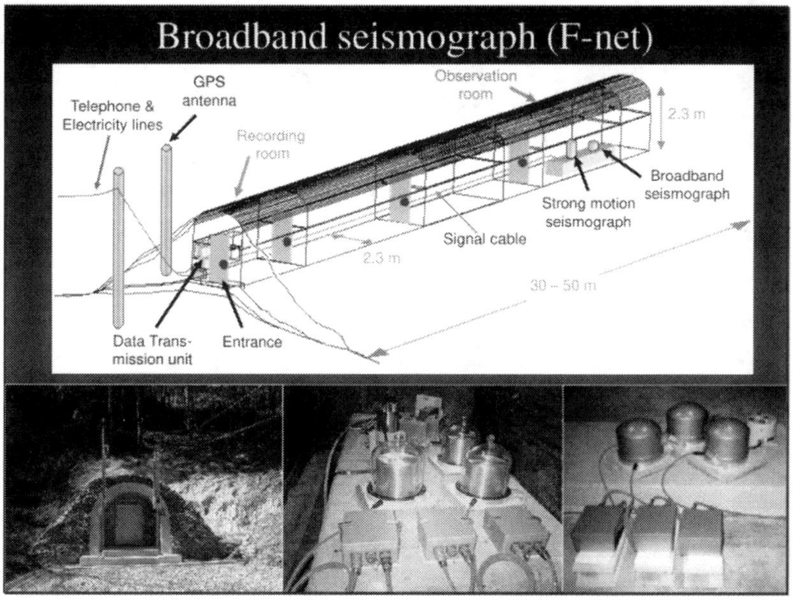

Figure 14. At first, broadband seismograph is settled in a backroom of an observation vault to avoid various meteorological disturbances.

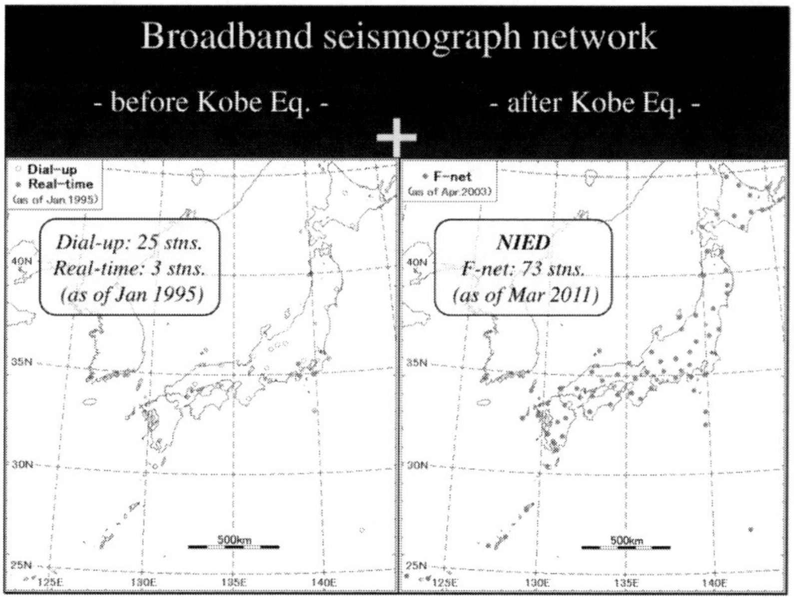

Figure 15. Before the Kobe earthquake, the distribution of broadband seismograph was very sparse and inhomogeneous, and almost all stations were not operated in real-time mode. In the KIBAN project, nearly 70 broadband seismograph stations were added with an average spacing of 100km and started operation in on-line real-time mode. This network is called F-net (F means full-range).

Journal of Physics: Conference Series **433** (2013) 012039 doi:10.1088/1742-6596/433/1/012039

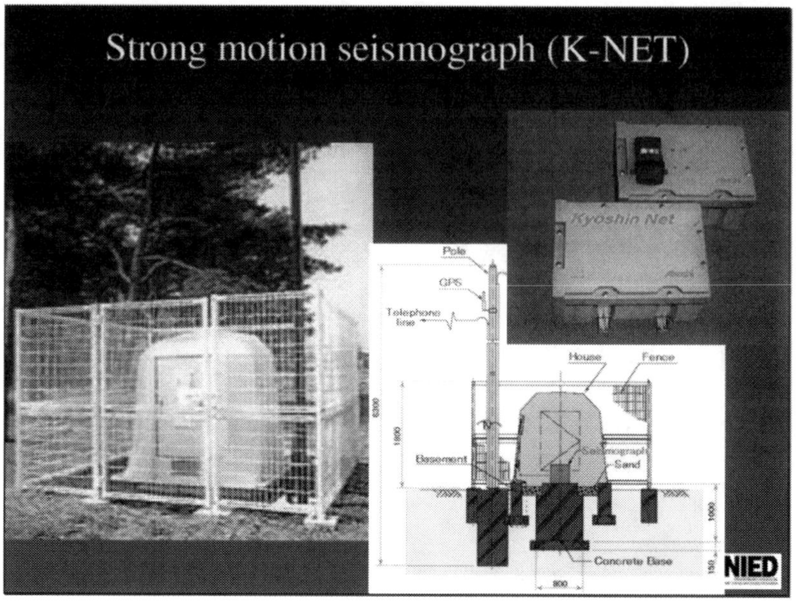

Figure 16. Next is the strong motion. After the Kobe earthquake, NIED has constructed a strong motion network called K-NET (K is an initial of Kyoshin which means strong motion in Japanese). At a K-NET station, three-component accelerometer is settled on the ground surface and the triggered records are transmitting via telephone line. With an average spacing of 20km, about 1,000 stations were constructed covering the whole of Japan.

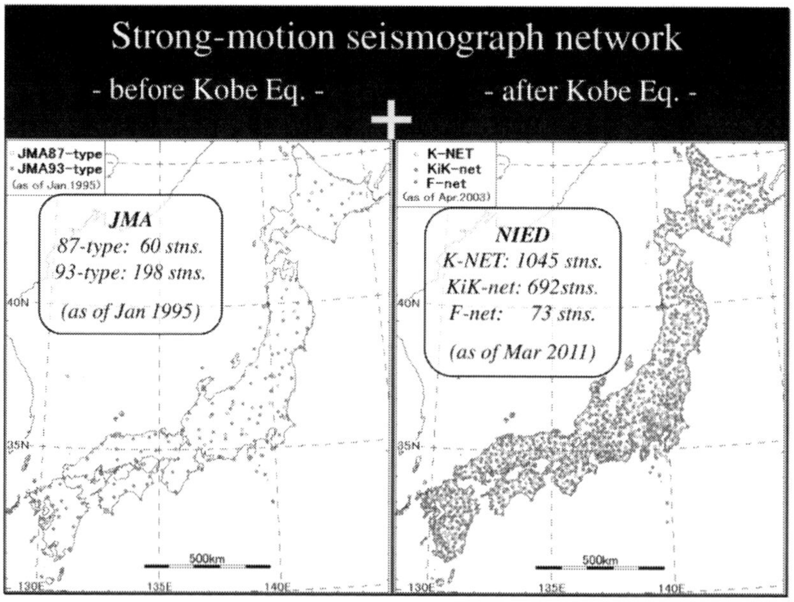

Figure 17. Before the Kobe earthquake, a nation-wide strong motion network was operated by JMA with nearly 250 stations. Since the density is insufficient to delineate precise shaking distribution, 1,000 K-NET stations were added after the Kobe earthquake. Moreover, through the KIBAN project, strong motion seismographs were also equipped in parallel at the F-net and Hi-net stations.

Figure 18. The last is the high sensitivity observation. To assure high sensitivity, sensors are installed at the bottom of borehole to avoid the near surface ground noises. Although the depth of borehole is usually 100-200m, more deep borehole reaching to 2,000-3,000m are used if necessary. This network is called Hi-net (Hi is the initial of High-sensitivity).

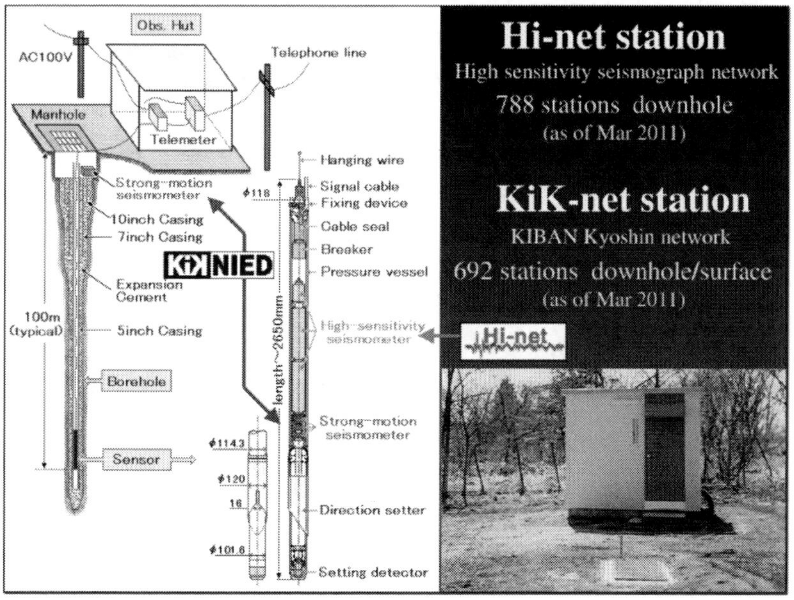

Figure 19. Now, about 800 Hi-net stations are in operation covering the whole of Japan with an average spacing of 25km. At the majority of the Hi-net stations, a strong motion seismograph is installed both at the ground surface and at the bottom of borehole. This network of a paired strong motion seismograph is called KiK-net (Kiban Kyoshin network).

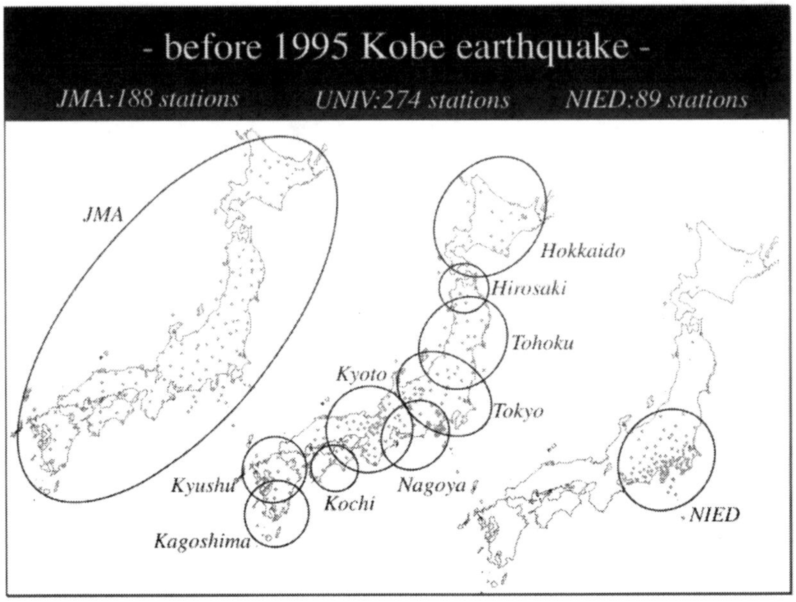

Figure 20. Before the Kobe earthquake, high sensitivity seismograph networks in Japan were independently operated by JMA, universities, and NIED. JMA was operating a sparse but nation-wide network, while universities and NIED were operating local networks covering the each area of concern.

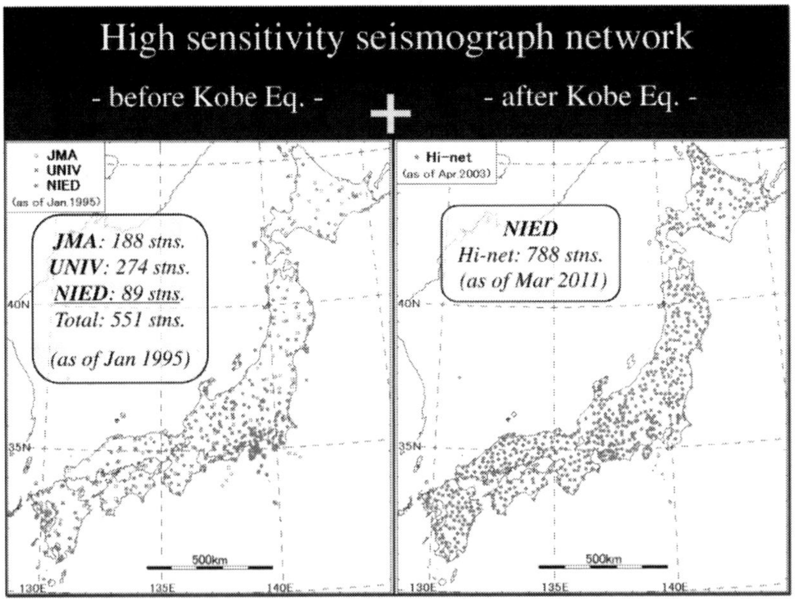

Figure 21. Thus, nearly 550 high sensitivity seismic stations were operated in Japan before the Kobe earthquake. After the Kobe earthquake, all these data were tentatively concentrated to JMA and a unified data processing was started in October 1997. Still the distribution of high sensitivity seismic station was inhomogeneous and its density was not enough. To improve this situation, about 800 Hi-net stations were newly constructed.

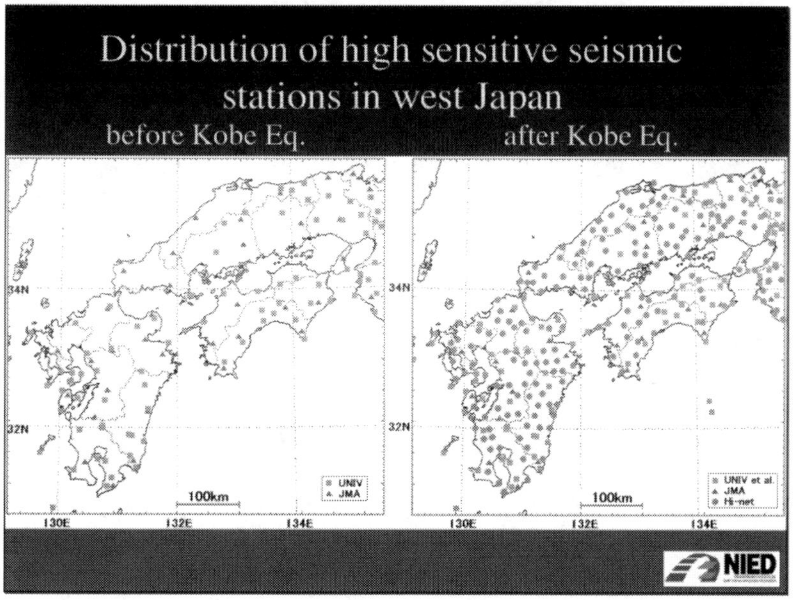

Figure 22. This figure shows how high sensitivity seismograph network was intensified in the western Japan.

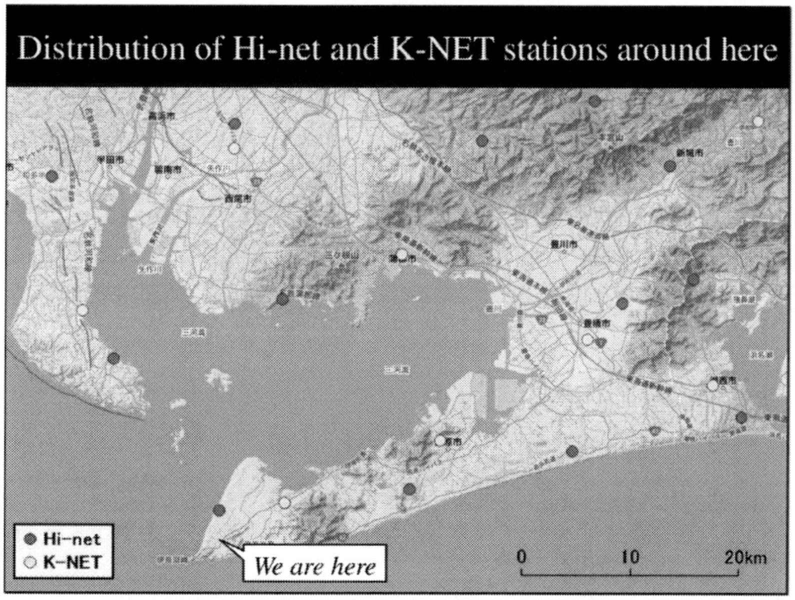

Figure 23. This map shows the distribution of Hi-net and K-NET stations around here, the Irago Cape. Red points show the Hi-net stations, while yellow ones show the K-NET stations. Imagine the seismic network is covering the whole of Japan with such a high density.

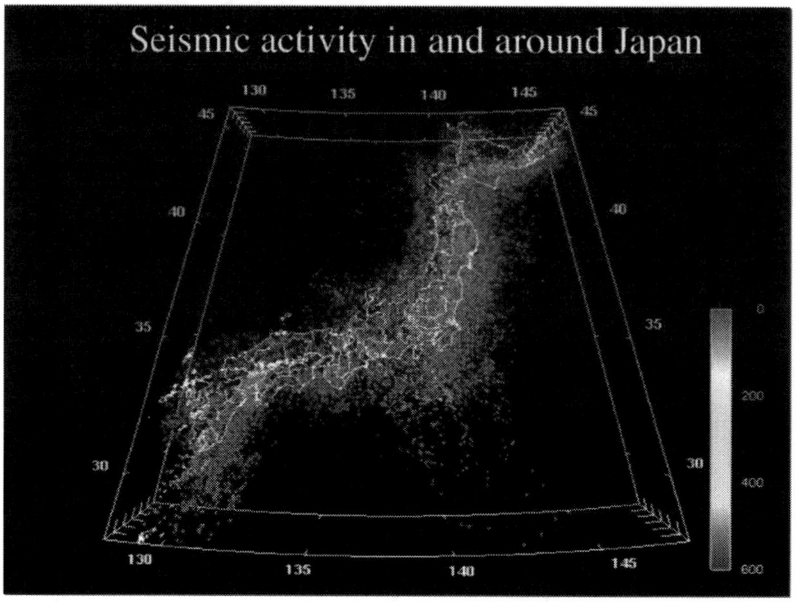

Figure 24. The data from Hi-net is primarily used for hypocenter determination. On average, about 20,000 earthquakes are located every year. This figure shows the hypocenter distribution around Japanese islands, where red points show the shallow events while blue ones correspond to the deep events.

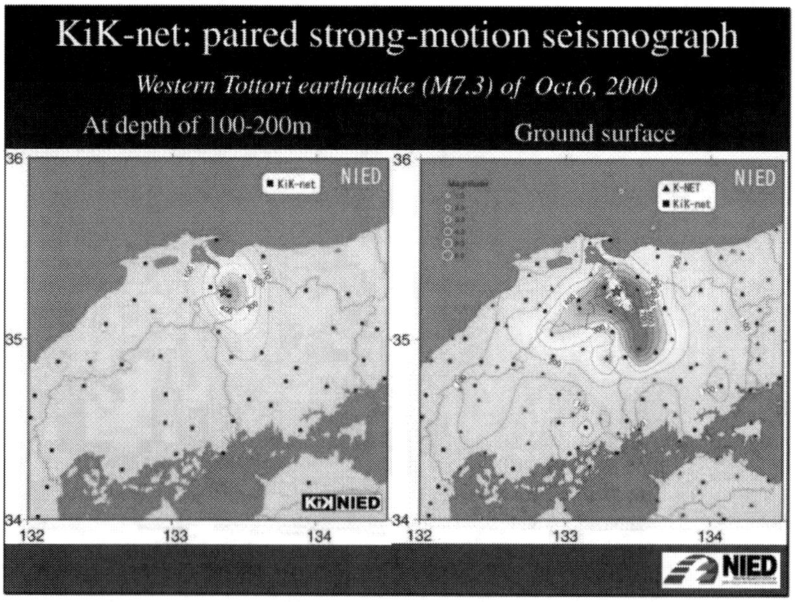

Figure 25. This figure shows an example of the results obtained from the KiK-net and K-NET. Associated to the occurrence of the western Tottori earthquake of October 2000, the distributions of the ground acceleration were obtained at both of the 100-200m depth level and the ground surface. Compared to the simple concentric pattern at depth, we can see a complex pattern at the surface reflecting the complexity of the ground conditions.

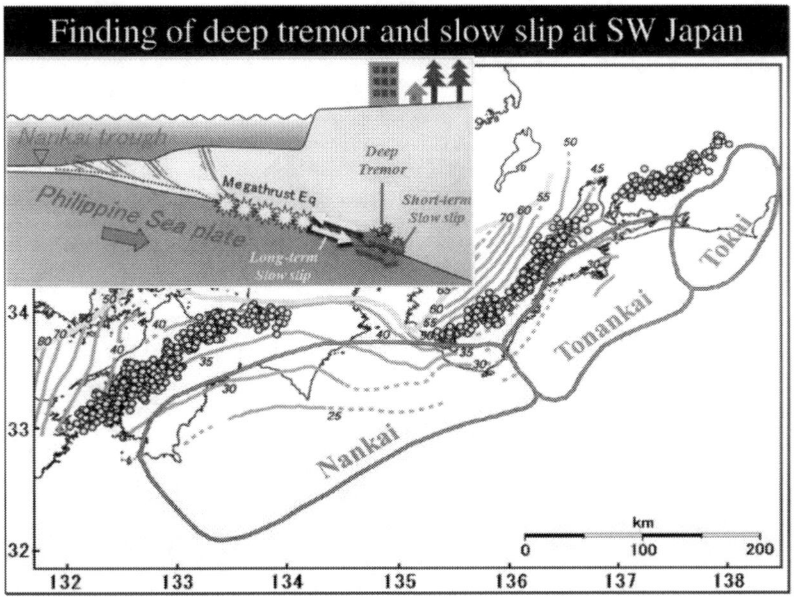

Figure 26. As the result of Hi-net observation, we could find the phenomena of deep tremor at southwest Japan for the first time in the world. They are generated along the depth contour of the Philippine Sea plate and correspond to the focal areas of the forthcoming mega-thrust earthquakes, Tokai, Tonankai, and Nankai. In a cross sectional view, the tremor is located at the deeper extension of the mega-thrust earthquake and is combined with slow slips of various time scale. We are expecting some new hints will be obtained as to the generation mechanism of big earthquakes.

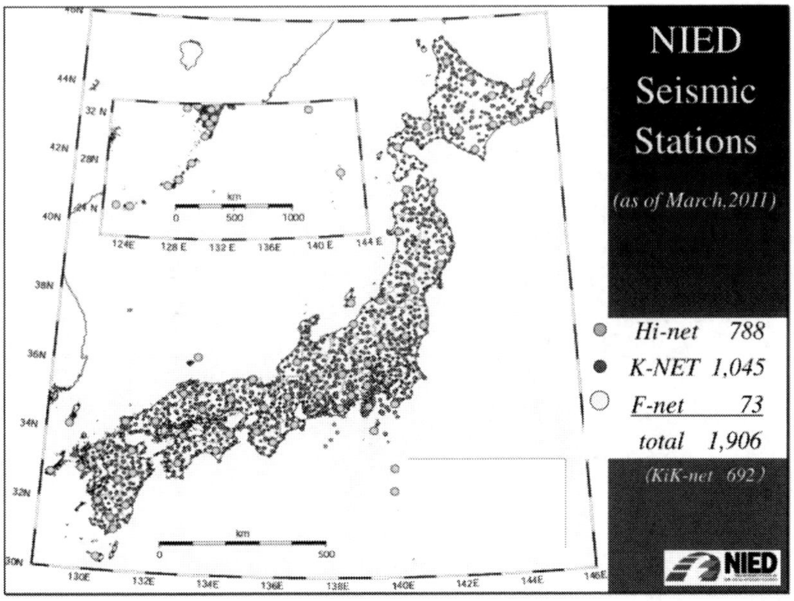

Figure 27. As a whole, NIED is now operating about 2,000 seismic stations in the Japanese islands.

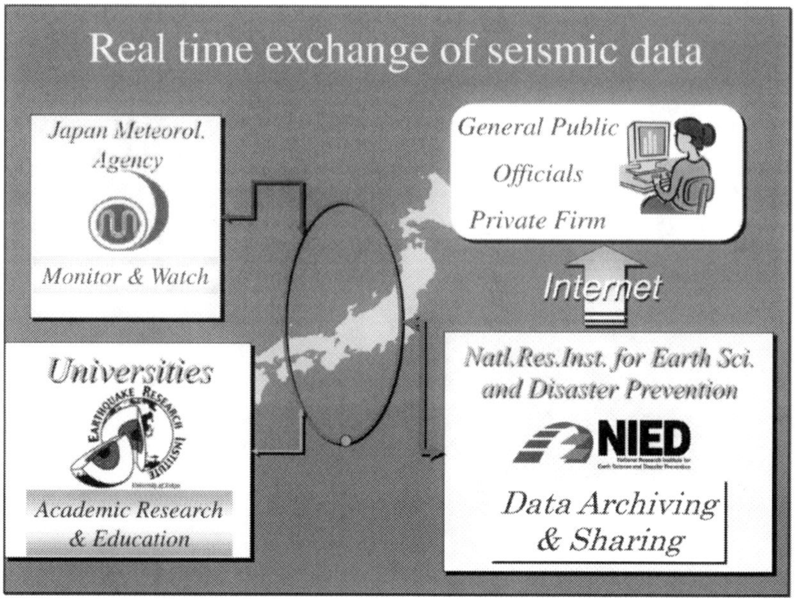

Figure 28. These seismic data are exchanged in real time among JMA, universities, and NIED. JMA is using them for monitoring and watching, while university groups are using them for the purpose of academic researches and education. NIED has a role of data archiving and sharing to provide them to the general public, officials, and private sector.

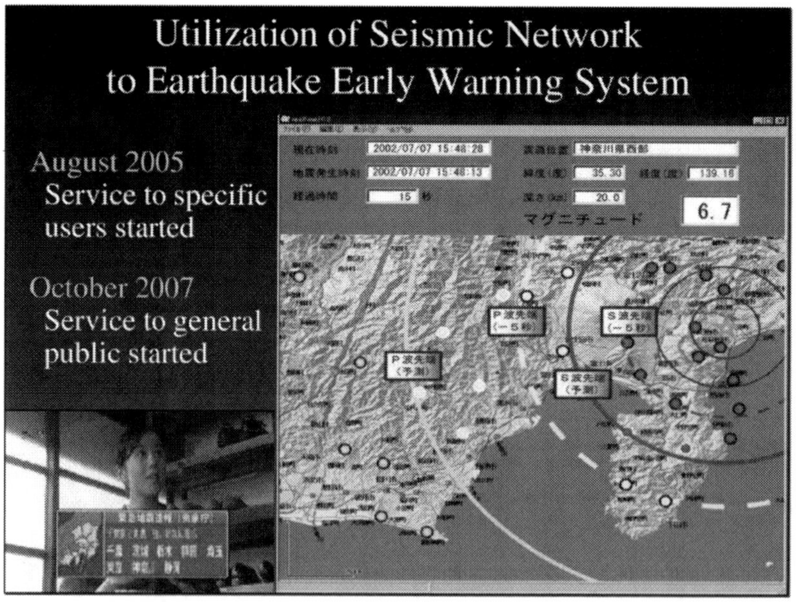

Figure 29. As a practical application of the dense seismic network, Hi-net data are utilized for realization of the Earthquake Early Warning System. In Japan, the service of Earthquake Early Warning to specific users was started in August 2005 and the service of them to general public was started in October 2007 for the first time in the world.

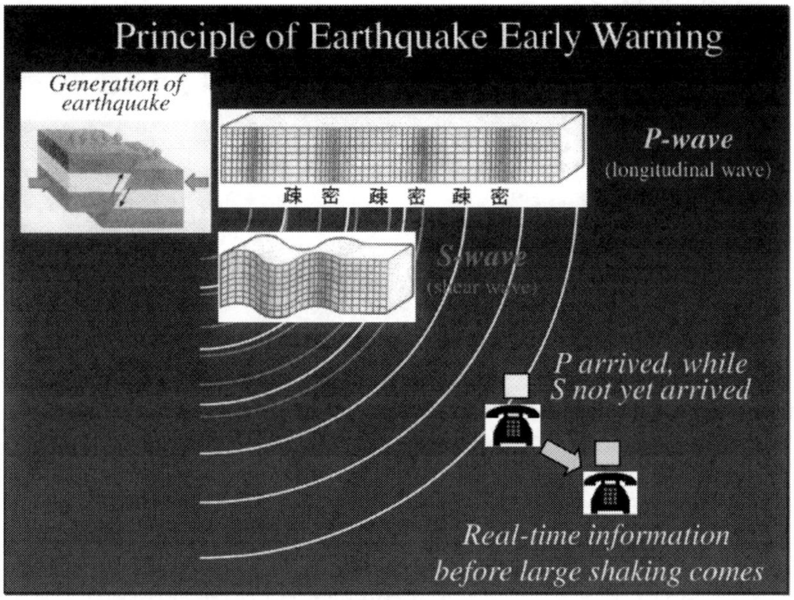

Figure 30. The principle of Earthquake Early Warning is as follows. When an earthquake occurs, it generates the elastic waves of two kind, P-wave and S-wave. P-wave is faster than S-wave but its amplitude is generally far small compared to that of S-wave. At this yellow point, P-wave is already arrived but S-wave is not yet arrived. If this information is transmitted to the remote place at once, they can prepare to the earthquake before the large shaking of S-wave comes or even before the arrival of the P-wave.

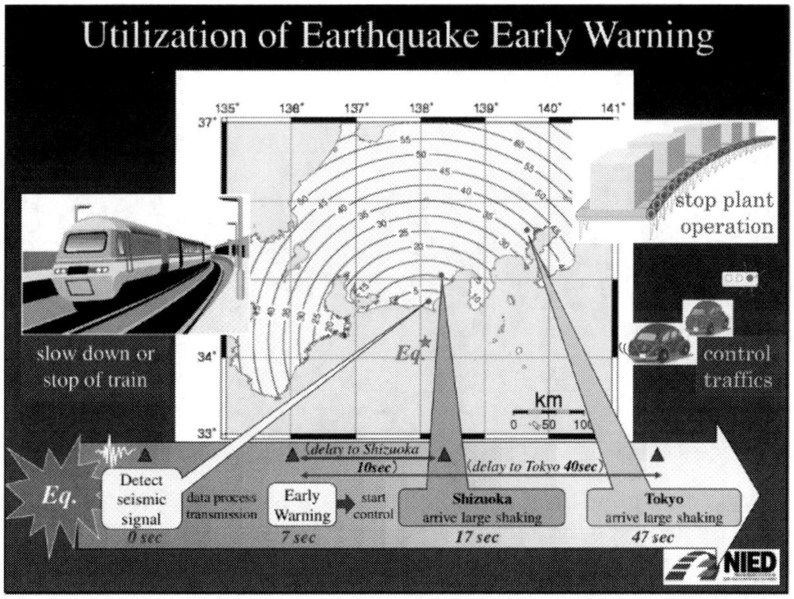

Figure 31. This is an example of scenario. When an earthquake happens here, first signal is detected at the nearest observation site. After some data processing and transmission, a quick Early Warning is issued within several seconds and the various controls are started to traffics, plant operation, and so on. At this Shizuoka City, a large shaking will arrive about 10 seconds later, and it will arrive at Tokyo 40 seconds later.

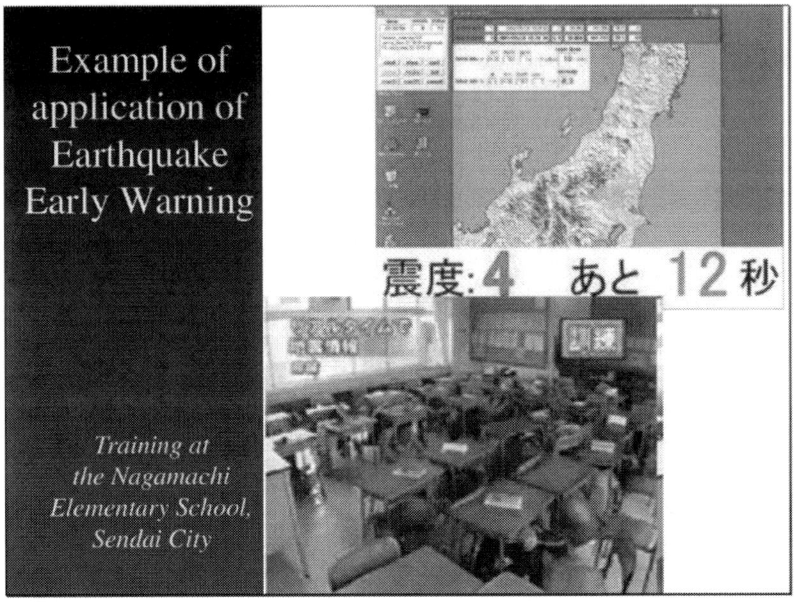

Figure 32. This picture shows a scene of the training at the Nagamachi Elementary School in Sendai City. Pupils are hiding under the desk receiving an Early Warning information that shaking of intensity 4 will come 12 seconds later.

Figure 33. Here, let us remind the great East Japan Earthquake of March 11 of the last year, which recorded M 9.0, the largest in the history of seismic observation in Japan. This picture shows the instance that tsunami arrived at the coast of Miyako City. As you can see, tsunami is not a big wave but the flood coming from the ocean.

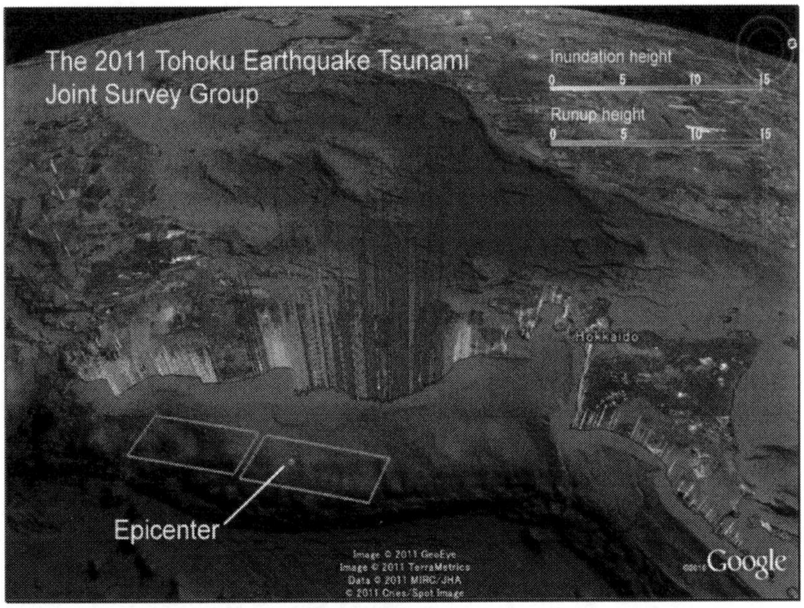

Figure 34. This diagram shows the tsunami height distribution. Most severe tsunami were concentrated in the Sanriku region with 30 to 40m run-up heights. When this earthquake occurs, the first information issued from the JMA was remarkably underestimated. The magnitude of the earthquake was guessed as M 7.9 and the estimated tsunami height level was announced only 3 to 6 meters in the first Warning. One reason for such an underestimation was the lack of data in the ocean area. Compared to the dense networks in the land area, seismic and tsunami observation in the ocean area was crucially insufficient.

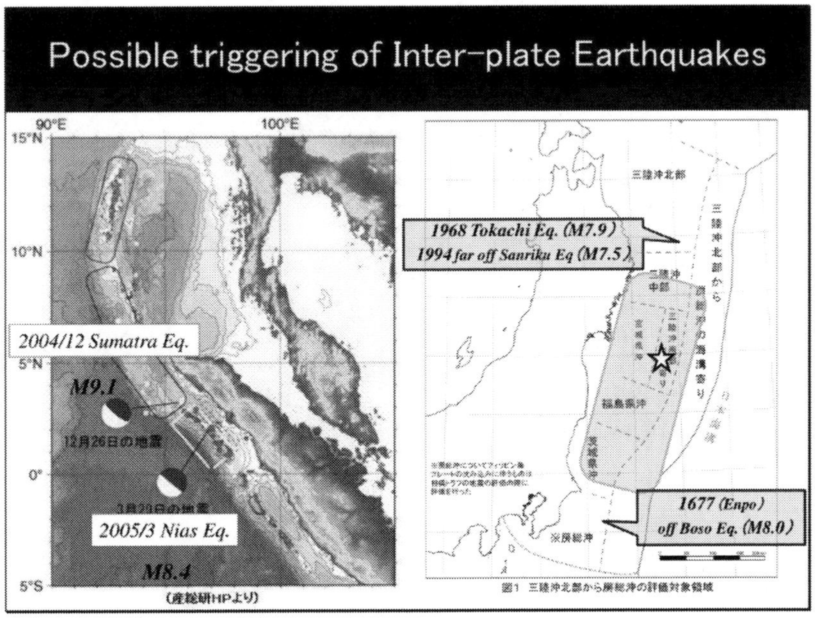

Figure 35. By the occurrence of the great East Japan earthquake, we are anxious about possible triggering of large inter-plate earthquakes in the surrounding region. In the case of 2004 Sumatra earthquake of M 9.1, a big thrust earthquake of M 8.4 was followed three months later at the neighboring area. We should be cautious to the similar phenomena for the East Japan earthquake,

(continued) especially at northern and southern extensions of the focal area of the main shock. In the northern extension, relatively large earthquakes were occasionally generated such as 1968 Tokachi and 1994 far off Sanriku earthquakes. On the contrary, in the southern extension we have only little records of the large earthquakes excluding 1677 off Boso earthquake of M8.0. It is said that this historical earthquake accompanied a big tsunami and gave considerable damages in the coastal areas of the East Japan.

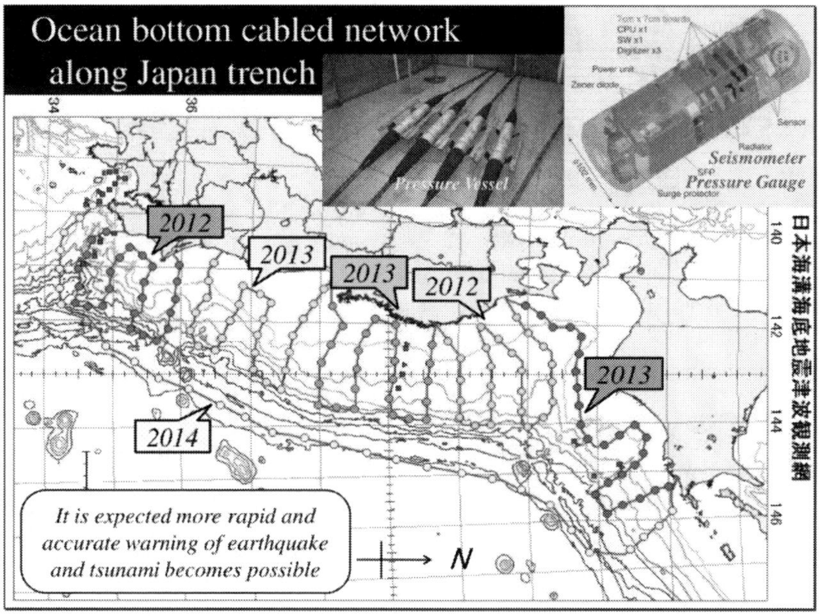

Figure 36. In order to improve the observation capability in this region and to prepare to the next big one, our institute, NIED has started the construction of an ocean-bottom cabled network along the Japan trench. Colored dots in the ocean show the locations of pressure vessel in which a set of seismometers and pressure gauges for tsunami detection are included. The network is consisted of 154 stations in total with a spacing of about 30km in the East-West direction and in a spacing of 50-60km in the North-South direction with a total cable length of 5,800km. In the fiscal year of 2012, red colored Boso network and green colored Aomori network are scheduled to start the operation. We are hoping that this network will contribute to monitor large earthquakes and tsunamis which may attack east Japan and to successfully transfer more rapid and accurate warnings to the general public.

Thank you
for
your kind attention
NIED

AUTHOR INDEX

AUTHOR INDEX